ARCTIC REGION AND ANTARCTICA ISSUES AND RESEARCH

THE ARCTIC

CURRENT ISSUES AND CHALLENGES

Arctic Region and Antarctica Issues and Research

Additional books and e-books in this series can be found on Nova's website under the Series tab.

ARCTIC REGION AND ANTARCTICA ISSUES AND RESEARCH

THE ARCTIC

CURRENT ISSUES AND CHALLENGES

OLEG S. POKROVSKY
SERGEY N. KIRPOTIN
AND
ALEXANDER I. MALOV
EDITORS

nova
science publishers
New York

Library of Congress Cataloging-in-Publication Data

Names: Pokrovsky, Oleg S., editor. | Kirpotin, S. N., editor. | Malov, A. I, editor.
Title: The Arctic: : current issues and challenges / edited by Oleg S. Pokrovsky, Sergey N. Kirpotin, Alexander I. Malov.
Other titles: Arctic (Nova Science Publishers)
Description: Hauppauge, NY : Nova Science Publishers, 2020. | Series: Arctic region and antarctica issues and research | Includes bibliographical references and index. |
Identifiers: LCCN 2020000179 (print) | LCCN 2020000180 (ebook) | ISBN 9781536173062 (hardcover) | ISBN 9781536173079 (adobe pdf)
Subjects: LCSH: Arctic Regions--Environmental conditions. | Arctic Regions--Climate. | Arctic Regions--Research.
Classification: LCC GE160.A68 A729 2020 (print) | LCC GE160.A68 (ebook) | DDC 363.700911/3--dc23
LC record available at https://lccn.loc.gov/2020000179
LC ebook record available at https://lccn.loc.gov/2020000180

Published by Nova Science Publishers, Inc. † New York

CONTENTS

INTRODUCTION

THE CHANGING ARCTIC: PEOPLE, RESOURCES AND ENVIRONMENT

Oleg S. Pokrovsky[1,*], Sergey N. Kirpotin[2] and Alexander I. Malov[3]

[1]Geoscience and Environment Toulouse, CNRS, Toulouse, France

[2]Bio-Clim-Land Centre, Tomsk State University, Tomsk, Russia

[3]N. Laverov Federal Centre for Integrated Arctic Research of Russian Academy of Sciences, Arkhangelsk, Russia

The importance of the Arctic in many fields of human activity strongly increased over past decades. The academic scientific research demonstrates a 3-fold increase in the number of journal articles dealing with "Arctic": from 1400 in 2000 to 4200 in 2018 (Fig. 1). This increase is not fortuitous but certainly stem from double importance of Arctic regions for humanity. The first importance is the role that Arctic plays in the on-going environmental changes, mostly linked to climate warming and environmental pollution. Here, the first key issues are the Arctic Ocean, ice melt, permafrost thaw, green house gases emission, and organic carbon mobilisation from soil to rivers. From the other hand, highly fragile Arctic ecosystems and biota are strongly affected by environmental pollution, be it organic compounds or toxic metals, or radionuclides. The rising concern of humanity to the key role of the Arctic in climate regulation on the planetrya scale and the extreme fragility of its ecosystem, biota and native population to on-going environmental change can certainly explain an explosive interest of scientific researchers to the Arctic in connection with 'climate change' (Figure 1).

The second big issue of the Arctic is its eminent role in problems of natural resources. The Artic shelf contains vast amount of hydrocarbons (gas and oil), whereas the terrestrial polar regions, now liberating from ice, may turn out to be highly important sites of future ore industry. The importance of possibly ice-free Arctic ocean as future maritime shipping routes will further enhance the accessibility of natural resources in this region. Taken together, this

* Corresponding Author's Email: oleg.pokrovsky@get.omp.eu.

can be the main driving factors of almost exponential increase in the interest to natural resources in the Arctic over past few years.

The present book adresses a wide variety of environmental, social and economic issues of the Arctic, in response to rising interest to this region in academic science, sociology and business. The 13 chapters represent state-of-the art reviews written by the experts on problems of native communities, climate change, political issues, implementation of large-scale projects, natural resources and conservation, environmental monitoring and assessment of pollution issues.

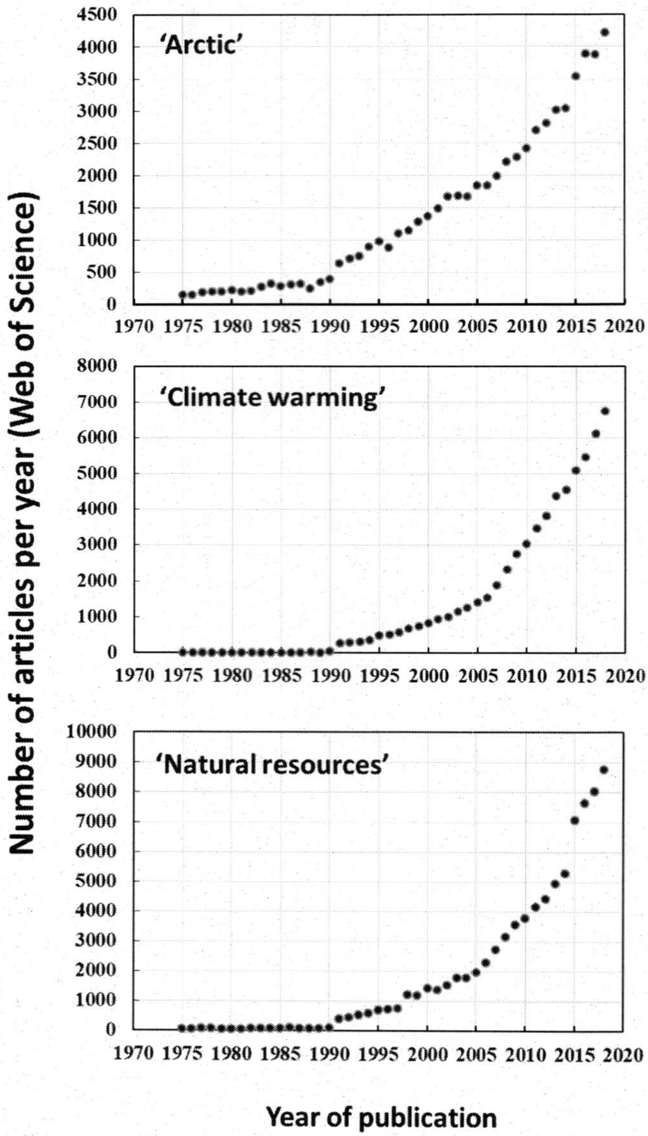

Figure 1. The amount of articles published per year in peer-reviewed journals (Web of Science) on the topic of "Arctic", "Climate Change" and "Natural Resources". Simultaneous increase of interest to these issues started in 1990th and demonstrate even stronger rise over past 5-10 years.

The first chapter looks at the needs of traditional communities and the ways in which they are impacted by climate change and the multiple uses of the Arctic which are made possible by climate change. This chapter explores unknown territory by bringing together Arctic and minority rights studies with social sciences and by putting the matter of learning in the Arctic into a theoretical framework from the social sciences. The chapter further demonstrates how climate change impacts the core aspects of societies in the Arctic, including the value of traditional knowledge and the societal status of Arctic people in local communities and it ends by discussion on reflective practice of learning about territorial significances in a changing arena.

The second chapter continues the discussion of the global process of climate change from the perspective of the local population of the Yamal-Nenets Autonomous district and adjacent regions. The chapter is based on empirical research on the territory of the Yamal and Khanty-Mansiysk district (Yugra), the corpus of interviews and observations. It explains the socio-economic, historical, infrastructural and natural linkages between the Arctic and subarctic zones, as well as the need to take into account the zonal changes throughout Western Siberia. It further shows the scenarios of overcoming the difficulties and adapting to the changing conditions caused by the new wave of development of the North, management decisions and natural disasters occurring during the life of the last two generations of the northerners. The chapter adresses vital questions for survival in the Arctic: to develop the ability to survive autonomously, using local knowledge and skills, community resources, or to focus on the operational supply and intensive transport and informational links in the Arctic?

The third chapter addresses the impact of climate warming on Arctic cities. This chapter characterizes large warm temperature anomalies and corresponding environmental changes in 11 Arctic cities. The authors unambiguously demonstrate that the Arctic urban climate is significantly warmer, especially during the coldest weather spells, than the corresponding regional climate. By introducing the concept of Urban Heat Island (UHI), a localized territory that still affects an area 2-3 times larger than the city itself, the authors show how the winter UHI during the polar night is supported by the anthropogenic heat flux and enhanced air pollution in cities. Further, they demonstrate that the summer UHI during the polar day is supported by stronger warming of better drained urban surface. The authors conclude that better understanding of the Arctic UHI may result in practical engineering and policy measures to counteract the detrimental consequences of the excessive warming on infrastructure and urban planning.

The fourth chapter addresses political and economical issues related to the delineation and delimitation of extended continental shelves, which is an emerging issue in international relations. The study identifies the critical variable which influenced the cooperation process on delineation. The authors present a novel methodology which is based on comparative study of two cases: the Arctic Ocean and the South China Sea. Overall this study contributes towards a better understanding of a legal nature of delineation and to demonstrate interrelations between international law and geopolitics in regard to the delineation in the Arctic Ocean.

The chapter 5 examines environmental research capacity in Western Siberia in relation to the global networking. Owing to the development of the oil and gas complex, Western Siberia among all Arctic regions has the most developed infrastructure: roads, oil and gas pipelines, pads, power lines, gas-compressor stations, cities and towns of oil and gas industry workers. This well-developed infrastructure (a good road network) provides unprecedented access to the region for researchers and field experiments, making them significantly cheaper than in other northern regions. This chapter demonstrates convincingly that the Yamalo-Nenets Autonomous District has unique unprecedented opportunities for the development of large-scale network projects based on transdisciplinarity, and is the most convenient site for studying both the fundamental questions of interaction of climate and landscapes and considering the practical aspects of these changes and evaluation of these social impacts. The chapter also provides a description of some bilateral and regional research networks and finding ways of their effective interactions with each other, with stakeholders and the state authorities in the Yamal-Nenets Autonomous District (YNAO).

The next chapters present an overview of natural resources of the Arctic with particular focus on Western Siberia and Russian European Arctic regions. Chapter 6 addresses the vulnerability of Arctic ecosystems on the example of Siberian vegetation. The current threats are associated with shifts in the ranges of various plant and animal species, threats to biodiversity and adverse consequences for the structure and composition of communities. The authors demonstrate that the vegetation of the polar tundra, by virtue of its boundary position in the biosphere, has a relatively low stability in the conditions of global climate dynamics. In particular, the Russian sector of the Arctic, due to its greatest length and typical character of the manifestation of zonality of tundra vegetation, presents special interest for monitoring climate-dependent changes. Based on the analysis of Russian publications that are not familiar to the English-speaking reader, and contemporary English-written literature as well as materials from authors' research, the chapter provides an overview of the main types of vegetation of the Subarctic and Arctic of Siberia, as well as the main approaches to its study.

Chapter 7 addresses conservation issues of migratory birds in the Arctic. The authors present their work carried out since 2010 in the Yamal-Nenets Region to improve the methods of monitoring and conservation of migratory waterfowl. The result of authors' approach provided a modern method of calculating the number and density of waterfowls. It is based on the following action algorithm: determination of extrapolation area; determining the type of habitat for each survey point for each species; determination of the area of all identified habitat types surveyed in the course of aerial surveys; determination of the total area of all identified habitat types in the work area; determination of the density of each species in each habitat; calculation of the number of each species in each type of habitat. According to data from counts in the YANAO, temporary hunting free zones during the spring hunt period have been created in key locations identified during the monitoring. The authors further demonstrate that this protection method has no analogues in Russia, has proven its effectiveness and deserves to be extended to other regions.

The Chapter 8 describes the results of a remote study of the long-term dynamics of lake areas in the permafrost zone of Western and Eastern Siberia. The authors present a synthesis of the data on changes in the number of extinct and newly formed lakes over a 40-year study period, depending on the geocryological zoning of the study area. Further, they combine it with the results of the analysis of climatic changes according to weather stations and reanalysis data in the permafrost zone of Western Siberia. Overall, this chapter presents a state-of-the-art knowledge of Arctic lake number and inter-annual dynamics and covers huge territory from western to eastern Siberia.

The last six chapters are devoted to Artic geology, geochemistry and geodynamics. Chapter 9 contains a comprehensive synthesis of groundwater resources in the Russian European Arctic (Arkhangelsk region). The author presents the dominant conditions of formation of the subterranean inflow into the rivers for the territory of the White Sea basin and demonstrates that underground runoff occurs from zones of active and relatively active water exchange in the ratio of 99:1. He further describes the inventory of groundwaters of the region and shows that the share of underground runoff in the total river flow in most of the territory is 10–20% whereas the maximum values (30 – 40%, and sometimes more than 50%) are characteristic of intensely karst areas.

The Chapter 10 characterizes the soil cover of poorly studied Arctic Island - the Novaya Zemlya Archipelago. The authors present the maps of the spatial distribution of natural and man-made radionuclides in the surface soil layer on the coast of Moller Bay. The obtained data are compared with the levels of specific activity of these radionuclides obtained earlier by other authors for the soils of the east coast of Novaya Zemlya. According to the data obtained, for ^{137}Cs and ^{40}K, the level of specific activity in the soils of the coast of Moller Bay in the west of the southern island of Novaya Zemlya is less than 3 and 2 times, respectively, compared to the east coast. The level of specific activity of $^{13}7Cs$ in the studied soils is ca. 2,6 times lower than in the soils of the northern part of the island. In a number of soil sections on the Moller coast, the presence of short-lived isotopes ^{134}Cs and ^{60}Co was established, which indicates the on-going fallout of man-made radionuclides into the soils of the Arctic. Finally, the authors demonstrate that atmospheric transport of pollutants to the territory of Novaya Zemlya is of greatest importance in winter, while in summer the main transport comes from water areas.

The Chapter 11 further elaborates on assessing the contamination of the Arctic ecosystems with radioactive elements. This contamination is associated with the presence of land and sea testing sites for modern weapons, radioactive waste disposal sites, and the operation of nuclear power and nuclear fleet. Along with natural and anthropogenic factors, transboundary transfers of radioactive components by air and water masses are of great importance for the state of the environment in the Western Sector of the Russian Arctic. The chapter presents the results of studies the radionuclides in the seas, coastal and island territories of the Western Arctic of Russia during sea expedition on research vessel «Professor Molchanov» in 2012. The authors show that ^{137}Cs activity in the bottom sediments of the Western Arctic seas of Russia has recently decreased, but in many samples a rapidly decaying ^{134}Cs radionuclide has been detected, which may indicate a continued flow of technogenic radioactivity into the Arctic seas. They demonstrate that the levels of radiation pollution are low and tend to decrease, but the picture may change due to increasing anthropogenic activity on the continental shelf and adjacent territories. Also, the active use of the Northern Sea Route with the arrangement of related infrastructure and the implementation of dredging in the mouths of large northern rivers could be cause the redistribution of natural and man-made radionuclides.

The Chapter 12 is devoted to seismic monitoring of the European Arctic and adjoining regions. This chapter elaborates on the results and analysis of the seismic monitoring of the European Arctic and adjoining regions obtained from the early 20th century and modern period from 2011 to 2018. The authors demonstrate that, with opening some new seismic stations in Russian sector of the European Arctic, it is now possible to register small-magnitude earthquakes within the Gakkel Ridge area and Barents-Kara Sea region including the "continent-ocean" transition zone. Based on the generalization and analysis of previously available and received new information of Arctic earthquakes distribution, the authors conclude that the modern Arctic geodynamics is determined by tectonic processes taking place at the divergent boundary of the North American and Eurasian lithospheric plates. Because the modern seismicity is indicators of geodynamic processes, the increased seismic activity of separate sites of the European Arctic and adjoining regions is defined by total influence of three main factors: (*i*) the partial discharge of stresses generated in the axial zones of the Arctic basins, (*ii*) the lithosphere reaction to the pressure of thick sediments and, (*iii*) the uploading from the ancient glaciation.

The Chapter 13 describes diamond deposits in the Russian European Arctic via addressing the problems of kimberlite exploration using conventional and novel approaches. The authors present a fundamental breakthrough in the identification of low contrast anomalies, associated with kimberlite pipes by geophysical methods via developing new innovative technologies for exploration of low-contrast diamond kimberlite bodies, hidden in the depth and to strengthen the forecasting-prospecting works. Specifically, they demonstrate that microseismic sounding method allows imaging diamond pipes at the depth greater than 1 km because diamond pipes are often located near faults zones. Further, via combined use of microseism sounding and radiometric's methods the reliability of interpretation can be greatly increased. Overall, results presented in this chapter show that, using of passive seismic and radiometric methods can increase effectiveness of prospecting and exploration of diamond pipes of the Arkhangelsk diamond province.

The last two chapters of the book deal with geochemical features of solid material in the Arctic. Chapter 14 presents the monthly, seasonal and annual vertical fluxes of sedimentary matter from the surface layer of the Arctic Ocean, including those directly determined near the North Pole. The composition of sedimentary material in the snow and ice cover was studied at the maximum distance from continents in the high latitudes of the Northern Hemisphere, and the vertical fluxes of material from the under ice layer were determined. To collect this unique material, the authors used sediment traps for water column particles; for snow and ice, they used large volumes of melt water. The author demonstrates that dispersed aeolian sedimentary material is of primary importance in snow and sea ice.

In: The Arctic: Current Issues and Challenges
Editors: Oleg S. Pokrovsky, Sergey N. Kirpotin et al.
ISBN: 978-1-53617-306-2

Chapter 1

LEARNING IN THE CHANGING ARCTIC

Nafisa Yeasmin*, PhD and Stefan Kirchner, PhD
Arctic Centre, University of Lapland, Finland

ABSTRACT

The Arctic is undergoing massive changes which impact local communities which rely heavily on traditional knowledge and local expertise, including indigenous peoples and minority groups. Globalization and climate change make it necessary that traditional forms and contents of learning are revisited in light of the challenges faced by communities in the Arctic. With a particular focus on the needs of traditional communities and the ways in which they are impacted by climate change and the multiple uses of the Arctic which are made possible by climate change, this chapter looks at Klob´s learning style how learning in the Arctic is changing as experiential learning today. While the Arctic and climate change are often the object of education, the focus of this text is on the impact of the latter on effective learning in the former. By bringing together Arctic and minority rights studies with social sciences and by putting the matter of learning in the Arctic into a theoretical framework from the social sciences, this research explores unknown territory. The chapter will show that climate change even impacts core aspects of societies in the Arctic, including the value of traditional knowledge and the societal status of Arctic people in local communities which lead to a discussion on reflective practice of learning about territorial significances in a changing arena.

Keywords: learning, theory, Arctic, communities, climate, knowledge

INTRODUCTION AND BACKGROUND

The Arctic region needs people to fill a shortfall in the labor market, due to declining local birth rate and increasing proportion of the elderly people. From a development perspective, the region would require 1.7 million new people for filling the gap in the labor

* Corresponding Author's Email: Nafisa.yeasmin@ulapland.fi.

market over the next 15 years in order to maintain the current standard of well-being and sustain its development (Vänskä 2015). “The region is full of resources e.g., oil, gas, mine and tourism sectors. Arctic region is sparsely populated and Arctic climate together with abundance of natural resources make it to stand out from other European regions” (Yeasmin, 2013). It is considered necessary to explore natural resources and make better use of them in the region. “The region itself can have a goal is to increase the number of capable labor force for future labor needs in industrial and economic sectors. There is also a need to attract labor from outside the region due to many new industrial projects concerning the extraction of raw material” (Yeasmin, 2013). The Arctic region, which is characterized by a low population density, needs at least a minimum level of diversity of the population and needs to tackle the problem of internal migration from rural to urban areas. It is difficult to attract people to stay in rural parts of the Arctic instead of moving towards urban centers in the south where they currently can expect better opportunities in terms of jobs or education. Extreme climate, cold weather and disparities in every sectors of integration and a lack of certain opportunities to some extent also pushed inhabitants like immigrants or local youth to move to the southern parts of countries such as Finland, Norway or Sweden (Yeasmin, 2013). Community viability has been approached through a variety of conceptual frameworks and tools (Burns 2012; Fiksel 2006; Zhang, Bakshi, 2010). It has value for sustaining small minority communities in the North, which are geographically isolated in global terms, and is thus crucial for the northern economy, which depends on human capital of the region (Arctic Human Development Report 2014). Diverse communities rely on their religious and spiritual norms, beliefs, and practices to cope with difficulties, challenges, and stressors in the host country (Hisham, Kenneth 2015; Diller 2014; Edgell 2012). Community practices can offer a sustainable framework for adjusting to many other important dimensions of a society, such as diversity, equality, social justice and intercultural competency (Kolan, Sullivan 2014). In this phase of transition, the Arctic and climate change are often the object of education. The focus of this text is on the impact of the latter on effective learning in the former. By bringing together Arctic and minority rights studies with social sciences and by putting the matter of learning in the Arctic into a theoretical framework from the social sciences, this research explores unknown territory. This chapter will show that climate change even impacts core aspects of societies in the Arctic, including the value of traditional knowledge and the societal status of Arctic people in local communities which lead to a discussion on reflective practice of learning about territorial significances in a changing arena.

The Impact of Climate Change on Local Communities in the Arctic

Climate Change as a Reality in the Arctic

The Arctic is particularly affected by climate change, which is already a tangible reality across the global high north. The direct effects of the emission of greenhouse gases were observed in early 1980 (Manabe and Stouffer 1980) as were the loss of sea ice, changing weathers patterns, flora and fauna (AMAP, 2017). The sudden rise of temperatures and loss of see ice is not normal. This can create an extreme effect on eco-system of the planet. The melting of glaciers, ice shelves and sea ice causes rising sea levels, which can create disasters such as floods in near-coastal areas. This has also an impact worldwide. Worldwide, more than a billion people in coastal and island nations will directly be affected by the sea-level raising (Storlazziy et al., 2013). The loss of ice leads to changes in the salinity of the sea water, which affects global ocean circulation patterns and biodiversity also far outside the Arctic (Norwegian Polar Institute, n/a). Changing weather patterns also lead to extreme snowfall, heat waves and drought in some places; they also lead to a new record of high and low temperatures (ibid.). Such changing weather patterns can be a threat to global food, climate change, economies and people. This change is gradual in nature. Some of the region might benefit from the changes in the short term, likewise high ocean temperature might increase profits from fishing, but on the other hand, is climate change already having an impact on the natural environment of the Arctic and the people who depend on it for their livelihoods (Kirchner, 2015). This Arctic reality causes great uncertainty regarding the impact of climate change on life in the far North. In the long run, the changes which are already underway today will be challenging for all.

Indirect Effects of Climate Change in the Arctic

Climate change in the Arctic is likely to reinforce socio-economic activities in the regions. Arctic governance systems and agreements related to Arctic economic activities respond to natural and human-generated stressors and challenges. This reactive nature of regulatory systems contributes to the discussion on sustainable development in the region. There is a need of traditional knowledge of Arctic communities for improving Arctic infrastructure, mining, tourism, Arctic marine ecosystem, aquaculture and agricultural activities challenges. The ways in which communities are impacted by climate change along with the ways they are surviving in the changing Arctic. Therefore it is potentially important to create strategies on multiple uses of the Arctic resources, which are made possible by climate change. In order to bring Arctic resources to market (Petrick et al., 2017), it is necessary to have a special support infrastructure for exploration and production activities, including in the Arctic Ocean. This leads to increasing risks for the natural environment and for the people who live in the Arctic. For a long time, the extreme characteristics of the Arctic, such as darkness, remoteness, lack of infrastructure had hindered exploitation for time being. Now, as the Arctic climate is changing, increased accessibility facilitates resources exploitation both offshore and onshore, as well as activities for supporting aquaculture, fisheries, mining etc. Climate change has made parts of the Arctic more accessible, which in turn has raised expectations for economic profits from extractive industries, marine transportation and tourism, as well as from the building of associated infrastructure (Dalsøren et al., 2013). Due to climate change, transport infrastructure is affected, which in turn impacts the safety of local communities, which can be seen in the increasing difficulties for commercial traffic on ice roads and transport of heavy loads (Furgal & Prowse, 2008), which are vital for supplying remote communities with food and other supplies. Climate change also means an increasing frequency of extreme weather events (Kolstad & Bracegirdle, 2008), more coastal erosion, a rising sea level and floods, all of which affects coastal ecosystems and infrastructures (Forbes et al., 2008). Such indirect, human-made, effects of climate change have a direct impact on the Arctic environment and the people who live in the Arctic. In fact, they have consequences for anthropogenic global warming (Corbett et al., 2010), local and regional air quality, and Arctic water, which is also under the risk of oil spills, pipeline leaks etc. (Wilkinson et al., 2017). The latter risk in particular increases as human activities in the Arctic increase, especially in the context of traffic and extractive industries, thereby providing additional challenges to the sustainability of development in the Arctic.

Increasing economic activities also have negative effects on air quality and the atmosphere. A case in point are sulphur emissions from increased shipping activities (Law et al., 2017). Other measures, such as the growth in extractive industries, leads to a rising risk of water pollution. The pollution not only affects the natural environment but also the people who live in the Arctic.

All these changes and socio-economic scenarios lead to a need for better knowledge and understanding among the people who live in the area, in particular in light of the goal of sustainable future development.

This changing Arctic provides challenges and opportunities for inhabitants, which raises questions regarding the current adaptive capacity (Hovelsrud et al., 2011) of local societies in the Arctic. On one hand, this can be seen as a reason for more immigration: "Economic activities in the region encompass diverse sectors, since regional development depends on Arctic fishing, underground technology, oil and gas extraction, mining, forestry, reindeer husbandry and tourism industry" (Röver 2014). The North is a developing region that needs extra human capital for the exploration and exploitation of gas and energy, as it is known that no less than 80 percent of the world's technology for underground mining is expected to come from the European, Norwegian and Russian North (Carl 2011). An influx of people means rising demand, innovations and diverse new ideas. Social networks and group thinking develop common values and comprehensive visions (blog posted to energy Blog on January 10, 2014). On the other hand, is there a need to devise ways towards sustainable development even under conditions of rapidly decreasing population numbers in rural areas. The traditional knowledge which has been created by indigenous peoples in the Arctic over thousands of years can play an important role in developing the knowledge necessary to engage in a sustainable development of the region, looking not only at short-term profits but also at ideas which require a long-term view as they will not promise immediate profits. The sensitive natural environment calls for a limited human footprint. Indigenous peoples in the Arctic have lived in this way for thousands of years.

Who Lives in the Arctic?

The Arctic is home to different ethnicities who bear different identities and cultures. Indigenous, non-indigenous locals, immigrants, refugees, students, spouses, migrant jobseekers and asylum seekers embrace different mixed identities and increase the awareness of various minority communities. In 2013, the overall population in the Arctic was around 4 million people (Heleniak & Bogoyavlensky 2015). It is assumed that the number of Arctic populations has stabilized at just over 4 million and is not projected to increase much further (Heleniak & Bogoyavlensky 2015), because the birth rate in the Arctic region is low. On the other hand, people are aging and local youth are migrating to the southern areas. Therefore, the region faces the problem of high unemployment rates in some Artic cities, rising pension costs and a reduced number of working-age residents. Together, these factors lead to a need to reform the economic situation in some Arctic countries. If the logic of economic stability is to be followed (as is the case in most Arctic areas), the region needs human capital to meet what is perceived as a demographic threat. Nowadays, the awareness of "mixed" identities in Arctic communities has increased. There are different kinds of local communities in the Arctic. Their cultural diversity can help to develop the Arctic into a creative and innovative region. The diverse ethnic indigenous group like Sámi, Inuit, Inuvialuit, Yupik, Iñupiat are not remants of the past but contemporary indigenous peoples which try to find a liveable combination of contemporaneousness and inflexible practice in the Arctic (Schweitzer et al., 2014). The indigenous peoples of the Arctic have their own traditional ecological knowledge, which is widely recognized. There is considerable variation in their cultural characteristics and "a trend towards cultural focus can be seen both in sense of commodity and external recognition, which can be an asset in the future" (Schweitzer et al., 2014). In addition to Arctic indigenous people, there are many newcomers, like immigrants from different countries, who live in the rural and urban regions of Arctic, but also non-indigenous minorities with long histories, such as the Romani people, as well as the members of the dominant settler society.

Cultural differences and diverse community demand new policies and legislation in which tend to rely on. Policies and legislation should also focus on old-fashioned identity categories for maintaining traditional experience to valuing differences. The accommodation of all diverse Arctic communities within the similar political and economic systems has often been challenging and there is a need to manage cultural diversity in order to sustaining the overall community. A demographical threat along with a trend to moving south - is instigating an unbalanced labor market in the rural North. To some extent, recognition of all small, dominant and minority culture led to different conflicts related to cultural pluralism. Cultural inclusion for prevailing the value systems of all culture is in a need of social institutional discussion for changing social construct. At the same time, the settlements have in some cases extinguished or pressed towards greater assimilation to cope with this changing Arctic environment rather than supporting cultural diversity. Now Arctic peoples' abilities, traditional knowledge, mind-sets, and social value could thrive and contribute to closing a knowledge gap in a challenging environment. Many communities in the urban area in the Arctic also lack the proper knowledge of northern indigenous peoples, minorities and their culture.

The role of cultural diversity management within the broader context of sustainable development. The value of diversity management in the north needs a sustainable learning framework to adjust to many other important and changing dimensions of the society, such as diversity, equality, social justice and intercultural competency (Kolan, Sullivan 2014) or environmental resilience.

The three major pillars of sustainable development are boosting the economy, protecting the environment and promoting social equality. Critics argue that Arctic policies are often best developed for the people of the North (Röver September 8-9, 2014). People in the changing Arctic require knowledge on Arctic policies, they need to know about social equality and develop a mind-set regarding diversity (and the changing forms of diversity, ranging from the indigenous/settler interaction to the role of modern day newcomers to the Arctic), development policies and continuous education and training on valuing differences and changes. Knowledge is potentially important for building resilience and capacity in a changing situation. A variety of development projects have been put forward to move the region towards more sustainable development (Atkinson 2004, 229-249) and communities need to learn the culture of organizations to develop likewise. Traditional knowledge adds value to organization processes. Social values of the communities, equity between communities can build the capacity for resilience (Common, 1995) for increasing integrity for surviving (Pearce et al., 1989). Collective engagement with other communities is necessary for the creation of conditions for a sustainable economy (Yeasmin 2014). Local communities need knowledge and information about what can be done at the local and regional levels to preserve and protect the environment and the natural and human capital in the Arctic region (Rasmussen, Olsen 2014).

LEARNING IN THE ARCTIC

In such a changing environment of complexity and uncertainty, constant learning can build resilience capacity and competencies of diverse communities in the Arctic. A combination of knowledge of rural, urban, ethnic indigenous and immigrants' communities could create a knowledge society (Singh, 2010). Awareness of the importance of learning in present times is essential to gain competitive advantages for surviving in a changing situation (Harvey and Denton 1999; Ulrich et al., 1995). A powerful learning environment is a valuable asset for enhancing the capacity of adaptability (Sun, 2003) which inspires, facilitates and empowers communities in a certain organizational territory (ibid.). Learning about the traditional knowledge of indigenous peoples or ethnic minorities can ease innovation and help society at large to assimilate new knowledge in several new ways (Dess & Picken, 2000).

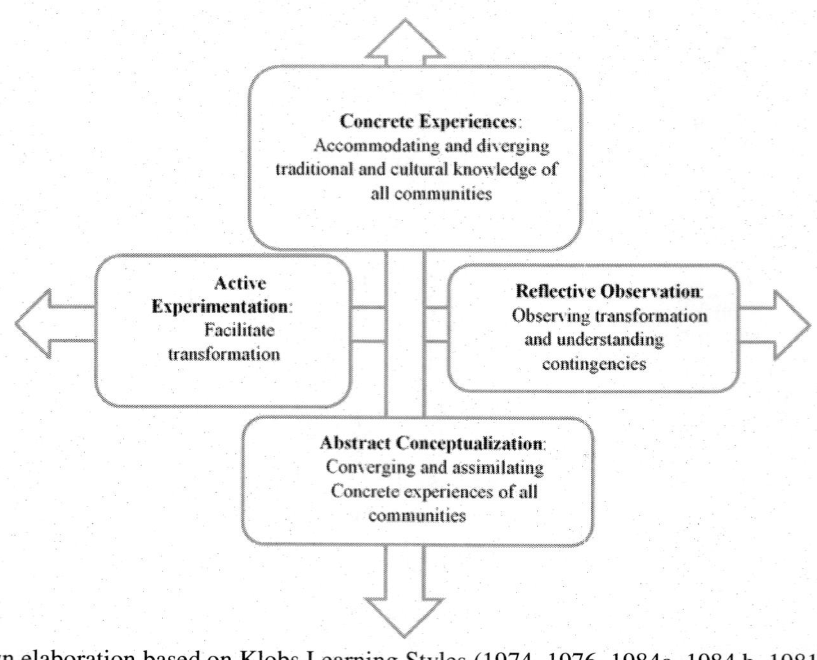

Source: Own elaboration based on Klobs Learning Styles (1974, 1976, 1984a, 1984 b, 1981).

Figure 1. Learning style in a changing Arctic.

In this study, we focus on Klob´s learning styles that the learning process consists primarily of the utilization of the learner´s previous experiences and practice and experimenting the new assimilated versions of logical approach, new ideas and concepts based on practical knowledge. And finally, accommodating different types of knowledges by observing the converging learning experiences for understanding each stage of experiential learning circle (see Figure 1).

METHODS AND CONTENTS

Learning in a changing Arctic demands accommodating feelings and watching others experiences. In this method of learning is to grasp information and identify certain triggers by watching diverse logical approaches. Klob´s learning theory (1974) is a four-step learning method. Klob argues that learning is a transformation of experiences by feeling, watching, thinking and practicing ways towards a new experience.

Learning of Traditional and Cultural Knowledge

Learning about others cultural and traditions accommodate different ways logic to accelerate pace of changes. Analysis on diverse communities and peoples support gaining information on their livelihoods, social values, economic access and cultural identity. Various environmental and socio-economic factors for surviving in a certain ecological environment are preserving in community traditions (Berkes, 1993). But to explore these hidden factors one needs people from diverse communities who can continually expand their interest to accommodate modes to learn and create new and expansive patterns of thinking regarding traditional and cultural knowledge (Singh, 2010). Sharing knowledge and concrete practices of different groups (Thurner et al., 2000) facilitates knowledge transmission. This makes the information accessible, which is required to sustain social relationships and suggests a new concept of practical and experiential approach, since supporting peoples by planning community management which is basically based on challenges and experiences could probably be more sustainable.

Since diverse communities are able to look at factors from different perspectives. Diverse communities bear several different viewpoints. Diverse viewpoints can create diverged ideas through, for example, brainstorming (Klob, 1984 a). The diverse cultural interests of different Arctic communities and their different lifestyles mean that different communities can teach each other food safety, human safety or the generation of income. Examples across the Arctic, in particular, but not exclusively, relating to indigenous peoples include reindeer herding, fishing and arts and crafts. The latter category raises important questions regarding intellectual property rights and the risk of cultural appropriation.

The creation of the term 'duodji' for specific Sámi-made items is an important step towards the protection of indigenous intellectual property rights which might inspire similar developments elsewhere. The introduction of reindeer husbandry in Canada and Alaska (which are also home to Caribou, which have long been hunted) in the late 1890s (with a herd from Russia (International Centre for Reindeer Husbandry, 2019), and again in the 1930s, provides an important example for such a transfer of skills and knowledge among indigenous peoples of the Arctic, in this case from the Sámi to the Inupiaq (International Centre for Reindeer Husbandry, 2019) and the Inuvialuit (International Center for Reindeer Husbandry, 2019a), respectively.

New communities in the Arctic are immigrants (refugees, asylum seekers, students, spouses etc.) bear different cultural skills. Diverging cultural and traditional knowledge can combine information and solve problems by using imaginations. (Ohmagari and Berkes, 1997). Collecting traditional and cultural knowledge which is scattered among diverse communities and stakeholders, though a difficult task, is an important source of effective climate adaptation (Aylett, 2014).

Reflective Observation

The changing Arctic demands collaborative learning styles between communities. Observing the contingencies between experience sharing and understating particular important factors need reviewing and reflection of knowledge. Assessing community behaviors and consequences is crucial in this reflective observation phase. Many communities are well-known and experienced by direct and indirect impact of changing Arctic likewise coastal communities and communities who are residing in industrial zone or nearby mining oil and gas extraction field, forestry, reindeer husbandry and in a tourist zone, gather empirical local knowledge which is valuable information by sharing it with other communities. To understand, whether or not they are bringing the empirical experiences, requires further observation between experiences and understanding. Ideally, their community concerns are adopted by the policy makers, researchers or stakeholders are whether or not credible enough need a reflective observation on reality for real learning of real issues. The rapid changes of climate and weather fluctuations affect the community and the ecosystem structure (AMAP, 2011). Acquiring knowledge through imitation learning is an effective approach for receiving examples of observation- action tuples (Liu et al., 2018). The tuple might be added some learning algorithm which could supervise the learning method of communities on changing ecosystem in the Arctic. In this reflective observation phases, communities can identify relevant proper actions, viewpoints, behaviors and other characteristics of other communities in similar regions. This reflective observation could constitute different enablers and varieties of interesting applications for environmental configuration. Behavioral changes of communities and peoples based on environmental challenges, such as climate change adaptation, need to be taken into focus in order to better understand the contexts in which such observational learning can occur (Ensor & Harvey, 2015). According to Ensor & Harvey, learning in a changing environment goes beyond individuals and includes communities, networks and systems (ibid.). New meanings of new experiences that support climate adaptation and embedded understanding of collaborative learning through observation in different perspectives are sense-making evidences for communities (Wals, 2007) and consider the ultimate impacts of community learning approaches.

Abstract Conceptualizing

In this part of the theoretical considerations, abstract conceptualizing means that sense-making on reflective observation and empirical experiences which will get a concise and logical framework. Specific evidences on what to change and how to change could be expected in this learning phase. Evidences are potentials of learning that can challenged and transformed into greater community views and values (Lebel et al., 2010). Clear explanations and logical frameworks support the understanding of information which has been gained through the observation of traditional and cultural knowledge. Assimilating all new skills and experiences within a regional governmental context for constituting a large -scale infrastructure framework based on practical values that can attract converging a common learning style for climate adaptation. Abstract conceptualizing process improves decision-making processes as well (Cundill & Rodela, 2012). This can influence natural resource management based on factual and empirical knowledge on the regional level.

Active Experimentation

Active experimentation of conceptualizing opportunities are significant for social learning, for example, face-to-face facilitation, network-centric activities, peer-to-peer facilitation support transformation. Collective awareness raising, established values protocols and other relationship during the learning process, up to abstract conceptualizing, can not only raise awareness but also reform kinship to climate change impacts (Pelling et al., 2014) that easily facilitate behaviors of institutions and other decision making-bodies by actively experimenting these findings and practices which could support adaptation needs of all communities as a whole who have been suffering from climate changes. The realistic and concrete knowledge of various communities is credible enough for national, regional and international organizations concerned with encompassing adaptation strategies or approaches to facilitate transformation. Different tools, approaches, methods can reinforce the need of careful consideration regionally and internationally. Therefore, finding facilitators (various actors and sectors) can accelerate those tools, and methods for active experimentation (Cundill et al., 2014). System management can demonstrate visible experiences among stakeholders create a new alternative method to support actors and sectors to adopt transformation. Active experiments of a logical framework facilitate the expansion of both debate and dialogue and can bring resources into use for critical development decisions. The emergence of new knowledge, skills-set, sense-making evidences reorganize conventional notion for experimenting mainstream climate change adaptation in the changing arctic. This type of assimilated and cross-sectoral knowledge can provide realization and increase understanding for various domains. Active experimentation also needs to be examined and observed in order to determine protected spaces for protecting the holes of the context of climate change which can provide benefits from transformation. Active experimentation facilitates greater transformative changes to respond adequately in changing conditions (Cumming, 2013). Experimentation of learning knowledge can build and accommodate capacities on incremental reforms of climate governance (Sabell & Zeitlin, 2011). Therefore, this active experimentation is not only the enactment of collecting evidences, perhaps it creates indeed perceptions for a broad understanding of cultural and traditional knowledge of communities (Matschoss & Heiskanen, 2017) and understanding agents of the communities who render local climate initiatives more operative and experimental (Hildén et al., 2017) for the benefits of all communities.

Conclusion and Application

In the Arctic, the dual forces of globalization and climate change are causing changes, which challenge the residents of the Arctic. These challenges affect in particular indigenous communities, which often are dependent on their close relationship for cultural and economic purposes. Under circumstances of radical change, knowledge is becoming more and more important as a key to a sustainable future life of Arctic communities. This increases the societal value and importance of learning. In order for Arctic societies to be able to limit the negative effects of the multi-faceted change currently underway in the Arctic and to adapt to an unclear future, constant learning is essential. The experience of fundamental change, which was not chosen by the people who live in the Arctic and which is often perceived as the result of outside forces beyond the control of Arctic residents, also impacts learning in the Arctic. The effective generation and transfer of knowledge in an Arctic context requires empathy for the learner and for others, openness to the validity of diverse experiences and a balancing act between culturally-induced respect for traditional knowledge and an objective evaluation of knowledge generated in the past. In a world which is not only new but which continues to undergo fundamental changes, many of which fall outside the frame of reference of human experiences, change is becoming the new normal. As learning is meant to prepare us for the future, learning in the changing Arctic will have to adapt and evolve as well in order to prepare future generations for the, still invisible, challenges which lie ahead.

References

Andrew, R. 2014. *Socio-Economic Drivers of Change in the Arctic*. Oslo: CICERO Center for International Climate and Environmental Research.

Arctic Monitoring and Assessment Program (AMAP). 2017. *Snow, Water, Ice and Permafrost in the Arctic (SWIPA) 2017*. Oslo: AMAP, p. 269.

Aylett. A. 2014. *Progress and Challenges in the Urban Governance of Climate Change: Results of a Global Survey*. Cambridge: Massachusetts Institute of Technology, MA.

Berkes, F. 1993. *Traditional ecological knowledge in perspective*. Traditional Ecological Knowledge: Concepts and Cases, Inglis, J. T., ed. Canadian Museum of Nature/ International Development Research Centre.

Berkes, F. 2008. *Sacred ecology: traditional ecological knowledge and resource management*. Second edition. Philadelphia, Pennsylvania, USA: Taylor and Francis.

Common, M. 1995. *Sustainability and Policy: Limits to Economics*. Cambridge, UK: Cambridge University Press.

Corbett, J. J., Lack, D. A., Winebrake, J. J., Harder, S., Silberman, J. A., and Gold, M. 2010. Arctic shipping emissions inventories and future scenarios. *Atmospheric Chemistry and Physics* 2010, 10: 9689–9704. https://www.atmos-chem-phys.net/10/9689/2010/acp-10-9689-2010.html.

Cumming, G. S. 2013. *Scale mismatches and reflexive law. Ecology and Society*, 18.

Cundill, G., Lotz-Sisitka, H., Mukute, M., Belay, M., Shackleton, S., and Kulundu, I. 2014. *A reflection on the use of case studies as a methodology for social learning research in sub Saharan Africa. NJAS-Wagen Journal Life Sci* 2014, 69:39–47.

Cundill, G., and Rodela, R. 2012. A review of assertions about the processes and outcomes of social learning in natural resource management. *Journal of Environmental Management* 2012, 113:7–14.

Dalsøren, SB., Samset, BH., Myhre, G., Corbett, J. J., Minjares R., Lack, D., and Fuglestvedt, JS. 2013. Environmental impacts of shipping in 2030 with a particular focus on the Arctic region. *Atmospheric Chemistry and Physics 2013, 13: 1941–1955.* https:// www.atmos-chem-phys.net/13/1941/2013/acp-13-1941-2013.html.

Dess, G. G., and Picken, J. C. 2000. *Changing roles: leadership in the 21st century. Organizational Dynamics*, 28(3), 18–33.

Ensor, J., and Harvey, B. 2015. *Social learning and climate change adaptation: evidence for international development practice.* https://onlinelibrary.wiley.com/doi/full/10.1002/wcc.348.

Forbes, D. L., Manson G. K., Mate, D., and A. Qammaniq. 2008. *Cryospheric change and coastal stability: combining traditional knowledge and scientific data for climate change adaptation. Ice and Climate News* 11: 17–18.

Furgal, C., and Prowse, T. D. 2008. *Northern Canada. From impacts to adaptation: Canada in a changing climate 2007.* Lemmen, D. S., Warren, F. J., Lacroix, J. and Bush, E., eds. 57–118. Ottawa, ON, Canada: Government of Canada.

Harvey, C., and Denton, J. 1999. To come of age: the antecedents of organizational learning. Journal of Management Studies 36(7), 897–918.

Heleniak, T., and Bogoyavlensky, D. 2015. *Arctic populations and migration.* In Larsen, J. N., and Fondahl, G, eds. Arctic human development report: regional processes and global linkages. Copenhagen: Nordisk Ministerråd.

Hovelsrud, G. K., Poppel, B., van Oort, B., and Reist, James D. 2011. Arctic Societies, Cultures, and Peoples in a Changing Cryosphere. AMBIO 40 (Suppl. 1): 100. https://link.springer.com/article/10.1007%2Fs13280-011-0219-4.

Hildén, M., Jordan, A., and Huitema, Dave 2017. Special issue on experimentation for climate change solutions: the search for climate change and sustainable solutions- the promise and the pitfalls of experimentation. *Journal of cleaner production*, 169:1-7.

International Centre for Reindeer Husbandry. 2019. *Inupiaq Eskimo.* http://reindeerherding.org/herders/inupiaq-eskimo/.

International Centre for Reindeer Husbandry. 2019a. *Inuvialuit/Sámi.* http:// reindeerherding.org/herders/inuvialuit-sami/.

Kolstad, E. W., and T. J. Bracegirdle. 2008. Marine cold-air outbreaks in the future: An assessment of IPCC AR4 model results for the Northern Hemisphere. London, UK: *Climate Dynamics* 30: 871–885.

Kolb, D. A. 1976. *The Learning Style Inventory: Technical Manual.* Boston, MA: McBer & Co.

Kolb, D. A. 1981. *Learning styles and disciplinary differences.* The modern American college, 232-255.

Kolb, D. A. 1984. *Experiential learning: Experience as the source of learning and development (Vol. 1).* Englewood Cliffs, NJ: Prentice-Hall.

Kolb, D. A., and Fry, R. E. 1974. *Toward an applied theory of experiential learning.* MIT Alfred P. Sloan School of Management.

Kolb, D. A., Rubin, I. M., and McIntyre, J. M. 1984. *Organizational psychology: readings on human behavior in organizations.* Prentice Hall.

Kirchner, S. 2015. Climate Change Effects on Snow Conditions and the Human Rights of Reindeer Herders. *Pace Environmental Law Review*, 33: 1-22. https://digitalcommons.pace.edu/pelr/vol33/iss1/1.

Law, K. S., Roiger, A., Thomas, JL., Marelle, L., Raut, J. L., Dalsøren, S., Fuglestvedt J., Tuccella, P. 2017. Local Arctic air pollution: Sources and impacts. AMBIO.

Lebel, L., Grothmann, T., and Siebenhüner, B. 2010. *The role of social learning in adaptiveness: insights from water management. Environ Agreements 2010, 10:333–353.*

Liu, Yuxuan, Gupta, A., Abbeel, P., and Levine, S. 2018. *Imitation from observation: learning to imitate behaviours from raw video via context translation.* Electrical Engineering and Computer Sciences University of California at Berkeley. Technical Report No. UCB/EECS-2018-37. http://www2.eecs.berkeley.edu/Pubs/TechRpts/2018/EECS-2018-37.html.

Manabe, S., and Stouffer, R. J. 1980. Sensitivity of a global climate model to an increase of CO2 concentration in the atmosphere. *Journal of geophysical research.* https://doi.org/10.1029/JC085iC10p05529.

Matschoss, K., and Heiskanen, E. 2017. Making it experimental in several ways: the work of intermediaries in raising the ambition level in local climate initiatives. Centre for Consumer Society Research: *Journal of Cleaner Production.*

Norwegian Polar Institute. https://www.npolar.no/en/themes/global-climate-change/.

Ohmagari, K., and Berkes, F. 1997. Transmission of Indigenous Knowledge and Bush Skills among the Western James Bay Cree Women of Subarctic Canada. *Human Ecology* (1997) 25(2):197-222.

Pearce, D. W., Markandya, A., and Barbier, E. 1989. *Blueprint for a Green Economy.* London, UK: Earthscan. ISBN 1853830666.

Pelling M., O'Brien, K., and Matyas D. 2014. Adaptation and transformation. *Climate Change.*

Petrick, S., Riemann-Campe, K., Hoog, S., Growitsch, C., Schwind, H., Gerdes, R., and Rehdanz K. 2017. Climate change, future Arctic Sea ice, and the competitiveness of European Arctic offshore oil and gas production on world markets. AMBIO. https://link.springer.com/article/10.1007/s13280-017-0957-z.

Sabel, C. F., & J. Zeitlin (Eds.). 2011. *Experimentalist Governance in the European Union. European Law Journal*, Oxford University Press.

Storlazziy Curt, et al. 2013. *Forecasting the Impact of Storm Waves and Sea-Level Rise on Midway Atoll and Laysan Island within the Papahānaumokuākea Marine National Monument—A Comparison of Passive versus Dynamic Inundation Models.* U.S. Geological Survey.

Singh, Kavita. 2010. An analysis of relationship between the learning organization and organization culture in Indian business organization. *Organizations and markets in emerging economies*, 2010, vol. 1, No. 1 (1).

Sun, He-Chuan. 2003. Conceptual clarifications for organizational learning, learning organization and a learning organization. *Human Resource Development International* 2003, 6(2), 153–166.

Turner, N. J., Ignace M. B., and Ignace, R. 2000. Traditional ecological knowledge and wisdom of Aboriginal peoples in British Columbia. *Ecological Applications* 10(5):1275-1287.

Ulrich, D., Jick, T., and von Glinow, M. A. 1995. High-impact learning: building and diffusing learning capability. *Organizational Dynamics*, 22(2), 52–66.

Wilkinson, J., Beegle-Krause, C. J., Evers, KU. Hughes, N., Lewis, A., Reed, M., and Wadhams, P. 2017. *Oil spill response capabilities and technologies for ice-covered Arctic marine waters: A review of recent developments and established practices.* AMBIO.

Wals, A. E. J., ed. 2007. *Social Learning towards a Sustainable World: Principles, Perspectives and Praxis*. Wageningen, the Netherlands: Wageningen Academic Pub.

Yeasmin, Nafisa. 2018. *The governance of migration manifest itself to those who are being governed: economic integration of immigrants in Arctic perspectives.* Doctoral dissertation: University of Lapland.

In: The Arctic: Current Issues and Challenges
Editors: Oleg S. Pokrovsky, Sergey N. Kirpotin et al.
ISBN: 978-1-53617-306-2

Chapter 2

CLIMATE CHANGE FROM THE ARCTIC PEOPLE'S POINT OF VIEW: RHYTHMS OF EVERYDAY LIFE, INFRASTRUCTURES AND LANDSCAPES

Lidia Rakhmanova[1,*], Larisa Kolesnichenko[2], Viktor Drozdov[2], Iurii Kolesnichenko[2] and Sergey N. Kirpotin[2]

[1]Laboratory of the ecosystems and climate change research,
National Research Tomsk State University, Tomsk,
Higher School of Economics, St. Petersburg, Russia
[2]Laboratory of biodiversity and ecology,
National Research Tomsk State University, Tomsk, Russia

ABSTRACT

The Chapter reveals to the reader a look at the global process of climate change from the perspective of the local population of the Yamal-Nenets Autonomous district and adjacent regions. Their destinies, way of life, professional plans and career trajectories unfold against the background of changing natural conditions, shifting seasons, unpredictable ice conditions, the collapsing Soviet and emerging post-Soviet infrastructure being built by new stakeholders. The first section explains the socio-economic, historical, infrastructural and natural linkages between the Arctic and subarctic zones, as well as the need to take into account the zonal changes throughout Western Siberia, as they form a transdisciplinary context of transformations in the Russian Arctic.

The second section shows the scenarios of overcoming the difficulties and adapting to the changing conditions caused by the new wave of development of the North, management decisions and natural disasters occurring during the life of the last two generations of the northerners. What is more important for survival in the Arctic: the ability to survive autonomously, using local knowledge and skills, community resources, or the bet made on the operational supply and close, intensive transport and informational links in the Arctic?

* Corresponding Author's Email: muza-spb@yandex.ru.

The third section raises the question of whether there are serious climate changes (that can be seen thanks to the research of landscapes, the state of permafrost, water resources, soils, ecosystems, ecology of regions), from the perspective of local residents, leading the economy and working in the far North? How do they feel them and what aspects of daily life are affected by climate change? The Chapter is based on empirical research on the territory of the Yamal and Khanty-Mansiysk district (Yugra) and northern part of Tomsk region, the corpus of interviews and observations.

Keywords: arctic, local communities, climate change, infrastructures, lifestyle

INTRODUCTION

When the ‘environment’ as a phenomenon emerges in social and human science research, it often presents itself as a context, being an object and not a subject of ongoing transformation processes. The fact that the environmental, air, water and soil conditions, as well as their balance, are one of the reasons that trigger migration flows in the Arctic and Sub-Arctic, drives changes in the social structure of local communities, changes in human potential indicators that alter employment patterns in different regions, and affect well-being, is confirmed through a comprehensive analysis of data and indicators from different types of empirical sources.

This chapter aims to show that the social research of the Arctic zone as whole and of the Arctic communities, when appropriate anthropological optics are applied, allows us to look at the Arctic environment not as a context in which the destinies of locals and outsiders are evolving, but as a subject or actor of change with which each locality and the community as a whole engages in a dialogue (Bolotova 2012).

We are facing an amazing fusion of two phenomena of different nature: vulnerable Arctic ecosystems and vulnerable human beings. A key characteristic of these complex and multi-layered relationships is that they are relationships of "mutually vulnerable actors," one of which has the capacity and a certain language to describe the process and consequences of ecosystem and human community’s coexistence. These descriptions are undoubtedly "pre-scientific", as they do not appeal to objective meteorological observations, or, for example, to the analysis of water in the Ob river basin or the analysis of permafrost processes, but it is this "pre-scientific" view that can fundamentally change scientists' perceptions of what unobvious links do exist between human and natural phenomena.

The key concept in this chapter is "viewpoint". And here we will talk about the point of view, about scientists' interpretation of how the local residents experience and perceive the undergoing changes. We should have been wary of such "double interpretation" if we had not had at our disposal the correlated data pools: the opinions of observers living in the Arctic environment with data on the state and statistics of variations in certain elements of the Arctic environment (vegetation, fauna, soils, permafrost, temperature regimes), the certainty of which is confirmed by the studies of colleagues presented in this book.

The methodological complexity of such a research needs to be specified, as the existence of a "pure narrative" of the local population on climate change, focused solely on them, is very questionable. It is difficult to speak of climate, because the 'climate' as a concept, although it appears in everyday life, is in fact a product of a fairly high level of abstraction. Our task was to find the foundation, the language of description and the appropriate optics to see through the tiniest details of the northern villages everyday life - certain elements that form a coherent mosaic of evidences. Evidences, focused on the climate change expressed "from within communities".

1. The Transformation of the Arctic Lifestyle from an Interdisciplinary Perspective: Drivers of Change in the Arctic Zone and Beyond

The first phenomenon by means of which we will try to identify the changes in climatic conditions in the West Siberian Arctic from the position of the local population is, undoubtedly, the way of life. It would be appropriate to raise the question of the extent to which the described way of life is 'Arctic' and where the boundary between the "existence" of the Arctic and non-arctic lifestyle and daily life lies. The question is particularly difficult if we take into account the fact that any transformations of climate and ecosystems do not only change the indicators at a particular point in the geographical space, but also shift the boundaries of the areas and zones associated with similar conditions, resources and properties.

In other words, to date, those territories that used to be part of the "Arctic" may no longer be formally considered the part of it. And yet, the inertia of local identity, the longevity of technology and tradition, leads to the fact that the elements and patterns that characterize the way of life in the Arctic extend to larger areas, where they do not seem to be necessary for survival.

On the other hand, we can observe enclaves of non-arctic urbanized culture, created by rotational settlements around the mineral deposits, which is becoming more and more a vital alternative and opportunity (Dudeck 2010), and is becoming more and more involved in the social and infrastructural landscape of the Arctic.

If we return to the thesis on the vulnerability of individuals and communities, a vulnerability, that is particularly characteristic for the Arctic territories, then, on the example of the creation of well-equipped settlements, with a well-thought-out structure, modern engineering networks and materials for building houses, we can see in it not the process of adaptation, but the creation of a barrier or even protective environment, which to a greater or lesser extent allows to live in Arctic in conditions that do not fully correspond to the Arctic (Arkema 2014): in terms of loads, risks and other indicators.

To live in the Arctic or subarctic zone as if you live in a mild climate means to live in a self-contained, autonomous system, which is connected with cycles, seasons, risks, and extreme phenomena of the Arctic environment only to a small extent, and, therefore, means not to have a critical, reflexive ability to observe and track the changes taking place. One of the key motives for observing and fixing any natural transformations is the close relationship between the way of life and the resources of the environment. In the modern context, this connection is manifested not only in such complex things as environmental management (hunting, fishing, gathering wild-growing herbs, etc.), but also in trade (food supply, shortages of certain goods), medicine, various types of business, education and cultural life.

Therefore, when we talk about the problem of preserving the traditional way of life among representatives of the indigenous peoples of the North (See: Dudeck 2015), the emphasis is made on non-material culture and oral tradition, as well as reindeer herding, hunting and fishing. However, change of climate and ecosystems, show us that the preservation of technologies and practices can sometimes mean the refusal to adapt to transformations (Cunningham et al. 2015), and can lead to absurd situations of imbalance. Such a case can be described as a "tradition of extracting a nonexistent resource". Despite the wisdom and flexibility of local cultures, which possess a multiresource nature and allow to switch from one resource to another, even in the face of extinction of a certain species of animals, changes in their habitat, changes in the migration routes of river fish (Huryn et al. 2005). The variability of such systems could be a source of endlessly living and flexible adjustment of human communities and ecosystems if another actor did not interfere with this dyad.

Historical "waves" of northern infrastructures formation have created a heterogeneous space on the territory of the Russian Arctic, characterized by transport and information gaps, gaps in supply systems and logistics (Pilasov, Galtseva, Atamanova, 2017). The role of the planned economy in the development of Western Siberia, its sub-Arctic and Arctic zones can now be estimated as a dual one. On the one hand, the history of exploration of the North shows us powerful leaps in development, construction of roads, new settlements, settlements and cities, forced exploitation and, on the other hand, - population settling in those areas that could not become self-sufficient and autonomous in terms of supply and production of intraregional products (Pelyasov 2016).. All these trends have led to the formation of a northern economy, supported and subsidized by infrastructure from outside the Arctic regions of the USSR (Bolotova, Vorobyev 2007).

This period was characterized by a multistage process of re-circulation and "transformation" of the extracted resources - minerals, oil, gas, fur, fish, caviar, wild reindeer, hunting products - through the network of supply, exchange, trade, planned preparations into social and economic benefits: medical care, subsidized transport flows (river transport, polar and circumpolar small aviation, construction of narrow gauge and new roads). The Arctic of the Soviet period, as well as the areas equated to the Far North regions, has formed a stable phenomenon of the "state of care and social responsibility".

Thus, the harsh climatic conditions of living and working in the Arctic and Subarctic, psychologically difficult conditions associated with a short astronomical day and polar night, partial isolation of settlements during the development of new territories, the lack of developed social and cultural environment - were leveled out by the system of alternative benefits and guarantees (Lukin 2014), which reduced social tensions in the northern regions and formed a balance between risks and available resources. The system of guarantees assumed:

- the increased "northern" wages,
- lowering the retirement age;
- increase in the wage coefficient;
- possibility to move to the chosen areas after the end of the legal term of work in the "North" (13 years) with compensation for the travel of the family and the employee to a new local residence;
- special food supply;

- compensation for electricity and heating costs;
- double paid vacation;
- organization of field medical services in hard-to-reach areas,

and many other benefits and activities aimed at alleviating hardship. Nevertheless, what exactly is the severe nature of the way of life in the territories described in this chapter? A.N. Pilasov and his colleagues note that the development of Arctic settlements is characterized by uncertainty, associated with the rapid deployment and rapid reduction of exploration waves in the North (Pilasov, Galtseva, Atamanova, 2017). The extreme living conditions noted by many authors are extremely important for this study precisely because after the collapse of the USSR and the change of the development model and management practices of northern territories, the administrative, planned socio-economic "envelope", which protected the population and mitigated the consequences of human contact with the Arctic ecosystem, was destroyed, exposing the most acute contradictions and showing the risk zones in the process of human-environment relations.

The absurdity of the situation was that the system envelope created in the Soviet period, was destroyed and died out gradually, forming on the map enclaves and conglomerates of settlements, isolated villages and entire districts that were uneven in terms of the level of development and the level of autonomy (Pelyasov 2016). Instant and complete destruction of the system of connections in the North could give an impulse to the total resettlement of local residents from the Arctic region after the collapse of the USSR. However, resources for such mobilization were not sufficient, so instead of active interregional migration, all the forces of local communities that found themselves on the ruins of Soviet structures were concentrated on the processes of adaptation to the environment that suddenly appeared before them directly as a "bare" reality of the Arctic (Ford et al. 2017).

Analyzing the situation in the late 2010s, we are dealing with the legacy of those adaptation processes that in the long run represent a long history of human development of the Arctic spaces, and, in the short term - the destruction of old and the emergence of completely new adaptation strategies in the 1990s. It was they who shaped the whole complex of approaches, institutions, techniques and practices that contributed to comfortable and fruitful living in a severe environment (See: Zamyatina, Pilyasov 2018a).

However, how do you determine where and when the adaptation started and where did it end? We can only measure its effectiveness, or we can say that adaptation has not been accomplished and the process has failed. In this case, the Arctic appears as a space of constant human trial and a space of constant adaptation failures, which provoke the birth of new logic of exploration and survival, inventions and institutions.

Human trial, as well as adaptation, is a process, but a 'process-event' that takes place within a limited time frame and is a crisis point in a person's biography or autobiography. Seriously, the key test is the middle position of the scales: at this moment it is decided whether to lean towards "survival" or towards smooth adaptation. This moment is directly related to the issues of "rooting" in this local area. What does it mean to be an indigenous resident of the Arctic and the North? How many times do we have to overwinter, how many years do we have to live in the Arctic region?

We dare to suggest that not only the elements of a person's self-identification as a member of a particular community or region are very important for Arctic identity (Rakhmanova 2014), but also the objective ability of community members to respond independently and together to the challenges of extreme Arctic conditions (Rakhmanova 2015).

During the 2018-2019 field season, we asked the informants how difficult it was for them to adapt to their environment, whether they wanted to move to more southern areas, to their homelands, and tried to find out what was the source of discomfort. Here are just a few quotes from the interview:

- Does the local climate suit you now?
- The climate here is very difficult.
- So you've adapted, but you're not used to it at the end?
- I have a serious illness because of my move, I am asthmatic.
- Is asthma stronger here?
- It wasn't there before, it was acquired here.
- And you didn't think about coming back despite the disease?
- There was nowhere to go back to (f., 63, Kargasok)

In order to continue to reflect on the peculiarities of human perception of the environment and its changes, it is necessary to show how multi-elemental and complex this environment is and how many factors influence the assessment of a person's degree of well-being, comfort or vice versa - the risk of living in a particular environment.

1.1. Geographical or Infrastructural Isolation of the Community/Locality

The linkage with the environment, the dependence on its condition (weather, climate in general, the presence of extreme natural phenomena), is more strongly felt and is lived through in remote villages. Moreover, the stronger and longer was the flourishing of the city-forming enterprise before the collapse of infrastructure ties, the better equipped were the social infrastructure facilities, medical care, food supply, transport communications with the regional center and neighboring villages, the more sharply the extinction and collapse of once strong ties is perceived.

Environmental challenges, which the community can easily overcome due to informational, market and transport links with other communities and key governing institutions, become more acute after the partial collapse of ties, revealing the specifics of Arctic living conditions and, sometimes, - even survival practices. Thus, the climate and health factors come to the forefront in the assessment of complex phenomena, that includes (See: Rakhmanova 2018):

1. Climatic conditions that imply instability of weather conditions, unpredictability of the emergence of temperature extremes.
2. Risky or almost hopeless efforts to maintain one's own household and to obtain the necessary vitamins and foodstuffs.
3. Demographic risks and difficulty in finding marriage partners both in the surrounding local communities (small population, sparse settlements), and the creation of a family with a spouse from non-arctic regions (difficulty of adaptation, risks)

4. The wide gap and contrast between the culture and specificity of the labor activity characteristic of the Far North and Arctic regions and the work in unified budget and state structures (education, medicine, trade, transportation), the complexity of switching between them and the need to master specific knowledge and skills.
5. Explicit influence of seasonal specifics, which dictates and determines the rhythm of human life.
6. Limited circle of communication, social forces of a closed community.
7. The problem of searching for one's "business of life" and raising the question of identity.
8. The impenetration of everyday life in the Arctic with disruptions and practices aimed at eliminating them and adapting to the changing life-as-trial environment (vie-epreuve) (Foucault, 2007: 477).

The specifics of the Arctic zone, the existence of hard-to-reach areas, the wealth of hidden resources, and the difficulty of controlling the remote micro-regions have determined the patterns of regulation and management strategies in the area. The management effect played a certain role during the Soviet period, on the one hand, being responsible for the excellent supply, and, on the other hand, for the leveling out of differences and uniqueness of communities.

In the post-Soviet period, the way to live and survive in the Arctic has become not only a technology, but also a significant symbolic resource for those communities (Rakhmanova 2014, Wallbridge 2012), which, through the preservation and sometimes even reconstruction of crafts, technologies and practices of environmental management and elements of everyday life, have gained their lost authenticity and involvement in the Arctic culture.

The boundaries of the "Arctic" in its flexible sense are expanding significantly beyond international classifications, beyond the boundaries of the Arctic Zone of the Russian Federation (AZRF), when we look at this phenomenon through the prism of social and human sciences.

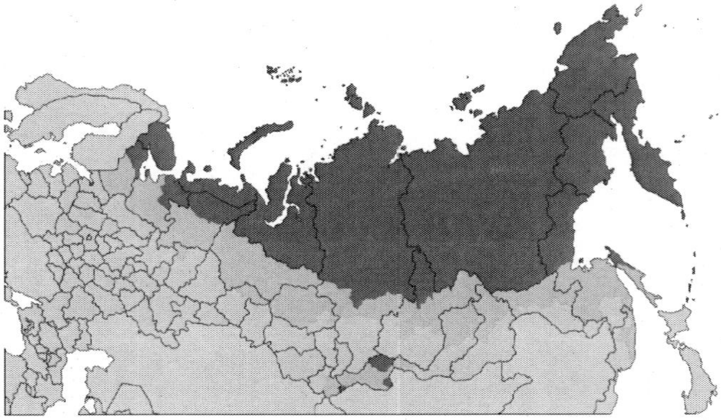

Figure 1. The map of the regions included or assigned to the High North.

On the example of the concept of the "Far North", which still coexists with the concept of the "Arctic zone", we observe the mutual overlapping of several principles of space categorization. In the case of the Arctic zone, the following indicators are the guiding principle (See: Lukin 2014):

1. "Latitudinal" principle (Arctic Circle - 66°33' North latitude)
2. Socio-climatic principle (Comfort of living in the area in terms of life support, labor activity, mobility, general state of health)
3. Climatic principle (natural parameters - temperature regime, specifics of vegetative period, etc.)
4. Landscape, or climate-geographical principle (zoning in terms of tundra, forest-tour, taiga, etc.) climate, presence of permafrost
5. Administrative (inclusion of the entire administrative unit within the Arctic zone if part of the territory meets the criteria)

The sophistication and complexity of the situation being analysed is that all the variables (except for the Polar Circle boundaries) underlying the principles described above are constantly changing, thus shifting the boundaries of the AZRF (de facto, but often not de jure). For example, the monthly averages of July and January change, but the macro trend in terms of temperature changes may be different from the changes observed on the basis of statistics over the last century. The depth of soil freezing, the promotion of vegetation in the Arctic zone, which is characteristic of more southern areas, also leads to a rethinking of the principles on which the territory is included in the Arctic. Changes in the boundaries of the subjects of the Russian Federation may also indirectly affect the geopolitical changes and the revision of governmental instruments for the allocation of the Arctic regions, the Far North regions.

The most important aspect of the change is the socio-climatic principle of the allocation of the Arctic territories: it extremely accurately reflects the interdependent changes in the environment, ecosystems, climatic indicators and people's livelihoods in northern communities.

If we compare visually the boundaries of the zones of the AZRF, areas of the Far North and areas equated to their status, we'll see the following:

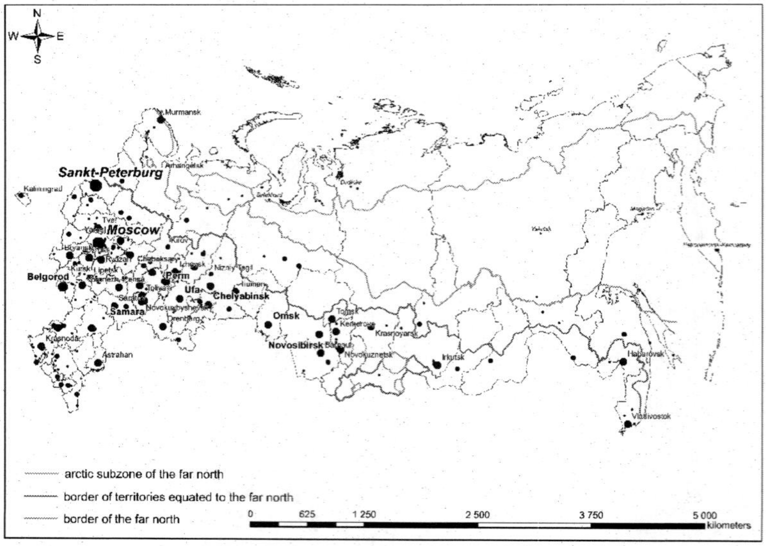

Figure 2. The map of the regions fully or partly included in Arctic Zone of RF (AZRF), included or assigned to the High North.

What does this map 'tell' us? The borders of the regions equated to the districts of the Far North are marked in red. What they are equated with means a metaphorization of the concepts of "North" and "Arctic", as well as the fact that the geographical principle is not a determinant, and such a classification is conditioned by a different factor. In other words, the territories of the Khanty-Mansiysk Autonomous District, part of the Republic of Karelia, Komi, Tomsk Oblast, Transbaikal Krai and Chita Oblast, and Amur Oblast are in an ambiguous position: these are regions where daily life, work, and life support are severe, but not so much because of the climate as for some non-natural reasons. What are these factors and reasons of a non-biological and unbiological nature?

2. The Arctic, Permeated By Connections: Infrastructure, Borders, Barriers and the Phenomenon of Autonomy in the North

This section is devoted to the identification of non-climatic factors that have a significant impact on the assessment of climate change by local residents of the examined areas of Western Siberia. Based on the main thesis that the phenomenon of "arcticity" is directly related to the phenomenon of spatial and logistics isolation, we will begin our analysis with transport infrastructures and their gaps. On the map below we see the road network of Russian Federation as whole and the transport infrastructure of the Tomsk region bordering the Khanty-Mansiysk Autonomous District in the northwest (highlighted with white background).

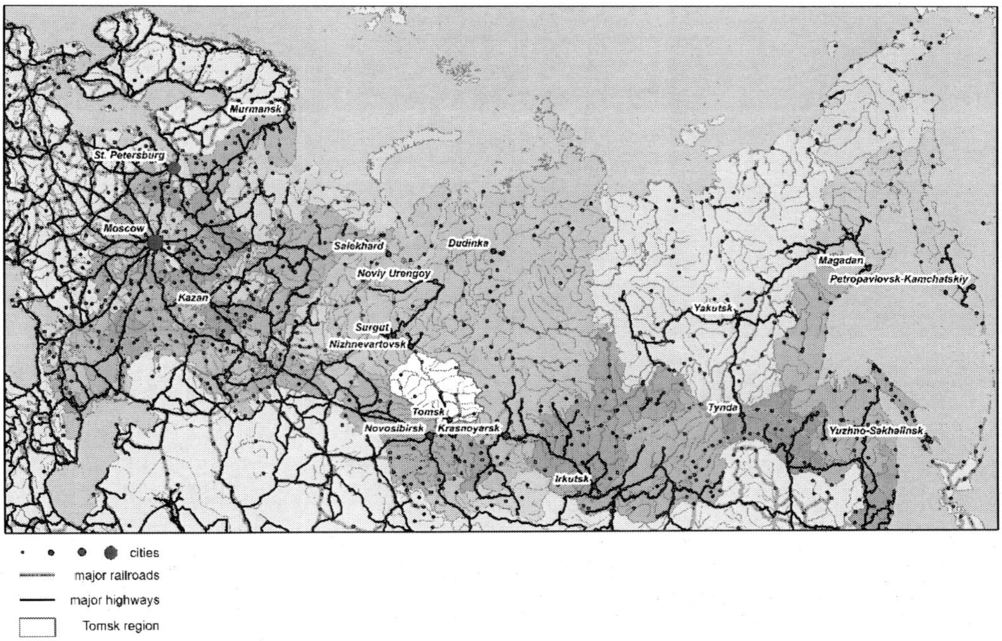

Figure 3a. Transport ways in Russia and Tomsk region.

We see significant differences between the density of different types of roads (asphalt, soil and railways) in the European part of the country, including the "Russian North" region, in the South and South-East. However, there are regions where the state roads are represented by small enclaves and fragments of roads connecting local territories with each other (Zamyatina, Pilasov 2018b). The map points to the fact that latitude is not always a defining dimension, and there is another dimension related to complicated logistics, life support and irregular connections within the micro-region. Such areas and areas are the focus of this study.

If we switch our attention to the next, more detailed map (Fig. 3b), we see that the geographical center of the Tomsk region is marked by a gigantic break of infrastructural links created by Human / State, while the only "natural" artery of the region is the Ob River and its inflows.

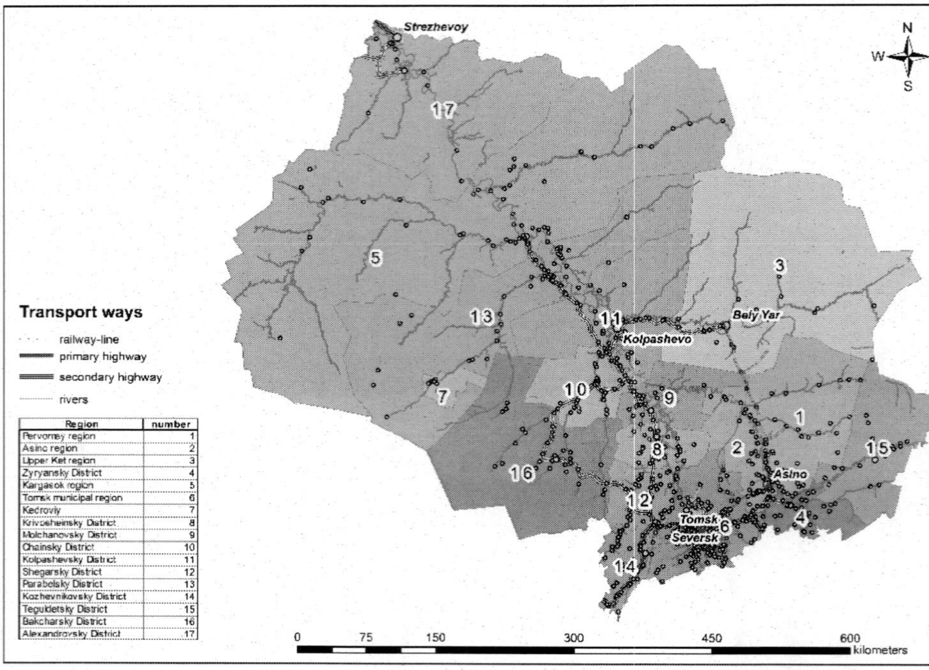

Figure 3b. Transport ways in Tomsk region.

The difference between the road network and the few railroad branches refers us to the chronology of the development of this space gradually, generation after generation. And if the railways indicate areas of intensive development of natural resources (e.g., timber industry), the route running along the Ob and then being interrupted, indicates a trend of strengthening inter-district ties within the region, which, however, has not been brought to a logical end.

If we compare this road map with the map of areas that are currently included in the list of areas equated to the districts of the Far North, we see that the nodes of transport infrastructure in the south and south-east of the region, the intensity and density of the road network is linked to the principles of border identification and zoning. Those areas, which are closely connected with the capital of the region in the south (Tomsk), and provided with a dense road network with asphalt pavement or year-round unpaved roads, as well as state winter roads, are not attributed to the areas of the Far North.

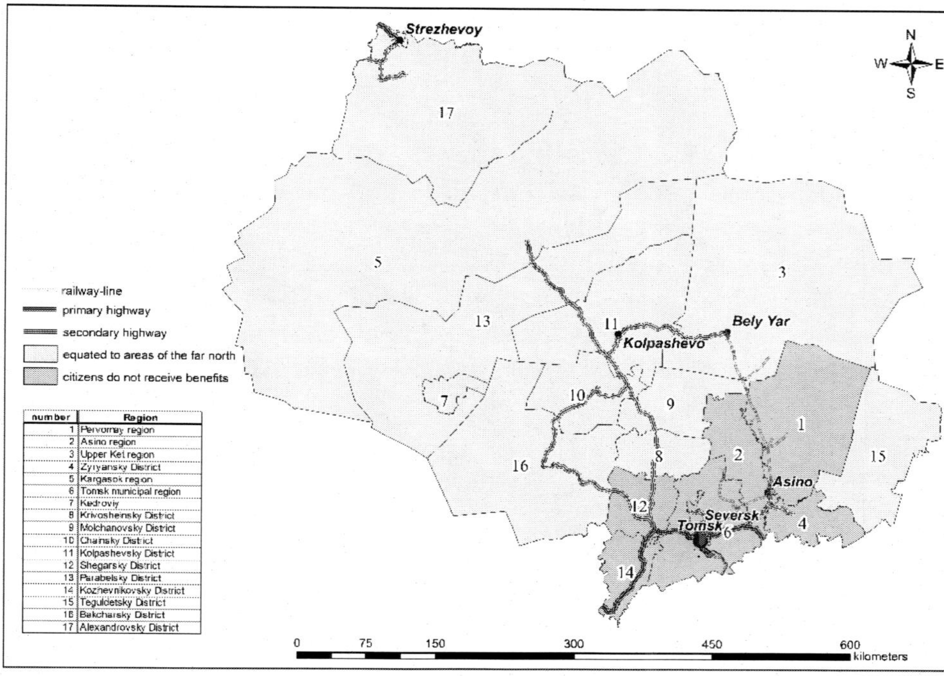

Figure 4. The districts of the Tomsk region equated to the districts of High North.

The federal highway Tomsk - Kargasok, leading to the North-West along the Ob riverbed, ends in Kargasok, the district center. The roadway in the North appears in Alexandrovsky - the center of Alexandrovsky district on the left bank of the Ob river and in Strezhevoy on the right bank. In fact, this gap and the presence of only water communication (ferry, "Voskhod" high-speed boats) between the districts of the region weakens the links between the center and the periphery both in terms of control and in terms of supply and life support.

Thus, the taiga and coastal settlements of the Ob River, the Tym River and other rivers of the Aleksandrovskiy and Kargasokskiy districts are, in fact, an "island territory", located in the heart of Eurasia. The "island status" of the inland area introduces a new indicator into the discussion of extreme living conditions - the degree of isolation, the duration of isolation and the degree of destruction/degradation of social and infrastructural links with regions outside the "island area".

It is in connection with this statement that it is necessary to study the Western Siberian Arctic in the context of the changes that occur not only within the AZRF, not only in the Far North, but also in the territories equated to them. On the one hand, this approach allows us to analyze the views of the population and their experience, taking into account the perspective of changing the boundaries of regions and the concept of "Arctic" as such. On the other hand, consideration of nonnatural factors in the analysis of informants' statements about changing climatic conditions should be correlated with subjective ideas about what do "harsh living conditions", "trial", "extreme conditions" mean.

Interviews with residents of relatively southern regions have shown that the negative properties of the purely northern latitudes and Arctic spaces are moving to the southern territories of Siberia and are manifested in the brightest way where warm, but interhuman and inter-settlement ties have been destroyed:

- And when the river 'stands still' (is frozen) and this means the isolation for the whole season, when neither roads work nor 'Voskhod' sails.

- Is this period of transport isolation longer now?

- Yes, it can be longer. It depends on how the river 'stands still', of course.

- How does the river 'stands still', is it earlier or later? Stable?

- Well, I don't know how it used to be. But before at least it was easier to get out of the village, but now only once a helicopter comes; once a week and not always takes people on board.

- Is it in the off-season?

- They manage to swim with 'obazki' (type of traditional boat) through the river, if there's no other chance to cross it.

- With the risk to life ...

- Yes, at the risk of a life! And there's no other chance to manage it in another way...

When the links are broken, the adaptive response of the community can be either an attempt to "fix" and reestablish the links (repair roads, find alternative modes of transport, organize partnerships with another region) or an attempt to make autonomous as many areas of life as possible and independent of external supply and life support systems. Any steps in this direction are inevitably linked to the climatic conditions of the micro-region.

2.1. Heating and Water Supply Autonomisation

In the northern regions, despite the rural character of most settlements, heating and gas supply has been centralized since the Soviet times. However, the old engineering systems, their gradual deterioration, played an evil joke on the residents when the population began to decline sharply. Since they were designed for a certain load and number of residents, the work of the housing and utility services became unprofitable. Tariffs increased, but there were no alternatives to survive the severe and long polar winters. The stoves are a legacy of villages that date back to the pre-war years, when wooden construction was durable and of good quality, or older settlements in the late 19th and early 20th centuries in the Arctic. The reverse side of the centralization of all resources is the strict control and, at times, the impossibility to improve their living conditions in a proactive manner:

> *- And people don't need to switch to autonomous provisioning?*
>
> *- And you can't.*
>
> *- That is, there are no sources?*
>
> *- It is forbidden by law to install autonomous heating in capital buildings, only with the permission of the administration. (m.,* 37, Pangody)

2.2. Arctic Diet, Food Supply Problems, Low-Quality and Expired Goods

An alternative to imported vegetables may be to run your own homestead farm in risky farming areas. Despite the permafrost, people are actively organizing greenhouses, inventing devices for heating the soil under the greenhouses, trying to "win back" a few degrees of heat in the North. And if earlier the appearance of fresh and useful products in the shops of the area was the result of the state governing and northern import, in the 21st century the dialogue with soil conditions, short vegetative period and other objective natural conditions falls on the shoulders of local residents themselves.

However, in some of the villages where we conducted interviews and surveys, it became clear that the plan of the settlement, its status and position do not allow the organization of homestead lands. Centralization and control, which were the key to stability and a high standard of living, here, on the ruins of infrastructure, suddenly became a significant limitation and barrier to achieving food autonomy.

2.3. The Siberian River as a Reflexive Macroobject Changing under the Influence of Climate Change and Ecosystem Shifts

The rapidly changing Arctic can be studied in depth not as a stand-alone macro-object, but in connection with adjacent regions and the changes that occur in them. The Ob River, which unites the three regions that are classified as Far North or at the core of the AZRF, is undergoing various changes throughout its course. Ice conditions, time shifts in the periods of freezing or flooding are extremely important factors that are persistently noted by local residents.

The river is one of the most understood elements of the ecosystem, with the eyes of traders, fishermen, entrepreneurs, families with sick old people and children chained to it. These are not only issues of supply, but also issues of health care, safety, life expectancy, psychological well-being of the population, access to cultural, educational and business resources, inter-settlement communication and communication with the regional center.

The river does not just act, but it is managed by actions or events, and thus it transmits meanings, transmits signals, saves information and destroys it, unites people, offers alternative ways of moving, and creates barriers, dividing the inhabitants of settlements on different banks. A member of the community who earns his living from fishing sturgeon constantly balances the danger of being caught by a fish inspector with the problem of losing contact with his competitors and allies in case of distress.

The inevitability and determination with which it is woven into many of the life scenarios of local residents creates counter expectations on their part and gives grounds to assign certain functions to the river. However, this process is not so smooth: we see the failure of the scenarios, on the one hand, due to the conflict of interests of different participants in this "river dialogue", which put different meanings and work differently with the river space and the space near the river. On the other hand, these failures are related to the river life cycle, early floods, water level surges and constant changes in the profile and properties of the riverbed.

Thus, having considered the significance of northern infrastructures, supply systems and the role of such macro-objects as the Siberian River, we have methodologically justified the need to consider the three Western Siberian regions as a single space of climate change and socio-economic transformation. A more detailed description, selection of settlements and territories for empirical research and its results are presented in the next section of the chapter.

3. Destinies and Community Life: Natural and Social Causes of Transformations

An empirical study was conducted in three regions: the Tomsk region, Khanty-Mansiysk and Yamalo-Nenets Autonomous Districts as a part of the socio-anthropological research on the Russian Foundation for Basic Research (RFBR) project "Comprehensive study of changes in vulnerable terrestrial ecosystems of the Western Siberian Arctic based on a megaprofile approach." A mixed research strategy combining quantitative (questionnaires) and qualitative (interviews and observations) methods was applied.

An appropriate toolkit was developed for the project, including three types of questionnaires and an interview guide. The first questionnaire was used to interview all respondents, regardless of the degree of rootedness and years of residence in the area. The key questions of this block were devoted to the assessment of climatic conditions in the area of residence, the level of comfort for living, work, commerce and fisheries; assessment of the radical climate changes occurring in the microregion as seen by local communities (Anisimov et al., 2017); identification of environmental factors that, according to respondents, have undergone the most significant changes in recent years and decades.

This survey also raised the questions about the key drivers of climate change in the opinion of local residents of rural areas, identified areas of life that were most affected by climate change. In addition, a question was posed about the view of the phenomenon of climate change from the everyday perspective in order to determine whether local residents of the studied regions perceive climate changes as a global, regional or local problem, or consider this phenomenon as a consequence of natural processes and macrocycles, occurring in the environment irrespective of the anthropogenic load and the impact of human activities.

The second questionnaire was based on the closed questions, in which the respondent had to assess the nature of changes in natural processes, weather conditions, temperature regime, seasons' duration, nature and level of precipitation, frequency of extreme weather events (hurricanes, thunderstorms, whirlwinds, extreme frosts, heat or drought), as well as the dynamics of the volume of wild-growing herbs, fish stocks in water bodies and animals in taiga and tundra.

The third questionnaire aimed at revealing the correlation between the degree of rootedness in the territory and the acuteness of perceiving climate change, which runs in parallel with the long-term human adaptation to the conditions of the given climatic zone and the specifics of the local microclimate. This questionnaire was offered only to respondents who do not consider themselves as local residents and have moved to the region from other climatic conditions due to work, watch keeping, or family circumstances, etc. Semi-structured interviews were conducted in institutions and organizations in rural settlements, as well as on the streets of the settlement at the gathering places: shops, markets, administration, post office, library, garages, port, ferry, near cafes, bars and hotels, infrastructure facilities, boiler houses, housing and public utilities departments, schools, kindergartens and hospitals. A total of more than 70 interviews were received in 16 localities.

The interview focused on the extent to which the transformations that affect the way of life and livelihoods of modern settlements are directly related to climate change, and to what extent the local people notice these transformations and perceive them as results of climate change, rather than other processes. The informant was asked to identify key markers and phenomena which, in his/her opinion, are signs and consequences of ongoing climate change or to recognize that the changes are so insignificant that it would be incorrect to associate this with the global climate change.

By asking in detail about the main activities of the informant - professional, daily, hobbies, subsistence fishing - the interviewer was able to identify areas of daily life in which the informant had some kind of expertise and traditional environmental knowledge (such as horticultural technologies in the northern territories or reindeer herding). Starting from these areas of knowledge and observation, on which the informant focused because of its importance and familiarity, we identified the processes that were most acutely felt, significant and constantly observed. After that the less important spheres of life were touched upon, and we identified the secondary processes and phenomena, which could be noted and observed by the local resident irregularly.

The interviews revealed a number of conceptual discrepancies between the research questions on which the research program was based and common interpretations of the interview questions and questionnaires, which split the phenomenon of "climate change" in the minds of respondents into three components: "climate change," "changes in weather conditions," "deterioration (changes with a single vector) of the environmental situation in the region."

During the initial operationalization of the basic research concepts, the environmental aspect was not part of the overall scope of climate change. However, thanks to the qualitative methods used to study the daily perception of climate change, new scenarios of perception of these processes were identified, including the establishment of causal links between climate change, the manifestation of unusual weather events and pollution of soils, reservoirs, air as a result of the development of oil and gas industry, construction and commissioning of industrial facilities, etc. Thus, we faced a challenge, which was formed in the framework of an interdisciplinary research project at the intersection of climatology, ecology, biology, soil science, medical sciences and social anthropology, focused on the process of climate change.

In the course of "assembling" the project it became necessary to bring together and analyse statistical data, observation series and people's opinions expressed in simple language and direct expressions. This is not only the task of translating everyday observations and impressions into academic and statistical language, but also the task of operationalizing concepts and triangulating data. During the pilot study, we encountered inconsistencies not only in the opinions of villagers in different latitudes, but also in the responses of single informants. Thus, a person may claim that "winters have become colder than before," but immediately adds that "in the 1950s, each winter was severe, but we did not notice it - we were young" (m., 63, Vyngapurovsky).

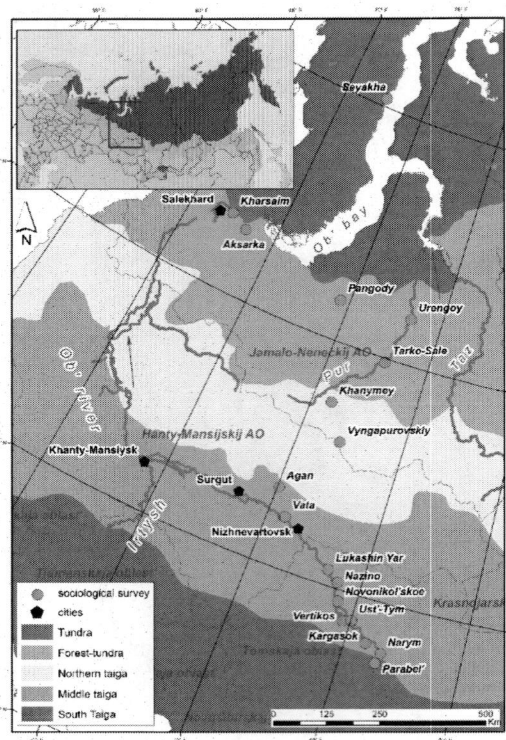

Figure 5. Research areas.

From a scientific point of view the questionnaires also gave incoherent and sometimes incomprehensible combinations of answers. The combination of qualitative and quantitative methods has led us to the need for a broader research context. If initially the focus of anthropologists' attention was on local interpretations and views, observations, and feelings of local people in northern Siberia about climate change - in other words, how they were perceived locally - by the end of the expedition it became obvious that it was impossible to obtain accurate data without the involvement of concomitant factors that in some way refracted the perception of climate change. This contextual expansion has been inspired by the locals themselves. In their discussions of climate change, they moved into the category of weather conditions, environmental conditions, water and air purity, and linked these phenomena to high mortality rates of cancer and cardiovascular disease. The discussion of climate shifted towards talking about the nearness of death and death of men of working age, who are the backbone of the community. Their sudden deaths and the weakening of community resources were the subject of an interview with one of the informants: "We live like a boiling pot and never know when we will lose the next man" (f., 57, Kargasok).

3.1. Are Climate Changes Experienced in the Region of Residence?

Before we present the results of quantitative and qualitative research in all its details concerning various aspects of climate change, we should present the results of the answer to the question: "Are there any subjective impacts of climate change in Western Siberia? Let us turn to Figure 6.

According to the answers of respondents from different regions, the most significant climate change is noticed by the residents of the North of Tomsk region - in Kargasoksky and Aleksandrovsky districts (46%), while the southern districts of the Khanty-Mansiysk autonomous district - Nizhnevartovsk and Surgutsky districts - experience and perceive climate change as insignificant (Figure 6).

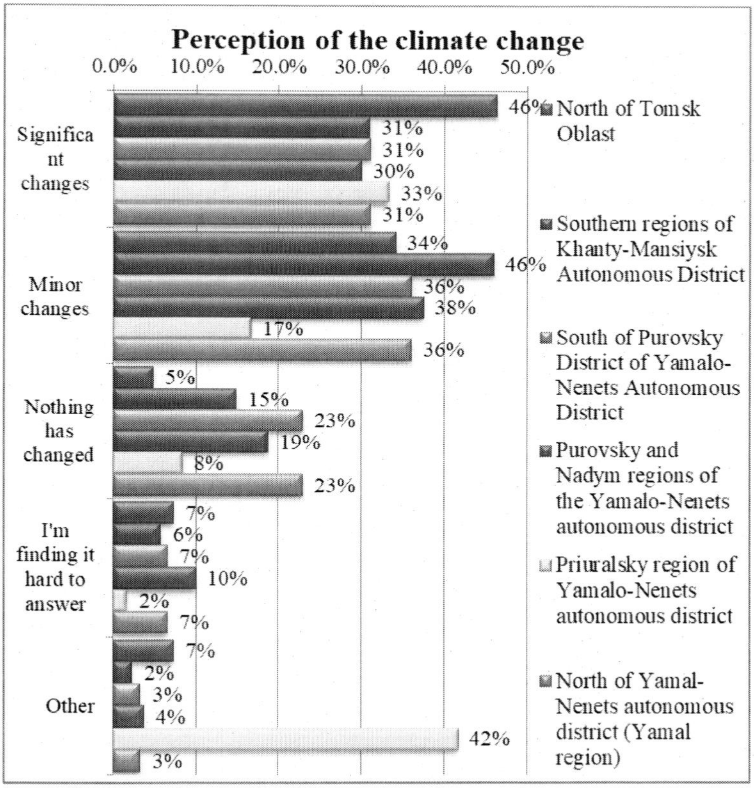

Figure 6. Degree of climate change in respondents' perceptions.

The views of respondents from the Central part of the Khanty-Mansi Autonomous Area and various districts of the Yamalo-Nenets Autonomous Area are similar and the answers are not unambiguous: 31-33% of respondents tend to believe that the changes are significant and obvious, while 36-38% of respondents say that the changes are not too noticeable and have little impact on everyday life. Regions in which one fourth of survey participants noted that there are no changes as such, or these processes can be assessed differently - the South of Purovsky Raion and Yamalsky Raion.

Non-standard distribution of answers in the Yamal district is worth noting separately, where 42% of respondents chose the answer "other," commenting especially on the nature of climate change. Respondents noted that there are local changes, such as changes in the structure of ice, shredding of rivers and lakes, the emergence of new species of insects and birds, the disappearance of fish in commercial quantities, but these changes are not complex.

3.2. Key Types and Directions of Climate Change in the Perception of Local Residents

In the quantitative survey devoted to subjective perception of climate change in different micro-regions, respondents could choose one or more answers to the question about the elements and processes in the environment that have undergone the most significant changes (Figure 7).

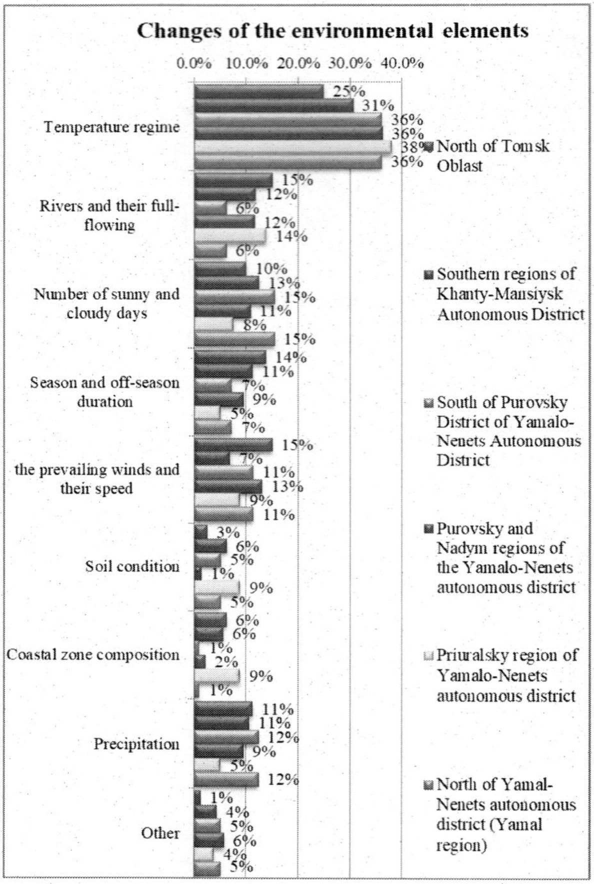

Figure 7. Environmental components that respondents consider to be most strongly/explicitly changing.

In order to check the distribution of answers to this question, respondents were offered a table in which they assessed all kinds of environmental elements and indicators based on the degree of radical change. Based on the results of two cross-surveys, we publish the following data. Now let us consider sequentially the elements and phenomena that the informants participating in semi-structured interviews noted and compare them with the results of quantitative research, highlighting the main elements and trends that were noted by respondents in the survey and interviews. It is worth noting that some of the following phenomena were not included in the quantitative research toolkit and were only revealed during the interviews, thus expanding the list of changing signs and observed phenomena that are influenced by climate change.

4. Temperature Regime and Season Duration

According to a quantitative study, only 26% to 42% of respondents in different parts of Siberia claim that the winter period is longer than before. In contrast, in the northernmost area of the study (Seyakha settlement), 34% of respondents are convinced that winters have become shorter (Figure 8).

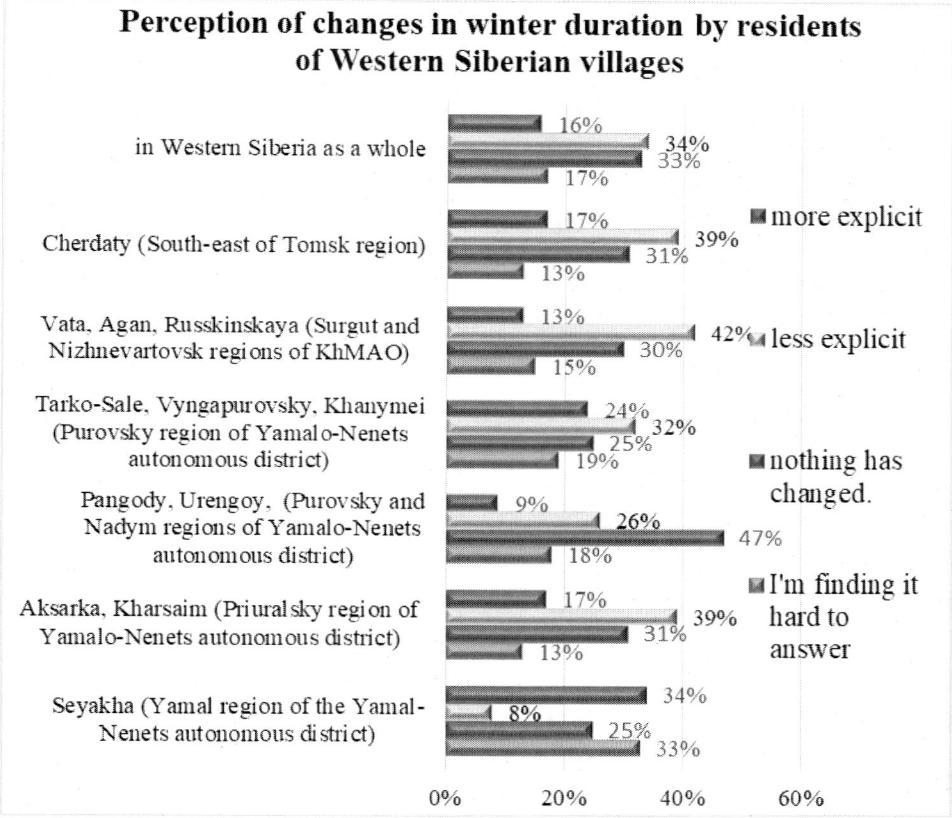

Figure 8. Perception of changes in winter duration by residents of Western Siberian villages.

Responses about winter temperatures are more similar - the majority of respondents in the northern regions are convinced that winters have become warmer (58-44%): "in winter there are abnormally warm weather for our places" (m., 49, Kargasok); "earlier it was colder. The frosts were much more frequent" (f., 42, Vata); "I remember we did not go to school for a month because of the frost like this, but now there is no such thing, maybe two days like this" (f., 38, Urengoy); "Well, of course, since the 83rd year, when the temperature was 53-55 we did not recognize it, we did not notice" (f., 72, Russkinskaya). In the southern regions it was noticed by a smaller number of respondents (37-35%), while in the village of Cherdaty and the village of Kargasok 35% of the population claim that it has become colder in winter: "Cold winters. It is getting colder, reaching minus 55 temperatures. This year's winter was very long, the snow was almost until May. Summer was delayed."

Often people noted that they did not see a certain trend and pattern in the temperature mode: “It is impossible to say at all, because everything is not standard. This winter is like this, next time it is like that, next year is the following, and there is no stability” (f., 43, Russkinskaya).

The fact that the summer period has become longer is noted only by residents of Yamal (Seyakha settlement), residents of other districts either do not notice any changes, or think that summer has become shorter, so in the area of Russkinskaya and Agan villages it was noted by 60% of respondents, in Cherdaty village - 48% (Figure 10, 11).

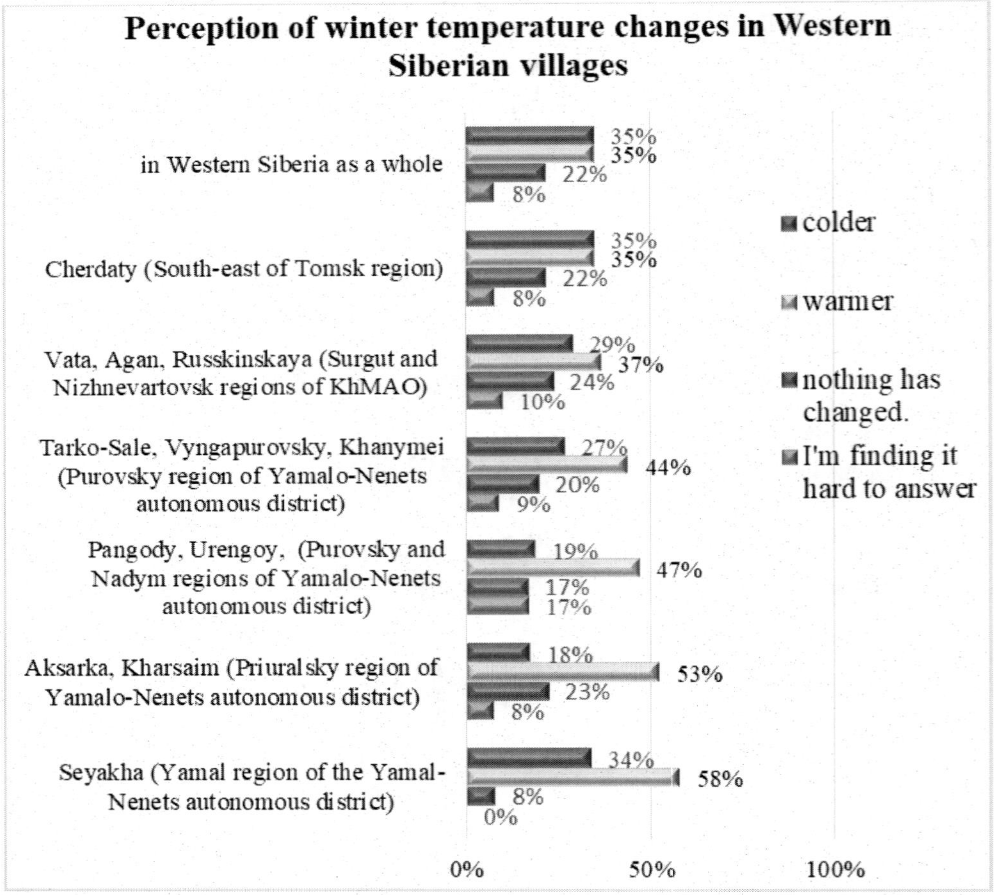

Figure 9. Perception of winter temperature changes in Western Siberian villages.

Interestingly, the opinions of the inhabitants of the subarctic and arctic zones of western Siberia differ dramatically. Whereas in the North and Central part of Yamal local residents believe that summer has become much warmer (up to 75% in Seyakha and up to 42% in Urengoy and Pangody), for the more southern districts of the Khanty-Mansi Autonomous Area and Tomsk region it is obvious that summer has become cooler (from 48 to 57% of answers). The changes in spring and autumn (Table 1) seem insignificant for residents of Western Siberia.

The residents of the central part of the Purovsky district of the Yamalo-Nenets autonomous district and the south of the Khanty-Mansiysk autonomous district point out that the spring period has become colder. The warming in autumn is noted by the majority of residents of Seyakha, Aksarka, Kharsaim villages, as well as Cherdaty settlement.

The increase in precipitation was noted by 33% of respondents in summer and 41% in winter. The majority of residents of the Far North note an increase in the number of unusually hot days (Sayakha - 83%, Aksarka, Kharsayim -33%, Urengoy, Pangody - 38%), while 48% to 50% of respondents from the southern regions, on the contrary, believe that the number of hot days has decreased (Figure 13). When asked if the number of unusually frosty days has changed, the majority of residents of the surveyed regions (36% to 54%) respond that the number of frosts has decreased (Figure 9). The exception is the most southern Cherdaty settlement (43%). There were recorded homogeneous answers to the question "has the intensity of sharp temperature changed": from 35% to 67% of respondents believe that sharp temperature changes have become more pronounced (Figure 10). Answers to the question about the change in the number of extreme weather events in different surveyed areas are not so homogeneous (Figure 11).

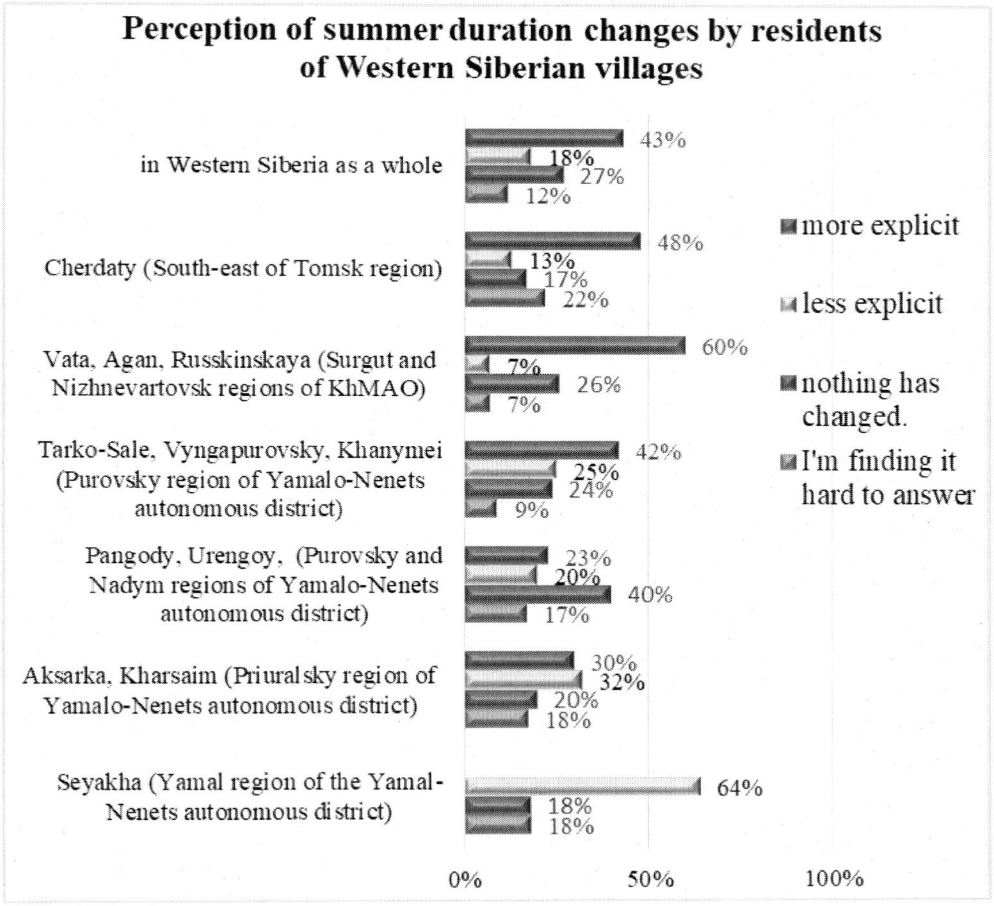

Figure 10. Perception of summer duration changes by residents of Western Siberian villages.

Table 1. Perception of changes in autumn and spring by the population of Siberia (%)

Question	Answers	Seyakha	Aksarka, Kharsaim	Urengoi, Pangodi	Trako-Sale, Khanymei, Vyngapurovsky	Russskinskaya	Cherdaty
Spring period became	Shorter than before.	34	27,5	27	25	15	26
	longer than before	2	22,5	20	27	47	22
	hasn't changed.	3	20,0	30	24	21	35
	I'm finding it hard to answer.	33	25,0	23	16	17	17
	other	0	5	0	8	0	0
In Spring it became	colder	17	35,0	29	**39**	**45**	**39**
	warmer	41	32,5	16	21	15	4
	Nothing has changed.	17	12,5	36	19	30	48
	I'm finding it hard to answer.	25	12,5	19	21	10	9
	other	0	0	0	0	0	0
Atumn period became	Shorter than before.	0	15,0	13	25	20	13
	longer than before	75	30,0	32	**41**	33	39
	hasn't changed.	17	30,0	36	19	29	39
	I'm finding it hard to answer.	8	17,5	19	15	18	9
	other	0	0	0	0	0	0
In autumn it became	colder	34	22,5	26	43	53	17
	warmer	**50**	**37,5**	17	14	9	**39**
	Nothing has changed.	8	17,5	29	17	22	35
	I'm finding it hard to answer.	8	17,5	17	20	16	9
	other	0	0	11	6	0	0

4.1. Increasing Number of Unpredictable and Extreme Weather Events

Perception of changes in the number of extreme weather events by residents of Western Siberian villages

- "if you take it ten years ago, the weather conditions were much better, of course. There were no such sharp changes in temperature, as we have now, say, happening. We can fall asleep, even in winter, when you fall asleep - you have there -16, in the morning you get up - you have a window of minus 40." (f., 36, Kargasok).
- "You know, 28 in the morning, -10. There can be drops up to 30 degrees" (f., 72, Russkinskaya).
- "Not so much before. But the weather is changing. It used to be not so abrupt. There, in two days, it can get so slightly warming or cooling. And now - sharp. Yesterday it was warm, but today - the frost hit. I also feel it. Who are sick - they feel it it all! Such weakness. That time - it was warm, the northern wind blew sharply, it got cold, and the body already feels it too. And the old people are lying at home, feeling it very badly. All sorts of joints hurt there..." (m., 59, Russkinskaya).
- "The heat wave was in 2012. Rains were low, and there were a lot of forest fires" (f., 42, Vata).
- "It was just 2016 when all anthraxrap started, it still burned here, but we had smoke right there, the smell of burning right there" (f., 38, Urengoy).

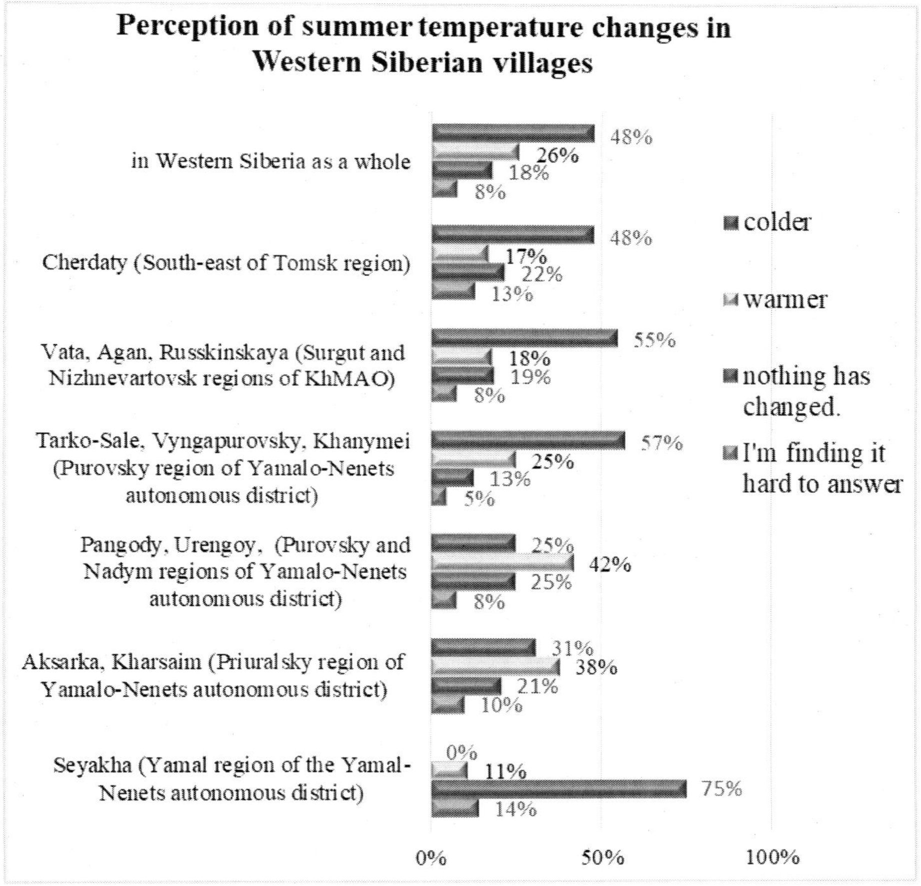

Figure 11. Perception of summer temperature changes by residents of Western Siberian villages.

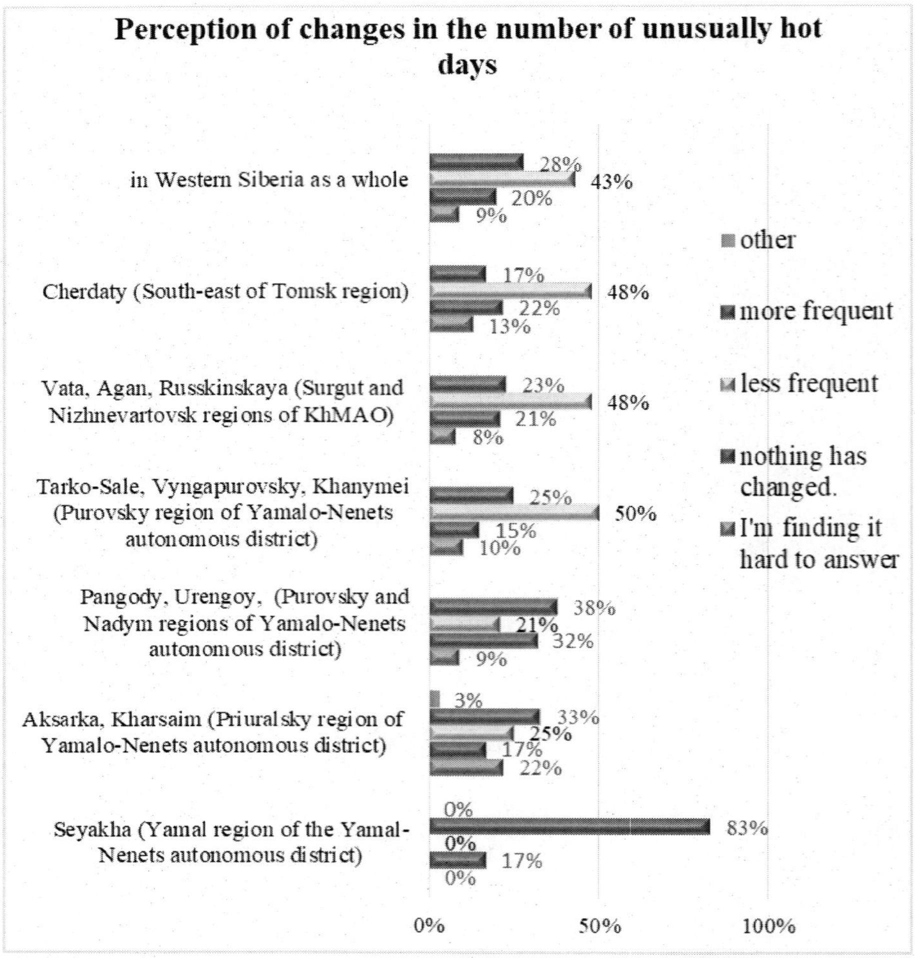

Figure 12. Perception of changes in the number of unusually hot days by residents of Western Siberian villages.

- "In 2014, the whole village was on fire, and everything went to the forest area, almost everything burned out, the inhabitants of indigenous nationality of their huts, bathhouses they did not suffer much, the reindeer that they kept run away" (f., 72, Russkinskaya).
- "Weather changes are abrupt: from warm to cold, from cold to warm" (m., 49, Kargasok).
- "But there was some time, in my opinion, 2012, I begin to remember, I remember, there was such a strong wind that the roofs blew, all day blew, impulses up to 25-26 m/sec" (f., 38, Urengoy).
- "The difference was not so obvious, but now it is very sharp, yesterday +27, and tomorrow +15" (m., 38, Agan).
- "Recently, the event has taken place, verything in the world burned, especially on the mountain and the phone burned down. Thunderstorm" (f. 70, Novonikolskoye).

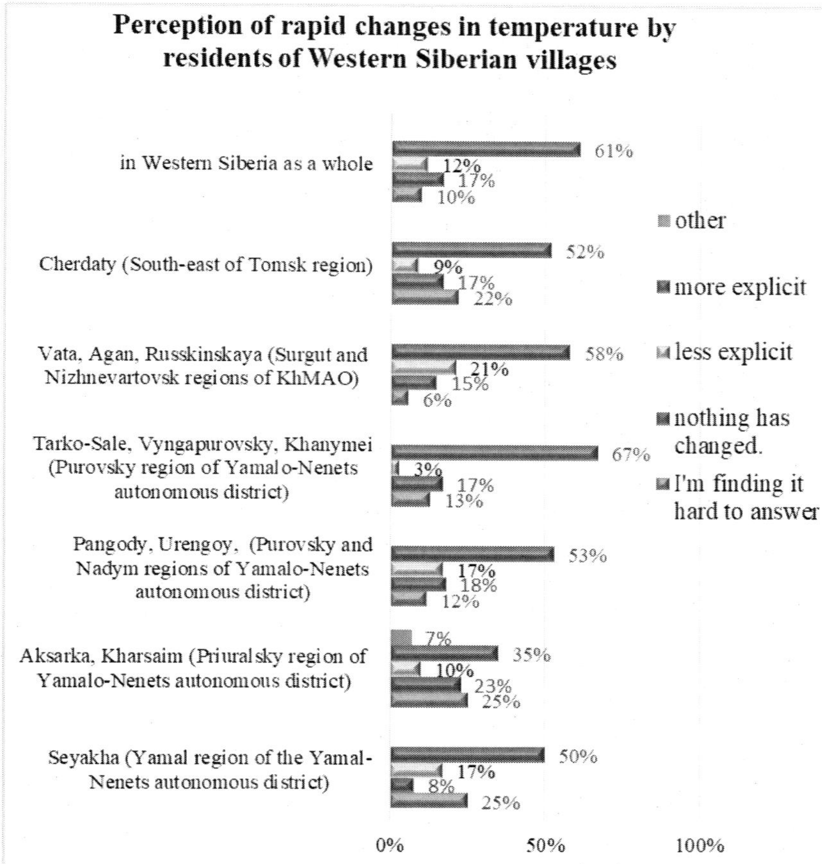

Figure 13. Perception of rapid changes in temperature by residents of Western Siberian villages.

4.2. Deterioration of Public Health Due to Climate, Weather and Environmental Factors

The phenomenon, which is directly related to sudden changes in temperature during the day, is the deterioration of health, the frequency of cardiovascular disease and the risk of complications for older people. Health is the most sensitive area of life that responds to climate change. People in the north of Yamal react particularly acutely to climate change with their health condition (Figure 18-1). Informants note in interviews:

> "The ecology is terrible. What can I say? We have swamps... Everyone, as they say, just point a finger - cancer, cancer"; "maybe we have some kind of pollution... or something else, but in short, we are all, as they say, all our Kargasoki district is the most elevated for oncology" (Zh, 57, Kargasok).

- Did climate change affect health?
- Health has been affected, especially the heart sufferers. I work in ritual services, and I can say that mortality has increased.
- Different ages?

– Mostly sick people.
– So the stronger people have started to survive, the healthier people, and the weaker people don't survive like they did in the 70-80s?
– Yes, many men in our country will not survive to the new retirement age of 65. (m., 49, Kargasok)

"my husband even died of, say, oncology. Also here. My acquaintance, not long ago, here, went to bury him... so... Tensions are very high, our men. We are sitting on coals, so to speak. We have oncology everywhere" (f., 57, Kargasok).

"I mean, such an abnormal variations! Judging by my parents - the feeling, that it affects their health! Here are these sharp changes" (f., 36, Kargasok).

"And now it may be warm, and than - "bang!" Well, today it is "27," tomorrow' - "15," roughly speaking. Big sharp differences" (f., 47, Agan)

4.3. Change of Rhythm and Duration of Floods, Change of Ob River-Bed

According to the quantitative survey (Figure 7) rivers, their condition, full-flowing, riverbed transformation - are marked as the most important and pronounced concomitant signs of climate change. The significance of changes in the river system in the areas of the river mouth and on the banks of the Pur River is noted by 11 to 15% of respondents. In the central part of the Surgut district of the Khanty-Mansiysk Autonomous District and on the bank of the Gulf of Ob (North Yamal) only 6% noted such changes as significant.

The interview revealed the implications of changes in river stream and their full-flowing nature:

Sure, it washes away. You can go out and have a look: dams are already being built and everybody ....walks along the path - there was suuuuch a passage to the shop." (railway, 36, Kargasok)

"It is the same with the river now... All the streams are drying up. It is necessary [to go fishing] only in spring . Nowadays, everything is dried up: the lakes have become shallow. Oblazki (traditional type of a boat) used to float there before, but now... The lakes are drying up. Water runs somewhere away. (m., 59, Russkinskaya)

I think that a lot of fish else die because of changes in the riverbed, sometimes the rolls change the riverbed for 2-3 times per navigation. Because nobody deals with this river, there are all the same excuses "due to the difficult financial situation," neither dredges, nothing works, of course, these pits are being washed away. .... Well, it is a pity about the river. Every year there are such changes, anomalies. Now, well, 20-30 cm of water falls per day. Nowadays the rise of water was good, it held on." (m., 59, Alexandrovskoye)

The flood has become longer, it used to be like that- it lasted untill the pentecoast, and that's all, the water goes down, but now, out, it still stands for 7 meters. In August, it will be time for her to swell over again, and now this one hasn't left yet. The form of riverbed changes every year, the river works every day, both in summer and in winter: it washes away the sand in one place, and in another one dredges it and the ravines falls down. (m., 67, Kargasok)

"The ice comes later, later, much, much, much later, so we remember it [as it used to be before] well. I know that +1 cold in fur coats in a hat is like this. And now in May there is nothing of it, thus it is said that there is a warming." (g, 63, Khanymei)

"Where I've been fishing, there was a high water at one time. Then for seven years it was gone and the duck began to leave, now it is normalizing, the level is the same as it used to be before. It started to rumour that "there is no water" - nothing like that." (m., 32, Agan)

"Of course, the river is changing, the riverbed form has changed, it shallows, the navigation passed here before, but now it's not the Ob', it's the duct of the Ob'." (m., 62, Wata)

"...usually flood waters, melting snow... And this year - I have never seen anything like this before! It's already warm, people are almost bathing, the rains have passed and Ob' has risen strongly and flooded the swamps. Ground waters have risen very much, I don't remember anything like that." (m., 32, Agan)

"The river has changed a lot: the riverbed, the river shallows, there was a port, a steamboat came. And now it is only a duck. It used to be a navigable one. Now it is an old river-bed. This part was an island. And now everything is flooded with water." (m., 58, Vata).

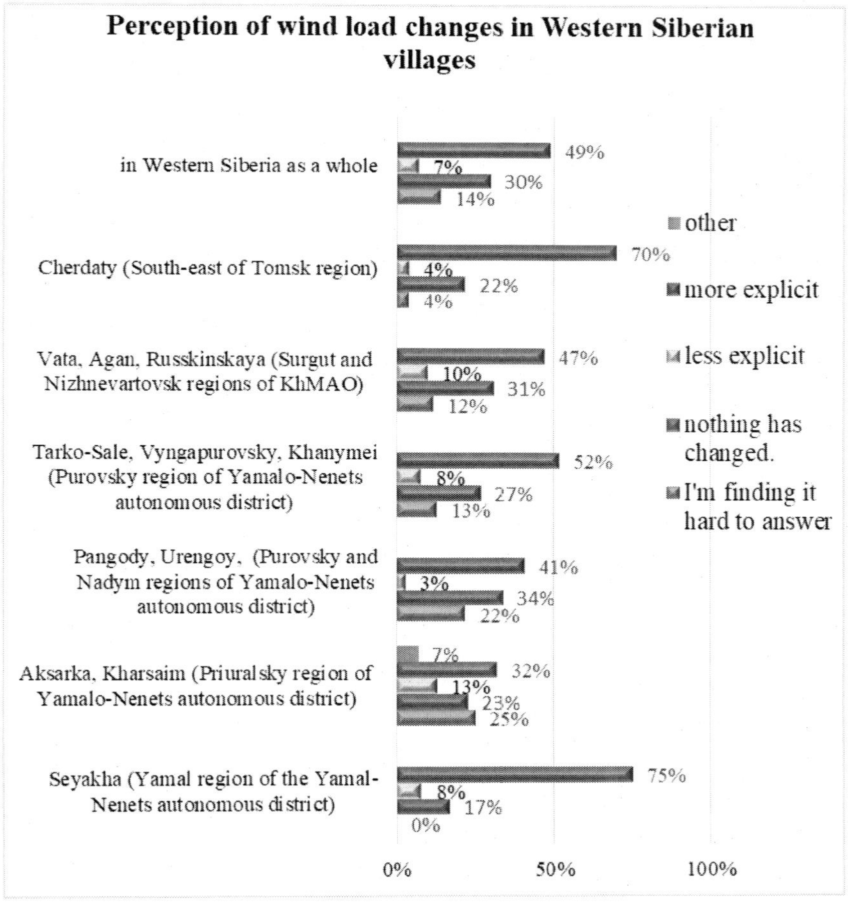

Figure 14. Perception of wind load changes in Western Siberian villages.

4.4. Establishing Links between Atmospheric Phenomena and Forest Development in the Course of Construction, Transport Infrastructure Development and Forest Industry

Strong winds are not so destructive for the health of residents, but they affect the management of the household economy, safety of urban infrastructure and transport, preservation of urban development, and wildfire risks. Except for the Priuralsky district of the Yamalo-Nenets autonomous district, a significant increase in wind power is observed in all districts (from 41% to 75% - Figure 14).

> "in the last few years - you know that - the climate has changed, and every year not for the better, of course. Like the wind. Why are the winds massive? - Because of the loss of forest... Mostly deforestation. More felling. And fires. Both. Everything as a complex"; "In winter these winds are also terrible, in summer the winds are strong, in winter too." (f., 36, Kargasok)
>
> "Three years ago, I even ate apples here, local apples, and now they are disappearing: you plant them, but they disappear for some reason, die, and why - it is unclear." (m., 67, Kargasok)

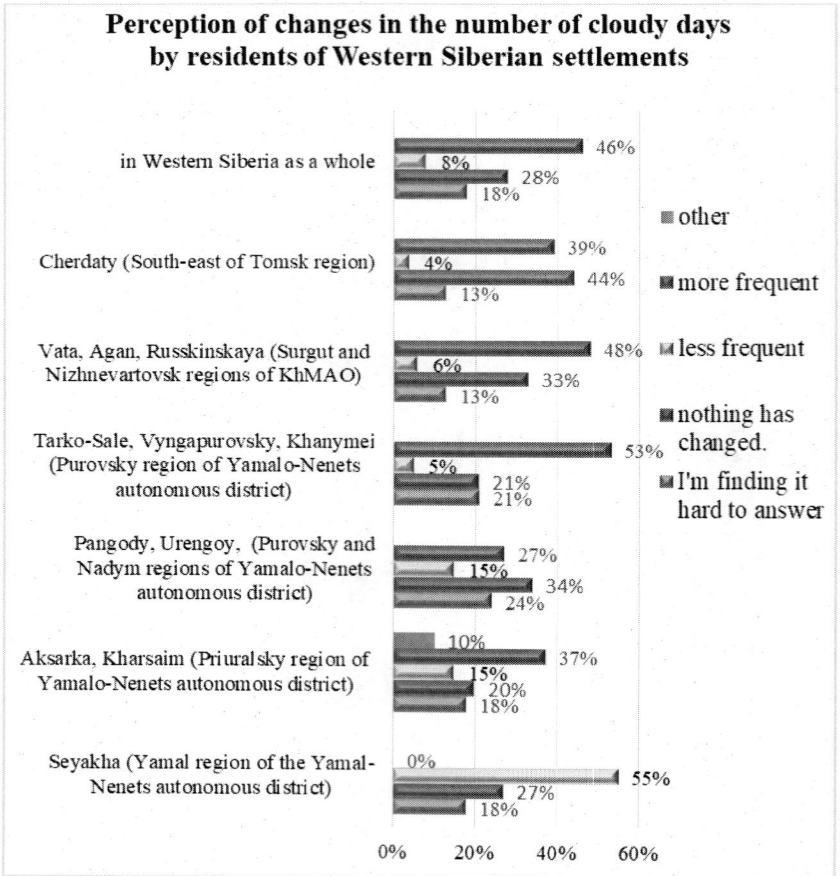

Figure 15. Perception of changes in the number of cloudy days by residents of Western Siberian settlements.

4.5. Unpreparedness of Housing and Communal Services Systems, Private Homestead Households, Housing Stock of Rural Settlements to Radical Changes in Seasons Rhythm, Duration and Temperature Regime

The interviews revealed unobvious links between climate change and the housing stock, as well as the autonomy of life support for each household: it concerns heating, energy and heat sources, housing materials, access to clean water, gas, firewood, and the ability to take care of their own homestead lands. Quantitative survey has shown that from 8 to 14% of respondents see this connection. Informants participating in the interview additionally noted:

"Every year you keep trying, because you don't know what to expect from winter to winter. Somewhere you get warmer, somewhere you buy something. Even those pensioners who live in brick houses, they try to buy these fans every year, that is, to install these electric heaters. Additionally. [Interviewer: Well, these are temporary measures...] Yes, they are, because our housing and utilities sector is also failing. Central heating in houses. It can be very cold in winter" (f., 36, Kargasok).

"We spent most of January at home. First of all, it was cold outside and cold inside the building. The winds are so strong that they blow into the walls" (f., 32, Lukashkin Yar).

"Electricity on its own... is a song! ...It's like drinking, eating. Every day without light at work for 6-8 hours, you don't have to go out... It happens both in summer and in winter: it's a constant blinking, constant all this is this. There is no stability. There is no stability in anything" (f., 36, Kargasok).

Table 2. Perception of changes in other weather phenomena by the population of Siberia (%)

Question	Answers	Seyakha	Aksarka, Kharsaim	Urengoi, Pangodi	Trako-Sale, Khanymei, Vyngapurovsky	Russskinskaya	Cherdaty
There are times when there's ice events	more often than before	59	45	31	36	29	26
	less often than before	33	12	9	9	19	0
	unchanged	8	10	44	37	33	52
	I'm finding it hard to answer.	0	22,5	16	16	17	22
	other	0	0	0	2	2	0
blizzard, snowstorm happen	more often than before	59	45	16	35	21	48
	less often than before	33	20	12	21	32	13
	unchanged	0	17,5	53	34	30	30
	I'm finding it hard to answer.	8	12,5	19	8	15	9
	other	0	0	0	2	2	0

4.6. Impact of Climate Change on Fishing and Other Commercial Activities

The residents of Siberia gave the most homogeneous answers to the questions that are directly related to their everyday activities. Thus, in the majority of the surveyed districts respondents complain about the decrease in the yield of berries and other wild herbs in combination with low productivity of taiga soils, and the complication of the legislative framework with regard to natural management.

- "I think there is less fish. Berries too. Because now everything is banned, you know. Everything is banned completely..." (f., 57, Kargasok).
- "Well, the berries are gone now, it used to be somehow calmer, you collect some wildberries, and then it either burned out, or everything was overgrown with rosehip" (m., 52, Novonikolskoe).
- "Things have changed, I don't forage for berries and mushrooms as much as I do. I am very angry, how can I say, whether it is rage or just a shame, before Yangiyakhinskiy fishing after about 35 km there is a huge hill, and there were blueberries, cowberries, mushrooms, the whole mass of people traveled continuously and everything was enough. And now for several years, there is nothing at all" (f., 63, Khanymey)
- "No, nothing has changed. Because: this year there are mushrooms. It alternates: now there are mushrooms here, next year mushrooms will be in another place where they were, say, five years ago. At the same time, no one will say that, because there were no bumps, yes, long time ago. The cone was there for one year once, but it... needs warm weather. If it is a warm summer, there will be bump" (m., 32, Agan).
- "And there was less fish, my father used to bring a pike with a head hanging on one side of the table and a tail on the other side, but now it's not like that" (f., 42, Vata).
- "Well, they disappeared, because they died. I say, where we live in spring the fish is dead, I already told my bosses, and it is everywhere so" (f, 53, Russkinskaya).
- "In general, they say nothing about the valuable fish, there was a lot of valuable fish before, but now there is no fish... Before there was a lot of sturgeon, people saytalk, now there is no fish" (f., 70, Novonikolskoe).
- "Previously, in every district center there was a fish factory, and a hell of the fish were caught, but now I do not know, much less. I just know the receivers and there is not so many of it. A lot of these receivers have closed down, just a lot of them. And they don't take this fish with them, well, 2-3 tons, well, they will take 2-3-4 tons to Altai, well, 5-7 tons a month, that's all" (m, 47, Kargasok).
- "They used to do a lot with it previously, handed it over, processed..." (m., 57, Wata).
- "No, well, that's it, but there's a lot of bream. There are few sturgeons. As a child, my grandfather caught pike, which did not fit in the bath, but now it is out of the question" (m., 35, Kargasok).

4.7. Reduction of Fish Stocks in Water Bodies

Almost everywhere there is a decrease in the quantity of fish stocks. It is remarkable that this problem is of more concern to residents of the Ob basin, where from 52 to 95% of respondents noted that the number of fish has decreased. In the Pura basin, this problem is not that significant (Figure 16).

- "The fish is like... There used to be a lot of fish. Now there is less. I don't know what is there, maybe it is poisoned, maybe that... I don't know, it dies, maybe there is no water in winter. It suffocates. It goes into streams where water is low and suffocates there. In spring the stream is full of it: there is a pike, big fish and everybody swims like this" (m., 59, Russkinskaya).
- "On the one hand - they say - there are fewer fish, but on the other hand - they are fishing more! I have a small lake at 13th, we haven't been fishing there for five years, there is a sea of fish there after the winters. And then this year I did not fish there only. Karaia - has become a sea! He had already started to get out, the water had arrived, it was going into the river, and the fish started to appear in the river. So we just get more" (m., 32, Agan).

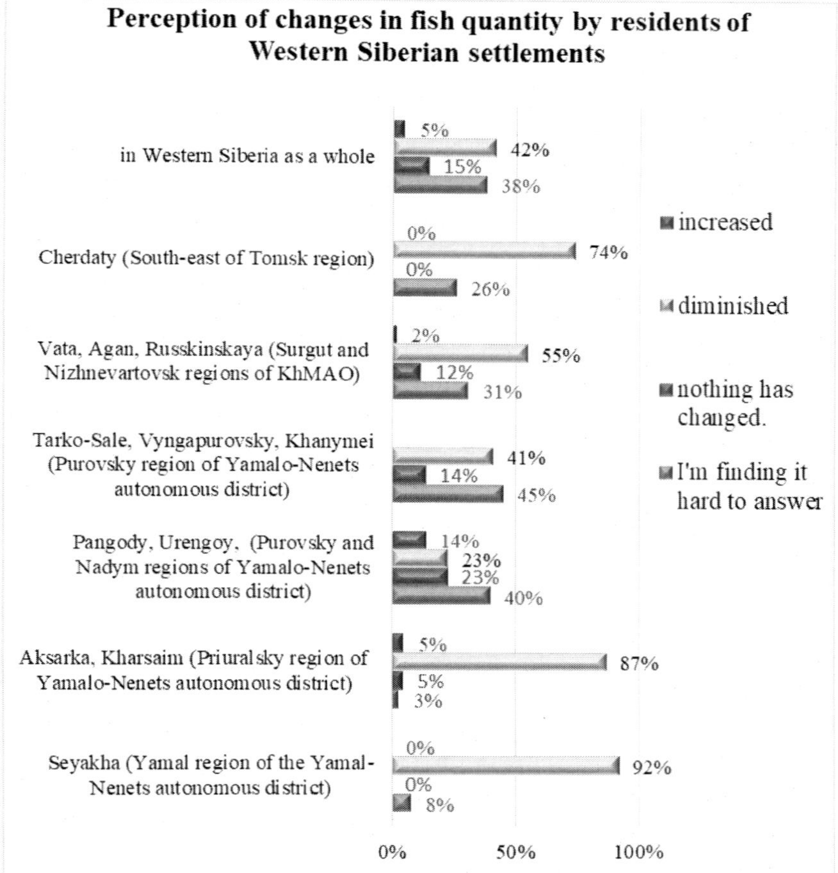

Figure 16. Perception of changes in fish quantity by residents of Western Siberian settlements.

Increase in precipitation in the off-season, which worsens the predictability of the state of household farms and their yields:

> "Take this summer, how it all flooded in the spring. Let's say I didn't even plant a vegetable garden. Because I have a high water level there. In the garden. You can't hope for it now. I have a vegetable garden: I haven't even planted any potatoes now. My neighbors planted it, but it was all wet and rotten. It's work and money - it's bought seeds - it's time and it's time to hire someone to dig up this garden. That is... Here we go. Rains were heavy today, of course" (f., 36, Kargasok)
>
> "Here, as if it were a pit, it rains all the time" (m., 58, Vata)

The decrease in the yield of wild herbs and vegetables on homestead lands is accompanied by the growth of noses and increase in the diversity of insect species:

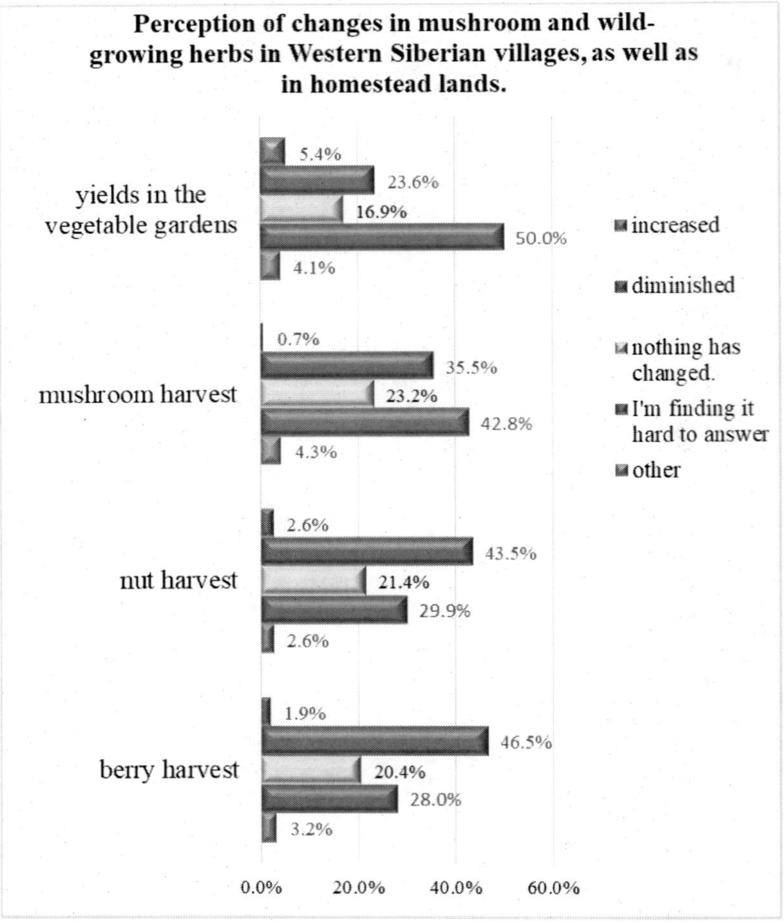

Figure 17. Perception of changes in mushroom and wild-growing herbs in Western Siberian villages, as well as in homestead lands.

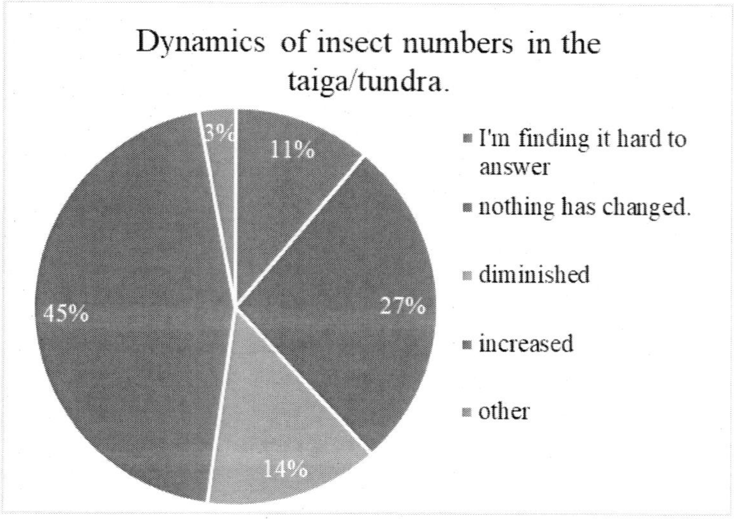

Figure 18. Dynamics of insect numbers in the taiga/tundra.

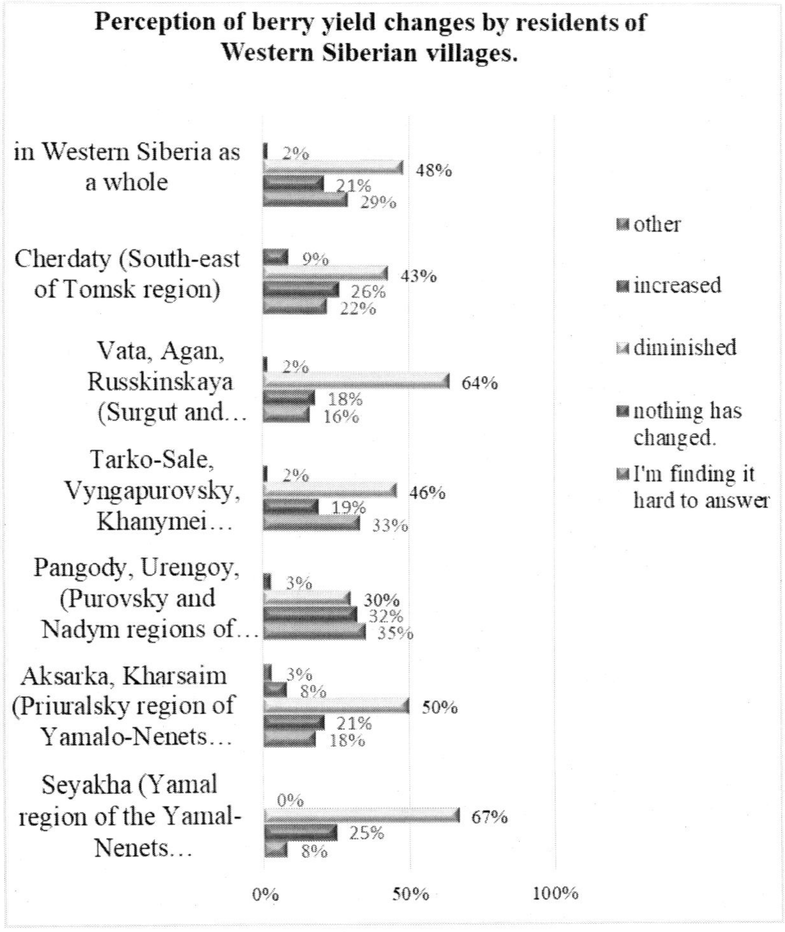

Figure 19. Perception of berry yield changes by residents of Western Siberian villages.

Table 3. Perception of changes in precipitation by the population of Siberia (%)

Question	Answers	Seyakha	Aksarka, Kharsaim	Urengoi, Pangodi	Trako-Sale, Khanymei, Vyngapurovsky	Russskinskaya	Cherdaty
Rainfall in summer	increased	33	12,5	23	28	39	26
	diminished	33	32,5	34	21	19	13
	unchanged	17	27,5	31	30	21	48
	I'm finding it hard to answer.	17	17,5	6	12	17	13
	other	0	0	6	9	4	0
Snow cover thickness	more than ever before.	8	27,5	31	36	17	35
	less than before.	42	32,5	22	17	33	13
	unchanged	33	10,0	19	29	22	30
	I'm finding it hard to answer.	17	22,5	14	15	17	22
	other	0	0	14	3	11	0

4.8. The Change of a Balance in Taiga Fauna (the Ratio of Wild Animals) Due to Anthropogenic Impact

As it was noted above, the impact of climate change on life in a particular settlement may be related not only to phenomena occurring in the region of residence, but also to those originating in other regions. Thus, for example, fires in the European part of Russia, according to informants, influenced the migration of lynxes to Western Siberia, and fires in Krasnoyarsk region influenced the growth of the bear population.

"Bears come, they aren't afraid of roads anymore, so they walk. They're fed, attracted, and else the waste is thrown out of the canteen..." (m., 59, Ruskinskaya)

-"I do not really notice it. Although there was one winter, 60 degree below zero, and after this winter there is almost no muskrat and squirrels" (m., 32, Agan)

"it has become more snakes, bears and ticks. We were not afraid to go to the woods previously, but if you go to the cemetery now then you will take off 2-3 ticks from yourself" (m., 35, Kargasok)

Oh, there are also changes in biodiversity, birds have appeared, that did not exist before: chaffinches, willows, gorgeous tails, kingfishers, oak-breeders" (m., 48, Kargasok)

"the lizards were running all the time: we were catching them, now I do not see them at all" (f., 42, Vata)

"It has become less ducks or they have changed the route, they are saying, the duck changes it, this one, they have the flight somewhere else" (m., 52, Novonikolskoe)

"Once, 50 cars of hunters came for hunting in the Northen Procachi at a time. All right, we shoot a little bit there, Khanty, but we don't choose everything, we leave it. Well, then they've came, shooted a lot, there were a couple of females there. But that's not the point. It has becomes no game in the pine wood. It will take 10-15 years till this population will resume and the situation will normalize" (m., 32, Agan)

"A tower has been built somewhere, a road has been built, bundles of gaz and oil wells are being built, the duck do not understand it immediately, change the route of migration." (m., 32, Agan)

4.9. Emergence of New Species of Insects, Birds, Animals, Fish

The increase in summer temperatures and the mildness of winters (Figure 3-6) lead to the advance of insects characteristic of the southern sub-Arctic zone to the south of the Arctic. Informants, first of all, mention dangerous species of insects, such as ticks, which are infection carriers:

"And now they're showing up, the ticks are all over the place. Earlier, when my father was a little boy, he used to say, we didn't even have snakes. Snakes had just appeared, and in Muravlenko they began to appear, too" (m., 59, Russkinskaya).

"When the fires came here in the 2012, the wolves and lynxes came here. After 2012. Or rather, not after the 2012, but after the fires in central Russia - in 2010. From there the animals came here" (g, 47, Agan).

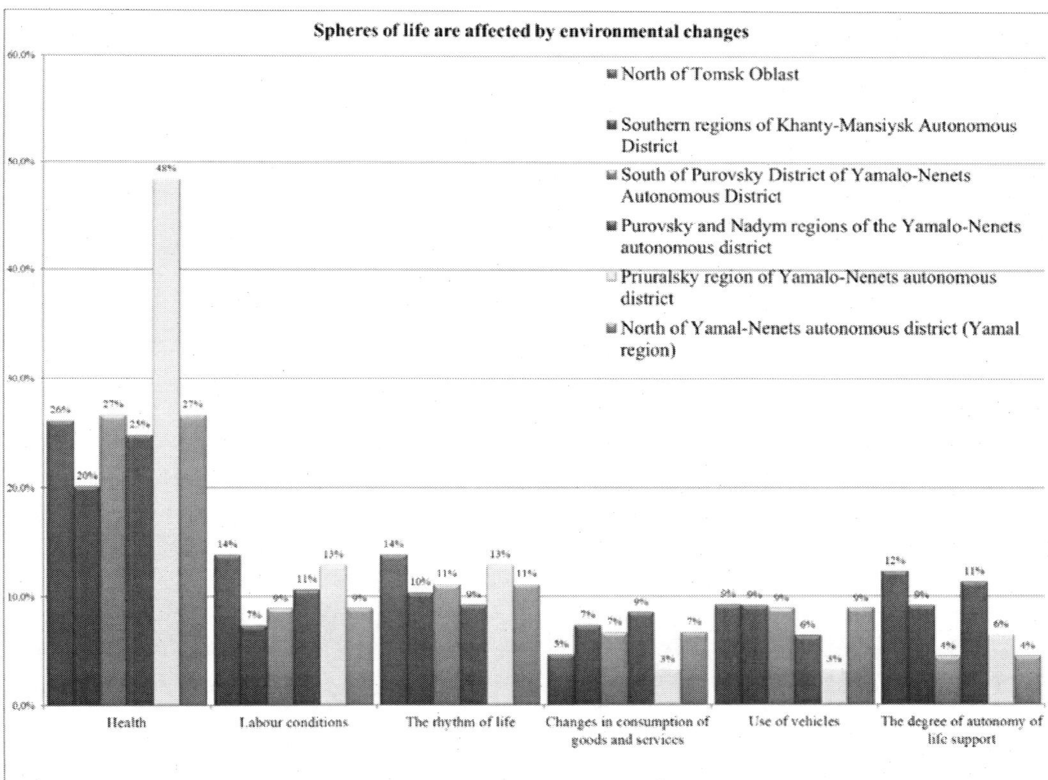

Figure 20. Impact of climate change on the livelihoods of local residents.

4.10. Relationship between Climate Change and All Spheres of Life

About half of the residents of the North of Yamal (Seyakha) are concerned about health problems caused by severe climatic changes and extreme weather events (48.4%), while for the southern regions of the Khanty-Mansiysk Autonomous Region (20.1%) health issues are not so common. From 10 to 14% of the population of different regions almost unanimously believe that climate change has an impact on the rhythm of life.

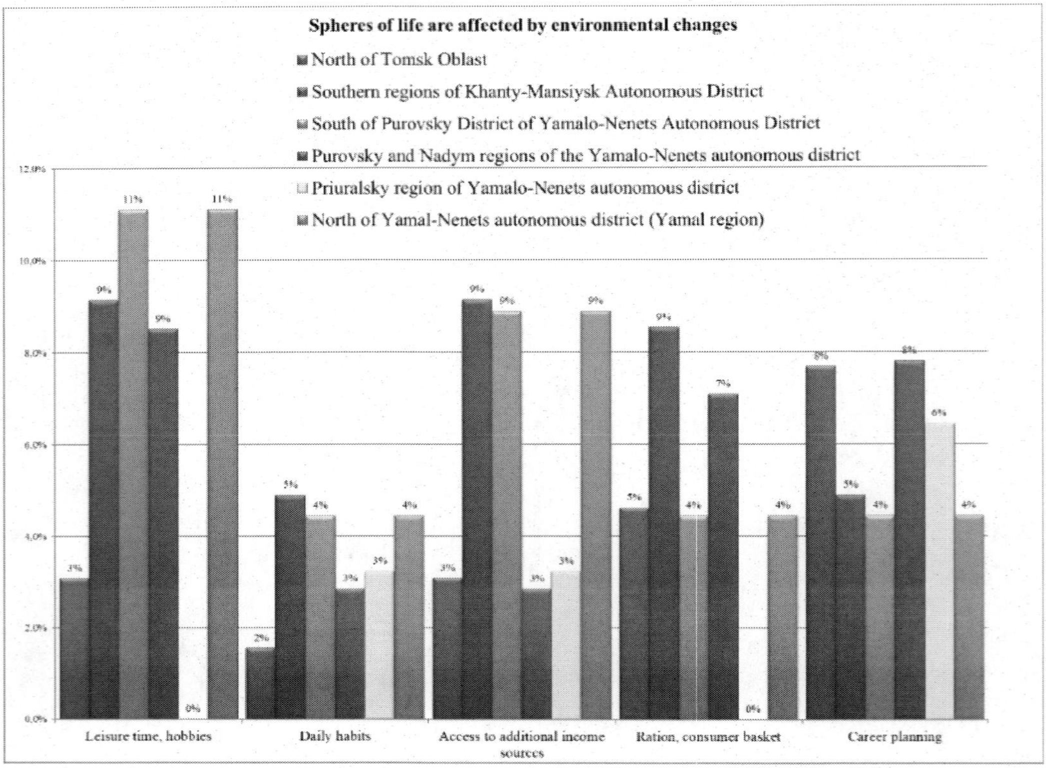

Figure 21. Impact of climate change on the livelihoods of local residents.

It is interesting to analyze how explanatory models that focus on a particular driver of change relate to areas of life that are affected by climate change in the opinion of respondents. Populations in different regions are concerned about different changes and transformations affecting their daily lives. An important aspect of everyday life in the light of ongoing climate change is the degree of autonomy of livelihood (especially in the North of the Tomsk region - 12%) and the central part of the Khanty-Mansiysk Autonomous District - 11.3%). This is especially important for individual households and rural settlements, whose infrastructure was built on the basis of the Soviet system, which was partially destroyed and lost its technical equipment, and is therefore not always ready for major changes, extreme weather events and shifting seasons.

"The climate has changed a lot. It's... bad. It's growing: there, this permafrost is melting, these tall ones, they're flowing down slowly. In winter, on the contrary, it is warm, they do not freeze, which are meant to be "eternal" and these are the big hillocks we have, these with reindeer moss. [Interviewer: Are there pastures on deer hillocks?] Yes, yes, a few years ago, eight years ago. At first - high; and now ... here only the rests stick out. Everything went down. There are no strong frosts, they do not freeze in winter, next year they still melt. Then again the snow blows up in winter and brings warmth with itself. Earlier, when we had severe frosts, then we were good, we went up then" (m., 59, Russkinskaya).

Supply and consumption patterns in the northern regions: "Everything has become more complicated, the cattle are very hard to hold now, there are a lot of needs for this now, and all this has become very expensive" (f., 32, Lukashkin Yar).

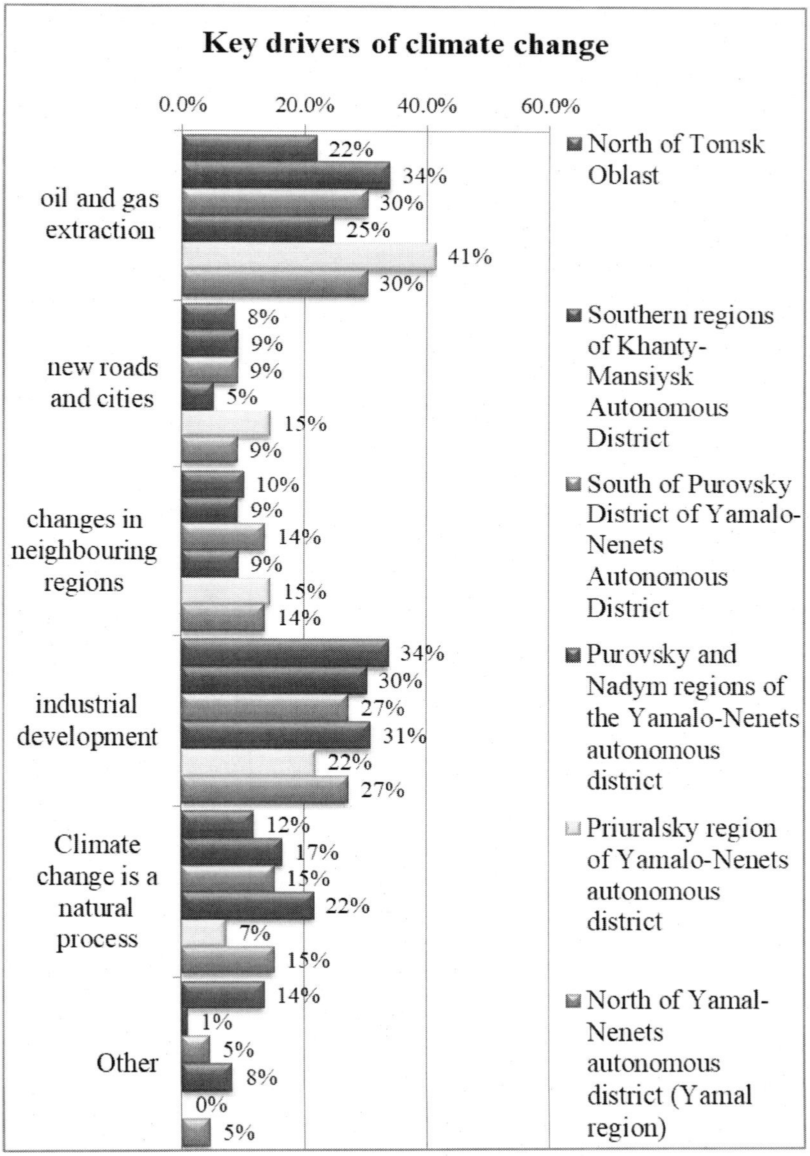

Figure 22. Key drivers of climate change.

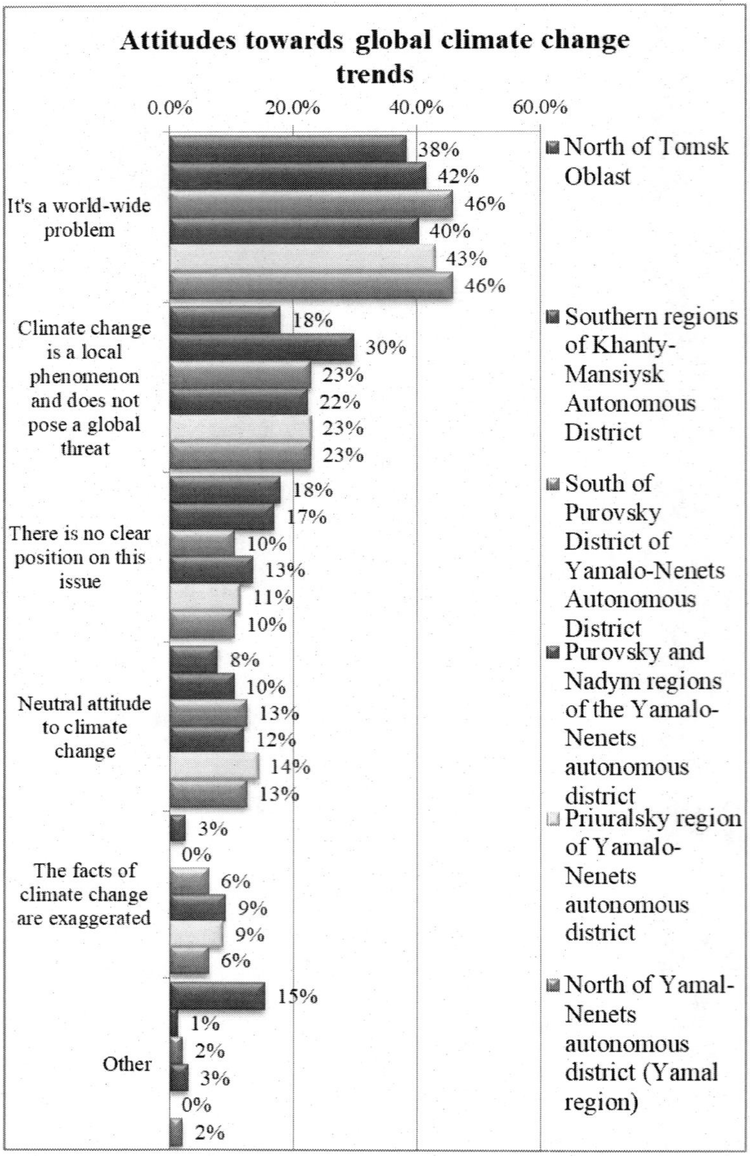

Figure 23. Respondents' attitudes towards global climate change trends.

CONCLUSION

According to a quantitative study, from 26% to 42% of respondents in different parts of Western Siberia claim that the winter period is longer than before. In contrast, in the northernmost area of the study (Seyakha village), 34% of respondents believe that winters have become shorter.

The answers about winter temperatures are more similar - the majority of respondents in the northern regions are convinced that winters have become warmer (58 - 44%), in the southern regions it was noticed by a smaller number of respondents (37-35%), while in the southernmost region (Cherdaty settlement) 35% of the population claim that winters have become colder. The fact that the summer period has become longer is noted only by residents of Yamal (Seyakha settlement): residents of other districts either do not notice any changes, or believe that the summer has become shorter, so in the area of Russkinskaya and Agan it is noted by 60% of respondents, in the settlement of Cherdaty - 48%.

The majority of residents of Seyakha (75%), Aksarka, Kharsaim (38%), Urengoi, Pangody (42%) noted the warming in summer, while 48% to 57% of respondents from more southern regions, on the contrary, believe that the summer has become colder. The changes in spring and autumn seem insignificant for the residents of Western Siberia, only in Seyakha 50% of respondents noted the increase in temperatures in autumn, and 45% of respondents from the village Russkinskaya and Agan noted a decrease in spring temperatures. The same is true for the perception of rainfall.

The majority of residents of the Far North note an increase in the number of unusually hot days (Sayakha - 83%, Aksarka, Kharsayim -33%, Urengoy, Pangody - 38%), while 48% to 50% of respondents from the southern regions, on the contrary, believe that the number of hot days decreased. When asked whether the number of unusually frosty days has changed, the majority of residents of the surveyed regions (36% to 54%) answered that the number of unusually frosty days has decreased. The exception is the interviewed residents of the southernmost village of Cherdaty (43%).

Uniform answers to the question "Has the intensity of rapid temperature fluctuations changed?" have been recorded; between 35% and 67% of respondents believe that sudden temperature changes have become more evident. Answers to the question about the change in the number of extreme weather events in different surveyed areas are not so uniform. The majority of respondents from different regions of Siberia believe that the number of cloudy days and wind pressure have increased. The residents of Siberia answer in the most uniform way those questions that are directly related to their daily life. Thus, in the majority of the surveyed districts respondents complain about the decrease in the yield of berries and other wild herbs, the increase in the number of mosquitoes. Almost everywhere there is a decrease in the number of fish stocks. It is noteworthy that this problem is of greater concern to the residents of the Ob basin, because here from 52 to 95% of respondents noted that the number of fish has decreased. This problem is not so significant in the Pur river basin.

It is important that, based on the results of semi-structured interviews and questionnaires, it seems that local residents have a well-reasoned and elaborated system of perceptions of the causes and main drivers of climate change. Views of respondents from different regions differ significantly. While for the residents of the Urals region of the Yamalo-Nenets autonomous district (Aksarka, Kharsayim) the development of oil and gas production is the dominant driver of significant climate change in the world and Western Siberia, the residents of the Northern settlements of the Tomsk region (Kargasok, Vertikos, Lukashkin Yar) believe that the responsibility for climate change lies primarily with the development of heavy industry and industrial production (34%), and only the second place is given to such a driver as mining and processing of minerals (22%).

A mixed perspective on the causes of climate change was evident in the interviews and questionnaires of respondents living in the Purovsky and Nadym districts of the Yamal-Nenets autonomous district (Pangody, Urengoy). A third of the sample (31%) noted the impact of industry on climate change, but 25% of local residents believe that the impact of the oil and gas industry is dominant. In addition, 22% of respondents from the region believe that climate change is a natural process that is part of the natural macrocycles.

In conclusion, we would like to note that the relevance of the study is justified, in particular, by the distribution of answers to the question about the degree of concern about climate change among local residents of the West Siberian Arctic and subarctic, as well as their views on the scale and context of these changes. From 38% (North of the Tomsk region) to 46% of respondents (South of the Purovsky Raion and North of Yamal) believe that climate change is global in nature and poses a threat to the future of the region. However, almost a third of the population in the southern regions of the Khanty-Mansiysk Autonomous Region believe that climate change is insignificant and only relevant to the micro-region. Although 8-13% of the respondents are neutral about climate change and 11-13% have not formed a clear position on this issue, more than 50% in each of the explored regions are concerned about the observed climate change trends.

ACKNOWLEDGMENTS

The study was carried out with the financial support of the Russian Foundation for Basic Research in the framework of scientific project No. 18-05-60264 and No. 18-00-01625.

We also express gratitude to Large-scale research facilities «System of experimental bases located along the latitudinal gradient» for fruitful cooperation (http://ckp-rf.ru/usu/586718/; http://www.secnet.online/megaustanovka-ru.html).

REFERENCES

Anisimov O. A., Ziltcova Ye. L., Zhegusov Yu. I. (2017) Public perception of climate change in the cold regions of Russia: an example of Yakutia. *Ice & snow*, 57(4). 565-574.

Arkema K. (2014) Key lessons for incorporating natural infrastructure into regional climate adaptation planning. *Ocean & Coastal Management*, 95. 189-197.

Bolotova A. (2012) Loving and Conquering Nature: Shifting Perceptions of the Environment in the Industrialised Russian North. *Europe-Asia Studies*, 64 (4).

Bolotova A., Vorobyev D. (2007). Managing Natural Resources at the North: Changing Styles of Industrialization in the USSR. *Patrimoine de l'industrie (Industrial Patrimony),* 9 (17): 29-41.

Chappin E. J. L., van der Lei T. *Modeling the adaptation of infrastructures to prevent the effects of climate change - an overview of existing literature.*

Cunningham, R., Cvitanovic, C., Measham, T., Jacobs, B., Dowd, A.-M., & Harman, B. (2015). Engaging communities in climate adaptation: the potential of social networks. *Climate Policy,* 16 (7), 894–908.

Davydov A. N. Mikhailova G. V. (2013) Climate change and conditions of life in the Arctic in perception of Nenets on island Vaigach. *Ecology of human*. 2013. 2. 29-34.

Dudeck S. (2010) "Indigenous Oil Workers between the Oil Town of Kogalym and Reindeer Herder's Camps in the Surrounding Area." *Biography, Shift-Labour and Socialisation in a Northern Industrial City – The Far North: Particularities of Labour and Human Socialisation Proceedings of the International Conference in Novy Urengoy, Russia, 4th – 6th December 2008*, 138-146.

Dudeck S. (2015) "Khanty Subsistence in a Changing Environment - Reindeer Herding and Indigenous Survival Strategies in Western Siberia." *Proceedings of the 29th International Abashiri Symposium: Environmental Change and Subsistance of Northern Peoples: Developement and Adaptation.* Abashiri: Association for the Promotion of Northern Cultures, 33-40.

Ford, J. D., Labbé, J., Flynn, M., Araos, M. (2017). Readiness for climate change adaptation in the Arctic: a case study from Nunavut, Canada. *Climatic Change*, 145(1-2), 85–100.

Foucault, M. (2007) Subject's hermeneutics: Course of lectures given at the Colleges de Française 1981-1982 academic year. *St. Petersburg. Science*, (In Russian).

Huryn, A. D., Slavik, K. A., Lowe, R. L., Parker, S. M., Anderson, D. S., & Peterson, B. J. (2005). Landscape heterogeneity and the biodiversity of Arctic stream communities: a habitat template analysis. *Canadian Journal of Fisheries and Aquatic Sciences*, 62(8), 1905–1919.

Lukin, Y.F. (2014) Status, composition, population of the Russian Arctic. *Arctic and North.* №15. 57-94. (In Russian).

Pelyasov, A. N. (2016) Russia's arctic frontier: Paradoxes of development. *Regional Research of Russia.* 6. №. 3. 227 – 239.

Pilasov, A.N. (2009) And the last will be the first. Northern periphery on the way to knowledge economy. *Moscow. URSS.* 542 p. (In Russian)

Pilasov, A.N., Galtseva, N.V., Atamanova, E.A. (2017) Economics of Arctic "islands" (on the example of the Nenets and Chukchi autonomous districts). *Economics of the region.* 13. № 1. 114–125.

Rakhmanova L. (2014) Memory and identity in the symbolic space of the island. *Language, Individual & Society.* 8. 322-330.

Rakhmanova L. (2014) Social identity as a space-marker of reformatory social activity (by the example of Solovky Islands' community). *SGEM International Multidisciplinary Scientific Conferences on Social Sciences and Arts.*

Rakhmanova L. Y. (2015) Interconnection of Socio-cultural Adaptation and Identity in the Socialization Process. RUDN Journal of Sociology.

Rakhmanova, L. (2018) Trial—adaptation—rooting: Understanding and embodiment of the concepts in the Arctic and Siberia. *Paths of Russia. 1917–2017: One hundred years of changes*, 227–244. Moscow, St. Petersburg: Nestor-History.

Wallbridge R. (2012) Climate change, need, and everyday life. *Conference paper, BSA Climate Change Study Group Event - Methodological and conceptual approaches for researching climate change at different societal scales.* 1-14.

Zamyatina, N. Y., Pilyasov, A. N. (2018a) A new interdisciplinary scientific field: Arctic regional science. *Regional Research of Russia.* 8. № 3. 215–224.

Zamyatina, N. Yu., Pilasov, A. N. (2018b) A new approach to the development of northern and Arctic territories of Russia: local transport system. *Problems of territory development.* № 4. 26–41. (In Russian).

In: The Arctic: Current Issues and Challenges
Editors: Oleg S. Pokrovsky, Sergey N. Kirpotin et al.
ISBN: 978-1-53617-306-2

Chapter 3

WARMER CLIMATE OF ARCTIC CITIES

Igor Esau*[1,*]*, Mikhail Varentsov*[2]*, Marlene Laruelle*[3]*,
Martin W. Miles*[4,5]*, Pavel Konstantinov*[2]*, Andrey Soromotin*[6]*,
***Alexander A. Baklanov*[7] *and Victoria V. Miles*[1]

[1]Nansen Center/Bjerknes Centre for Climate Research, Bergen, Norway
[2]Moscow State University, Moscow, Russia
[3]George Washington University, Washington D.C., US
[4]NORCE/Bjerknes Centre for Climate Research, Bergen, Norway
[5]University of Colorado, Boulder, US
[6]Tyumen State University, Tyumen, Russia
[7]The World Meteorological Organization,
Global Atmospheric Watch Programme, Geneva, Switzerland

ABSTRACT

The International Panel for Climate Change (IPCC) has recently envisioned a number of detrimental environmental effects and challenges that will emerge with the global warming in excess of the 1.5°C limit set by the Paris Agreement. The Arctic was mentioned as a particularly vulnerable region on the Earth with rapid and amplified warming affecting all components of the cryosphere and biosphere. This chapter characterizes even larger warm temperature anomalies and corresponding environmental changes in the urban Arctic. A large share of the Arctic population is currently living in 1.5°C and even 3.0°C warmer climate. We report the urban temperature anomalies – urban heat islands (UHIs) – in 11 Arctic cities. The Arctic urban climate is significantly warmer, especially during the coldest weather spells, than the corresponding regional climate. The winter UHI during the polar night is driven by the anthropogenic heat flux and enhanced air pollution in cities. The summer UHI during the polar day is driven by stronger warming of drier and better drained urban surfaces. We show that the UHI intensity depends on the land cover types of the surrounding environment. The Arctic UHI is localized, but still affects an area 2-3 times larger than the city itself. The warmer temperatures enhance the vegetation productivity creating "green belts" around cities and

[*] Corresponding Author's Email: igore@nersc.no.

towns. Sustainability studies revealed that better understanding of the Arctic UHI may result in engineering and policy measures to counteract detrimental consequences of the excessive warming.

Keywords: Arctic climate, urban heat island, meteorological observations, UHIARC, remote-sensing data, MODIS, urban sustainability

INTRODUCTION

A Special Report on Global Warming of 1.5°C by the International Panel for Climate Change (IPCC, 2018) has identified a number of detrimental environmental effects and unwelcome changes emerging if this threshold is crossed. The global warming over land has already reached this threshold (IPCC, 2019); and the annual average Arctic surface air temperature rose by 2.7°C from 1971 to 2017, with the 3.1°C increase in winter and the 1.8°C increase in summer (AMAP, 2019). Perhaps, the most prominent effects of this unprecedented Arctic warming are observed through sea ice retreat (Kinnard et al., 2011), permafrost thaw (Biskaborn et al., 2019), northward advance of boreal forest (Harsh et al., 2009) and rapid release of soil carbon into the atmosphere (Sitch et al., 2007).

Permafrost thaw, for example, may cross a tipping point starting a self-supporting feedback cycle of degradation if temperature anomaly increases beyond 2°C. Indeed, the regionwide increase of soil temperature has been already reported (Biskaborn et al., 2019). Degrading permafrost not only releases green-house gases but also weakens ground-bearing capacity that puts Arctic infrastructure at risk (Shiklomanov et al., 2016; Hjort et al., 2018). In a review of Arctic urban vulnerability, Streletsky et al., (2012) concluded that the permafrost degradation has already damaged 10% (in Norilsk) to 80% (in Vorkuta) of constructions in the Russian cities. Under the risk of total collapse, the whole towns like Newtok and Barrow (now Utqiagvik, Alaska) are planning for relocation on more stable ground, whereas many apartment buildings are abandoned in the Russian Arctic towns.

Another IPCC special report on climate change, sustainable land management, and changes in terrestrial ecosystems (IPCC, 2019) devoted significant attention to global urbanization and urban climate change. The report emphasized that urban areas are warming more rapidly than their natural background. It stated: "Both global warming and urbanization can enhance warming in cities and their surroundings (heat island effect), especially during heat related events, including heat waves (*high confidence*)." Bear in mind, however, that the Arctic cities, except Barrow, were not considered in this report. This Chapter addresses this knowledge gap. We report the observed intensity and environmental impact of the urban heat island (UHI) in 11 representative Arctic cities and towns (see Table 1).

Looking from a global perspective, this Chapter contributes to the World Meteorological Organization (WMO) work on promoting safe, healthy and resilient cities. The WMO develops the Integrated Urban Weather, Environment and Climate Systems and Services (IUS) initiative to improve sustainability of global cities under changing climate. A conceptually novel methodology for the Urban Integrated Hydrometeorological, Climate and Environmental Services (Grimmond et al., 2014; Baklanov et al., 2017, WMO, 2018) is in an implementation phase in many cities around the world. Using 30 demonstration cities, guidelines and best practice have been formulated for the IUS within distinct administrative frameworks (WMO, 2019). Up to date, however, the Arctic cities are not included in these coordinated efforts.

Although the amplified Arctic warming threatens urban infrastructure, logistics and traditional lifestyles, other effects of the warming might be beneficial for urban resilience. Sea ice retreat and advance of boreal forest ecosystems may eventually contribute to sustainable Arctic urbanization. Norman Pressman (1996), an outstanding urban geographer and visionar, described the Arctic cities as remote outposts where architectural solutions were aimed, at least in idea, to insulate people from harsh climate. A lack of connections with nature and more general a lack of sense-of-place (Stedman, 2003) made the northernmost cities vulnerable to swings in geopolitical agenda and market conjecture, both of which are defined from centers much further to the South. In this context, improving logistics and emergence of more familiar ecosystems within the urban reach may restore social connections with the environment. Urban afforestation (green space building) is one of the proven mitigation approaches. McBride and Douhovnikoff (2012) has disclosed a high symbolic value of urban trees in reconnection of people and nature. Arctic arboreta and botanical gardens are now found in Akureiry, Tromsø and Apatity; and urban greening programmes are running in Siberian cities.

Table 1. Socio-geographical and climate characteristics of 11 Arctic cities/towns considered in this study. Climate of the urban background is given according to the Koeppen-Geiger classification (Kottek et al., 2006) and a bioclimatic classification (http://www.globalbioclimatics.org)

City	Lat.	Lon.	Popul.	Climate/ Bioclimte	Soil	Biome
[1] Barrow (Utqiagvik), Alaska	71°17′N	156°47′W	4,200	ET/ POLAR continental	Permafrost	Tundra
[2] Fairbanks, Alaska	64°50′N	147°43′W	31,500	Dsc/BOREAL continental	Discontinues Permafrost	Boreal forest (Taiga)
[3] Tromsø, Norway	69°40′N	18°56′E	64,400	Dfc/ BOREAL oceanic	Unfrozen	Boreal forest
[4] Longyearbyen, Spitsbergen	78°13′N	15°38′E	2,200	ET/ POLAR oceanic	Permafrost	Tundra
[5] Murmansk, European Russia	68°58′N	33°05′E	295,400	Dfc/ BOREAL sub-continental	Unfrozen	Tundra-forest
[6] Apatity, European Russia	67°34′N	33°24′E	55,700	Dfc/ BOREAL sub-continental	Unfrozen	Boreal forest (Taiga)
[7] Vorkuta, Urals Russia	67°30′N	64°02′E	56,100	Dfc/ POLAR continental	Discontinues Permafrost	Tundra-forest
[8] Salekhard, West Siberia, Russia	66°32′N	66°36′E	49,300	Dfc/ POLAR continental	Discontinues Permafrost	Tundra-forest
[9] Nadym, West Siberia, Russia	65°32′N	72°31′E	46,600	Dfc/ POLAR continental	Discontinues Permafrost	Tundra-forest
[10] Novy Urengoy, West Siberia, Russia	66°05′N	76°41′E	104,100	Dfc/ POLAR continental	Discontinues Permafrost	Tundra-forest
[11] Norilsk, Taimyr, Russia	69°20′N	88°13′E	179,600	Dsc/ POLAR continental	Permafrost	Tundra-forest

The complex set of socio-environmental interactions and feedbacks unleashed by the Arctic warming and urbanization would require own monography; this broad topic is clearly beyond the scope of this chapter. Instead, this chapter is focused on documenting warmer climate of Arctic cities and accompanying vegetation changes. An extensive literature review is embedded into the Chapter. It shall help our reader to establish links between the warmer urban climate and multitude of current issues and challenges reported in the Arctic studies across different disciplines. This Chapter consists of four sections. Introduction is followed by the section presenting the cities and their individual issues. The third section presents and discusses results of our studies. The last section outlines conclusions and challenges.

THE ARCTIC CITIES

The modern Arctic is a very urbanized region. More than three quarters of the total Arctic population reside in cities (Laruelle, 2019). The share is smaller in Alaska and Scandinavia but exceeds 90% in the Russian Arctic where almost 98% of the Taymyr regional population is concentrated in the Norilsk agglomeration. Table 1 lists all 11 cities considered in this chapter. Their geographical location and climate types are given in Figure 1. In the following description, we group the cities by similarity of their physio-geographical conditions.

Barrow (*Utqiagvik*) is a town situated on the coast of the Arctic Ocean in northernmost Alaska, in the continuous permafrost zone. As other Arctic cities, Barrow heavily relies on artificial foundations of sand and gravel – materials by composition and thermal properties distinctly different from those of the natural landscape.

Longyearbyen is a town similar in size and biome to Barrow. It however has milder maritime Arctic climate. It is found in a narrow valley open to an ice-free sea inlet – a fjord. In this sense, it is similar to Tromsø, which is the largest city in northern Norway.

Tromsø is situated on an island surrounded by an ice-free narrow fjord. Mountain slopes around the fjord are rather steep. They frequently lock a shallow mixed boundary layer over the city creating proper conditions for aggravation of heat, moisture and chemical air pollution in city. Although Tromsø has been included in a "twin cities" climate programme – an exploratory study of future climate analogs for urban adaptation (Rohat et al., 2017), there are no proper urban climate studies in this city yet.

Murmansk – the largest city in the Arctic with its almost 300 thousand inhabitants – is in many climatic aspects similar to Tromsø. It is located along a narrow fjord surrounded by steep hills. The climate of Murmansk is better studied (Yakovlev, 1972; Varentsov et al., 2014; Demin et al., 2017) than climate of any other Arctic city with exception of Apatity. Unfortunately, those studies remain disconnected and fragmented. Moreover, the city has a complex relief with local elevation differences of more than 50 m.

Apatity is a mid-size city in the interior of Kola Peninsula. Its climate is considerably more continental than climate of other cities in Fennoscandia. The city occupies the top of a hill, while its meteorological station (WMO ID 22213) is located at lower ground near the lake Imandra. We emphasize the elevation differences as the Arctic climate is characterized by frequent and strong temperature inversions extending from the surface to 1-2 km height (Zhang et al., 2011). Such temperature inversions themselves could create an illusion of UHI as a statistical artifact. Temperature inversions and reduced turbulent mixing obstruct interpretation of local climate anomalies. Varentsov et al., (2018) have addressed this challenge of urban climatology in details. Intercomparison of ground-based and satellite observations with modeling results indicated that not less than 50% of the wintertime urban temperature anomaly in Apatity must be attributed to the anthropogenic heating and land use change factors. High concentration of research and education institutions in the city facilitates detailed studies of urban environment and climate here. Moreover, the studies of the city are often trans-disciplinary and consider socio-environmental interactions (e.g., Johansen and Skryzhevska, 2013).

The last group of 6 Arctic cities (Fairbanks, Vorkuta, and the Siberian cities: Salekhard, Nadym, Novy Urengoy, Norilsk) have severe continental climate with very low winter temperatures and frequent episodes of calm anti-cyclonic weather causing cold air spells. Winter temperatures can drop down to -45°C – -55°C; and remain that cold for weeks. All these cities incorporate patches of frozen ground and permafrost.

Fairbanks is a mid-size city in the interior of Alaska, US. The city hosts the University of Alaska and the International Arctic Research Centre with their extensive environmental programmes. It has a long (>100 years) series of meteorological observations. Because of this, it was actively studied as an archetypical example of cold boreal climate in high northern latitudes. The UHI was defined as a synchronous temperature difference between the urban Fairbanks International Airport and rural Air Force Base meteorological stations.

In Fairbanks, an interplay between the UHI and air quality issues becomes an important issue. Bowling (1986) indicated that air quality issues are challenging high latitude communities. Even moderate amounts of pollutants released into an atmosphere with extremely poor dispersion creates air quality hazards, often localized in one or a few neighborhoods. High correlation between the UHI intensity and poor air quality has been found in many mid-latitude cities as well (e.g., Li et al., 2018). But in the Arctic cities, the persistent air inflow into the UHI center can accumulate pollutants in downtown over several days.

Vorkuta, *Salekhard*, *Nadym*, and *Novy Urengoy* are Russian cities to the West and East of Urals mountains. The latter three cities are found on a flat and palludified plain. They have rapidly grown during 1970s and 1980s so that distant effects of the created UHI are now emerging. The knowledge is needed to adjust the construction standards to the warmer climate and predict geotechnological risks. By contrast, Vorkuta has been growing since 1940s. Today, about 80% of buildings are damaged in the city (Streletskiy et al., 2012; Streletskiy et al., 2019).

Norilsk is one of the most remote and the second largest city in the Arctic. It harbors the coldest winter temperatures; and to the large degree it is built-up on permafrost. Use of cold-ground protecting technologies is widespread as well as damage from UHI negligence (Shiklomanov et al., 2017). Norilsk is a mining city with widespread and heavy environmental pollution from metal smelters. This pollution (hydrogen sulfide, in particular) altered or destroyed ecosystems on a large swath of land around the city (Telyatnikov and Prystyazhnyuk, 2014). Although over nine months of the year, the surrounding ecosystems are dormant and insensitive to air pollution, accumulated ground and water pollutants act in summertime. Researchers reported reduced biodiversity in all systematic groups of higher and lower plants, but especially in mosses, lichens, dwarf shrubs, and larch trees. On this degraded background, ecosystem and landscape changes due to climate warming are emerging over the last 2-3 decades. Steady greening has reclaimed 21,000 km^2 by 2017. The observed transition from tundra to taiga and a slight decrease in abundance of lakes are robust signs of higher soil temperatures and substantial permafrost degradation (Nyland et al., 2018). The permafrost degradation is much more advanced in the city itself as damages to buildings and infrastructure indicate (Shiklomanov et al., 2017).

By contrast, Apatity and Vorkuta agglomerations are less damaging to the nature, albeit their environmental footprints are also known. Rigina et al., (1999) reported the environmental damage from a metal smelter in Monchegorsk – a city north of Apatity. They concluded that combination of mountains and temperature inversions shelters Apatity from severe air pollution. Virtanen et al., (2002) reported pollution damages for tundra vegetation around Vorkuta city. The damage is concentrated within 10 km from the urban industrial zone.

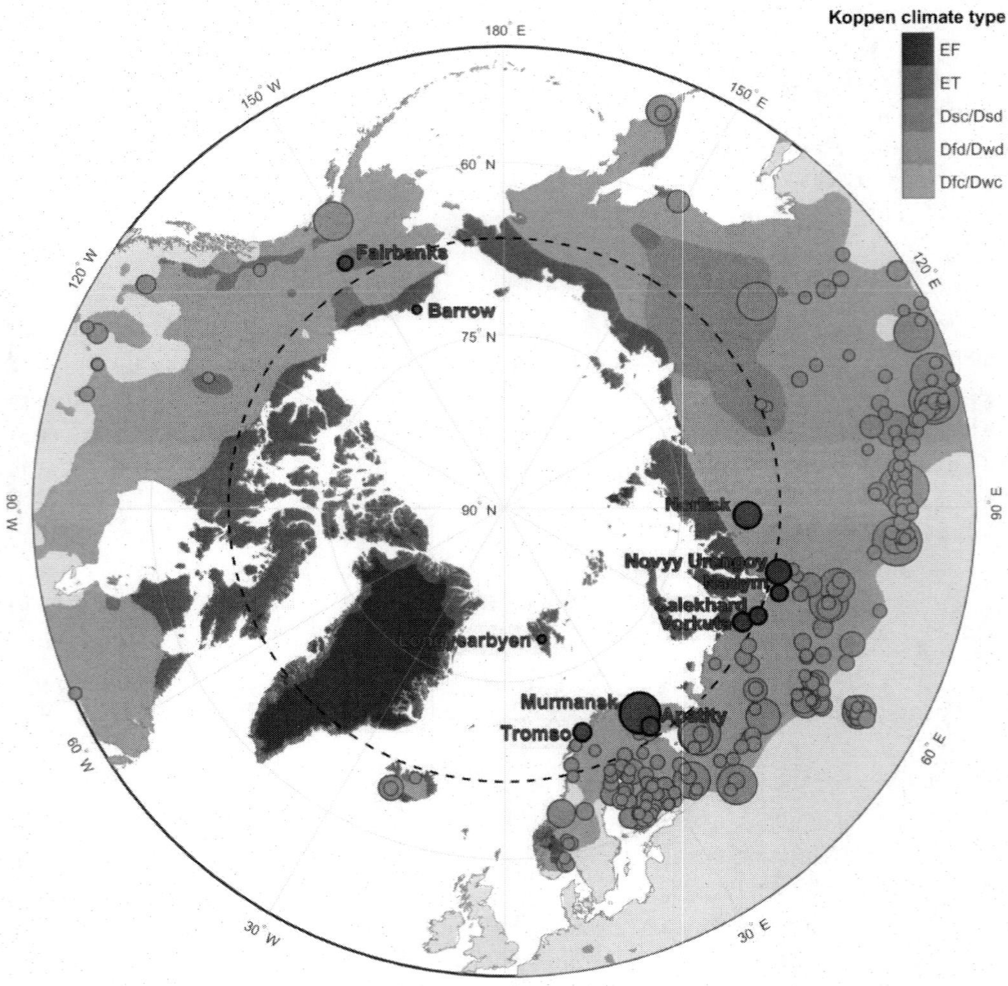

Figure 1. Geographical locations of the Arctic cities (red circles) considered in this study. Green circles identify other "winter" cities with more than 15,000 inhabitants in boreal and polar bioclimatic zones. The size of circles shows the urban population.

METHODOLOGY OF URBAN CLIMATE STUDIES

Urban climate studies seek to identify climate anomalies induced by specific urban factors. This knowledge is needed to help urban planning in design of sustainable, resilent and comfortable city environment with minimal intervention to nature, thus, implementing a strong sustainability concept (Ekins et al., 2003; Wu, 2013). An extensive list of climate and socio-environmental factors to assess the urban environmental quality could be found in e.g., Musse et al., (2018). The list of the most influential factors is still open and debated, however, it usually includes the anthropogenic heat flux from urban metabolism, primarily from heating (Dong et al., 2017). The other influential factors include changes in the surface heat budget, soil moisture and thermal conductivity, sky view factors, urban aerodynamical roughness and form factors. Those factors are, however, hard to get as it would require running costly research projects. Therefore, a common practice is to deduce the urban factors from their observable effects comparing local climates of urbanized and natural landscapes. The differences then could be attributed to generalized land use – land cover types so to allow urban climate characterization using solely geomorphological features. This approach is adopted, e.g., in the World Urban Database and Access Portal Tools (WUDAPT; www.wudapt.org) project utilizing a Local Climate Zone (LCZ) types by Stewart and Oke (2012).

Surface air temperature (SAT) and, more recently, land surface temperature (LST) are the most accessible climate variables that characterize the urban-rural differences. In recent years, high resolution remote sensing data products have taken over the more traditional in situ meteorological observations. The satellite remote sensing is advantageous in its homogenies and dense spatial coverage, whereas the meteorological stations provide certified and regular but sparse, fragmented and often non-representative observations (Stewart, 2011). Typical accuracy of the SAT is 0.5°C for temperature loggers and 0.1°C for the certified stations (Konstantinov et al., 2018). Typical accuracy of remotely sensed LST is 1.0°C to 3.0°C depending on land cover type and weather conditions (Urban et al., 2013). Frequently, the UHI based on the LST data is referred to as the surface urban heat island (SUHI).

The UHI intensity is determined as a urban-rural temperature difference. It strongly depends on fractional area distribution of land use – land cover (LULC) types around the city (Esau and Miles, 2018) or on immediate environment of the meteorological stations (Stewart, 2011). Figure 2 shows the diversity of local climates characterized by the mean seasonal LST around Fairbanks. The winter and summer LST for shrubland considerably differ from that of forest and the urban areas. For instance, the SUHI in Fairbanks with respect to shrubland south of the city has an intensity of 0.6°C (DJF) and 2.4°C (JJA), while the SUHI intensity with respect to forest is 2.1°C (DJF) and 2.6°C (JJA) correspondingly. Thus, the SUHI, and hence, the UHI intensity depends on composition of the LULC types around the city. Our study includes both the UHI and SUHI data. Moreover, we compare the SUHI obtained by a popular buffer-based algorithm (BF-UHI) with one obtained by a simplified urban-extent algorithm (SUE-UHI) from Chakraborty and Lee (2019).

The SUHI intensity, ΔT, in the BF-UHI is the difference between the maximum temperature of the city poligon, T_{Umax}, and the mean temperature, $\langle T_R \rangle$, of minimally developed land (natural LULC types) outside the city, i.e.,

$$\Delta T = T_{Umax} - \langle T_R \rangle, \qquad \text{(Eq. 1)}$$

where the angular brackets represent the spatial averaging. The obtained SUHI intensity is insensitive to the number of urban and rural pixels (Tan and Li, 2015). The SUE-UHI intensity is defined as the difference between the averaged LST over all urban and all non-urban LULC pixels (excluding water pixels) within the same densely populated "urban" area. Thus, it is expected that the SUE-UHI will be of lower intensity than the BF-UHI.

MODIS Data Products for Urban Climate Studies

The Arctic cities are insulated enclaves embedded into largely pristine environment. Such insulation greatly facilitates identification and attribution of anomalous urban climate features. A popular buffer-based method (e.g., Zhou et al., 2015) is rather suitable for such cities. This method, abbreviated here as BF-UHI, identifies buffers around an urban polygon. The urban polygon may include vegetated and water patches but is dominated by bare and urbanized (artificial) surfaces. To justify the BF-UHI application, we observe that small-scale LULC heterogeneity does not reveal a visible impact on the averaged LST anomalies. *A posteriori* analysis (see Figure 3) shows that the averaged LST is rather spatially smooth.

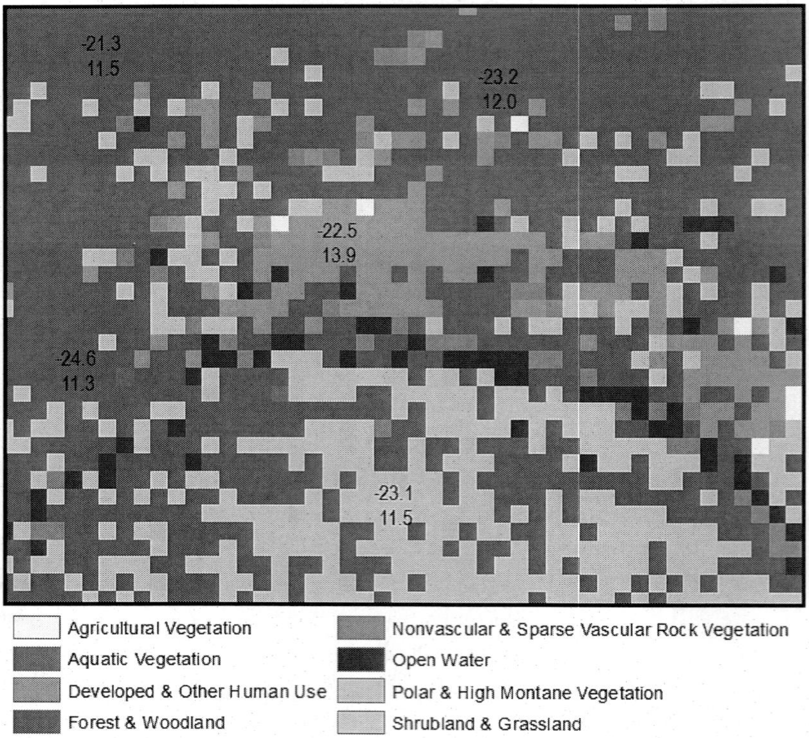

Figure 2. An example of the local climate diversity related to the different LULC types around Fairbanks. The urban area corresponds to the Developed and Other Human Use LULC type. Colors represent the LULC types combined into 8 major classes. The numbers show the mean LST over the corresponding LULC type for the summer (JJA) and winter (DJF) months over 2000-2017. See Esau and Miles (2018) for more details.

We utilized several data products from the MODerate-resolution Imaging Spectroradiometer (MODIS) onboard of the Terra/Aqua satellite platform. Although the MODIS data are available for clear sky conditions, spatial data composites and interpolation allow for data extension to partially cloudy conditions. Zhang et al., (2016) study indicated that seasonally averaged characteristics are less sensitive to the cloud cover issue. The mean MODIS LST could be recovered even for time periods with 60% of images contaminated by clouds. Nevertheless, severe cloud contamination of the MODIS data in the Arctic remains as an issue for remote sensing analysis (Ostby et al., 2014).

A detailed description of data processing can be found in Esau et al., (2016) and Miles and Esau (2017). In this study, we utilize the MODIS data over 2000 – 2017. The LST data were obtained from the MOD11A2 8-day average, clear sky data product for the local day and night slices. Quality of the MODIS LST products was assessed in several studies. For example, Urban et al., (2013) assessed this product on the pan-Arctic scale. The mean difference between the MODIS LST and the SAT does not exceed 1.6°C (the correlation coefficient is 0.89). Although both the difference and strength of the LST-SAT correlations depend on atmospheric stability (e.g., Zilitinkevich et al., 2005), the high correlation suggests that this stability factor has somewhat limited influence of the climate statistics.

The remote sensing data were further processed to obtain seasonally averaged diurnal (averaged 10:30 and 13:30 local solar time) and nocturnal (averaged 22:30 and 01:30 local solar time) LST. We consider the mean LST values for the polar night winter season (DJF) and for the polar day summer season (JJA).

We compare our BF-UHI values with the SUHI values from the SUE-UHI data. The global SUE-UHI dataset utilize the same MODIS data products to obtain the SUHI in 9500 urban clusters. The SUE-UHI data do not include small urban poligons, areas with significant elevation difference and areas with small fraction of the urban LULC. In result, only a few Arctic cities are in the SUE-UHI set. In addition, we excluded Tromsø and Nadym from that dataset as these cities were geographically misrepresented.

The environmental quality could be approximated through the Normalized Difference Vegetation Index (NDVI) (e.g., Fung and Siu, 2010; Liang and Weng, 2011). The NDVI represents a measure of canopy 'greenness' where values below 0.2 are generally non-vegetated surfaces. The MODIS NDVI 16-day composites with a 250 m spatial resolution (MOD13Q1) were analyzed over 2000-2018. We choose to work with the annual maximum NDVI or NDVImax, reached by the end of July, in order to eliminate to seasonality of the vegetation development in that index. The processing procedure has been described in Esau et al., (2016) and Miles and Esau (2016). We used the NDVImax in the range between 0.3 and 1.0 to exclude water, bare soil and other non-vegetated pixel from the analysis. As we are focused on ecosystem responses, each pixel-value NDVImax is considered as characterizing a surrounding ecosystem. The near-neighbor interpolation covers gaps in the NDVImax composite's values over the same LULC types.

The Urban Heat Island Arctic Research Campaign (UHIARC)

Before the Urban Heat Island Arctic Research Campaign (UHIARC), Arctic urban climate was largely unexplored. Detailed UHI studies were known for Fairbanks (Bowling, 1986; Magee et al., 1999), Barrow (Hinkel and Nelson, 2007; Klene et al., 2013), Murmansk

(Yakovlev, 1972; Demin et al., 2015; Demin et al., 2017) and Apatity (Rigina et al., 1999; Konstantinov et al., 2015; Demin et al., 2016; Shumilov et al., 2017). In the high Arctic, Climate in Longyerbyen on Svalbard was studied by Nordli (2010), and fragmentary, by O. Humlum (http://www.climate4you.com). But, up to our knowledge, urban meteorological network has not been deployed in any Eurasian city above the polar circle prior to UHIARC.

The UHIARC network is a dense observational network of temperature loggers (iButton) and automatic weather stations (Davis Vantage Pro2). The UHIARC was deployed in several mid-sized cities (Apatity, Salekhard, Vorkuta, Nadym, Novy Urengoy) in the winter 2016–2017 (Konstantinov et al., 2018). Earlier preliminary installations had been tried in Apatity, Murmansk, and Norilsk. In the winter 2018-2019, the network was run in Apatity, Nadym and Salekhard. The most long-term and comprehensive study was conducted in Apatity where Varentsov et al., (2018) compared the UHI as obtained from the network with the BF-UHI data and with results of meteorological modeling in the area. The UHIARC dataset is freely available from http://urbanreanalysis.ru/uhiarc.html.

The UHIARC six cities have grown out from small settlements during the Soviet era. Similarity of urban architectural solutions and surface morphology determines similarity of the urban LCZs there. The concentration of population in mid-rise apartment blocks (LCZ5) distinguishes the Russian Arctic cities from sprawled low-rise neighborhoods (LCZ6) of Scandinavian, American cities. The heavy industry LCZ10 and the large low-rise LCZ8 are also found in the cities. The open low-rise LCZ6 is found in Salekhard, but dominates in Barrow, Fairbanks, Tromsø and Longyearbyen.

RESULTS AND DISCUSSION

Urban Heat Islands

All 11 Arctic cities of this study harbor persistent UHI and SUHI (Table 2). Larger SUHI intensity as compared to the UHI intensity from the UHIARC data could be easily understood when we recall that LST and SAT are physically different. The SUHI is sensitive to the surface heat budget as it is based on the LST. Both the anthropogenic heat flux in winter and absorption of solar radiation in summer rise the LST. The surface exchanges heat with the lower atmosphere by turbulent mixing. As the turbulent mixing is often suppressed in the stably stratified atmosphere, where the LST is lower than the SAT, this difference between surface and air temperatures could be significant. Konstantinov et al., (2018) have revealed that the strongest UHI are found in the calm days with the lowest SAT. A lack of mechanically generated turbulent mixing under the calm conditions exacerbates this temperature difference supporting significantly stronger SUHI.

A physical mechanism supporting the UHI in summertime could be described as follows. The surface absorbs incoming short-wave solar radiation. Although the urban and other disturbed surfaces generally absorb less radiation, they are also better drained, and therefore dryer, so a larger share of available energy can be used to rise temperature instead of driving evapotranspiration. The summer Arctic landscape is wet. Wetlands and open water bodies surround the cities. The local LST of wetlands is lower as wetlands are covered with more active vegetation (larger NDVImax). The largest NDVI value is related to the forest vegetation around the city edge and along rivers where drainage is better. Figure 3 reveals reasonable collocation between the lower LST and the higher NDVImax in summertime around Nadym. It shows that a "green belt" (NDVImax > 0.78) around Nadym also corresponds to a "cold belt" in the LST rendering the SUHI over artificial surfaces.

The role of solar radiation, vegetation cover and surface wetness become clearer when we look at the airport area to the south-east of Nadym. The landing strip does not have any anthropogenic heating so that its winter LST remains lower than that of the city core. By contrast, its summer LST is as high as in the city core. The surface of the landing strip is rather small to explain its large SUHI footprint. Our direct field measurements on the sandy surface around the strip (August 2018) confirm that surrounding sparsely vegetated sands are warm and dry biasing the energy budget towards the sensible heat flux.

Table 2. The UHI/SUHI intensity in the Arctic cities. NA stays not "Not Applied" specifying absence of data

City	UHI			Comment
	Summer	Winter	Annual	
Barrow	0.1	2.0	2.0	BUHIS (Hinkel and Nelson, 2007)
	9.6*	0.2	4.9*	BF-UHI
Fairbanks	0.0	0.7	0.3	Magee et al., (1999)
	0.4	0.0	0.1	SUE-UHI
	1.2	2.4	1.8	BF-UHI
Tromsø	2.2	3.8**	3.0**	BF-UHI
Longyearbyen	0.6	0.8	0.7	Nordli (2010)
Murmansk	0.2	0.9	0.6	Yakovlev (1972)
	NA	3.0	NA	UHIARC
	0.6	1.4	0.9	SUE-UHI
	3.5	5.2**	4.4**	BF-UHI
Apatity	NA	1.0	NA	UHIARC
	2.5	1.1	1.8	BF-UHI
Vorkuta	NA	1.1	NA	UHIARC
	1.2	0.7	0.8	SUE-UHI
	5.0	7.2	6.1	BF-UHI
Salekhard	3.3	3.7	3.5	BF-UHI
	NA	1.4	NA	UHIARC
Nadym	1.4	1.8	1.6	BF-UHI
	NA	0.9	NA	UHIARC
Novy Urengoy	2.0	2.3	2.1	BF-UHI
	NA	1.1	NA	UHIARC
Norilsk				UHIARC
	0.6	0.3	0.5	SUE-UHI
	4.8	4.7	4.8	BF-UHI
Snow climate type of the Northern Hemisphere	1.78	0.35	NA	SUE-UHI

* The very large SUHI is due to cold sea/lake ice surfaces in the buffer zone.

** The large SUHI is due to warm open water sea surfaces in the buffer zone.

Barrow is already at risk due to the Arctic warming, sea ice retreat and coastal line erosion (Lynch and Brunner, 2007). Despite large uncertainties of climate projections and extreme weather events, the Barrow city council is urged to prepare and implement a climate adaptation plan. The UHI knowledge should be part of these efforts. Since 2001, the Barrow Urban Heat Island Study (BUHIS) has monitored a 150 km^2 area with about 70 temperature loggers (Hinkel and Nelson, 2007). The UHI was calculated as an averaged group difference between reading within urban and rural patches. The winter UHI averages to 2°C, occasionally exceeding 6°C under cold and calm conditions. The summer UHI is minimal with averaged intensity of 0.1°C when referred to wet tundra. Besides the weather factors, the UHI has strong correlation with natural gas consumption.

This UHI have already resulted in additional permafrost thaw, thicker seasonally active soil layers and greening. Klene et al., (2013) reported urban soil temperatures 0.3–2.3°C higher and the active soil layer 15–40 cm thicker than those in the surrounding tundra. Accelerated snowmelt in Barrow, up to 2 weeks earlier than in tundra, have been known since the mid-1970s. These changes raise an issue of inappropriate infrastructure design and construction criteria being implemented – the development compromising the soil bearing capacity and increasing vulnerability to the global warming.

Longyearbyen has a very long history of meteorological observations since 1911 (Nordli, 2010). Several stations were simultaneously working at a relatively small area for at least a few months (often longer). Hence, it is feasible, at least in principle, to infer the UHI from the observations. The mean multi-annual temperature series adjustment between the town and the airport is 0.7°C. But complex relief, varying mountain winds, and very large temperature trends (2.4°C per 100 years) make such estimations unreliable.

Murmansk has been studied by climatologists since 1960s (Yakovlev, 1972). Meteorological observations from several urban stations over 1930s–1970s clearly show heat trapping in the lower parts of the city and along the fjord. Although data are fragmented, they also suggest that the most intensive UHI (0.8°C) is found in the densely built areas. Yakovlev relates this UHI amplification with a high aerosol load. He pointed out that the urban aerosols (soot) can block outgoing thermal radiation. Since 1960s, Murmansk has strongly expanded. The UHIARC data (preliminary) revealed more intensive UHI of about 3°C (Konstantinov et al., 2014). The BF-UHI data support the SUHI of large intensity (5.2°C for DJF) in the city. Demin et al., (2015) questioned the large UHI using the SAT records from regular meteorological stations. Although their idea to look at the SAT changes in relation to urban growth (1935-2014) is interesting, it has not been developed enough in the published analysis.

Apatity is another well-studied city in Fennoscandia. The most detailed study is published in Varentsov et al., (2018). It also addresses the role of temperature inversions and katabatic drainage flow in the UHI formation. The drainage can lower the SAT by a few degrees in relief depressions. Even small elevation differences can cause hazardous meteorological phenomena such as road icing, fog etc., challenging the urban management. The UHI significantly moderate surface-layer inversions which become stronger again at 100 m and higher levels. These numbers agree with the unmanned helicopter data obtained for the UHI in Nadym (Varentsov et al., 2019).

The six Arctic cities in severe continental climate can have very low winter temperatures and frequent episodes of calm anti-cyclonic weather (cold air spells) with temperatures down to -45°C – -55°C. All these cities reveal significant UHIs in winter and summer seasons. Figure 3 illustrates the UHI and SUHI in those cities.

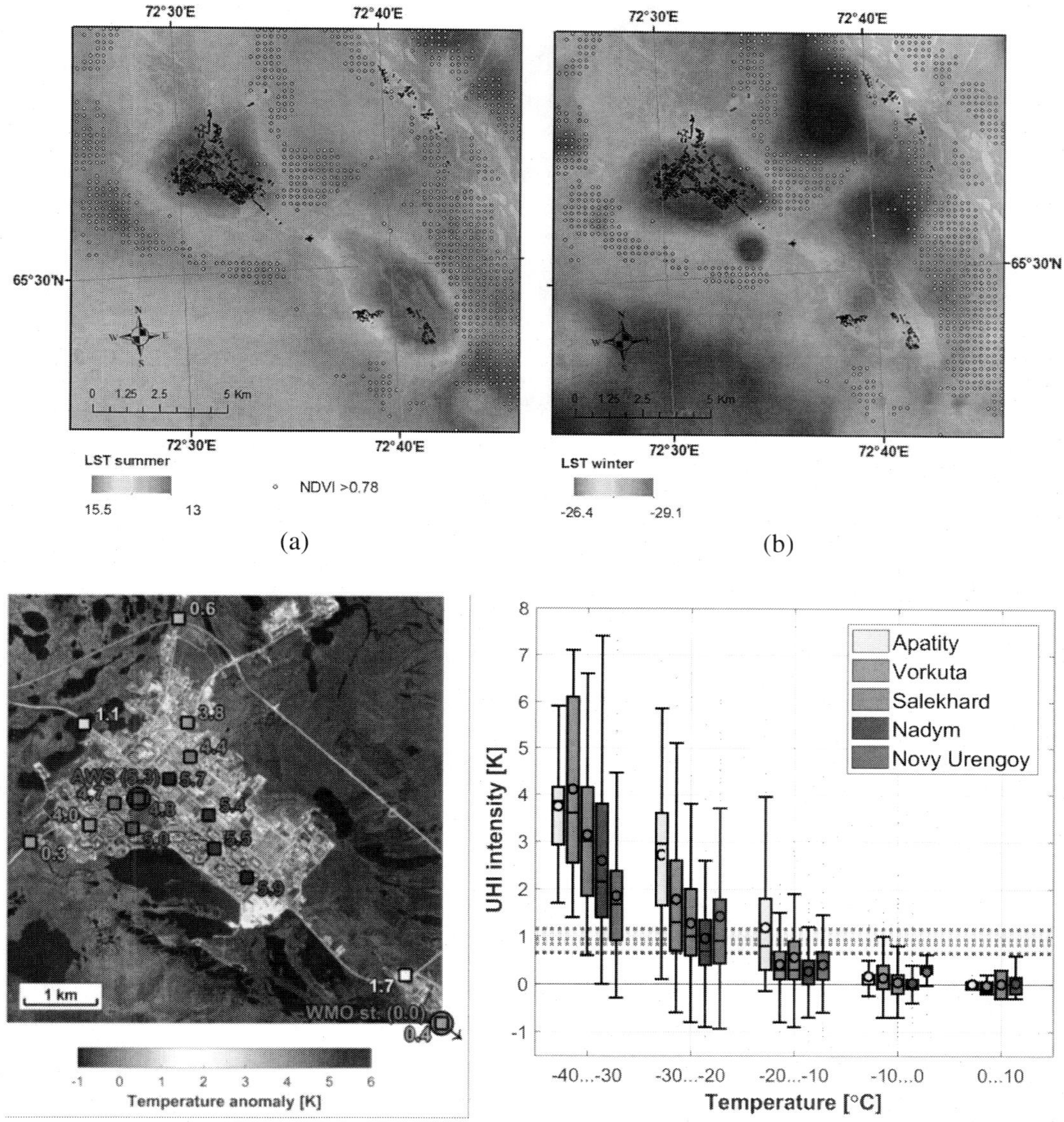

Figure 3. (a) The SUHI intensity shown as the LST in the BF-UHI data averaged over summers (JJA) 2000 – 2018 in Nadym. Patches of highly productive vegetation (NDVImax > 0.78) are highlighted with yellow dots. (b) The SUHI intensity as in (a) but for winters (DJF). (c) The UHI intensity from the UHIARC during 14.12.2016 – 01.03.2017 in Nadym. Colors show the SAT anomalies relative to the rural station (WMO) in Nadym. Circles show locations of the urban AWS and rural WMO stations. Squares show locations of the temperature loggers. (d) Dependence between the UHI intensity and the SAT in five Arctic cities as identified by colors. The data are binned in 10°C intervals. The box-and - whiskers show the statistics over 02.02.2017 – 15.03.2017 for Vorkuta, Salekhard, Nadym and Novy Urengoy (Konstantinov et al., 2018) and 09.12.2015 – 01.03.2016 for Apatity (Varentsov et al., 2018). The dashed colored lines show the averaged winter UHI intensity in the corresponding cities.

Fairbanks was actively studied as an archetypical example of cold boreal climate in high northern latitudes. Frequent strong surface temperature inversions in the area (Malingowski et al., 2014) require accounting for elevation difference here. The first brief study by Bowling and Benson (1978) looked at plausible causes of the UHI such as of variable albedo, the daily temperature cycle, snow cover, and wind speed. They reported a very large UHI intensity of around 10°C, but cautiously attributed it to the extreme temperature inversions. Magee et al., (1999) studied the UHI intensity, seasonality and trends with almost 50 years (1947-1997) of data. They report significantly lower values with the annual average of 0.33°C. Some aggravation of the UHI could be seen from 1970 to 1990, when the largest increase in population occurred.

In Fairbanks, as in other cities as well, an interplay between the UHI and air quality issues becomes an important issue. Bowling (1986) indicated that air quality issues are challenging high latitude communities. Even moderate pollution might lead to localized air quality hazards when the atmospheric dispersion is weak. High correlation between the UHI intensity and poor air quality has been found in many mid-latitude cities (e.g., Li et al., 2018). But in the Arctic cities, the persistent air inflow into the UHI can accumulate pollutants in downtown over several days.

Vorkuta, *Salekhard*, *Nadym*, and *Novy Urengoy* are UHIARC cities. The largest city – Novy Urengoy – supports the UHI of 1.0°C (UHIARC; Konstantinov et al., 2018) and the SUHI of 2.1°C (BF-UHI; Miles and Esau, 2017). The mean winter UHI intensity in the other cities ranges from 0.8°C to 1.4°C with the 95th percentile of about 5°C; while the 99th percentile exceeds 7°C.

The major issues of the cities on permafrost are related to geohazards due to weakening ground bearing capacity (Hjort et al., 2018). All these cities are hot spots of permafrost thaw and ground weakening. One third of the urban infrastructure there is located in areas with high vulnerability to the temperature rise. On the background of global warming, mitigation measures to reduce the local SUHI intensity and impact are urgently needed. Such measures might also reduce wintertime fuel consumption and green-house gas emissions. As highly relevant illustration of the infrustrature damages by permafrost, one can refer to problems with the Global Seed Vault in Longyearbyen. The Washington Post (K. Epstein) reported on 27 March 2019 that the structure is already threatened by warming, possible flooding and the risk is projected to increase.

Emerging Ecosystem Changes in Warmer Urban Climate

Perhaps the most prominent and radical changes due to locally warmer climates are observed in the biological production and composition of emerging ecosystems. Many studies (e.g., Kumpula et al., 2014; Yu et al., 2015) have demonstrated that naturally or anthropogenically disturbed land does not recover vegetation composition typical for the regional ecotone but tends to be occupied by an alternative ecosystem with a larger share of shrubs and trees (Figure 4). Warmer urban climate finds its climate analogs as far south as 600 km in the northern and middle taiga bioclimatic zones. Soil tending, fertilization and plant care may help urban ecosystem to tolerate the Arctic climate. Greening experiments in botanical gardens and in cities such as Nadym have successfully introduced blossoming trees and a range of other invasive species.

Local climate differences across fragmented LULC patches provide opportunities for gradual ecosystem transformations. Temperature variability on kilometer-scale provides buffering to mitigate transient effects of inter-annual weather variability (Lenoir et al., 2013).

The buffering capacity peaks in forest-tundra ecotone at about 65°N. It gives us a reason to expect that the Arctic cities, located at about this latitude, may host strongly modified ecosystems. The UHI footprint typically covers an area 2-3 times larger than the urban area itself. The free evolution of the interacting suburban ecosystems provides more accurate knowledge about environmental processes in warmer climates than the popular plot-scale experiments (e.g., Elmendorf et al., 2012).

Figure 4. Advance of alternative ecosystems on disturbed soils under climate change. Two photos present an unintended heating experiment – afforestation of warmer sandy mound covering the gas pipeline Nadym-Punga (Siberia). The natural surrounding biome is dwarf-shrub tundra on hummocks with lichens on frozen peatlands. Permafrost is discontinues in this areas. Reproduced from Martyshak et al., (2014).

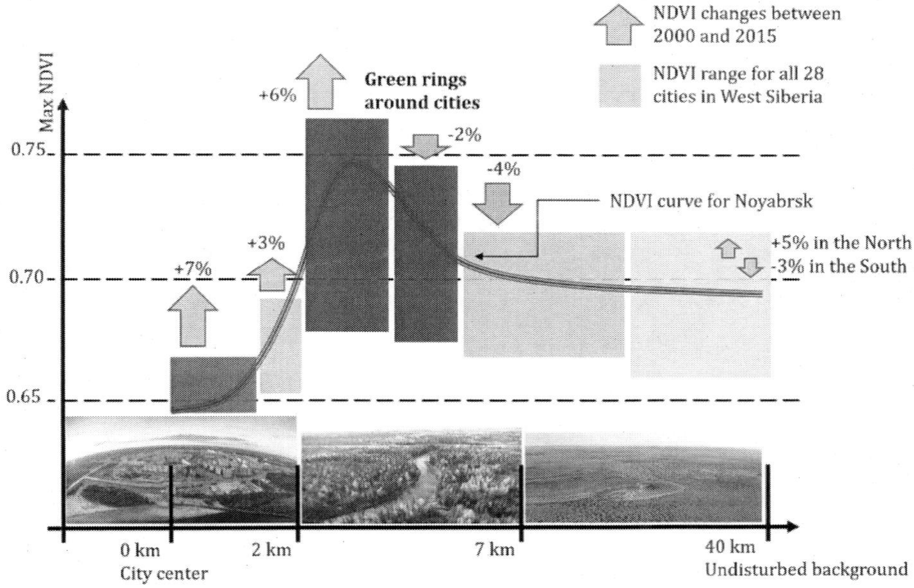

Figure 5. Summary of the observed changes in the biological production in and around the Arctic cities. The data obtained from the NDVImax analysis in Esau et al., (2016).

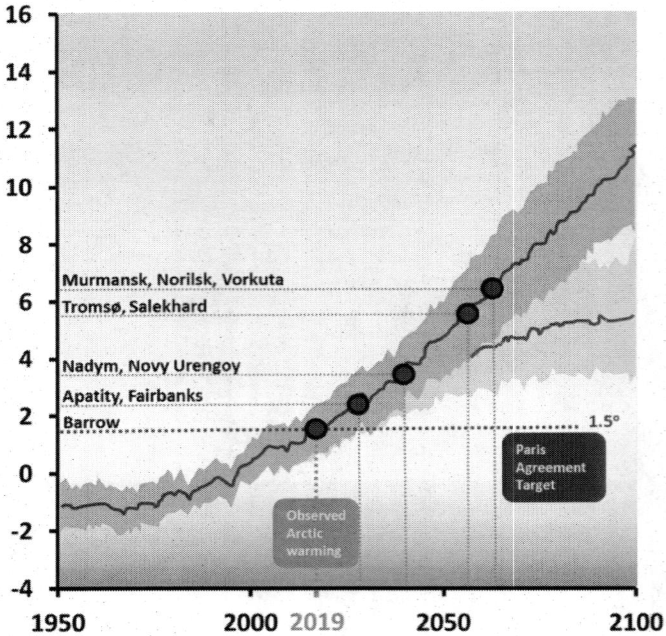

Figure 6. Winter season future temperatures for the Arctic (60–90°N) averaged over 36 CMIP5 global climate models and expressed as departures from the means for the 1981–2005 period. The red line is the ensemble mean for RCP8.5, the blue line is for RCP4.5. Shaded areas denote ± one standard deviation from the ensemble mean (Overland et al., 2014; and Fig. 2.15 of AMAP, 2017). The observed SUHIs from the BF-UHI are shown as red dots collocated with the expected future Arctic temperature anomalies, e.g., the observed wintertime urban temperature anomaly in Nadym corresponds to the regional warming as expected to be reached by 2060. Observe that the present Arctic climate is already 1.5°C warmer than the means.

Extensive NDVI study of 28 cities in northern West Siberia supplied data to generalize an urban greening pattern (Esau et al., 2016). Figure 5 shows that the urban core, which is densely built-up and has low NDVI, is surrounded by a green belt of more productive vegetation communities. This green belt is more or less fragmented in each individual case. Figure 3 shows the green belt and the SUHI in Nadym. Since the Arctic cities are relatively young – their extent and population has stabilized in 1990s – their green belts are not as visible in the NDVI itself as in the recent NDVI trends. The urban and sub-urban ecosystems are becoming greener more rapidly than the natural background.

Conclusion: Current Issues and Challenges

The Arctic cities are found in harsh climate and fragile environment. Historically, the urban development in the region was focused on insulation from "winter" – as Norman Pressman put it – and was disconnected from nature. Global warming, which is more rapid and amplified in the Arctic, opens opportunities to change that attitude. No, the polar night, cold weather, ice and snow will not disappear. But warmer summer temperatures will make possible and meaningful city transformation into living environment with green space and urban forest as elements of design; and warmer winter temperatures will facilitate outdoor recreational activities very much as it is now in the rest of the snow climate zone. An extensive meta-analysis covering studies during 1999-2007 (Shaposhnikov and Revich, 2016) has indicated that severe cold weather spells are hazardous for human health and induce higher mortality in the age group ⩾65. The warmer urban climate decreases weather-related health issues, especially due to frost damage and cardiovascular cases.

The warmer urban climate in the 11 Arctic cities/towns is documented in this study as well as in the referred literature collection. Even on the background of the strong Arctic warming, persistent temperature anomalies in the Arctic cities are impressive. Their socio-environmental impact cannot be overstated. According to Miles and Esau (2017), 70% of the population in northern West Siberia is currently living in the areas experiencing more than 2.5°C warming where the urban temperature anomalies add 1.5°C to the background regional warming of about 1.0°C relative to the 1960-1990 period. Among them, almost 50% or about one million people witness environmental and climate changes at 3.5°C warming.

The warmer urban climate could be seen in a climate change perspective using the climate analogue methodology (Fitzpatrick and Dunn, 2019; Bastin et al., 2019). Figure 6 presents the observed annual mean SUHI intensity (from the BF-UHI dataset) as projected onto the future Arctic warming prediction from 36 models in the Climate Model Intercomparison Project phase 5 (CMIP-5). Even expecting strong Arctic warming in the RCP8.5 scenario, the present-day climate of many Arctic cities corresponds to the regional climate conditions projected by the mid- or end of the 21st century. In fact, these persistent warm temperature anomalies in the Arctic cities allows for observations of early harbingers of changing environmental processes and ecosystem transformations, e.g., shrubification and afforestation (Figure 4), that will unfold in the warming Arctic in future decades.

Our data and the conclusion of the Arctic climate studies provide a general picture of the future Arctic UHI that is different from the conclusions of the IPCC (2019) report. We project that the UHI intensity in the region will decrease as the Arctic is warming and the coldest temperatures with the strongest UHI will be less frequent. The socio-economic trends work in the same direction. Improved building insulation and cold-ground preserving technologies, e.g., thermosypones and ventilated basements (Melnikov et al., 2014; Shiklomanov et al., 2017), reduce the SUHI and UHI in winter, while urban greening reduces them in summer.

The Arctic urban climate will change the urban environment and will be changed by the socio-economic transformations and urban development. Orttung and Reisser (2014) highlighted two main issues affecting sustainability of the Arctic cities: economy of resource extraction and climate change. Society responds to these drivers evolving decision-making processes, mitigation policies, and internationalization of the Arctic. Up to date, however, economy remains prevalent over environmental protection diverting the efforts from strong sustainability approach. Perhaps the most significant trend in sustainability is a gradual relocation of the urban population to smaller, mid-size settlements (Laruelle and Hohmann, 2017). Population in Murmansk, Vorkuta and Norilsk are in decline, whereas population of mid-size cities (Fairbanks, Apatity, Salekhard, Nadym) is stable or increasing. Sustainability of the Arctic cities dramatically depends on their abilities to diversify production and to increase the level of local educational institutes.

Urban ecology can elucidate connections to biogeophysical environment in which city dwellers reside (Grimm et al., 2008). The Arctic cities are found in harsh climate and fragile environment. Environmental damage is widespread around many of those cities. Nevertheless, urban dwellers begin to show a significant degree of care for green space.

Urban afforestation, greening and conservation of sub-urban ecosystems are reflected in strong positive trends of the NDVImax. This greening is supported by the warmer soil temperatures and deeper seasonally active soil layer as many field studies have reported. Seen from ecosystem perspective, the warmer urban climate provides a climate buffering – the fine-grained thermal variability – increasing biodiversity and climate change resilience of ecosystems (Lenoir et al., 2013). The climate buffering effects peaks at about 65°N where most of the Arctic cities are found. It explains their exceptional role in northward proliferation of boreal and temperate ecosystems with a urban-rural biodiversity gradient unobserved anywhere else.

Eventually, the warmer urban climate in some Arctic cities can increase urban and peri-urban food production – a resilience approach advocated by IPCC (2019). This perspective is as much fantasy as might look like. Stammler and Sidorova (2014) described a phenomenon of peri-urban Russian summer cottages – dacha – in the Arctic. This social phenomenon has its roots outside the north but has been introduced by incomers to improve their subsistence, food balance and livelihood. The authors concluded: "The dacha has become an indispensable counterpart of life in the urban concrete housing blocks for most Russian northern inhabitants… dacha has come to stand for a human-environment relationship that gradually re-introduces rurality to urban life in the Russian Arctic so permanently that dacha places start losing their seasonal character."

The strong SUHI is however damaging to infrastructure and buildings in the Arctic cities. As frozen grounds warm and thaw, the risk of geocryologic hazards increases (Nelson et al., 2001). Although there are risks that the global warming will weaken the soil bearing capacity and aggravate the hazards on the regional scale (Hjort et al., 2018), so far, evidence indicates the largest vulnerability in relation with the SUHI and disturbed surfaces. Model-based estimations suggest decline of the ground bearing capacity by 10% to 40% in Siberia. Taken SUHI into account, those estimates should be corrected significantly upward. Unfortunately, the SUHI is still poorly studied and not accounted for in construction projects.

So, what makes the warmer climate in cities special for the current Arctic issues and challenges? *First*, in the Arctic, many aspects of human life, economy and surrounding environment are very sensitive to warmer temperatures, especially to the warmer summertime LST. This sensitivity makes the urban heat island more influencial than in other locations over the globe. Moreover, even relatively small settlemets as well as larger infrastructure objects, e.g., landing strips and sand quaries, induce significant temperature anomalies as the atmospheric stability trapps the anomanies near the ground (Zilitinkevich and Esau, 2005; Davy and Esau, 2016). Without active turbulent diffusion, both heat and air pollutants are transported horizontally and only slowly dispersed. The atmospheric stability explains persistent and concentrated impact of the urban factors on the local environment and climate.

Second, the warmer urban climate forces the changes in other components of the environment. The UHI drives accelerated permafrost thaw and vegetation changes; followed by transformations in human behavior, urban design and approaches to improve urban resilience. In some places, such changes are already challenging and dominant issues. Today, as much as two thirds of the total Arctic population leaves in 2.5°C warmer world. They are witnessing afforestation and challenged by permafrost degradation. Many approaches for the UHI mitigation from temperate latitudes are not working in the Arctic; at the same time, the Arctic UHI knowledge is insufficient to develop effective regional guidelines.

Finally, it is unlikely that the urban climate change in the Arctic will follow trajectories similar to those envisioned for the cities in temperate climates. The UHI will likely decrease, but generally more friendly environmental conditions will support sustainable and diverse urban development in the region. Table 3 summarize the issues and challenges of the warmer climate of the Arctic cities.

Table 3. The overview and qualitative assessment of the current issues and challenged related to the warmer climate of the Arctic cities

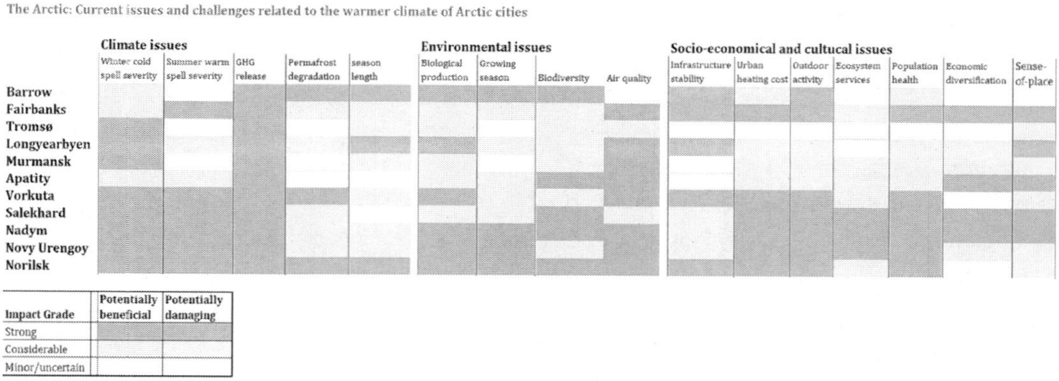

ACKNOWLEDGMENTS

This summary is based on the research efforts funded by the Belmont Forum project *Anthropogenic Heat Islands in the Arctic: Windows to the Future of the Regional Climates, Ecosystems, and Societies*, funded by the Research Council of Norway, the Russian Foundation for Basic Research and the US National Science Foundation.

REFERENCES

AMAP (2019). *Arctic Monitoring and Assessment Programme climate change update 2019: An update to key findings of snow, water, ice and permafrost in the Arctic (SWIPA)*. Oslo, Norway. 12 pp.

Baklanov, A., Grimmond, C. S. B., Carlson, D., Terblanche, D., Tang, X., Bouchet, V., et al., (2017). From urban meteorology, climate and environment research to integrated city services. *Urban Climate*, *23*, 330-341, https://doi.org/10.1016/j.uclim.2017.05.004.

Bastin, J.-F., Clark, E., Elliott, T., Hart, S., van den Hoogen, J., Hordijk, I., et al., (2019). Understanding climate change from a global analysis of city analogues. *PLOS ONE*, *14*(7), e0217592. https://doi.org/10.1371/journal.pone.0217592.

Biskaborn, B. K., Smith, S. L., Noetzli, J., Matthes, H., Vieira, G., Streletskiy, D. A., et al., (2019). Permafrost is warming at a global scale. *Nature Comm.*, *10*(1), 264. https://doi.org/10.1038/s41467-018-08240-4.

Bowling, S. A., Benson, C. S. (1978). Study of the subarctic heat island at Fairbanks. *Environ. Protection Agency Report*, EPA-600/4-78-027, 7 pp.

Bowling, S. A. (1986). Climatology of high-latitude air pollution as illustrated by Fairbanks and Anchorage, Alaska, *J. Clim. Applied Meteorol.*, *25*, 22–34.

Chakraborty, T., & Lee, X. (2019). A simplified urban-extent algorithm to characterize surface urban heat islands on a global scale and examine vegetation control on their spatiotemporal variability. *Int. J. of Applied Earth Observ. Geoinform.*, *74*, 269–280. https://doi.org/10.1016/j.jag.2018.09.015.

Davy, R., & Esau, I. (2016). Differences in the efficacy of climate forcings explained by variations in atmospheric boundary layer depth. *Nature Comm.*, *7*(1), 11690. https://doi.org/10.1038/ncomms11690.

Demin, V., Kozelov, B., Elizarova, N., Menshov, Y. & Konstantinov, P. (2016). The role of relief in the heat island phenomenon in the city of Apatity, *Fundamental and Applied Climatology*, *2*, 95-106. (in Russian).

Demin, V., Anziferova, A., Mokrotovarova, O., (2015). Air temperature change in Murmansk since the beginning of XIX century, *Proc. Kola Science Center RAS*, *2015*(20), 113–125. (in Russian).

Demin V., Anziferova A., Chaus O. (2017). Influence of meteorological conditions on calculation of the urban heat island in Murmansk, *Proc. Hydrometeorol. Res. Center of the Russian Federation*, *363*, 160-175 (in Russian).

Dong, Y., Varquez, A. C. G., & Kanda, M. (2017). Global anthropogenic heat flux database with high spatial resolution. *Atmos. Environ.*, *150*, 276–294. https://doi.org/10.1016/j.atmosenv.2016.11.040.

Ekins, P., Simon, S., Deutsch, L., Folke, C., & De Groot, R. (2003). A framework for the practical application of the concepts of critical natural capital and strong sustainability. *Ecol. Economics*, *44*(2–3), 165–185. https://doi.org/10.1016/S0921-8009(02)00272-0.

Elmendorf, S. C., Henry, G. H. R., Hollister, R. D., Björk, R. G., Boulanger-Lapointe, N., Cooper, E. J., et al., (2012). Plot-scale evidence of tundra vegetation change and links to recent summer warming. *Nature Climate Change*, *2*, 453–457. https://doi.org/10.1038/NCLIMATE1465.

Esau, I., Miles, V. V., Davy, R., Miles, M. W., & Kurchatova, A. (2016). Trends in normalized difference vegetation index (NDVI) associated with urban development in northern West Siberia. *Atmo. Chem. Phys.*, *16*(15), 9563–9577. https://doi.org/10.5194/acp-16-9563-2016.

Esau, I., & Miles, V. (2018). Exogenous drivers of surface urban heat islands in northern west Siberia. *Geography, Environment, Sustainability*, *11*(3). https://doi.org/10.24057/2071-9388-2018-11-3-83-99.

Fitzpatrick, M. C., & Dunn, R. R. (2019). American urban areas in the late 21st century. *Nature Comm.*, 1–7. https://doi.org/10.1038/s41467-019-08540-3.

Fung, T., & Siu, W. (2000). Environmental quality and its changes, an analysis using NDVI, *Int. J. Remote Sensing*, *21*(5), 1011-1024. https://doi.org/10.1080/014311600210407.

Grimm, N. B., Faeth, S. H., Golubiewski, N. E., Redman, C. L., Wu, J., Bai, X., et al., (2008). Global change and the ecology of cities. *Science*, *319*(5864), 756–760. https://doi.org/10.1126/science.1150195.

Grimmond, C. S. B., Tang, X., Baklanov, A. (2014). Towards integrated urban weather, environment and climate services. *WMO Bull.*, *63*(1), 10-14.

Harsch, M. A., Hulme, P. E., McGlone, M. S., & Duncan, R. P. (2009). Are treelines advancing? A global meta-analysis of treeline response to climate warming. *Ecol. Letters*, *12*(10), 1040–1049. https://doi.org/10.1111/j.1461-0248.2009.01355.x.

Hinkel, K. M., & Nelson, F. E. (2007). Anthropogenic heat island at Barrow, Alaska, during winter: 2001–2005. *J. Geophys. Res.*, *112*, D06118. https://doi.org/10.1029/2006JD007837.

Hjort, J., Karjalainen, O., Aalto, J., Westermann, S., Romanovsky, V., Nelson, F., et al., (2018). Degrading permafrost puts Arctic infrastructure at risk by mid-century. *Nature Comm.*, *9*(1), 5147. https://doi.org/10.1038/s41467-018-07557-4.

IPCC (2018). Special Report on Global Warming of 1.5°C, *Intergovernmental Panel on Climate Change* (IPCC). Assessed August 30, 2019, https://www.ipcc.ch/sr15/.

IPCC (2019). Special Report on climate change, desertification, land degradation, sustainable land management, food security, and greenhouse gas fluxes in terrestrial ecosystems. *Intergovernmental Panel on Climate Change* (IPCC), Assessed August 30, 2019, https://www.ipcc.ch/report/srccl/.

Johansen, H., & Skryzhevska, Y. (2013). Adaptation priorities on Russia's Kola Peninsula: climate change vs. post-Soviet transition. *Polar Geogr.*, *36*(4), 271–290. https://doi.org/10.1080/1088937x.2013.788577.

Kinnard, C., Zdanowicz, C. M., Fisher, D. A., Isaksson, E., de Vernal, A., & Thompson, L. G. (2011). Reconstructed changes in Arctic sea ice over the past 1,450 years. *Nature*, *479*(7374), 509–512. https://doi.org/10.1038/nature10581.

Klene, A. E., Nelson, F. E., & Hinkel, K. M. (2013). Urban–rural contrasts in summer soil-surface temperature and active-layer thickness, Barrow. *Polar Geogr.*, *36*(3), 183–201. https://doi.org/10.1080/1088937X.2012.706756.

Konstantinov P., Baklanov A., Varentsov M., Kukanova E., & Repina I. (2014). Experimental Urban Heat Island Research of Four Biggest Polar Cities in Northern Hemisphere. In *Geophys. Res. Abst. EGU General Assembly, 16*, 10699.

Konstantinov, P. I., Grishchenko, M. Y., & Varentsov, M. I. (2015). Mapping urban heat islands of arctic cities using combined data on field measurements and satellite images based on the example of the city of Apatity. *Izvestiya, Atmospheric and Oceanic Physics*, *51*(9), 992–998. https://doi.org/10.1134/S000143381509011X.

Konstantinov, P., Varentsov, M., & Esau, I. (2018). A high density urban temperature network deployed in several cities of Eurasian Arctic. *Environ. Res. Letters*, *13*(7). https://doi.org/10.1088/1748-9326/aacb84.

Kottek, M., Grieser, J., Beck, C., Rudolf, B., & Rubel, F. (2006). World Map of the Köppen-Geiger climate classification updated. *Meteorologische Zeitschrift*, *15*(3), 259–263. https://doi.org/10.1127/0941-2948/2006/0130.

Kumpula, T., Pajunen, A., Kaarlejarvi, E., Forbes, B. C., & Stammler, F. (2011). Land use and land cover change in Arctic Russia: Ecological and social implications of industrial development. *Global Environ. Change*, *21*(2), 550–562. https://doi.org/10.1016/j.gloenvcha.2010.12.010.

Laruelle, M., & Hohmann, S. (2017). Biography of a polar city: population flows and urban identity in Norilsk, *Polar Geogr.*, *40*(4), 306-323, https://doi.org/10.1080/1088937X.2017.1387822.

Laruelle, M. (2019). The three waves of Arctic urbanisation. Drivers, evolutions, prospects. *Polar Record*, 1–12. https://doi.org/10.1017/S0032247419000081.

Lenoir, J., Graae, B. J., Aarrestad, P. A., Alsos, I. G., Armbruster, W. S., Austrheim, G., et al., (2013). Local temperatures inferred from plant communities suggest strong spatial buffering of climate warming across Northern Europe. *Global Change Biology*, *19*(5), 1470–1481. https://doi.org/10.1111/gcb.12129.

Li, H., Meier, F., Lee, X., Chakraborty, T., Liu, J., Schaap, M., et al., (2018). Interaction between urban heat island and urban pollution island during summer in Berlin. *Science of the Total Environ.*, *636*, 818–828. https://doi.org/10.1016/j.scitotenv.2018.04.254.

Liang, B., & Weng, Q. (2011). Assessing urban environmental quality change of Indianapolis, United States, by the remote sensing and GIS integration. *IEEE J. Selected Topics in Applied Earth Observ. Remote Sensing*, *4*(1), 43–55. https://doi.org/10.1109/JSTARS.2010.2060316.

Lynch, A. H., & Brunner, R. D. (2007). Context and climate change: An integrated assessment for Barrow, Alaska. *Climatic Change*, *82*(1–2), 93–111. https://doi.org/10.1007/s10584-006-9165-8.

Magee, N., Curtis, J. & Wendler, G. (1999). The urban heat island effect at Fairbanks, Alaska, *Theor. Applied Climatol.*, *64*, 39-47.

Malingowski, J., Atkinson, D., Fochesatto, J., Cherry, J., & Stevens, E. (2014). An observational study of radiation temperature inversions in Fairbanks, Alaska. *Polar Science*, *8*(1), 24–39. https://doi.org/10.1016/j.polar.2014.01.002.

Matyshak, G., Goncharova, O., Bobrik, A., Tarkhov, M., Ogneva, O., & Moskalenko, N. (2014): The influence of gas pipeline on functioning of frozen peatland soils (north of West Siberia, Russia), presentation on *Arctic Change conference*, Ottawa, Canada.

McBride, J. R., & Douhovnikoff, V. (2012). Characteristics of the urban forests in arctic and near-arctic cities. *Urban Forestry and Urban Greening*, *11*, 113–119. https://doi.org/10.1016/j.ufug.2011.07.007.

Melnikov, V. P., Melnikova, A. A., Anikin, G. V., Ivanov, K. S., & Spasennikova, K. A. (2014). Engineering solutions for building on permafrost in perspective energy-efficient enhancement. *Earth's Cryosphere*, *18*(3), 82–90.

Miles, V. V, & Esau, I. (2016). Spatial heterogeneity of greening and browning between and within bioclimatic zones in northern West Siberia. *Environ. Res. Letters*, *11*(11), 115002. https://doi.org/10.1088/1748-9326/11/11/115002.

Miles, V., & Esau, I. (2017). Seasonal and spatial characteristics of Urban Heat Islands (UHIs) in northern West Siberian cities. *Remote Sensing*, *9*(10). https://doi.org/10.3390/rs9100989.

Musse, M. A., Barona, D. A., & Santana Rodriguez, L. M. (2018). Urban environmental quality assessment using remote sensing and census data. *Int. J. Applied Earth Observ. Geoinformat.*, *71*, 95–108. https://doi.org/10.1016/j.jag.2018.05.010

Nelson, F. E., Anisimov, O. A., & Shiklomanov, N. I. (2001). Subsidence risk from thawing permafrost. *Nature*, *410*, 889.

Nordli, Ø. (2010). The Svalbard Airport. *Bull. of Geogr. – Phys. Geogr. Ser.*, *3*(3), 5–25.

Nyland, K. E., Gunn, G. E., Shiklomanov, N. I., Engstrom, R. N. & Streletskiy, D. A. (2018). Land cover change in the lower yenisei river using dense stacking of landsat imagery in Google Earth Engine, *Remote Sensing*, *10*, 1226.

Orttung, R. W., & Reisser, C. (2014). Urban sustainability in Russia's Arctic: lessons from a recent conference and areas for further investigations. *Polar Geogr.*, *37*(3), 193–214. https://doi.org/10.1080/1088937X.2014.919362.

Ostby, T. I., Schuler, T. V, & Westermann, S. (2014). Severe cloud contamination of MODIS Land Surface Temperatures over an Arctic ice cap, Svalbard. *Remote Sensing of Environ.*, *142*, 95–102. https://doi.org/10.1016/j.rse.2013.11.005.

Overland, J. E., Wang, M., Walsh, J. E., & Stroeve, J. C. (2014). Future Arctic climate changes: Adaptation and mitigation time scales. *Earth's Future*, *2*(2), 68–74. https://doi.org/10.1002/2013EF000162.

Pressman, N. E. P. (1996). Sustainable winter cities: Future directions for planning, policy and design. *Atmos. Environ.*, *30*(3), 521–529. https://doi.org/10.1016/1352-2310(95)00012-7.

Rigina, O., Baklanov, A., Hagner, O., & Olsson, H. (1999). Monitoring of forest damage in the Kola Peninsula, Northern Russia due to smelting industry. *Science of the Total Environ.*, *229*(3), 147–163. https://doi.org/10.1016/S0048-9697(99)00078-9.

Rohat, G., Goyette, S., & Flacke, J. (2017). Twin climate cities—an exploratory study of their potential use for awareness-raising and urban adaptation. *Mitigation and Adaptation Strategies for Global Change*, *22*(6), 929–945. https://doi.org/10.1007/s11027-016-9708-x.

Shaposhnikov, D., & Revich, B. (2016). Toward meta-analysis of impacts of heat and cold waves on mortality in Russian North. *Urban Clim.*, *15*, 16–24. https://doi.org/10.1016/j.uclim.2015.11.007.

Shiklomanov, N. I., Streletskiy, D. A., Swales, T. B., & Kokorev, V. A. (2016). Climate change and stability of urban infrastructure in Russian permafrost regions: prognostic assessment based on GCM climate projections. *Geogr. Rev.*, 1–18. https://doi.org/10.1111/gere.12214.

Shiklomanov, N. I., Streletskiy, D. A., Grebenets, V. I. & Suter, L. (2017). Conquering the permafrost: urban infrastructure development in Norilsk, Russia, *Polar Geogr.*, *40*, 273-290.

Shumilov, O. I., Kasatkina, E. A., & Kanatjev, A. G. (2017). Urban heat island investigations in Arctic cities of northwestern Russia. *J. Meteorol. Res.*, *31*(6), 1161–1166. https://doi.org/10.1007/s13351-017-7048-8.

Sitch, S., Mcguire, A., Kimball, J., Gedney, N., Gamon, J., Engstrom, R., et al., (2007). Assessing the carbon balance of circumpolar Arctic tundra using remote sensing and process modeling. *Ecol. Applicat.*, *17*(1), 213–234. https://doi.org/10.1890/1051-0761(2007)017[0213:ATCBOC]2.0.CO;2.

Stammler, F., & Sidorova, L. (2014). Dachas on permafrost: the creation of nature among Arctic Russian city-dwellers. *Polar Record*, *51*(06), 576–589. https://doi.org/10.1017/s0032247414000710.

Stedman, R. C. (2003). Is it really just a social construction?: The contribution of the physical environment to sense of place. *Society and Natural Resources*, *16*, 671–685. https://doi.org/10.1080/08941920390217627.

Stewart, I. D., & Oke, T. R. (2012). Local climate zones for urban temperature studies. *Bull. Am. Meteorol. Soc.*, *93*, 1879–1900. https://doi.org/10.1175/BAMS-D-11-00019.2.

Stewart, I. D. (2011). A systematic review and scientific critique of methodology in modern urban heat island literature. *Int. J. Climatol.*, *31*, 200–217. https://doi.org/10.1002/joc.2141.

Streletskiy, D. A., Shiklomanov, N. I., & Grebenets, V. I. (2012). Changes of foundation bearing capacity due to climate warming in northwest siberia. *Earth Cryosphere*, *16*(1), 22–32.

Streletskiy, D. A., Suter, L. J., Shiklomanov, N. I., Porfiriev, B. N., & Eliseev, D. O. (2019). Assessment of climate change impacts on buildings, structures and infrastructure in the Russian regions on permafrost. *Environ. Res. Letters*, *14*(2), 025003. https://doi.org/10.1088/1748-9326/aaf5e6.

Tan M. & Li X., (2015): Quantifying the effects of settlement size on urban heat islands in fairly uniform geographic areas. *Habitat Int., 49*, 100-106.

Telyatnikov, M. Y., & Prystyazhnyuk, S. A. (2014). Anthropogenous influence of Norilsk industrial area on plant vegetation cover of the tundra and forest tundra. *Contemporary Problems of Ecology*, *7*(6), 654–668. https://doi.org/10.1134/S1995425514060158.

Urban, M., Eberle, J., Hüttich, C., Schmullius, C., & Herold, M. (2013). Comparison of satellite-derived land surface temperature and air temperature from meteorological stations on the Pan-Arctic scale. *Remote Sensing*, *5*(12), 2348–2367. https://doi.org/10.3390/rs5052348.

Varentsov, M., Konstantinov, P., Samsonov, T. & Repina, I. (2014). Investigation of the urban heat island phenomenon during polar night based on experimental measurements and remote sensing of Norilsk city, *Current Problems in Remote Sensing of The Earth from Space*, *11*, 329-337.

Varentsov, M., Konstantinov, P., Baklanov, A., Esau, I., Miles, V., & Davy, R. (2018). Anthropogenic and natural drivers of a strong winter urban heat island in a typical Arctic city. *Atmos. Chem. Phys.*, *18*(23), 17573–17587. https://doi.org/10.5194/acp-18-17573-2018.

Varentsov, M., Konstantinov, P., Artamonov, A., Repina, I., Pashkin, A., Samsonov, T., et al., (2019). intensive urban heat island research campaign in the Arctic: The first results and application for model verification. *EMS Annual Meeting Abstracts*. *16*, EMS2019-840-2.

Virtanen, T., Mikkola, K., Patova, E., & Nikula, A. (2002). Satellite image analysis of human caused changes in the tundra vegetation around the city of Vorkuta, north-European Russia. *Environ. Pollution*, *120*(3), 647–658. https://doi.org/10.1016/S0269-7491(02)00186-0.

Wu, J. (2013). Landscape sustainability science: Ecosystem services and human well-being in changing landscapes. *Landscape Ecol.*, *28*(6), 999–1023. https://doi.org/10.1007/s10980-013-9894-9.

WMO (2018). Guidance on Integrated Urban Hydro-Meteorological, Climate and Environmental Services. Volume 1: Concept and Methodology, April 2018, Online: https://elioscloud.wmo.int/share/s/Hq6CMCrqRHCt0_05oVqGbQ.

WMO (2019). Guidance on Integrated Urban Hydro-Meteorological, Climate and Environmental Services. Volume 2: Demonstration Cities, June 2019, Online: https://elioscloud.wmo.int/share/s/Rf3EW264RZWGJuLrCuzo9w.

Yakovlev, B. A. (ed.), (1972). *Climate of Murmansk*, Leningrad, Hydrometeoizdat, 108 p. (in Russian).

Yu, Q., Epstein, H. E., Engstrom, R., Shiklomanov, N., & Streletskiy, D. (2015). Land cover and land use changes in the oil and gas regions of Northwestern Siberia under changing climatic conditions. *Environ. Res. Letters*, *10*, 124020. https://doi.org/10.1088/1748-9326/10/12/124020.

Zhang, H., Zhang, F., Zhang, G., He, X., & Tian, L. (2016). Evaluation of cloud effects on air temperature estimation using MODIS LST based on ground measurements over the Tibetan Plateau. *Atmos. Chem. Phys.*, *16*(21), 13681–13696. https://doi.org/10.5194/acp-16-13681-2016.

Zhang, Y., Seidel, D. J., Golaz, J.-C., Deser, C., & Tomas, R. A. (2011). Climatological characteristics of Arctic and Antarctic surface-based inversions. *J. Clim.*, *24*, 5167–5186. https://doi.org/10.1175/2011JCLI4004.1.

Zhou, D., Zhao, S., Zhang, L., Sun, G., & Liu, Y. (2015). The footprint of urban heat island effect in China. *Scientific Reports*, *5*, 11160. https://doi.org/10.1038/srep11160.

Zilitinkevich, S., & Esau, I. N. (2005). Resistance and heat-transfer laws for stable and neutral planetary boundary layers: Old theory advanced and re-evaluated. *Quar. J. Royal Meteorol. Soc.*, *131*(609). https://doi.org/10.1256/qj.04.143.

In: The Arctic: Current Issues and Challenges
Editors: Oleg S. Pokrovsky, Sergey N. Kirpotin et al.
ISBN: 978-1-53617-306-2

Chapter 4

A COMPARATIVE STUDY ON THE COOPERATION IN THE ARCTIC OCEAN AND THE SOUTH CHINA SEA

Ekaterina Antsygina[1,*], LLM, Lassi Heininen[2], PhD and Nadejda Komendantova[3], PhD
[1]Faculty of Law, Queen's University, Kingston, Canada
[2]University of Helsinki, Helsinki, Finland
[3]International Institute for Applied Systems Analysis (IIASA), Laxenburg, Austria

ABSTRACT

Delineation and delimitation of extended continental shelves is an emerging issue in international relations. Submissions to the Commission on the Limits of the Continental Shelf (CLCS) triggered disputes among the coastal states with the overlapping claims. However, some of them managed to reach non-objection agreements which allowed the CLCS to proceed. The aim of this study is to identify the critical variable which influenced the cooperation process on delineation. The methodology is based on comparative study of two cases: the Arctic Ocean and the South China Sea. The original hypothesis of this research was that the pre-existing international, much functional, cooperation was the critical variable which influenced the adoption of non-objection memoranda of understanding in the Arctic Ocean, and that it might have potential to influence in the South China Sea. The main objectives of this study are to contribute towards a better understanding of a legal nature of delineation and to demonstrate interrelations between international law and geopolitics in regard to the delineation.

Keywords: Arctic, South China Sea, delineation, cooperation, extended continental shelves, comparative study

* Corresponding Author's Email: ekaterina.antsygina@gmail.com.

INTRODUCTION

Economic benefits from the exploitation of continental shelves' natural resources are the major drivers of the rising attention to extended continental shelves. Emerging procedures of the extended continental shelves' delineations enhanced disputes in respect to overlapping claims between the coastal states in different parts of the world.

Delineation is the establishment of the extended continental shelf's outer limits (Rybkina and Reisell, 2017), and the obligation to delineate arises only if the state intends to claim sovereign rights over the continental shelf that extends beyond 200 nautical miles (nm). According to Article 76 of the United Nations Convention on the Law of the Sea (UNCLOS), in order to delineate, states parties to the UNCLOS must provide scientific data on the extended continental shelf[1] to the Commission on the Limits of the Continental Shelf (CLCS).[2] Outer limits established based on the CLCS recommendations are considered final and binding.[3] It is a relatively new practice of states: the first submission was made in December 2001, and the most recent one - in May 2019.[4]

The CLCS must refrain from consideration of the submissions involving disputes unless it has the consent of all the parties involved (CLCS 2001), a condition that is difficult to comply, though not impossible (Antsygina and Perez-Salazar, 2020). States with overlapping claims tend to object to the consideration of submissions on the disputed areas by the CLCS. Thus, many submissions will be left without expert analysis and coastal states will not be able to establish final and binding outer limits of their respective continental shelves. Though delineation is a unilateral act of a coastal state, establishment of the final and binding limits on the basis of the CLCS recommendations confirms states' exclusive rights against other state-parties to the UNCLOS. In addition, preceding delineation can be useful if there are overlapping claims and delimitation of maritime boundaries is necessary because the states get a clear picture of their respected entitlements supported by the expert opinion, and it is easier to avoid disputes on the status of maritime features (for example, submarine elevations or mid-ocean ridges).[5]

Currently, Canada, The Kingdom of Denmark, Norway, The Russian Federation, and the USA, – as the Arctic littoral states - are in the process of delineation of their extended continental shelves. Despite the existing overlapping claims, states were able to reach non-objection memoranda of understanding so the CLCS can proceed with their submissions. To identify the critical variable which influenced the process of negotiations, we conducted the

---

[1] A submission to the CLCS is a set of particulars of the outer limits of a continental shelf beyond 200 nm along with supporting scientific and technical data, prepared by a state. Such a submission cannot be rejected by the CLCS, whether or not state included all the necessary evidence of the geomorphological shelf's existence.

[2] The Commission on the Limits of the Continental Shelf is a technical body for evaluation of coastal states' submissions on the delineation of the outer limits of the continental shelf in accordance with the requirements of Article 76(8) of the UNCLOS. The Commission issues recommendations to states on the basis of their submissions.

[3] The procedure is established by the Article 76 (paras. 7-8) of the UNCLOS.

[4] The Russian Federation submitted on 20 December 2001, and Canadaa made a submission on 23 May 2019. See http://www.un.org/depts/los/clcs_new/commission_submissions.htm.

[5] As a general rule, the outer limit of the continental shelf shall not either exceed 350 nm or shall not exceed 100 nautical miles from the 2,500 metre isobaths (Article 76 (6). However, this rule does not apply to submarine elevations that are natural components of the continental margin, such as its plateaux, rises, caps, banks and spurs. Hence, proving a status of some geological features, such as submarine elevations under Article 76 (6), can help to expand continental shelves outside the maximum limits established by Article 76 (6).

case study analysis of the Arctic Ocean (AO) and the South China Sea (SCS), including overlapping claims and the risk of a dispute.

In both cases, there are overlapping claims on extended shelves and submissions to the CLCS originally got negative reactions by neighboring states. Unlike in the Arctic, the coastal states of the SCS were not able to reach non-objection agreements allowing the CLCS to proceed with the submissions. Both regions are rich with the natural resources. Also, in the selected cases the coastal states have common collective goals that should enhance at least a functional cooperation.

Our original hypothesis was that the pre-existing cooperation was the critical variable which influenced the adoption of non-objection memoranda of understanding in the AO. We also analyzed whether delimitation in the AO could trigger a dispute between the coastal states. The research question was whether the pre-existing cooperation has positive influence on cooperation on delineation and delimitation of extended continental shelves, and if cooperation is not a critical variable then what are the factors contributing to the cooperation on delineation in the AO. The study was conducted within the Arctic Futures Initiative (AFI) and the Risk Program of the International Institute for Applied Systems Analysis (IIASA).

Background

Sovereign rights for continental shelves include exclusive rights to exploitation and exploration of natural resources, scientific research, and construction of artificial islands and installations. Hence, they touch upon such matters as oil mining, the authorization and regulation of drilling, the protection of environment in the case of oil spills, and many more. Natural resources that belong to the continental shelves include both non-living resources (oil, gas, metals, etc.) and living sedentary species (crabs, lobsters, pearl oysters, etc.) (Rothwell and Stephens, 2017).

But there is a difference between continental shelves within 200 nautical miles (nm),[6] and shelves beyond that limit, which are also called extended shelves. Though states establish sovereign rights for the whole continental shelf, the processes for parts within 200 nm and beyond that limit (extended shelves) are different. This difference is linked to the procedure of establishment of a title for continental shelves beyond 200 nm that is based on the natural prolongation criterion.

States-parties to the UNCLOS have to provide scientific data supporting their claim over the extended continental shelf to the Commission on the Limits of the Continental Shelf (CLCS) in order to establish final and binding outer limits. The procedure of delineation was agreed to prevent excessive claims by coastal states. Parties claiming jurisdiction over the continental shelf under the distance criterion (within 200 nm) are not bound to submit any information. But if a state-party to the UNCLOS bases its claims on the natural prolongation concept, in order to establish a final and binding outer limits of the extended continental shelf it must prove the existence of the prolongation of the landmass into the sea.

Thus, the fundamental differences between the distance and natural prolongation criteria are the requirement of the existence of a continental margin outside 200 nm, submission of

[6] Here and further refers to 200 nautical miles from the baselines from which the breadth of the territorial sea is measured.

the scientific data proving such an existence to the CLCS, and delineation of the outer limit on the basis of the CLCS recommendations.

It is not clear yet whether a non-party to the UNCLOS can follow the procedure with the participation of the CLCS, and if not, how such a state can demonstrate geological and geomorphological continuity confirming that its claim is not excessive without disclosing commercially valuable information.

The CLCS must refrain from considering submissions involving boundary disputes unless it has the consent of all the parties involved. Rule 46 establishes a procedure for submissions where there is a dispute (Annex I to the Rules of Procedure), in these cases the CLCS will only consider the submission of a coastal State with the prior consent of all States that are parties to the dispute.

That's why cooperation in delineation is important for states: they first can collect scientific data, and then all the efforts on the collection of data are not wasted if there are no objections and the CLCS can proceed. Second, it's impossible to establish final and binding outer limits without the recommendations of the CLCS, so non-objection agreements allow states to legitimize their continental shelf's claims.

METHODS

We apply the case study analysis method of two regions as well as the method of conceptual analysis for legal issues to examine existing legal framework and legal issues arising in respect to the disputes on maritime zones in the AO and the SCS.

In the Arctic region only four out of five coastal Arctic states (the so-called "Arctic Five": The Russian Federation, Canada, USA, The Kingdom of Denmark, and Norway) are parties to the UNCLOS, which means that the USA, which is not a party to the UNCLOS, does not have an obligation to delineate on the basis of the CLCS recommendations. Nevertheless, the USA considers the whole of the UNCLOS, with the exception of Part XI, as customary law (Churchill 2015), and has initiated the process of delineation (U.S. Department of State 2019). It is not yet clear whether the USA will submit data to the CLCS. In the South China Sea, relations of China, Brunei Darussalam, Vietnam, the Philippines, and Malaysia are considered. Taiwan has maritime claims and has presence on the disputed islands but only Chinese claims and relations with other states of the SCS were considered since they include those of Taiwan. Due to the limited timeframe of research, the influence of other stakeholders, though it can be significant, is outside of the study's scope. We nevertheless acknowledge that indigenous communities in the High North play an active role in the shaping of the policy of the Arctic states (Heininen 2012).

The rationale for selecting the AO and the SCS cases is linked to the different approaches of the coastal states to their neighbors' submissions involving overlapping claims. In both regions, the original reactions to submissions were not friendly, however subsequent actions of the coastal states followed different patterns: the Arctic states gave a prior consent to the CLCS' consideration of submissions and the coastal states of the SCS with overlapping claims objected to the submissions, so the CLCS will not be able to proceed according to the Rule 46 of the Rules of Procedure of the Commission and Annex 1 to the Rules of Procedure of the Commission (CLCS/40/Rev.1).

The hypothesis which we tested in frames of case studies was that the pre-existing cooperation on such issues as environmental protection, climate change, rescue at sea, etc. and presence of fora where stakeholders can discuss problems as a critical factor influencing cooperation on delineation. The theoretical basis of this research is in neoliberalism and complex interdependence theories.

We compare two case studies according to the following criteria which were identified by Heininen (Heininen 2018):

- Relevant treaties and soft law instruments
- Membership in international and regional organizations
- Functional cooperation on environmental protection and climate change
- Extended continental shelves' claims, reactions by the coastal states
- Types of disputes in the regions
- Foreign interest in the regions
- Military presence and security issues
- Land reclamation activities
- Allocation of power in the regions
- Presence of natural resources in the zones of overlap
- Transport
- Population and economic development

The empirical data for this research is derived from publicly available executive summaries of submissions of the coastal states to the CLCS, communications to the CLCS, recommendations of the CLCS, MOUs, treaties, policies of the coastal states and reports of various institutions.

CASE STUDIES

The Arctic Ocean

The Arctic Ocean is the smallest and the shallowest of the world's oceans. It covers an area of about 14 million km^2 and has a vast continental shelf, which would be very attractive for exploitation in the absence of the present ice conditions [10]. By the year 2040 the Arctic Ocean may permanently become ice-free in summer (Holland, Bitz, and Tremblay, 2006). The size of the population in the circum-Arctic region is about four million people and 10% of the present circum-Arctic population belongs to indigenous groups.

The majority of the Arctic states belong to the high-income economies (Canada, The Kingdom of Denmark, Norway and the USA), only Russia belongs to an upper-middle income economy. The Russian Federation is a major player in the region due to the length of its relevant coasts, infrastructure, extended continental shelf claims, and military presence. However, it is hard to say that Russia is a hegemon of the region, there is a balance of power that exists due to the established cooperation. The Arctic Five are equal in terms of the participation in the shaping of the region's future (Ilulissat Declaration (Arctic Ocean Conference, 2008) can serve as a good example). The Arctic states also have a common interest to control the AO and to keep the "outsiders" away. Establishment of the sovereign rights for extended continental shelves for most of the AO will help to prevent other states from engaging in the mining activities in the AO region.

As stated by Heininen (Heininen 2018), today the Arctic is a peaceful region with high stability based on a wide international and regional cooperation. Majority of the borders is defined and though there are several disputes, there are no serious conflicts over sovereignty. In respect to the delimitation of the continental shelf in the AO, The Russian Federation and Norway have reached an agreement on the delimitation of the areas comprising the continental shelf beyond 200 nm in the Barents Sea as well as in the western Nansen Basin (Treaty between the Russian Federation and the Kingdom of Norway Concerning Maritime Delimitation and Cooperation in the Barents Sea and the Arctic Ocean 2010). There is also an agreement between Russia, as the successor of the USSR, and the USA on the delimitation of the maritime zones in the Chukchi and Bering Seas, as well as in the Arctic and Pacific Oceans (The Agreement between the United States of America and the Union of Soviet Socialist Republics on the Maritime Boundary 1990). This agreement is applied provisionally as The Russian Federation has yet to ratify it. There is a territorial dispute between Canada and The Kingdom of Denmark over the Hans Island, but it's handled quite well, and the states announced negotiation on the status of the island in 2018. Also, there are maritime areas in the Beaufort Sea disputed by the USA and Canada. There is a difference in the position of states on the status of the Northwest and Northeast Passages as well, but those differences are not that significant while the Arctic is covered with ice and shipping is not well developed. This situation might change in the future when the Arctic shipping becomes commercially viable.

Most of the overlapping extended shelves in the Arctic Ocean are not yet delimited. As recommendations by the CLCS are expected within the decade, the Arctic states will probably start consultations concerning the maritime delimitation of continental shelves in the near future. Nevertheless, the Arctic coastal states will probably postpone delimitation until the CLCS delivers recommendations that are suitable for them. The agreements will be effected on the basis of international law. States are free to use any delimitation method they consider suitable to achieve an equitable solution (Tulupov 2015).

Legal stability is also important for the region. The law of the sea is applicable to the Arctic Ocean. During the Arctic Ocean Conference, held in 2008 in Ilulissat, Greenland, the representatives of the Arctic states made an explicit statement in regard to the applicable law:

> *"the law of the sea provides for important rights and obligations concerning the delineation of the outer limits of the continental shelf, the protection of the marine environment, including ice-covered areas, freedom of navigation, marine scientific research, and other uses of the sea. We remain committed to this legal framework and to the orderly settlement of any possible overlapping claims".*

Only four of the five coastal Arctic states are parties to the UNCLOS, which means that the delimitation of extended shelves in the Arctic will require an application of both treaty and customary rules. The Russian Federation ratified the UNCLOS on 12 March 1997, Canada on 7 November 2003; Norway on 24 June 1996; The Kingdom of Denmark on 16 November 2004. The USA took an active part in the negotiation on the UNCLOS but is not a party to it. Thus, it does not carry an obligation to submit information to the CLCS. Newertheless, the USA considers Article 76(7) as customary and is collecting data on the extended shelf (U.S. Department of State 2019); there is no information whether the USA intends to make a submission to the CLCS (which is probably possible since there is no prohibition by the UNCLOS (Baumert 2017). The non-party status of the USA adds several nuances to the delimitation and delineation. The Arctic states use both Soft and Hard Law instruments, and apart from the disagreement on the status of the Northeast and Northwest passages, and some straight baselines, there are no different interpretations of the UNCLOS or the law of the sea rules by the Arctic states.

With the global warming the foreign interest for the Arctic is growing. The European Union is highly interested in the Arctic matters but sometimes lacks legal standing. But the law is one of the tools that the EU can use to expand influence in the High North. It can legislate on a limited number of the Arctic-related matters, such as environmental protection, transportation, scientific research (Antsygina, Czarski, Konopleva, and Jirova, 2017).

The Russian Expedition to the North Pole, including planting of the Russian flag on the Arctic seabed close to the North Pole, and in particular strong reactions by other Arctic states manifested (Kuersten 2015), influenced the rising interest towards extended continental shelves in the Arctic Ocean, which had started earlier. Though Russian officials stated that this action did not have the character of territorial acquisition (Kuersten 2015, at 603), placing the flag close to the North Pole, in the area of overlapping claims for the extended continental shelves of Russia, The Kingdom of Denmark and Canada, was considered as provocative (Kuersten 2015, at 605). Though the submission of Denmark expanding till the 200 nm limit of the Russian continental shelf was also very irritating. The CLCS submissions of Russia (CLCS 2015), Norway (CLCS 2005), The Kingdom of Denmark (CLCS 2014), and the submission of Canada (CLCS 2019) make it clear that their claims overlap in some areas and there is a need for the maritime delimitation in the Central Arctic Ocean. Several Arctic actors would like to include the North Pole into its continental shelves. It is highly unlikely that this zone contains any valuable natural recourses, but the status of the North Pole "owner" can be a geopolitical victory.

The Russian Federation and The Kingdom of Denmark have overlaps in the Amundsen Basin, on the Lomonosov Ridge, and in the Makarov and Podvodnikov Basins; and Russia and Canada - in the Lomonosov Ridge, Makarov Basin and on the Mendeleev Rise (CLCS 2015). The southern extended continental shelf of Greenland (Denmark) includes an overlap in the south-western part with Canada. The Northern Continental Shelf of Greenland includes unresolved disputes with Norway, the USA, and Canada (CLCS 2014). Norway has overlapping claims in the Southern Banana Hole with Iceland and Denmark/the Faroe Islands, and in the maritime areas between Svalbard and Greenland. There is a possible overlap between Canada's and the USA's entitlements in the AO.

The USA is the only state participating in the outer continental shelves' delimitation in the Arctic Ocean that is not a party to the UNCLOS. According to Eugene Petrov, the outer limits of the US's continental shelf will probably include the Chukchi Plateau, and part of the Canada and Podvodnikov Basins (Petrov 2015). There is not yet information about the claim of the USA in the Arctic Ocean, but, most likely, there will be overlaps with the extended shelf of Canada.

Despite there are overlaps of claims, the Arctic states were able to agree not to object to the consideration of submissions by the CLCS. The Russian Federation and The Kingdom of Denmark have agreed that will not object to the consideration by the CLCS of their respective submissions concerning their disputed areas and that they will mutually take CLCS's recommendations without prejudice to the rights of the other party connected to the CLCS submission, nor to the delimitation of the continental shelf between them. Canada has overlaps with proposed extended continental shelves of The Russian Federation, The Kingdom of Denmark and the USA. Canada and Denmark reached an agreement that, upon submission of data on the limits of the continental shelf, by one of the parties to the CLCS, the other would promptly send a diplomatic note to the UN Secretary General. The note would advise that it does not object to the consideration of the submission to the CLCS and would specify that the respective recommendations by the CLCS will be without prejudice, both to the consideration by the CLCS of its own submission and to matters relating to delimitation. Russia and Canada also reached a similar non-objection agreement.

Norway, The Kingdom of Denmark and Greenland, signed an Agreement on the delimitation of the continental shelf and the fishery zones in the area between Greenland and Svalbard. The treaty establishes that the parties will decide on the delimitation of the extended continental shelves upon the establishment of their respective outer limits. The government of Denmark together with the Government of Greenland, did not object to the Commission considering Norway's submission or to making recommendations. In reciprocity, Norway made no objections to the CLCS considering and making recommendations on the partial submission by The Kingdom of Denmark in respect to the North-Eastern continental shelf of Greenland (CLCS 2013). The USA did not object to the consideration of the submissions, noting that this consideration would be exercised without prejudice both to the delineation of its own shelf and to matters relating to delimitation.

In order to analyze the possibility of a dispute over the extended shelves and delimitation scenarios, it is important to understand objectives of the coastal states. Global warming, depletion of the world's mineral reserves and rapid technological development, allowing commercially viable marine mining, are contributing factors driving the Arctic states to assert their sovereign rights for continental shelves outside 200 nm. The presence of the natural resources in the areas of overlap can obstruct the process. The question is whether there are resources in the Central Arctic Ocean outside states' 200 nm entitlements. The US Geological Survey estimates the Arctic resources that can be extracted with some feasible profit. There are very limited data to make any estimates on the potential value of resources but given the likelihood of ferromanganese crusts, ferromanganese nodules, gas hydrate deposits, and petroleum reservoirs, it is likely the resources are worth many billions if not trillions of dollars (U.S. Departent of State 2019). However, the probability to find reserves in the area of overlaps is low. The majority of estimated oil and gas recourses are located offshore within 200 nm or on the extended shelves areas that are not contested.

The sovereign rights for exploitation and exploration also include sedentary species that are either immobile in their harvestable stage or are under the seabed or are unable to move without constant physical contact with the seabed or subsoil. Pearl oysters, lobsters and snow crabs fall into this category. All these species have a so-called thermal window. Temperature affects the geographic distribution and abundance of ectothermic organisms (Terblanche, Hoffmann, Mitchellet al. 2011). Snow crabs can potentially be living resources of the continental shelves in the Arctic. They prefer cold water (- 1 to 11°C) and live at a range of depths from 20 to 2000 metres (Government of Canada 2016). Canada produces about two-thirds of the global supply of the snow crab (Government of Canada 2016). It is still hard to predict the effects of global warming and patterns of sedentary species in the Central Arctic, but there is a possibility that these species will move up north (CAFF 2018).

In the absence of more detailed information, it is impossible to conclude that interest in natural resources is the major driving force of the future delineations and delimitations of the extended shelves. It must be noted that coastal states not only enjoy rights but also undertake certain obligations. Thus, they have to develop and to enforce regulations preventing pollution connected with seabed activities and to concern and to protect environment (McDorman 2015).

Geopolitical ambitions and the battle for the North Pole has been interpreted to be another factor influencing the relations between the Arctic states and Arctic cooperation, as was seen in the above-mentioned case of strong reactions of a few Arctic littoral states after the Russian Expedition to the North Pole (Kuersten 2015, at 603). The crisis in Ukraine and sanctions imposed mutually by Russia and the Western powers can also contribute to the destabilization of the relations in the Arctic. In spite of concerns of a possible competition and conflicting interests on the North Pole between The Russian Federation, The Kingdom of Denmark and Canada, the five littoral states of the Arctic Ocean were able to agree on the principles of governance over the Arctic Ocean in their ministerial meeting in May 2008 in Ilulissat, Greenland. Though the Arctic is still geostrategically important area with heavy military structures present, a benefit of the achieved stability and existing cooperation are valued more important. The contributing fators are common interests between the Arctic states and, to a lesser extend, ad-hoc ministerial meetings. Therefore, the globalized Arctic is with high geopolitical stability and constructive cooperation, as was stated by the eight Arctic states at the latest ministerial meeting on the Arctic Council in May 2019 (Joint Ministerial Statement 2019).

Another possible reason for the delineation and delimitation is a construction of artificial islands and installations. Global warming brings lots of negative consequences, but also the accessibility of new sea routes, commercially viable mining in the Arctic and possibility to construct new installations. It is likely that the growth of Arctic shipping will require major infrastructure developments, but these projects will be effected mostly within 200 nm, where the Arctic states already established their sovereign rights. Nevertheless, the right to construct installations in respect to the extended shelves means the possibility to establish 500 metres safety zones and to exercise and to exclusive jurisdiction with regard to customs, fiscal, health, safety, and immigration laws and regulations (UNCLOS, Articles 60 and 80; McDorman 2015). Such an opportunity can be used for imposing some limitations on the freedom of navigation (Rothwell and Stephens, 2017). With the development of Arctic shipping, this can become an important consideration, especially if Article 234 of the UNCLOS, allowing coastal states to impose domestic regulations within 200 nm, will no longer be applicable. Since Article 234 regulates only areas ice-covered for most of the year, it cannot be put in application if there is no ice for at least 183 days a year.

So far there are no major projects on the construction of artificial islands, installations, constructions, due to the high costs and the absence of need for such developments. According to Poussenkova, new oil-production projects in the Arctic were frozen because of commercial unviability, technological, financial, institutional challenges, the absence of infrastructure and human resources shortage (Poussenkova 2018). Building infrastructure for the Arctic shipping is also not feasible soon, although the Arctic routes can decrease the shipping time from Rotterdam to Yokohama for 2 weeks, it is still not very popular due to high costs. To become an effective shortcut between Europe and Asia, the Arctic Ocean will require large investments in improving infrastructure and rescue services in the area

(Antsygina and Perez-Salazar, 2020). As competitive shipping depends on the use of large-scale vessels the Arctic ports need to adjust their services to respond to the needs of larger vessels.

Scientific research on the continental shelf is another prerogative of a coastal state. Usually, access for scientists is granted on a certain condition such as access of domestic scientists to the board of the vessel, sharing of data and, etc. According to Article 246 (2), marine scientific research in the exclusive economic zone and on the continental shelf shall be conducted with the consent of the coastal state. The coastal state may in their discretion withhold its consent to the conduct of a marine scientific research project under some conditions. This power of discretion, however, is not granted to the coastal state in respect of marine scientific research projects to be undertaken beyond 200 nm. Nevertheless, it gives additional power to a coastal state and the Arctic Five is interested in minimizing the influence of other states in the region. This common interest might contribute to cooperation on delineation.

The last, but not the least factor can be a fulfilment of a treaty obligation. A coastal state party to the UNCLOS willing to establish the outer limits of its continental shelf has to submit data to the CLCS within 10 years of the entry into force of the UNCLOS for that state. All the Arctic states-parties to the UNCLOS (Russia, The Kingdom of Denmark, Canada, and Norway) have submitted their data.

The objectives of the Arctic states differ, and some analysis will be required to consider the possibility of a dispute. If no agreement is reached, states must resolve disputes by peaceful means in accordance with the Charter of the United Nations. Compulsory dispute-settlement mechanisms established by the UNCLOS include submission of a dispute to the International Court of Justice (ICJ), the International Tribunal for the Law of the Sea (ITLOS), an Arbitral Tribunal, or a Special Arbitral Tribunal. Pursuant to Articles 286, 298 of the UNCLOS and the declarations of the coastal states, the possible dispute resolution methods might include only negotiation and conciliation. Thought the UNCLOS established a compulsory dispute-settlement mechanism, states-parties had a choice to opt it out for the delimitation issues. Article 298 establishes that if negotiations on delimitation are not successful, states must accept submission of the matter to conciliation, and if conciliation is not resultative the parties shall, by mutual consent, submit the question to the ICJ, ITLOS, an Arbitral Tribunal, or a Special Arbitral Tribunal. Most likely, the Arctic states will try to avoid resolving any possible dispute in the international dispute resolution institutions. Negotiations allow states more flexibility: states are free to establish any borders they consider suitable, but a tribunal must "…balance up the various considerations which it regards as relevant in order to produce an equitable result" (ICJ 1982). Such a result can be unpredictable for states and even not desired since it will be harder to give up some positions for a state when there is a favorable decision by an international court.

The presence of fora where states can meet and discuss their matters is important for cooperation. The Arctic states are members to various international and regional organizations (the EU is included into Table 1 as it has scientific, energy, environmental, geopolitical, transportation interests in the Arctic and is an economic partner of the Arctic states).

The most important body for Arctic cooperation is the Arctic Council. Though the decisions of the Arctic Council do not have a binding power, it is hard to deny its complex influence on a stability building and cooperation in the Arctic. It facilitates dialogue between the Arctic states and concentrates on policy-shaping, since its establishment in 1996 by the Ottawa Declaration (Arctic Governance Project, 2010). The Arctic states also share such common policies as the Norther Dimension and Concerning the Arctic Environmental Protection Strategy. Concerning the Arctic Ocean, the AC does not play that important role, being neither an international organization nor having a mandate for legally-binding decisions, on behalf of the member-states, and having a consensus principle, the Council has not been able to discuss on issues dealing with fisheries and oil and gas drilling. This is the main reason why the legally-binding agreements on the Arctic – for example, the Search and Rescue Agreement - are under the auspices of the AC but been negotiated between the Arctic states. Also, the latest agreement, "International Agreement to Prevent Unregulated High Seas Fisheries in the Central Arctic Ocean", signed in October 2018, was negotiated between the five Arctic littoral states, Iceland, the EU and China, Japan and South Korea.

The South China Sea

The South China Sea is almost 4 times smaller than the Arctic Ocean. It is about 3.5 million km^2. However, it is a home to more than 270 million people (Wilkinson, DeVantier, Talaue-McManus, Lawrence, and Souter, 2005), that is about 67 times more than the population of the Arctic. The density of population influences the economic, ecological and environmental situation in the region. The low and middle-income economies in the region are generally resource-dependent and experience the growing demand in foo, clean water and other resources, as well as adequate infrastructure and jobs (Ibid, at 24). This can lead to overfishing, use of bombs and other illegal practices for fishing, coral reefs degradation, pollution, and other negative consequences for the region (Ibid, at 31).

The SCS can be characterized as a semi-enclosed sea according to Article 122 of the UNCLOS. Article 123 of the UNCLOS calls for cooperation among states bordering semi-enclosed seas in respect to management, conservation, exploration and exploitation of the living resources, protection and preservation of the marine environment, and scientific research. However, this provision is rather aspirational and does not impose any binding obligations on the states of the region. Nevertheless, it enhances cooperation among the coastal states in the region. Cooperation is also required due to the environmental degradation that might influence food security, public health, economic situation, and, etc.

The SCS is surrounded by China, Brunei Darussalam, Indonesia, Malaysia, the Philippines, and Vietnam. Indonesia is not participating in the delineation procedures and is excluded from the study. The SCS is relatively shallow, with wide continental margins, and is rich with both non-living and living natural resources. Unlike in the Arctic, exploitation of the natural resources is not affected by the harsh weather conditions. The SCS is globally important due to its natural resources, shipping routes with more than $5.3 trillion in trade that passes through it (Glaser 2012), and the security situation. The risk of an armed conflict in the SCS is considerable, though there are indicators that states of the region are interested in the effective management of the dispute. With the support of the ASEAN, a non-binding Declaration on the Conduct of the Parties in the South China Sea was adopted, and negotiations on the binding code of conduct are ongoing. According to some prognoses, destabilization of the situation in the region might lead to a new world war (Keyuan 2017).

As specified by the World Bank, the Philippines and Vietnam belong to lower-middle-income economies, China and Malaysia – to the upper-middle income economies, and only Brunei Darussalam is considered as the high-income economy (The World Bank 2019). China is the major player in the region, with ambitions to become a marine power. While there is a balance of power in the AO, the dominance of China in the SCS region might add difficulty for reaching compromise solutions. The dominance of China in the SCS is contested by the USA in respect to the freedom of navigation of military ships to enter 200 nm exclusive economic zone. The USA is also interested in ensuring the military stability in the region (the 1951 Mutual Defense Treaty and the 2014 Enhanced Defense Cooperation Agreement with the Philippines) since 1.2 trillion of its trade (Glaser 2012) passes through the SCS. It adds additional tensions to the region and requires efforts on the management of the dispute.

There are several types of the disputes in the region:

1. Territorial disputes over the Spratly Islands between China (including Taiwan), Vietnam, The Philippines, Malaysia, and Brunei; over the Paracel Islands between China and Vietnam; and Scarborough Shoal between China and The Philippines. The disputes over land are not the most important, in our opinion. Resources that belong to the EEZs and continental shelves generated by the contested islands are the real reasons for the confrontation.
2. The dispute over the overlapping maritime zones (EEZs and continental shelves) generated by the mainland coasts and islands is the most significant. Here the classification of maritime features as islands is of high importance. Rocks can generate only 12 nm territorial sea and artificial islands cannot generate maritime zones, so land reclamation activities do not facilitate the expansion of sovereign rights. There are also overlaps of extended shelves and shelves within 200 nm in the area that contribute to the enhancement of the conflict.
3. The dispute over the interpretation of the freedom of navigation between the USA and China. Positions of China, Vietnam and Malaysia are inconsistent with the US approach to the freedom of navigation. China and Malaysia attempt to regulate foreign military activities within 200 nm and Vietnam requires notification of foreign military ships exercising a right of innocent passage in its territorial sea (O'Rourke 2018).

Another serious problem of the region is piracy; despite the anti-piracy efforts, the region remains vulnerable (ReCAAP 2015). Shipping is one of the most important matters in the SCS. All the SCS states are interested in the safety of the sea routes. This sphere, where the coastal states have a common interest, can become a basis for the functional cooperation that might lead to an increase in trust among the SCS coastal states. About 40% of the global trade passes through the SCS (Morton 2016). Though the shortest way from Europe to Asia is through the Arctic, the absence of the adequate infrastructure for supertankers, harsh weather conditions, costs make the SCS route the most attractive shipping artery. Rights to the continental shelves do not influence the freedom of navigation apart from the 500-meter safety zones established for artificial islands and constructions. However, the shipping business does depend on the stability in the region, that's why territorial and maritime disputes might affect the attractiveness of the SCS for shipping. Due to the globally important shipping routes, the situation in the SCS is different from the AO: states of the region cannot keep others away like the Arctic state; the foreign interest in the SCS (especially of the USA) is too significant.

The law of the sea is applicable to the region, all five coastal states participating in the extended shelves delimitation are parties to the UNCLOS. China ratified the UNCLOS on 7 June 1996, Vietnam – 25 July 1994, Malaysia - 14 October 1996, Brunei Darussalam - 5 November 1996, The Philippines - 8 May 1984. Thus, the UNCLOS governs all the relations of the SCS states in respect to the maritime claims. However, implementation of the UNCLOS provisions by the SCS coastal states could be inconsistent. Apart from the different view on the freedom of navigation, China proclaimed that it enjoys sovereign rights and jurisdiction over the relevant waters, seabed and subsoil within the "9-dash line." The "9-dash line" is based on the rights "formed over a long course of history" and it was used by China previously, but it was officially presented to the international community as a response to the Joint Submission by Malaysia and Vietnam and submission by Vietnam to the CLCS. This line is a line of allocation (Smith 2010) and, according to legal scholars (Morton 2016; Beckham 2010; Nong 2012) and the Arbitral Tribunal for the South China Sea Arbitration (PCA 2016), is not based on the rules of the law of the sea. The Philippines asked the Tribunal to clarify the source of maritime rights and entitlements in the South China Sea. The Tribunal concluded that China is not entitled to exercise "historic rights" within the "nine-dash line," so all the maritime zones shall be established according to the UNCLOS. Vietnam, Malaysia, the Philippines, and Indonesia objected to the "9-dash line." Indonesia stated that "…thus far, there is no clear explanation as to the legal basis, the method of drawing, and the status of those separated dotted-lines" (Republic of Indonesia, 2010).

China follows the UNCLOS in other seas and oceans and has this ambiguous position only in the SCS. Legal certainty and compliance with the UNCLOS might facilitate the dispute management in the region, as well as delineation and delimitation procedures.

So far only Vietnam and Malaysia made submissions in respect to the extended shelves in the SCS. China and the Philippines made extended shelves' submissions to the CLCS in relation to entitlements in other seas but left open the possibility to submit data concerning the SCS (Elferink 2014). Brunei filed a Preliminary information where it stated the intention to submit data on the extended continental shelf for the areas known as the Northwest Borneo Shelf, the Northwest Borneo Trough and the Dangerous Grounds to the edge of the deep ocean floor of the South China Sea Basin (Brunei Darussalam, 2009).

On 6 May 2009, Malaysia and Vietnam submitted jointly an information on the limits of the continental shelves beyond 200 nm in respect of the southern part of the SCS. The submission mentioned unresolved disputes in the area. Though the submitting states did not use the contested islands for defining the outer limits of the continental shelves and considered their mainland coasts for construction of the limits, they could not secure non-objection of other relevant coastal states. China and the Philippines requested the CLCS not to consider the submission (The People's Republic of China, 2009). The next day after the Joint submission was presented, Vietnam submitted to the CLCS information on the limits of the continental shelf beyond 200 nm in respect of the North Area appurtenant to Vietnam. This submission also received objections by China and The Philippines.

The submissions triggered hostile reactions by the states with the overlapping claims. The situation with the competing claims is very different from the AO: in the SCS, continental shelves within 200 nm are overlapping with the extended ones, thus, there is a conflict of distance and natural prolongation criteria. This issue will be discussed in detail in the results section, but the sovereignty over the contested islands gets higher importance in this case.

As a reaction to the submissions, China presented it's "9-dash line" and stated that it had indisputable sovereignty over the islands in the SCS and the adjacent waters and enjoyed sovereign rights and jurisdiction over the relevant waters as well as the seabed and subsoil (The People's Republic of China, 2009). Interesting, that China has different positions in the SCS and the East China Sea: it made a submission on the extended shelves in the latter and is basing the sovereign rights on the UNCLOS rules in respect of that submission. While in the SCS it adheres to the "9-dash line." In the present situation, it is hard to say that states will be able to resolve the stalemate. The recommendations by the CLCS might be useful as represent an independent expert opinion on the geological and geomorphological state of affairs.

While the motivation of the Arctic states is not totally linked to the resources of the AO, the situation in the SCS is different and resources of the continental shelves can be a major objective for the states in the dispute. The SCS is very rich with both living and non-living natural resources. There is no data for the central SCS though. According to the 2013 report of the US Energy Information Administration, 11 billion barrels of oil and 190 trillion cubic feet of gas are estimated to be found in the SCS. Most likely, some areas of overlap contain reserves of oil and gas.

Non-living natural resources are not the only treasures of the SCS. It is also rich with living resources. About 5 million tons of fish are caught here annually (Wilkinson, DeVantier, Talaue-McManus, Lawrence, and Souter, 2005). The population of the coastal states is dependent on the fishing activities: it's a source of food and money. Fish stocks belong to the EEZs, so delineation and delimitation of extended shelves do not influence the fishing rights. However, sedentary species, such as lobsters and oysters, are referred to the sovereign rights on continental shelf.

Another objective is the construction of artificial islands. The land reclamation activities are ongoing in the SCS; China, Vietnam and the Philippines are building military facilities on the maritime features of the SCS (Morton 2016). Though they cannot help states in enlarging maritime zones, these constructions are strategically important for the military purposes.

Though the situation with security is quite serious in the SCS, there are no indicators of the intention to escalate the conflict among the parties. It is harder to reach compromise solutions as China is a dominant power in the region, but China is not interested in the military conflict since it is open for the negotiations on the obligatory code of conduct for the states in the SCS (NDTV, 2018). Political will is necessary for the resolution of the territorial and maritime disputes, and though so far there are significant improvements in the situation, there is a hope that the SCS dispute will not become a wicked problem. Also, political leadership is required for the cooperation on environmental protection, management of fisheries, combat with piracy, etc. Functional cooperation might become a basis for the wider cooperation on management of the resources in the zones of overlaps. Joint development programs, the creation of the national parks can be a good tool for the conflict management.

The SCS states are members to various international and regional organizations, ASEAN being the most important one. ASEAN plus three is a forum for the SCS states' negotiations on the conflict management. It might play a role that the Arctic Council is taking in the Arctic. There are several common regional policies: Sustainable Development Strategy for the Seas of East Asia (SDS-SEA); Resolution and Plan of Action on Sustainable Fisheries for Food Security for the ASEAN Region Towards 2020; the Putrajaya Declaration of Regional

Cooperation for Sustainable Development of the Seas of East Asia; Maritime Cooperation Fund.

Many efforts are made in the sphere of conflict management and confidence building in the SCS, thus, other important areas receive less attention. At the same time, there was pre-existing cooperation in the SCS, such as the PEMSEA and SDS-SEA; and the original theory about the pre-existing cooperation as a major variable for the cooperation on delineation is falsified.

Legal stability might contribute to the resolution of the dispute. There are both hard and soft law instruments applicable by the coastal states, though soft law instruments tend to be a preferable form. Some authors think that the use of the soft law instruments is not beneficial for the parties as such tools lack a binding force (Kao, Sifford Pearre, and Firestone, 2012). We consider, that the soft law is as good as hard law in the case of the SCS. First, even hard law instruments lack enforceability in many cases, for example, China is not abiding by the UNCLOS in the SCS. Second, there are many examples that a successful cooperation can start with the adoption of the soft law (Proelss, and Muller, 2008). Third, soft law instruments might be preferable due to the cultural differences. Thus, the shape of the legal instrument is not that important as the presence of the political will to abide by the agreed provisions.

RESULTS OF COMPARATIVE ANALYSIS

The cases of the Arctic Ocean and the South China Sea were compared, and the following results were achieved (Table 1).

First, the original hypothesis was falsified, and the preexisting cooperation is not the only critical variable influencing approach to the delineation procedures. There is pre-existing cooperation in the SCS (see Table 1) on environmental protection, conflict management and confidence building, but there is no cooperation on delineation. Strong sovereignty, legal certainty, and common interests in the AO might have a greater impact on cooperation on the delineation. Certainly, the achieved stability creates a support for the further cooperation, but the absence of major territorial and maritime disputes in the zones of overlaps in the AO and the low possibility to find natural resources in the contested areas contribute to the willingness of the Arctic states to follow the UNCLOS-established order for the delineation of continental shelves.

Second, the type of overlapping claims also influences the willingness of states to involve the CLCS. Article 76 of the UNCLOS introduced a new definition of continental shelf and two different criteria for establishment of a title for a continental shelf: distance and natural prolongation.

Emerging delimitations of extended continental shelves might include two scenarios: delimitation of two extended shelves and delimitation of an extended shelf with a continental shelf within 200 nm. There is no practice on the correlation of distance and natural prolongation criteria in respect to extended shelves that might appear in the second scenario and which is applicable to the South China Sea case. In the AO the overlapping claims are based on the natural prolongation criteria (all overlaps are between the extended shelves).

Table 1. Comparison of two cases according to the identified criteria

	Arctic Ocean	South China Sea
Existing cooperation enhances international cooperation	Yes fisheries, climate change, CLCS submissions, environmental protection	Yes Many efforts in the sphere of conflict management/confidence building
Claims for extended continental shelves supported by international law	Yes legal stability, Ilulissat Declaration	Partially China-legal ambiguity of the "9-dash line"
Overlapping claims for continental shelves	Only extended shelves Coastal states respect established sovereign rights for 200 nm entitlements	Extended shelves and shelves within 200 nm (possibly influence the importance of the disputed islands and can be a trigger for escalation of a conflict)
Disputes	Minor Hans Island: Canada- The Kingdom of Denmark (announced negotiation 2018) Beaufort Sea: USA-Canada Northwest and Northeast Passages Norway-Russia resolved	Major Spratly Islands (multiply claims) Paracel Islands (multiply claims) Maritime zones (apart from extended shelves) Freedom of navigation: China-USA
Natural resources	Yes insignificant in the disputed areas?	Yes significant in the disputed areas
Joint development of the resources	Yes, possible	Possible
Security issues	Minor	Major
Need in confidence building	Not significant	Significant
Land reclamation (artificial islands construction) and militarization	No/yes	Yes/yes
Political will for cooperation	Yes	Yes, in confidence building and conflict management, ? in delineation and delimitation
Third party interest	EU – positive, scientific research (*seals)	USA – negative, freedom of navigation, different positions with China
Joint management of straits	Bering Strait: USA- The Russian Federation	Strait of Malacca: Singapore, Malaysia, Indonesia
Shipping	Insignificant, with potential to be become globally significant	Globally significant
Freedom of navigation and innocent passage	UNCLOS general interpretation	UNCLOS general interpretation and minor position as well. A special interpretation of the freedom of navigation: some states want to restrict foreign military activities within its EEZ and some want prior notification
Other limitations on navigation	Article 234 of the UNCLOS	No
Semi-enclosed sea (cooperation is necessary, Article 123 UNCLOS)	No	Yes
Climate change (cooperation is necessary)	Yes	Yes
Piracy	No	Yes
Population	Few	Dense
Economies	Lower-middle-income economies: none Upper-middle income economies: The Russian Federation High-income economies: Canada, The Kingdom of Denmark, Norway, USA	Lower-middle-income economies: The Philippines, Vietnam Upper-middle income economies: China, Malaysia High-income economies: Brunei Darussalam
Allocation of power	Almost equal in terms of cooperation	Hegemon-China
Change in allocation of power	Yes, after the Cold war	Yes, China as an emerging marine power
International agreements, Soft Law	UNCLOS: all except the USA London C. (London Convention on Dumping): Canada, The Kingdom of Denmark, Norway, USA MARPOL 73/78+ Protocol 97: all (+EU) POPs: Canada, The Kingdom of Denmark, Norway (+EU) United Nations Framework Convention on Climate Change (UNFCCC): all (+EU) Kyoto Protocol: The Kingdom of Denmark,	UNCLOS: all London C. (London Convention on Dumping): China, The Philippines MARPOL 73/78+ Protocol 97: China, Malaysia, The Philippines, Vietnam POPs: China, The Philippines, Vietnam United Nations Framework Convention on Climate Change (UNFCCC): all Kyoto Protocol: all

	Arctic Ocean	South China Sea
	Norway, The Russian Federation (+EU) Paris Agreement: Canada, The Kingdom of Denmark, Norway International Whaling Commission: The Kingdom of Denmark, Norway, The Russian Federation, USA CUES (Code for Unplanned Encounters at Sea, non-binding): Canada, The Russian Federation, USA Convention on the International Regulations for Preventing Collisions at Sea: all Agreement to prevent unregulated High Seas Fisheries in the Central Arctic Ocean: all (+EU) Convention on Biological Diversity (CBD): Canada, The Kingdom of Denmark, Norway, The Russian Federation (+EU) Convention on International Trade in Endangered Species (CITES): all (+EU)	Paris Agreement: Malaysia, Vietnam, Brunei Darussalam International Whaling Commission: China Military Maritime Consultative Agreement: China (+USA) CUES (Code for Unplanned Encounters at Sea, non-binding): all Convention on the International Regulations for Preventing Collisions at Sea: Brunei Darussalam, Malaysia, China, Vietnam Convention on Biological Diversity (CBD): all Convention on International Trade in Endangered Species (CITES): all
Regional agreements, Soft Law	ILO 169: The Kingdom of Denmark, Norway Svalbard Treaty: all International Agreement on Polar Bears: all NEAFC (The North East Atlantic Fisheries Commission): The Kingdom of Denmark, Norway, The Russian Federation (+EU), Canada as Co-operating Non-Contracting Party NAFO (The Northwest Atlantic Fisheries Organization): all (+EU) AMEC (The Arctic Military Environmental Cooperation) Norway, The Russian Federation, USA	ASEAN-China Free Trade Agreement: all SEAFDEC: Brunei Darussalam, Malaysia, The Philippines, Vietnam Treaty of Amity and Cooperation in Southeast Asia (TAC): all Declaration on the Conduct of the Parties in the South China Sea (non-binding): all
Dispute resolution	UN Charter, UNCLOS	UN Charter, UNCLOS, TAC
Membership in International Organizations	UN: all EU: The Kingdom of Denmark NATO: Canada, The Kingdom of Denmark, Norway, USA EAPC (The Euro-Atlantic Partnership Council): all IMO: all EEA (The European Economic Area): The Kingdom of Denmark, Norway (+EU) EAEU: The Russian Federation EFTA: Norway NAFTA: Canada, USA	UN: all IMO: all APEC (Asia-Pacific Economic Cooperation): all Asia-Pacific Parliament. Forum: China, Malaysia, The Philippines, Vietnam ACD (Asia Cooperation Dialogue): all Asian Development Bank: all East Asia Summit: all
Membership in Regional Intergovernmental Organizations	The Arctic Council: all the Arctic Contaminants Action Program (ACAP) Arctic Monitoring and Assessment Programme Working Group Conservation of Arctic Flora and Fauna (CAFF) Emergency Prevention, Preparedness and Response (EPPR) Protection of the Arctic Marine Environment (PAME) Sustainable Development Working Group (SDWG) BEAC (Barents Euro-Arctic Council): The Kingdom of Denmark, Norway, The Russian Federation (+EU) CBSS (Council of Baltic Sea States): The Kingdom of Denmark, Norway, The Russian Federation (+EU) NCM (Nordic Council of Ministers): The Kingdom of Denmark, Norway	ASEAN (Association of Southeast Asian Nations): Brunei Darussalam, Malaysia, The Philippines, Vietnam ASEAN plus three: includes China APFIC (Asia-Pacific Fishery Commission): China, Malaysia, The Philippines, Vietnam BIMP-EAGA: Brunei Darussalam, Malaysia, The Philippines SEAFDEC: Brunei Darussalam, Malaysia, The Philippines, Vietnam WPNS (Western Pacific Naval Symposium): all
Common Regional Policy	The Northern Dimension: The Kingdom of Denmark, Norway, The Russian Federation (+EU) AEPS (Arctic Environmental Protection Strategy): all, AEPS incorporated into Arctic Council	Sustainable Development Strategy for the Seas of East Asia (SDS-SEA): all, PEMSEA Resolution and Plan of Action on Sustainable Fisheries for Food Security for the ASEAN Region Towards 2020: Brunei Darussalam, Malaysia, The Philippines, Vietnam The Putrajaya Declaration of Regional Cooperation for Sustainable Development of the Seas of East Asia: all Maritime Cooperation Fund: ASEAN + China

Arguably, the Arctic states shall delineate to know the limits of their estimated shelves, so the collection of data and submission to the CLCS is the first step to the future delimitation. They are all interested in the establishment of the final and binding outer limits to "prove" to the rest of the world that their claims are not excessive and then they can decide on delineation between themselves.

In the SCS, the objectives of the states differ. The type of overlaps here will include: 1) extended shelves of the mainland coasts with shelves within 200 nm of the contested islands, and 2) extended shelves of the mainland coasts with extended shelves of the contested islands. The second situation is not considered as no one has claimed extended shelves of the contested islands, Vietnam and Malaysia did not use the islands for their submissions. Thus, only the first situation is analyzed. There is no interest for all of the SCS states to know the exact limits of the extended shelves in the zones of overlaps, as they apply different criteria for the title establishment: The Philippines would use distance criterion, China uses it's "9-dash line", Brunei might use both distance and natural prolongation, and Vietnam and Malaysia use the natural prolongation criterion. Since the "9-dash line" does not correspond to the UNCLOS, only the analysis of the distance versus natural prolongation criteria will be presented.

It shall be noted that there is a gap in the legal scholarship with respect to this issue. Yet there is no jurisprudence resolving the issue of the relationship of distance and natural prolongation criteria, and delimitation of extended shelves is a recent trend in jurisprudence. As the International Court of Justice (ICJ) noted in the Judgment on Territorial and Maritime Dispute (Nicaragua v. Colombia), "…the jurisprudence … for continental shelf delimitation involves no case in which a court or a tribunal was requested to determine the outer limits of a continental shelf beyond 200 nautical miles" (ICJ, 2012). Since the number of disputes concerning the delimitation of the extended continental shelf increases, the issue becomes a growing challenge for the international system of dispute resolution. The concurrent ICJ cases on this topic include Nicaragua v. Colombia (ICJ, 2016) and Somalia v. Kenya (ICJ, 2017). Other cases have been presented before the ITLOS including Ghana/Côte d'Ivoire (ITLOS, 2017) and Bangladesh/Myanmar (ITLOS, 2012). In the next decades, this field will definitely develop through the subsequent decisions of international courts, arbitrators, and agreements of states.

In the Libyan Arab Jamahiriya /Malta case, where the distance and the natural prolongation criteria were opposed to each other as grounds for a title for continental shelves within 200 nm, Libya was relying on the natural prolongation and Malta on the distance from the coast. Though both states were parties to the UNCLOS, customary law applied to the dispute because the UNCLOS was not in force at that time. In the Libyan Arab Jamahiriya/Malta case, the ICJ recognized the relevance of a distance criterion and did not give a role to "geological facts." It concluded: "the law applicable to the present dispute, that is, to claims relating to continental shelves located less than 200 miles from the coasts of the States in question, is based not on geological or geomorphological criteria, but on a criterion of distance from the Coast or, to use the traditional term, on the principle of adjacency as measured by distance" (ICJ, 1985). Thus, in the delimitation of continental shelves within 200 nm the distance criterion prevailed. In a situation like the SCS, in the case of delimitation of continental shelves between states when one party establishes title on the basis of the natural prolongation criterion (for continental shelf outside 200 nm), and another uses the distance criterion (for continental shelf within 200 nm), we argue that the latter will trump all the pretentions for the extended shelves. This hypothesis is supported not only by the Libyan Arab Jamahiriya/Malta case, but also by the intentions of states negotiating the UNCLOS, provisions of Article 82, and the convenience in administration of maritime zones (Antsygina, and Perez-Salazar, 2020).

Hence, except for Malaysia and Vietnam, the SCS states are not interested in the establishment of final and binding outer limits, and if distance criterion prevails, the qualification of contested maritime features as islands gets high importance as they would generate maritime zones and trump extended shelves. The contested islands might also generate extended shelves and delimitation of this types of overlaps will be required. Thus, the submissions of Malaysia and Vietnam and Vietnam just added an additional point of conflict over the contested features. Also, states might be reluctant to involve the CLCS in such a politically vulnerable situation. The question is whether coastal states shall first resolve their territorial disputes over islands and then proceed with the delineation. In this regard, there is a problem how the 10-year deadline affects sovereign rights on extended shelves.[7]

Third, the presence of natural resources in the zones of overlap in the SCS and high population density contribute to the significance of contested islands which, in its turn, might escalate the conflict. As it was already mentioned, there is a low possibility to find natural resources in the Central AO. Also, exploitation of these resources, if they are found, is not commercially viable and it's hard to predict when it will be profitable. In the SCS, however, the climate conditions are not harsh, there are infrastructure and transportation, and the exploitation of the resources can be commercially attractive. Most likely, there are natural resources in the zones of overlaps in the SCS. This contributes to the hardship of finding compromise solutions for the coastal states.

Fourth, since the Arctic states have a firm sovereignty over their territories and national waters in the Arctic, it is their common interest in keeping the Arctic region and its resources under their control and to be in charge of environmental protection. This includes regulating or limiting an access of non-Arctic states to the region, though constructive multilateral cooperation, mostly in the context of the Arctic Council, has been dominating. Since the multilateral cooperation is mostly functional cooperation on environmental protection due to long-range pollution as the first trigger, and climate change, the environment, as originally a field of low politics, has played an important role here. The SCS states cannot enjoy the same as there are globally important shipping routes in the SCS and foreign presence is already significant. This common interest of the Arctic Five, as well as the eight Arctic states, might be a contributing factor to a smooth process of delineation and delimitation. The existing cooperation as such, being able to solve a few disputes and leaning on UNCLOS may also play a lesson to learn/learned lesson, or at least an example, for the SCS coastal states (Storey 2013).

Fifth, legal certainty in the Arctic Ocean and adherence to the same legal rules provide for predictability and stability. Legal certainty and compliance with the UNCLOS might

[7] Under the Article 4 of the Annex II to the UNCLOS a coastal state party to the UNCLOS willing to establish the outer limits of its continental shelf has to submit data to the CLCS within 10 years of the entry into force of the UNCLOS for that state. Later, states-parties agreed that the obligation under Article 4 of Annex II and Article 76(8) could be satisfied by submitting a preliminary information indicative of the outer limits of the continental shelf beyond 200 nm. Decision regarding the workload of the CLCS and the ability of states, particularly developing states, to fulfill the requirements of article 4 of annex II to the United Nations Convention on the Law of the Sea, as well as the decision contained in SPLOS/72, paragraph (a), SPLOS/183, (June 20, 2008).

facilitate the dispute management, delineation and delimitation in the SCS. The first step to the legal certainty might be the functional cooperation on environmental protection and piracy, where all the actors would have a common interest. When there is a certain level of trust achieved states can strive to agree on the application of the same legal framework for the SCS.

Sixth, unbalance of power in the SCS contributes to the possibility to reach compromise solutions. In the AO states are equal in cooperation and it's easier for them to meet halfway. Hegemonic position of China allows it to dictate own conditions.

Seventh, the prognosis for delimitation of extended shelves in the AO is optimistic for the compromise solutions. Behind are existing cooperation, equality of states and the absence of such factors as contested islands and natural resources in the zone of overlap. Mutual trust grown up due to the cooperation and common goal to establish sovereign rights on extended continental shelves opposable to the rest of the world can contribute to the smooth delimitation process. There is a possibility of a dispute over the North Pole, positions of Russia, Canada and The Kingdom of Denmark indicate a strong will to establish sovereign rights for it as the symbolic point at the top of the world. Scenarios include establishment of sovereign rights of The Russian Federation, Canada or Denmark, a possibility to establish joint management area, or delimitation in a way where the delimitation lines start from the North Pole, so no one actually "owns" it.

Conclusion

The delimitation of extended continental shelves is understudied and will gain momentum in this decade, especially taking into consideration the process of ice melting in the Arctic. Delineation involving the CLCS in respect to the contested maritime areas might be complicated by the presence of different types of claims (continental shelves within and beyond 200 nm as opposed to the overlaps of the extended shelves only) and unresolved territorial disputes. As the overlapping claims in the AO are of the same legal nature, it was easier for the states to reach the agreements not to object to the submissions. The situation is different in the SCS where types of overlaps include both shelves within and beyond 200 nm. The strong sovereignty is thus more important for cooperation on delineation than the pre-existing cooperation.

The presence of natural resources in the zones of overlap contributes to the significance of the overlapping claims. At the same time, joint development of the resources can be a beneficial compromise solution for the coastal states. Or, due to growing pressure by rapid climate change/global warming, functional cooperation on environmental protection, such as to refuse to take wastes from the developed countries, could play an important role within the SCS countries. This way to resolve the dispute can be applied by both the Arctic states and states in the South China Sea. Some level of trust and confidence might be required for the joint development activities, so pre-existing cooperation serves as a basis for the joint development agreements. In the SCS point of common interest can be a combat with piracy: all the SCS states are interested in the safety of the sea routes. This sphere, where the coastal states have a common interest, can become a basis for the functional cooperation that might lead to an increase in trust among the SCS coastal states.

The process of delineation and delimitation can be smoother if the coastal states adhere to the same legal framework. Consistent application of the UNCLOS might facilitate the dispute management in the SCS, as well as delineation and delimitation procedures. The prognosis for delimitation of extended shelves in the AO is optimistic for compromise solutions. Behind are existing cooperation, equality of states and the absence of such factors as contested islands and natural resources in the zone of overlap. The further research will analyze the delimitation scenarios in the AO.

REFERENCES

Antsygina, Ekaterina, and Pérez-Salazar, Bernardo. 2020. "Sovereign Rights on Extended Continental Shelves: The Case of the Nicaragua Rise in the Western Caribbean." *International Journal of Marine and Coastal Law* (accepted).

Antsygina, Ekaterina, and Perez-Salazar, Bernardo. 2020. "Maritime Territorialisation and Governance: Geopolitical and Legal Issues Concerning the Delimitation and Sovereign Rights on Extended Continental Shelves in the Caribbean Sea and the Arctic Ocean." In *Frontiers: Law, Theory & Cases*, edited by Jairo Becerra, Jaime Cubides Cárdenas, and Dimitri Endrizzi, 56-104. Bogota: University Catolica of Colombia [forthcoming in 2020]. (Antsygina and Perez-Salazar, 2020).

Antsygina, Ekaterina., Czarski, Miriam., Konopleva, Svetlana. & Jirova, Anna. (2017). "The European Union's Role in a Changing Arctic." *A Changing Arctic (blog)*, August 2017. https://achangingarctic.wordpress.com/group/the-european-unions-role-in-a-changing-arctic/. (Antsygina, Czarski, Konopleva, and Jirova, 2017).

Arctic Governance Project. 2010. "*Arctic Governance in an Era of Transformative Change: Critical Questions, Governance Principles, Ways Forward.*" at pp. 3-6. https://tinyurl.com/ljqzvl7. (Arctic Governance Project, 2010)

Arctic Ocean Conference. (2008). "*The Ilulissat Declaration.*" 27-29 May. http://www.oceanlaw.org/downloads/arctic/Ilulissat_Declaration.pdf.

Baumert, Kevin A. 2017. "The Outer Limits of the Continental Shelf under Customary International Law." *AJIL, 111*, 4, 827-872. (Baumert 2017).

Beckham, Robert. 2010. "The UN Convention on the Law of the Sea and the maritime disputes in the South China Sea." *American Journal of International Law, 107*, 142, at p. 154 (Beckham 2010); Nong Hong, UNCLOS and ocean dispute settlement: law and politics in the South China Sea (Abingdon and New York: Routledge, 2012), p. 66. (Nong 2012).

Brunei Darussalam's Preliminary Submission to the Secretary-General concerning the Outer Limits of its Continental Shelf 12 May 2009. http://www.un.org/depts/los/clc s_new/submissions_files/preliminary/brn2009preliminaryinformation.pdf. (Brunei Darus-salam, 2009).

CAFF. 2018. "*State of the Arctic Marine Biodiversity Report.*" https://www.caff.is>orYouTube<https://www.youtube.com/watch?list=PLvo6-hWkBAzZyOB5wf1hE_02hZrVe7fKJ&time_continue=283&v=Fk3dW7eqACE at 4m:43s-5m:30s. (CAFF, 2018).

Churchill, Robin R. 2015. "The 1982 United Nations Convention on the Law of the Sea" in *The Oxford Handbook of the Law of the Sea*, edited by Donald R. Rothwell et al. Oxford: OUP, at 37. (Churchill 2015).

CLCS. 2005. "*Executive Summary of the Continental Shelf Submission of Norway in Respect of Areas in the Arctic Ocean, the Barents Sea and the Norwegian Sea.*" http://www.un.org/Depts/los/clcs_new/submissions_files/nor06/nor_exec_sum.pdf (CLCS 2005).

CLCS. 2013. "*Executive Summary of the Partial Submission of the Government of the Kingdom of Denmark together with the Government of Greenland to the Commission on the Limits of the Continental Shelf, The North-Eastern Continental Shelf of Greenland.*" http://www.un.org/depts/los/clcs_new/submissions_files/ dnk68_13/DNK 2013_ES.pdf. (CLCS, 2013).

CLCS. 2014. "*Executive Summary of the Partial Submission of the Government of the Kingdom of Denmark together with the Government of Greenland to the Commission on the Limits of the Continental Shelf, The Northern Continental Shelf of Greenland.*" http://www.un.org/depts/los/clcs_new/submissions_files/dnk76_14/dnk2014_es.pdf. (CLCS 2014).

CLCS. 2015. "*Executive Summary of the Partial Revised Submission of the Russian Federation to the Commission on the Limits of the Continental Shelf in Respect of the Continental Shelf of the Russian Federation in the Arctic Ocean.*" http://www.un.org/depts/los/clcs_new/submissions_files/rus01_rev15/2015_08_03_Exec_Summary_English.pdf. (CLCS 2015).

CLCS. 2019. "*Executive Summary of the Partial Submission of Canada to the Commission on the Limits of the Continental Shelf regarding its continental shelf in the Arctic Ocean.*" https://www.un.org/Depts/los/clcs_new/submissions_files/can1_84_2019/CDA_ARC_ES_EN_secured.pdf. (CLCS 2019).

CLCS/3/Rev.3, 6 February 2001. "*Annex I to the Rules of Procedure of the Commission on the Limits of the Continental Shelf*", para. 5. (a). Accessed May 16, 2018. http://www.un.org/depts/los/clcs_new/commission_rules.htm.

Continental Shelf (Libyan Arab Jamahiriya/Malta), Judgment, I.C.J. Reports 1985, p.13, para. 61 at p. 46. (ICJ, 1985).

Continental Shelf (*Tunisia/Libyan Arab Jamahiriya*), Judgment, I.C.J. Reports 1982, p. 60, para. 71. (ICJ, 1982)

Delimitation of the maritime boundary in the Bay of Bengal (Bangladesh/Myanmar), [2012] ITLOS Rep 4. (ITLOS, 2012).

Dispute Concerning Delimitation of the Maritime Boundary between Ghana and Côte d'Ivoire in the Atlantic Ocean (Ghana/Côte d'Ivoire), [2017] ITLOS Rep. (Not yet published). Available at: https://www.itlos.org/cases/list-of-cases/case-no-23/. (ITLOS, 2017).

Elferink, Alex Oude. 2014. "Do the coastal states in the South China Sea have a continental shelf beyond 200 nautical miles?" (2014) In *The South China Sea disputes and law of the sea*, edited by Jayakumar, S., Koh, Tommy, Beckman, Robert, pp. 164 – 191. (Elferink 2014).

Glaser, Bonnie S. 2012. *Contingency Planning Memorandum no. 14, Armed Clash in the South China Sea*, New York: Council on Foreign Relations, at 4. https://www.cfr.org/report/armed-clash-south-china-sea. (Glaser 2012).

Government of Canada. *Fisheries and Oceans Canada.* 2016. "Snow Crab." http:// www.dfo-mpo.gc.ca/species-especes/profiles-profils/snow-crab-crabe-neiges-atl-eng.html. (Government of Canada, 2016)

Heininen, Lassi. 2012. *Arctic Strategies and Policies: Inventory and Comparative Study, 2nd ed.* Stell: The Northern Research Forum & The University of Lapland. Accessed November 19.

Heininen, Lassi. 2018. "Artic Geopolitics from Classical to Critical Approach-Importance of Immaterial Factors." *Geography, Environment, Sustainability, 11* (1), 171-186. Accessed July 9, 2019. doi:10.24057/2071-9388-2018-11-1-171-186. (Heininen 2018).

Holland, Marika M., Bitz, Cecilia M., and Tremblay, Bruno. 2006. "Future abrupt reductions in the summer Arctic sea ice" *Geophysical Research Letters, 1*, 33, 23. (Holland, Bitz, and Tremblay, 2006).

https://www.rha.is/static/files/NRF/Publications/arctic_strategies_7th_draft_new_20120428.pdf. (*Heininen 2012*).

Kao, Shih-Ming., Sifford Pearre, Nathaniel. & Firestone, Jeremy. 2012. "Regional Cooperation in the South China Sea: Analysis of Existing Practices and Prospects." *Ocean Development & International Law, 43*, 3, 283-295, doi:10.1080/00908320.2012.698926. (Kao, Sifford Pearre, and Firestone, 2012).

Keyuan, Zou. 2017. "Navigation in the South China Sea: Why Still an Issue?" *The International Journal of Marine and Coastal Law, 32*, 2, 243. (Keyuan 2017).

Kuersten, Andreas. 2015. "Imagining the Arctic: International Law, Governance, and Relations in the High North." *Mich St Int'l Rev, 24*, 599. (Kuersten 2015).

Maritime Delimitation in the Indian Ocean (Somalia v. Kenya), *Preliminary Objections of 2 February 2017, Judgement*, [2017] ICJ Rep. (Not yet published). Available at: http://www.icj-cij.org/docket/files/161/19330.pdf.

McDorman, Ted L. 2015. "The Continental Shelf" in *The Oxford Handbook of the Law of the Sea*, edited by Donald R Rothwell et al., at 188. Oxford: OUP (McDorman 2015); see also Articles 208(1) and 240 UNCLOS.

Morton, Katherine. 2016. "China's ambition in the South China Sea: is a legitimate maritime order possible?" *International Affairs, 92*, 4, 909–940. (Morton 2016)

NDTV. 2018. "*China, ASEAN Agree On Draft South China Sea Code of Conduct.*" August 02. https://www.ndtv.com/world-news/china-asean-agree-on-draft-south-china-sea-code-of-conduct-1894310. (NDTV, 2018).

Note Verbale of the Republic of Indonesia No 480/POL-703/VII/10 July 8, 2010. Unofficial translation available at: www.un.org/depts/los/clcs_new/submissions_files/mysvnm33_09/idn_2010re_mys_vnm_e.pdf. (The Republic of Indonesia, 2010).

Permanent Court of Arbitration Case N° 2013-19, *The South China Sea Arbitration Award of 12 July 2016.* https://pca-cpa.org/wp-content/uploads/sites/175/2016/07/PH-CN-20160712-Award.pdf. (PCA, 2016).

Petrov, Eugene. 2015. "Political Geology in the Arctic." *Geo ExPro, 12.* Accessed May 17, 2018. http://www.geoexpro.com/articles/2015/03/political-geology-in-the-arctic. (Petrov 2015).

Poussenkova, Nina. 2018. "*Russia's Arctic Energy" Paper presented at the Russia Conference 2018 "Cold peace in the Arctic?"* Norwegian Institute of International Affairs, Oslo, Norway, September 14. (Poussenkova, 2018).

Proelss, Alexander, and Müller, Til. 2008. "The Legal Regime of the Arctic Ocean", *Heidelberg Journal of International Law*. (Proelss and Muller, 2008).

Question of the Delimitation of the Continental Shelf between Nicaragua and Colombia beyond 200 Nautical Miles from the Nicaraguan Coast (Nicaragua v. Colombia), Preliminary Objections, Judgment, [2016] ICJ Rep 100. (ICJ, 2016).

Rothwell, Donald R., and Stephens, Tim. 2017. *The International Law of the Sea*, 2nd ed. Oxford: Hart Publishing, at 124. (Rothwell and Stephens, 2017)

Rovaniemi Joint Ministerial Statement of the 11th Arctic Council Ministerial Meeting in May 7, 2019 in Rovaniemi.

Rybkina, Alena, and Reissell, Anni. 2017. "Continental shelves beyond 200 nm. Arctic Territory – Data and Modeling." *Geoinformatics Research Papers*, *5*, BS1002. July 9, 2019. doi:10.2205/CODATA2017, 1 of 172. (Rybkina and Reisell, 2017)

See O'Rourke, Ronald. 2018. *China's Actions in South and East China Seas: Implications for U.S. Interests—Background and Issues for Congress*. https://fas.org/sgp/crs/ro w/R4278 4.pdf citing Navy Office of Legislative Affairs email to CRS, June 15, 2012 and Department of Defense, Asia-Pacific Maritime Security Strategy, undated but released August 2015, pp. 7-8. (O'Rourke 2018).

See Regional Cooperation on Combating Piracy and Armed Robbery against Ships in Asia (ReCAAP). 2015. *Annual Report*. Singapore http://www.recaap.org/DesktopMod ules/Bring2mind/DMX/Download.aspx?Command=Core_Download&EntryId=421&Port alId=0&TabId=78. (ReCAAP, 2015).

Smith, Robert W. 2010. "Maritime Delimitation in the South China Sea: Potentiality and Challenges." *Ocean Development & International Law Journal*, *41*, 3. (Smith 2010).

Storey, Ian. E. g. 2013. "*Arctic Lessons: What the South China Sea Claimants can Learn from Cooperation in the High North*". ISEAS Perspective #65/2013. Singapore 16 Dec 2013. (Storey 2013).

Terblanche, John S., Hoffmann, Ary A., Mitchell, Katherine A., Rako, Lea, le Roux, Peter C., and Chown, Steven L. 2011. "*Ecologically relevant measures of tolerance to potentially lethal temperatures*" *J. Exp. Biol.*, *214*, 22. (Terblanche, Hoffmann, Mitchell, Rako, le Roux, and Chown, 2011)

Territorial and Maritime Dispute (*Nicaragua v. Colombia*), [2012] ICJ Rep 623 at 668 para 125. (ICJ, 2012).

The Agreement between the United States of America and the Union of Soviet Socialist Republics on the Maritime Boundary, Washington 1 June 1990. (The Agreement between the United States of America and the Union of Soviet Socialist Republics on the Maritime Boundary, 1990).

The People's Republic of China, Note Verbale CML/17/2007 May 7, 2009. http://www.u n.org/depts/los/clcs_new/submissions_files/mysvnm33_09/chn_2009re_mys_vnm_e.pdf. (The People's Republic of China, 2009).

The World Bank. "*World Bank Country and Lending Groups*." https://datahelpdesk. worldbank.org/knowledgebase/articles/906519-world-bank-country-and-lendin g-groups. (The World Bank).

Treaty between the Russian Federation and the Kingdom of Norway concerning maritime delimitation and cooperation in the Barents Sea and the Arctic Ocean. Murmansk 15 September 2010, in force July 7, 2011. UNTS, vol. 2791, No. 49095, p. 3.(Treaty between the Russian Federation and the Kingdom of Norway concerning maritime delimitation and cooperation in the Barents Sea and the Arctic Ocean, 2010).

Tulupov, Dmitry. 2015. "Amusing Delimitation: How to Neatly Split the Arctic Shelf." *RIAC*. Accessed May 17, 2017. http://russiancouncil.ru/en/analytics-and-comments/analytics/amusing-delimitation-how-to-neatly-split-the-arctic-shelf/ (Tulupov 2015); Also see Article 83 (1) of the UNCLOS.

U.S. Department of State. 2019. "*U.S. Extended Continental Shelf Project*." Accessed November 19. https://www.state.gov/about-the-u-s-extended-continental-shelf-project/. (U.S. Department of State, 2019).

United Nations Convention on the Law of the Sea, 10 December 1982, 1833 UNTS 3 (entered into force 16 November 1994). (UNCLOS).

Wilkinson, C., DeVantier, L., Talaue-McManus, L., Lawrence, D. & Souter, D. (2005). *Global International Waters Assessment, South China Sea, GIWA Regional assessment 54*. Kalmar: UNEP, University of Kalmar, at 78. https://wedocs.unep.org/bitstream/handle/20.500.11822/8816/GIWA_regional_assessment_54_South_China_sea.pdf?sequence=3&isAllowed=y. (Wilkinson, DeVantier, Talaue-McManus, Lawrence, and Souter, 2005).

In: The Arctic: Current Issues and Challenges
Editors: Oleg S. Pokrovsky, Sergey N. Kirpotin et al.
ISBN: 978-1-53617-306-2

Chapter 5

DEVELOPING ENVIRONMENTAL RESEARCH CAPACITY IN THE YAMAL-NENETS AUTONOMOUS DISTRICT WITHIN A GLOBAL NETWORKING FRAMEWORK

T. S. Kolesnikova[1,*], S. N. Kirpotin[1], L. G. Kolesnichenko[1], E. N. Morgun[2], O. M. Shaduiko[3], L. P. Borilo[3], A. I. Sinitskiy[2], A. L. Titovskiy[4], Oleg S. Pokrovsky[5], A. V. Kouraev[6], P. J. Mitchell[7] and S. N. Vorobiov[1]

[1]Bio-Clim-Land Centre, Tomsk State University, Tomsk, Russia
[2]Arctic Research Centre of YANAO, Salekhard, Russia
[3]TSSW Centre, Tomsk State University, Tomsk, Russia
[4]Department of Science and Innovation of YANAO, Salekhard, Russia
[5]GET UMR 5563 CNRS, University of Toulouse, France
[6]LEGOS UMR 5566 CNES - CNRS - IRD - UPS, University of Toulouse, France
[7]Faculty of Foreign Languages, Tomsk State University, Tomsk, Russia

ABSTRACT

Based on an analysis of modern literature and understanding of the reasons behind the existing trends in the development of fundamental science the chapter examines environmental research capacity in Western Siberia and global networking. There are tendencies increasingly acquiring the character of mega-science and organization as a network. Mega-science (first and foremost including Arctic Science) requires mega-resources for the implementation of large-scale projects, which, as a rule, involve tens and hundreds of leading research groups from different countries. Such large-scale interaction can be based on fundamentally new concepts and principles of organizing joint activities. The main driving force behind the development of institutions is not competition, but rather integration and mutually beneficial cooperation between partners.

* Corresponding Author's Email: kirpota@inbox.ru.

In this situation, big science must always prove to society the expediency of its existence and its usefulness for society, especially in terms of allocating huge funds and resources for its development. The concept of transdisciplinarity comes to the fore, which implies building bridges between fundamental science and society: shaping public opinion and policy, companies and other diverse groups of stakeholders.

Landscapes of the Siberian Arctic, including the continental shelf, regulate many natural processes on a global scale. For example, they have an impact on carbon balance, and they influence global climate change. Owing to the development of the oil and gas complex, Western Siberia among all Arctic regions has the most developed infrastructure: roads, oil and gas pipelines, pads, power lines, gas-compressor stations, cities and towns of oil and gas industry workers. This well-developed infrastructure (a good road network) provides unprecedented access to the region for researchers and field experiments, making them significantly cheaper than in other northern regions, which exploit small aircraft or helicopters (Kirpotin et al., 2018).

This chapter demonstrates convincingly that the Yamalo-Nenets Autonomous District (Okrug) has unique unprecedented opportunities for the development of large-scale network projects based on transdisciplinarity, and is the most convenient site for studying both the fundamental questions of interaction of climate and landscapes and considering the practical aspects of these changes and evaluation of these social impacts. The chapter also provides a description of some bilateral and regional research networks and finding ways of their effective interactions with each other, with stakeholders and the state authorities in the Yamal-Nenets Autonomous District (YNAO).

Keywords: mega-science, transdisciplinarity, networks, Arctic, mega-transect, stakeholders, Yamal

INTRODUCTION

The modern social world is in transition from a hierarchical form of organization to a networking one. Although it retains a hierarchical form, it is no longer viewed as dominant. Connections are gaining more and more weight; networking, covering the most diverse spheres of public life, and cooperation links, not subordination, come to the fore. Today's social world is a world of intersecting, overlapping networks of a very different nature. At their core, these networks are nothing more than an indicator of the transition from a society focused on the production of goods and services to a society where information and interaction have become the main values. The Internet has played an impressive role in this. Over its relatively short period of existence, the Internet has become involved in all spheres of life. This occurred precisely due to its non-hierarchical network structure. Over a couple of decades, it has managed to impose a significant imprint on the activities of all other institutions. The Internet is the most important element, tool and factor of the modern information revolution.

Trends towards interaction are observed in all social institutions. The problems that contemporary science will address are so great that society and the state allocate huge money for the development of scientific and technical research. This means that they have the right to expect an increasing contribution from science and technology to solve society's problems. But here another difficulty arises – the difficulty of making an objective decision regarding development and financing, the priority and importance for society of various research areas, and scientific, technical and innovative projects. One thing has become clear: our technologies seriously violate, and perhaps would irrevocably destroy, the ecological systems, on which our very existence rests. It turns out that these negative consequences are impossible to accurately predict, and we can only assume some development scenarios. However, it is very difficult to predict which of these scenarios are being implemented and how. Risk assessment is now of particular importance because it must take into account all the aspects, not only scientific and technically specialized but primarily social, environmental, ethical, etc. With the development of up-to-date technologies, new types of risks and dangers are emerging. The development of nuclear energy is one of the most vivid examples of the unpredictability and uncontrollability of the consequences of scientific and technological progress. Today it is obvious that such "aggressive" decisions must not be taken. The role of science in modern society, on the one hand, is increasing. Science forms a basis for political lobbying and also for political, social and economic decision-making. On the other hand, its role is decreasing, since no one else believes in its omnipotence. Now science itself has to prove to society the practical effectiveness and necessity of its existence. This makes it impossible to restrict science to a narrow academic framework. The contemporary scientist is no longer in an "ivory tower" and cannot afford to dismiss humanity any more (Gorokhov, 2011, Gorchitskaya, 2011, Stepin, 1996). The "paradigm" of the relationship between science and practice, science and politics, as well as the mechanism of knowledge production, is changing. Now it is not produced somewhere "above" before "coming down." Knowledge is constructed, built, and formed in the course of a dialogue of scientists, authorities and the public.

Policies should help translate normative ideas into action, but be connected with a scientific foundation. Today science takes upon itself what previous policy did, and modern politics, for example, in the field of climate issues, cannot even formulate its problems without science. Thus, nowadays science is becoming a tool and at the same time an important component of modern politics that solves specific problems for society. It is in this sense that we speak of so-called transdisciplinary science, which goes beyond not only the framework of individual disciplines but also disciplinary science in general, which is part of a broad social sphere. "In the modern world, biological, psychological, social and environmental phenomena are closely interrelated. In order to properly describe this world, we need an ecological perspective that the Cartesian worldview cannot provide. Therefore, we need a new paradigm. The beginning of such changes, the transition from a mechanistic to a holistic concept of reality is already noticeable in all areas" and is called "transdisciplinarity" (Capra, 1983). The new vision of reality is based on an understanding of the deep interconnection and interdependence of physical, biological, psychological, social and cultural phenomena, and exceeds the conceptual boundaries of modern scientific disciplines. Individual researchers, communities and networks join together according to new principles: theories, models; and organizations should not have advantages over others, they do not contradict each other and go beyond the generally accepted disciplinary differences.

Siberia is of global importance, it is a mega-system that largely determines the climate of the planet, it decisively affects all global processes (Kirpotin et al., 2015, Kirpotin, Callaghan, 2016; Kirpotin et al., 2018). Numerous Russian and international scientific groups have worked and are working in Siberia, some of whom conduct research at the highest level. However, the studies are fragmented and not coordinated; they are not carried out according to unified programs. As a result, there is no comparable data that could become the property of the world scientific community. However, “strong” research teams actively working in Siberia and actively interacting with international colleagues could not continue to ignore the need for unification. So, as an initiative “from below,” attempts to create a scientific network began. Later, as a solution to the problem of isolation of research, the concept of Siberia as a unique natural mega-facility appeared (Kirpotin et al., 2015, Kirpotin, Callaghan, 2016; Kirpotin et al., 2018). In 2016, Siberian Environmental Change Network (SECNet) – an open community of educational institutions, research organizations, scientific groups and individual scientists, was created. It was united by the goal of promoting the sustainable development of Siberian regions. The project at the very beginning was transdisciplinary, and one of its most important tasks was to synthesize knowledge about environmental changes in Siberia, assess the consequences of these changes on the population of Russia and the whole world through the application of uniform methodologies, approaches and principles in research. It was planned that the knowledge that would become available through the collaboration of various research centers would be applied to enable the world's population to adapt to the effects of climate change. To implement the project, the infrastructure of the SECNet requires substantial growth, development and large-scale modernization as part of the implementation of a large system infrastructure project.

Undoubtedly, the availability of developed infrastructure (equipping new and existing stations with necessary equipment, residential and laboratory facilities) and the development of unified protocols will make Siberia attractive to domestic and international partners at all levels. Moreover, it will stimulate and attract whole fields of science to engage in cooperation, and therefore obtain new sources of financing. At the same time, the implementation of such a large-scale project implies a set of actions that require significant financial and organizational resources, as well as specific legal regulation for the preparation, creation and subsequent operation of a global project necessary for the implementation of the strategic objective of SECNet, namely, a qualitative and positive change in the social-economic situation of the Siberian region. At the same time, the creation of such a large-scale project gives rise to a number of consequences, among which the risk of a funding shortage for constant support of the entire infrastructure is of particular importance. The practice of solving such global problems shows that the greater the number of stakeholders involved in the process of working on them, the more complete the statement of the problem is. Therefore, a transdisciplinary approach is optimal.

1. Transdisciplinarity and the Network Form of Communication as a Consequence of the Adaptation of Fundamental Science to the Realities of the Information Society

The production of scientific knowledge in modern realities essentially depends on political, economic and social factors, in particular, on the role that civil society plays in decision-making. There are acute issues of trust in science, politically influential scientific communities in shaping the political agenda, the impact of scientific knowledge on the way of thinking and behavior of ordinary citizens. Thus, a new concept has emerged: public research (transdisciplinary) (Skocpol, 1984; Rueschemeyer, Stephens, 1992; Strezhneva, 2012), where all stakeholders are involved in making a decision and assessing its risks. This tendency to level scientists, politicians and the public in the decision-making process leads to the abolition of hierarchical relations and the formation of a new form of interaction – the network.

Transdisciplinarity and the network form of communication is a consequence of the adaptation of fundamental science, on the one hand, to the realities of the information society, and on the other hand, to the new forms of science itself, which is increasingly focused on solving large-scale interdisciplinary projects carried out by international scientific consortia (Capra, 2003, Makarova, 2013). The transdisciplinary approach implies the involvement of all actors in the decision-making process since each participant in a problem situation considers the issue from his/her local position and sees only one of its sides and the details of this side that are contained in his/her model of reality (Kiyaschenko, 2009, Nicolescu, 2006). Only by taking into account the interests of all actors, can one get closer to an improving intervention (as the classicists of systems analysis call a change in a situation that does not entail negative consequences for any of the actors) (Ackoff, 1982, Tarasenko, 2013). “Sustainable development,” about which there is so much talk nowadays, can be recognized as a matter of fact an improving intervention. In addition to the stated individual goals of each specific network structure, there is a common “unwritten” goal for all: the result of functioning should be a contribution to the attempt by modern society to develop a strategy of “sustainable development” (an improving intervention), a resource use model that aims to meet human needs while preserving the environment, so that these needs can be met not only for the present but also for future generations.

All human technologies are built on a short-term linear perspective. In the course of observations of the surrounding natural environment (living system), consisting of ecosystems inhabited by countless organisms, we can see that all organisms have developed together over billions of years. Existing within their ecosystem, they did not cease to use and process the same molecules of soil, water and air. This is a rational use of resources, a constant desire to establish cooperative ties, which contributes to the harmonious integration of parts of the system at all levels of its organization (Jantch, 1972, 1980; Beck, 1992; Capra, 1996).

Returning to evaluating the effectiveness of scientific, technical and innovation activity, based on the above, we can conclude that it is a complex problem and the solution of such a problem is beyond the capabilities of any particular science. First of all, it is not within the scope of the problem statement, but inside it is “extra-scientific” – societal, which means all stakeholders must make decisions. Secondly, it is interdisciplinary; systems must interact to reach a non-linear solution with a long-term perspective.

The problems that are proposed to be solved by modern science are so global that society and the state allocate a significant share of budgetary funds for the development of scientific and technical research, and therefore, they have the right to expect an increasing contribution from science and technology to solving the problems facing society. But here another difficulty arises – the difficulty of making an objective decision regarding development and financing, the priority and importance for society of various scientific directions, scientific, technical and innovative projects. A tool is needed to assess the expected effectiveness and scientific justification for making specific decisions. Risk assessments and studies of the positive and negative consequences of the implementation of the results of a scientific and technical solution are required. Risk assessment is now of particular importance because it must take into account the most diverse aspects, not only scientific and narrowly technical but primarily social, environmental, ethical, etc. With the development of modern technologies, new types of risks and dangers are emerging. They set tasks for the state-related not so much with the elimination of the damage already inflicted, as the task of preventing possible damage. So the need becomes long-term planning, which should calculate and eliminate possible risks. To become relatively independent, the assessment must be not only interdisciplinary but international, i.e., disinterested experts from other countries should be involved in the assessment. In addition, representatives of regional authorities and the public should take part in the assessment, especially when it comes to evaluating scientific, technical, innovative and economic projects, the implementation of which affects their vital interests. The development of atomic energy is one of the most vivid examples of the unpredictability and uncontrollability of the consequences of scientific and technological progress. No one foresaw the consequences of possible accidents at nuclear power plants (not only environmental ones that had an impact on human health and its ecosystems, but also financial ones). Today it is obvious that such "aggressive" decisions cannot be taken (Brown, 1981;Capra, 1983; Gorokhov, 2006). However, if the decision is made by all stakeholders in compliance with the above conditions, it will involve comprehensive and, primarily, public research study and social assessment at the international level. Such a study will be problem-oriented, not only in the sense of its interdisciplinarity but also because it does not fit into the traditional disciplinary organization of science, although it complements it organically (Capra, 1983; Gorokhov 2010).

The mechanistic paradigm must be replaced by an ecological one. The beginning of these processes in science, in personal and collective relationships, and in the nature of organizations is already noticeable. New alliances and coalitions, already linked by hundreds of groups and networks, are based on a non-hierarchical, non-bureaucratic and non-violent nature. Some of them are very efficient at the global level. These multiple networks and coalitions have not yet established themselves in the political arena, but actively interact with stakeholders at all levels, continuing to develop a new vision of reality, creating a critical mass (Capra, 1996).

In this sense, it is very important to form the ecological thinking of the whole of society and not its individual structures. The main prerequisite for conscious public participation in the discussion of problems is regular awareness of the population about the real state of affairs, the dissemination of scientific knowledge in a modern and accessible form to a wide range of people. This implies a broad dialogue between the creators of technology, its consumers, state structures and the public, experts and those whom its regular or abnormal operation affects or may affect, and all parties involved in the complex process of public mutual learning, forming their collective understanding (Callaghan et al, 2015, Callaghan et al, 2019, Knyazeva, 2011).

The network nature of the organization of science is manifested in the erosion of the boundaries between its previously clearly defined parts. This is expressed in the growth of interdisciplinary research areas, in the transition from thematic to problematic types of construction and implementation of scientific research. Methods and norms of social and humanitarian knowledge penetrate natural science and even technical science. The realities understudy can no longer be connected hierarchically since such a connection does not fulfil the requirement of simultaneity and consistency (information dissemination and transformations are always delayed). It is the network ("web") that is an adequate form of organization of modern social being (Brysina, 2006, Kirpotin, 2005, Kirpotin, 2007, Kirpotin, 2009, Kirpotin, 2011, Kirpotin, 2013).

As can be seen, all network interactions more easily line up in all areas of modern science and they are becoming more urgent and indispensable. However, there are such areas of scientific activity where, due to their specificity, network work has long and firmly established itself as the most optimal approach in conducting research. The history of the development of polar observations speaks for itself (Allison, 2007, Barr, 2008, Corby, 1982, Gerson 1958, Kotlyakov, 2011, Mazzoli, 2016).

In the harsh conditions of the Arctic, researchers do not compete with each other, but build their relations on the basis of mutual aid, solidarity and integration, being an example of cooperation. The Arctic is a unique region in which nature is sometimes "broken up" in accordance with generally accepted scientific doctrines. Natural selection, competition as a driving factor of evolution, does not work in the harsh conditions of the Arctic. "... there are possible (and observed) cases when species that have better developed mutual aid (the so-called competition in altruism) have won the struggle for existence" (Kropotkin, 1907, 2011).

What is the driving force of modern science? The competitive struggle of ideas according to Darwin or mutually beneficial cooperation – cooperation according to Petr Kropotkin (Barzun, 1958; Malthus, 2008; Asmolov et al., 2013)? The answer to this question we get when analyzing the modern approach to the study of the Arctic (research station networks phenomena). Each "scientific unit" (as in this case we call the research station) strives for interaction and networking, each network for integration with other similar networks. All this allows us as soon as possible to master the main value of modern society as a whole and the scientific community in particular, which is information. There is a need for interdisciplinary, international research of the problem, with the possibility of including all stakeholders in the definition of research goals and strategies. This cooperation of all stakeholders is unique; all ways of understanding the world are involved in the decision-making process in an attempt to create a comprehensive, complete picture of the issue being studied in order to use the results of research practically and purposefully. Such a striving towards network interaction with the departure of science beyond the academic framework to one degree or another is observed in all scientific organizations. The transformation process proceeds at different speeds. The scale of the transformation depends on the goals of a particular specific research station, network, or scientific organization. However, the general trend is noticeable at all levels.

2. French-Russian Research Network Car-Wet-Sib – Bilateral Level and Transdisciplinary Essence

The International Research Network (GDRI) CAR-WET-SIB is unique in that it was born thanks to transdisciplinarity (presentations of the Tomsk Region at the Embassy of France in Moscow). This is a rare case when a scientific consortium was formed due to a strategic solution from above to solve an urgent task of global significance – the study of the climate-regulating function of West Siberian mires.

In December 2007, one of the 7 International Research Networks (GDRI) "Biogeochemical Carbon Cycle in Wetlands of Siberia" (CAR-WET-SIB) was established between Russia and France. CAR-WET-SIB is an example of a bilateral Russian-French scientific association/network, which was focused on the study of several corresponding scientific disciplines within the framework of one scientific field – Earth sciences.

CAR-WET-SIB (2008-2011) and its successor CAR-WET-SIB II (2015-2018) successfully realised all its planned activities. Currently (2019), the preparation of a new IRN (International Research Network) "From Tundra to Steppe (TTS): fluxes of carbon, water and metal in Siberia"is underway. IRN TTS is a logical continuation of activities of CAR-WET-SIB and CAR-WET-SIB II, with an enlarged consortium and scientific scope.

2.1. Scientific Context of the GDRI Car-Wet-Sib II

The main two goals of the GDRI Car-Wet-Sib II project were: i) to provide new and crucially important data for the quantitative prediction of the direction and magnitude of changes of fluxes of matter (water, chemical elements) in the regions of high latitudes under the ongoing climate warming and ii) to constrain a new hydro-geochemical model of carbon and related element transport capable of operating in permafrost-bearing environments. This mechanistic modelling will require quantitative calibration of elements, pools and fluxes assessed from highly seasonally and spatially resolved natural observations.

The scientific approach of the project was based on:

a) a combination of field (observations and measurements, samples collection) and laboratory (analytical) studies, remote sensing techniques and numerical modelling.
b) a selection of key sites within Central, Western and North-Western Siberia.

Within this GDRI, fundamental research was developed, based on solid and long-term existing collaboration between French and Russian (Siberian) teams. Indeed, this collaboration has existed for more than 15 years and was formalized during the period 2008 to 2011 by the creation of the GDRI Car-Wet-Sib I, initially headed by Bernard Dupré (France) and Sergey Kirpotin (Russian Federation). During this period, we benefited from the financial support of the INSU (Institut National des Sciences de l'Univers), the French Embassy in Moscow, the University Paul Sabatier, Toulouse, and the RFBR (Russian Foundation for Basic Research). The scientific expertise of all the partners encompasses the fields of chemistry, geology/(isotope)geochemistry, (micro-) biology/(microbial-) ecology, forestry/biodiversity, agronomy, eco-toxicology, remote sensing; which is, in our opinion, relevant for an integrated study on transport of waters and chemical elements in the Arctic and subarctic environment. Also, this GDRI was an important tool in the teaching programme of the research community involved in the GDRI because the network allows both the training of students and improving of fundamental knowledge. The knowledge developed is available for the entire scientific community through the publication of scientific articles, books, and participation in local and international conferences. Also due to the high priority research of the Arctic Ecosystems, this GDRI helped to attract scientists from other countries.

The purpose of the GDRI "Car-Wet-Sib II" was to provide support and coordination for scientific exchanges between the laboratories or teams involved and for activities related to the scientific program.

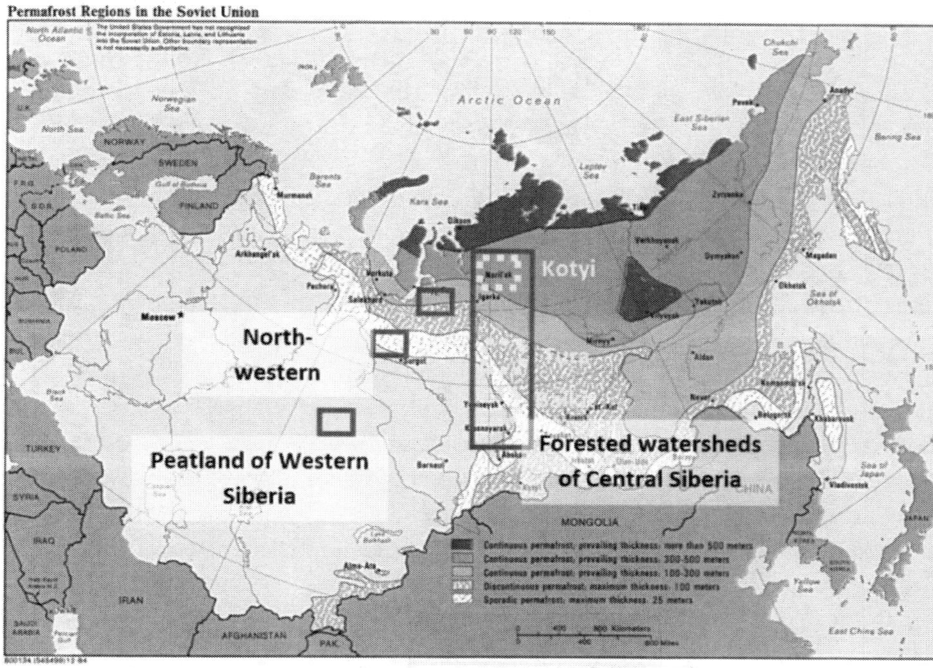

Figure 1. Map showing the studied sites of the CAR-WET-SIB II.

For this purpose, the GDRI aimed to:

- facilitate and encourage contacts and exchanges of information;
- encourage cooperation initiatives;
- coordinate and structure multi-partner research programs in order to reply to invitations to tender for supporting research and technological development;
- ensure harmonization and the complementary nature of the programs of conferences, congresses, etc. devoted to its scientific theme;
- encourage permanent training initiatives and promote pedagogical operations.

Qualitatively new understanding of biogeochemical processes controlling the terrestrial cycle of carbon and related elements in aquatic and soil environments of Western and Central Siberia were produced, combined with novel satellite approaches calibrated for the first time using direct ground observations.

The novelty of the project consisted in approaching the consequences of climate warming and permafrost thaw on hydrogeochemical fluxes in high latitudes via studying the dissolved organic carbon and related elements transfer along a latitude transect of three regions (from north to south Kotyi, Tura and Baikit) having identical lithology and vegetation but a different air temperature regime and permafrost thickness and coverage.

2.2. Results and Outcomes of the GDRI Car-Wet-Sib-II

In October 2008, a two-sided Russian-French seminar "Interdisciplinary Methods for Studying the Biogeochemical Carbon Cycle in Wetlands of Western Siberia in the Context of

Climate Change" was held in Toulouse with the support of the RFBR and the CNRS (Centre National de la Recherche Scientifique). In April 2010 RFBR together with CNRS and DFG (explain) was successfully organized and held the trilateral Russian-German-French seminar "Climate regulating role of forest bog ecosystems: landscape-ecological and social aspects" at the site of the University of Hamburg.

Within the CAR-WET-SIB project, with the assistance of the Embassy of France, three PhD students – Vera Stepanova, Maria Kolmakova and RinatManasypov – defended their theses under the joint supervision of TSU and the University of Toulouse.

In the framework of the "Colloque CNRS-Russie" held in 2010, following the Year of France in Russia and Russia in France, the activities of the GDRI CAR-WET-SIB were rated as the most successful. In 2012, the Siberian Branch of the Russian Academy of Sciences and the National Center for Scientific Research of France (CNRS) signed an agreement on the creation of the "French-Siberian Center for Science and Education", with one of the priority areas "Environment, climate, continental surface, biosphere" coordinated by Tomsk State University. As the Science and Technology Attaché at the Embassy of France, Michelle Tararin, noted in his interview in 2010: "Tomsk has become a recognized world leader in the field of ecology and climate".

The coordinated research efforts of Oleg S. Pokrovsky (Midi-Pyrenees Observatory) and his colleagues from Tomsk State University, aimed at studying the thermokarst lakes of Western Siberia, starting in 2009, discovered conceptually new features of the exchange of greenhouse gases between lake water and the atmosphere. In the course of detailed fieldwork during 2010–2012, it was discovered that small thermokarst reservoirs (<100 m2), which had not previously received research attention, formed at the very beginning of permafrost thawing and not represented on existing maps and databases of world lakes, characterized by concentrations of carbon dioxide and methane, are 3–10 times higher compared to the large (> 0.01 ha) lakes, which have been considered up to now. The contribution of these small lakes to the total surface coverage of the region is significant, and their consideration can drastically change our predictions relating to the response of ecosystems to climate warming and greenhouse gas emissions. The results of this work were published in the "Biogeochemistry" journal.

In 2013 a mega-grant "Biogeochemical cycles of arctic mire-lake landscapes of Western Siberia as an indicator of global climatic changes and the basis for environmental management of the region (BIO-GEO-CLIM)" was won according to the 220th Resolution of the Government of the Russian Federation (dated April 9, 2010) under an agreement with the Ministry of Education and Science of the Russian Federation No. 14.B25.31.0001. This project was successfully completed in 2015 under the leadership of Oleg S. Pokrovsky. For the development of French-Siberian scientific cooperation, the Car-Wet-Sib coordinator from the Russian side, Sergey Kirpotin, was awarded the highest academic award in France in 2015 and became a Chevalier l'Ordre des Palmes académiques.

Over the past ten years, more than 100 articles have been published jointly in high-level international journals such as Biogeosciences, Chemical Geology, Biogeochemistry, The Cryosphere, Hydrology and Earth System Sciences, Polar Biology, Water (Switzerland), Journal of Hydrology, Chemical Geology, Science of the Total Environment, Nature Geoscience and Nature Communications (Shirokova et al., 2013; Zakharova et al., 2014; Manasypov et al., 2015; Pokrovsky et al., 2015; Vorobyev et al., 2015; Pokrovsky et al., 2016; Vasil'chuk et al., 2016; Polishchuk et al., 2017; Morgalev et al., 2017; Loiko et al.,

2017; Shevchenko et al., 2017; Vorobyev et al., 2017; Ala-aho et al., 2018; Serikova et al., 2018; Vorobyev et al., 2019: Serikova et al., 2019).

From the first days of the start of its activities, CAR-WET-SIB paid great attention to the development of transdisciplinarity. It should be noted that the idea for the GDRI first appeared after a presentation of the Tomsk Region at the French Embassy in Moscow, when the Attaché of the Embassy for Science and Technology Lise Breuil was impressed with the report on the West Siberian swamps and their global role and decided to find partners in France for the Siberian side.

Thanks to Lise Breuil's initiative in 2007, the delegation of CNRS and the Observatoire Midi-Pyrénées (OMP) from Toulouse arrived in Tomsk to discuss the possibilities of organizing a large joint project. The delegation included fellow compatriots, Oleg S. Pokrovsky and Alexei Kouraev (graduates of Moscow State University), who later played a major role in promoting joint projects with TSU and France. They became the main executors and organizers of all projects in the framework of CAR-WET-SIB from the French side.

It should be noted that this topic has an interesting background. It was previously promoted by the not-purely scientific, but the popular science publication of Fred Pearce in the journal New Scientist "Climate warning as Siberia melts" about the landslide thaw of permafrost in Western Siberia in 2005 This publication had the effect of an exploding bomb and attracted the attention of not only the public but also scientists themselves to this problem.

The scientists who raised this problem in a popular scientific publication at the time were severely criticized and condemned by their colleagues. They were accused of populism and wanting to earn cheap popularity. However, time has confirmed the correctness of this decision, and today all leading research groups are forced to engage in such "popularization", often at great financial expense, hiring professional "boundary organizations" which specialize in building bridges between fundamental science and society.

Figure 2. Finnish journalist Marja Manninen interviews Oleg S. Pokrovsky.

The scientists involved in CAR-WET-SIB regularly invited cameraman Dmitry Gureev to their expeditions to produce popular science films about field research. In addition, representatives of leading television companies participated in the CAR-WET-SIB's expeditions in different years: BBC Russian Service (Oleg Boldyrev), France-2, Finnish Television (Marja Manninen) (Figure 2). Well-known Tomsk journalist Julia Ratomskaya prepared a 45-minute TV program about CAR-WET-SIB "Simplifying the Complicated." CAR-WET-SIB scientists have given dozens of interviews to regional, federal, and international newspapers and magazines.

Perhaps CAR-WET-SIB is one of the best bilateral International Consortiums in terms of using the principles of transdisciplinarity. It was a kind of pioneer and even a forerunner in this area.

2.3. The Next Stage IRN "from Tundra to Steppe"

The next stage of activities of CAR-WET-SIB is planned as International Research Network (IRN) "From Tundra to Steppe (TTS): fluxes of carbon, water and metal in Siberia" that will operate from 2019 to 2022. The IRN TTS is based on fruitful cooperation established during the GDRI Car-Wet-SIB I and II and several other national and international projects and relies on a group of complementary Russian and French partners. The previous cooperation has shown the ability of these different research groups to produce research of high scientific quality. The high number of international conferences and publications reveals that the GDRI was a powerful tool to promote international networking and collaborations and train a new generation of young scientists – post-docs and graduate students learning different but very complementary techniques.

The activities in the TTS biogeochemical framework rely on the studies of five different and complementary regions of Siberia. These sites are: 1) the forested watersheds of Central Siberia, 2) the thermokarst lakes of North-Western Siberia, 3) the large surface of pristine peatlands of Western Siberia, 4) the Yenisey river watershed, and 5) Central Yakutia (Eastern Siberia).

Within this region, the IRN aims to answer the following questions: a) What are the controlling factors of the permafrost hydrological regime on local (small watershed) and global (latitude transect) scale? b) How does the thickness of the soil active layer and permafrost coverage affect the organic carbon and related elements dynamics in the plant-soil-river continuum via leaching, colloidal transport, and transformation in the river channel? and c) In which direction and to which extent are the ongoing climate warming and permafrost thaw in high latitudes capable of modifying the hydrochemical/biogeochemical fluxes of carbon and related elements?

Field and laboratory studies will be focused on Western Siberia palsa bog mire complexes developed in the Central and Northern part and in the Altai Mountains (ecology, primary productivity and mineralization, botany). The thermokarst lakes of North-Western Siberia will receive particular attention, regarding both the water column and the sediment compartments.

The biogeochemical cycling and chemical elements transit in the ecosystems of the humid zone depend strongly on the water cycle. While the emission of gases to the atmosphere has been relatively well studied lately, transport of organic matter with surface water is still poorly quantified. Assessment of carbon losses with surface water is possible only if we know the hydrological regime of various landscapes. So it is of paramount importance 1) to evaluate the water regime and the water fluxes in different elements of the hydrological network of Siberian regions (bogs, thermokarst lakes, rivers etc.) and 2) to quantify the global contribution of the different ecosystems (palsa, forest, tundra) to the water outflow from the watersheds. Among the regions that will be studied are Western Siberia (the northern permafrost-affected part and the southern part with large presence of peatlands) and Central Siberia (permafrost-affected zones in Yakutia and Buryatiya), that represent hydrologically dynamic territories subjected to rapid climatic and ecological changes.

2.4. Meda-Transect Approach and Infrastructure

Car-Wet-Sib is also an infrastructural project. Since 2008, already in its first projects, a unique mega-transect for research and sampling was formed, with a length of 2500 km from the Altai mountains of southern Siberia to the Arctic zone. Later it became the basis for Siberian Environmental Change Network (SECNet) established in 2016 (Figure 3).

Today the international scientific community is invited to use in their research the network of unique experimental and observation sites and research stations located along the latitudinal gradient: Se-Yaha, Zapolyarny, Khanymey, Mukhrino, Kaybasovo, Vasjugan, Aktru, the modern ground-based research infrastructure of Western Siberia to study environmental changes, understand and predict future changes.

Unique concepts were developed as a methodological basis for both the Car-Wet-Sib and later SECNet activities: the concept of Western Siberia as a unique natural mega-facility and the mega-transect approach as its infrastructural axis for research.

The term 'mega-science' usually concerns physics. Extremely expensive and sensitive equipment like CERN's Large Hadron Collider is so expensive that no one country in the World, even the richest one, can pay for its installation and even work on it. Therefore, different countries and leading scientific centres pool their resources for the development of mega-science. Scientific consortiums are being formed to work on mega-facilities. For any research organization, it is incredibly encouraging to become a member of such a consortium. As a result of long-term multidisciplinary research in Siberia and taking into account the global significance of this region, the Eurasian midpoint, Tomsk State University developed the concept of Western Siberia as a unique natural mega-facility possessing exceptional global significance, attractive to the world scientific community and capable of being a research platform for organizing large network projects and research consortiums (Kirpotin et al., 2018).

The mega-transect approach is a unique concept and infrastructure, which was developed in Tomsk State University for environmental monitoring and landscape-ecological research. The approach can be followed in all seasons, for field sampling, ground-based research on the field stations in combination with remote sensing and ecosystem manipulations. The mega-transect was established at the framework of the concept of Western Siberia as a natural mega-facility, a kind of natural equivalent to CERN's Large Hadron Collider, to attract leading international research groups (Kirpotin et al., 2018).

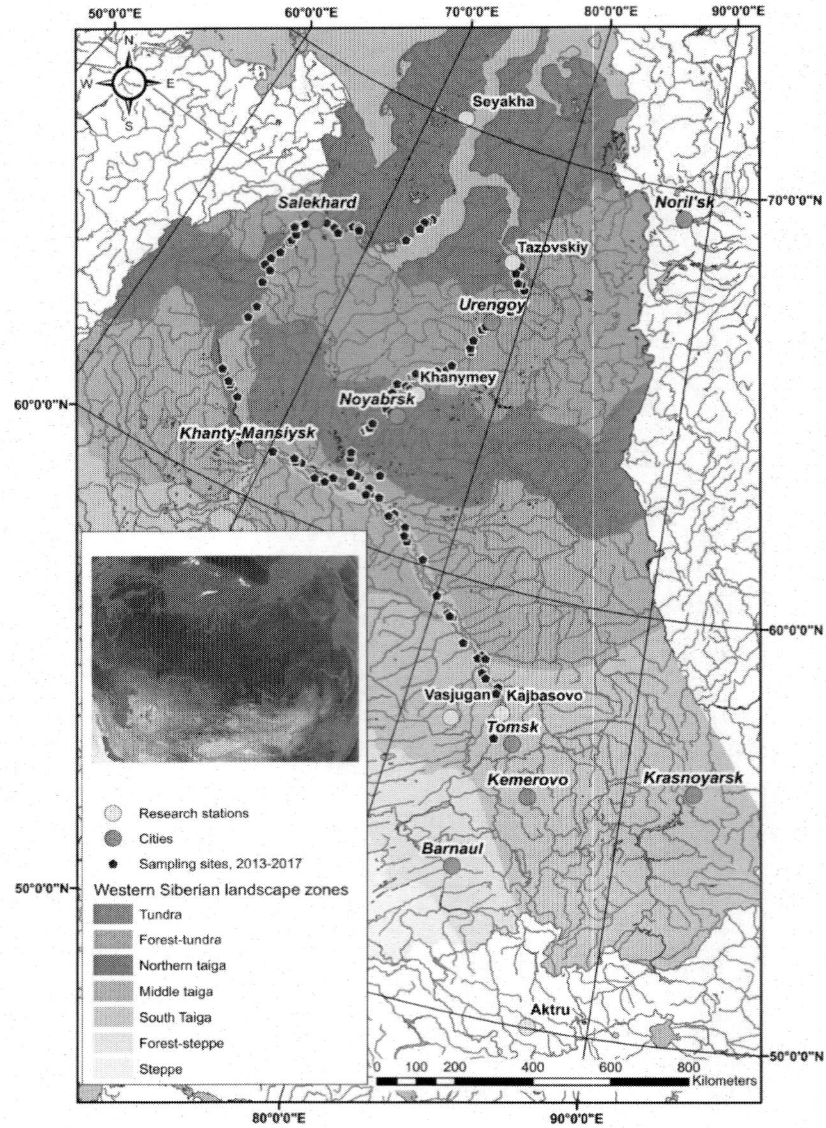

Figure 3. West-Siberian latitudinal mega-transect.

Within the confines of Western Siberia, a unique mega-transect (Figure 3) unparalleled anywhere in the world for conducting surveys, monitoring, sampling, live experiments, and manipulations was founded, extending 2500 km from the Altai Mountains in Southern Siberia at the border with Mongolia to the deep Arctic in Yamal peninsula. A cluster of research stations has been set up along the mega-transect: Aktru (North-Chuya Ridge, South-Eastern Russian Altay), Kaibasovo (the mid-course floodplain of the Ob River), Khanymey (the southern edge of the permafrost zone). Some of the research stations were included in the International Circumpolar network of research stations INTERACT-II, the largest project of the EU's program Horizon 2020. For the future, it is proposed to establish an even more extended trans-meridional mega-transect from the West to the East (about 7500 km) along the gradient of continentality from the Ob river basin to the Kolyma river basin in Yakutia, additionally including the basins of the biggest rivers' estuaries in the Arctic Ocean: Yenisei, Lena and Indigirka. This mega-transect will give a new dimension to the Trans-Siberian Scientific Way project (Kirpotin et al., 2018).

Figure 4. Sergey Kirpotin (left) and Warwick Vincent (Co-chair of T-MOSAiC - right) at Arctic Change 2017, Quebec.

The infrastructure of the West Siberian mega-facility and research mega-transect gives unprecedented opportunities, providing logistics for high-level field research at the framework of large-scale national and international projects (Kirpotin et al., 2018).

In December 2017, the Mega-transect project and successful experience of the TSU Bio-Clim-Land Center in organizing large-scale scientific research based on a mega-transect approach was presented by Sergey Kirpotin for the Steering Committee of the new pan-Arctic programme T-MOSAiC (Terrestrial – Multidisciplinary distributed Observatories for the Study of Arctic Climate), formed at the International Arctic Scientific Committee (IASC) (https://www.iasc.info/images/activities/T-MOSAiC-program.pdf). A similar transect for research, called the Gradient of Northern Territories, is being launched in Canada, but Russia's priority in organizing and successfully operating the world's first mega-transect for research and all-seasons environmental monitoring has been unconditionally recognized and approved by the Steering Committee of T-MOSAiC program as one of the key drivers of its development. In Quebec, a historic docking of two mega-structures: the Siberian latitudinal

Mega-transect and the Canadian gradient of the northern territories was carried out (Figure 4). It was decided to conduct joint ground research on a single agreed conceptual and methodological basis (http://www.tsu.ru/news/uchenyy-tgu-voshel-v-novyy-mezhdunarodny y-arktiche/; http: //www.tsu. com / news / na-severnom-polyuse-sostoitsya-stykovka-megapr ofil /). Today one of the key T-MOSAiC Action groups "Arctic Transects" (Chair: Sergey Kirpotin – Russia) completely based on this methodology (https://www.t-mosaic.com/action-groups.html).

Since its founding, Mega-transect has been focused on large-scale interaction with society, regional authorities and indigenous people on the basis of a trans-disciplinary approach. It actively cooperates in this context with other networks and has conducted several joint activities with SECNet, INTERACT, NEON, T-MOSAiC and NASEC.

3. Search for Ways of Establishing Research Networks' Effective Interaction among Themselves, with Stakeholders and State Authorities in the Territory of Yamal-Nenets Autonomous District

The task of enabling real interaction of networks, various in scale, specificity, and subject specialization, in the territory of the Yamal-Nenets Autonomous Okrug/District seems incredibly difficult. Moreover, it seems completely impossible without the support of the Okrug Government, effective interaction with local and indigenous people, as well as with different groups of stakeholders. At the same time, the main problem is that scientists, as a rule, cannot formulate their recommendations, forecasts and estimates in a form that is convenient and acceptable for officials and managers. In this regard, the discussion of this issue is convenient and correct, to begin consideration of the expectations of the Okrug Government from scientists.

At this point, it is necessary to clarify the nature of the Yamal-Nenets Autonomous Okrug. This is a special, even unique situation that has been developed in comparison with other northern territories of Russia. It is a key region of Russia in terms of oil and especially gas production. The extraction infrastructure is actively developing; new grandiose objects are rapidly being built, such as the biggest LNG plant in Russia. On the other hand, the region is the territory of residence of the largest Nenets population of the indigenous peoples of the north, which may fall outside the category of small peoples. Reindeer herding also creates powerful pressure on the environment; its uncontrolled development leads to overgrazing and degradation of reindeer pastures. At the same time, the interests of large hydrocarbon producing companies and indigenous people are often in conflict. Finally, large scientific projects, both Russian and international, are active in the Okrug. In this situation, the authorities are simply forced to balance between the interests of different actors, to find a compromise and reach a consensus in the search for sustainable development of this complex region.

The Yamal-Nenets Autonomous District occupies 769.3 thousand sq. km, while the number of the resident population as of January 1, 2019, is 541479 people, which is the highest figure compared to other Arctic regions, not only of Russia but also of the World (http: / /tumstat.gks.ru/wps/wcm/connect/rosstat_ts/tumstat/ru/statistics/jnoStat/). Industrial progress in recent decades has contributed to the rapid growth of the population of the district due to the influx of labour migration.

The region is rich in natural resources – oil, gas and condensate, and solid minerals. The sustainable socio-economic development of the Russian Federation is ensured, in many respects, by the functioning of the oil and gas sector of the Yamal-Nenets Autonomous District. This subject of the Russian Federation is the guarantor of energy security not only in our country but also in other countries of Europe and Asia. More than 30 enterprises produce oil and gas condensate, the main oil-producing enterprises in the district are subsidiaries of JSC Gazprom Neft and JSC NK Rosneft. Today, mining accounts for more than 88% of industrial production in the region, however, the economy of the Yamal-Nenets Autonomous District is supported by other main activities: industry, construction, trade, transport and communications, agriculture and forestry, agro-industrial complex (reindeer husbandry, fishing, hunting, processing of fur raw materials, cattle breeding, fur farming, industrial processing of meat and fish). In the largest cities of Yamal – Novy Urengoy, Noyabrsk and Nadym –in the last few years various sectors of the economy, from agriculture to the production of consumer goods, have been successfully developing. Although the enterprises of the oil and gas industry remain the main taxpayers, these municipalities cannot be called single-industry: in each, 5–10 large diverse enterprises are represented. Since 2014, this status has been removed from all cities of the YNAO.

The territory of the district is inhabited by representatives of many peoples, however, it is the indigenous people who give it uniqueness, the Nenets, Selkups, Komi, Zyrians and northern Khanty are among them. This fact attracts additional attention to the region. Indigenous peoples and their habitats are under the protection both of the Russian Federation and regional authorities. Special attention is paid to the protection of the habitat environment of indigenous peoples, and the priorities of the right to use natural resources, which is associated with the active mining of minerals in the region; and the preservation of indigenous identity and culture, which are endangered by the urbanization of the district. The development of indigenous backgammon in different regions of the planet, in general, has become a global problem of our time. An indicator of its relevance and scale is, in particular, the fact that the United Nations has proclaimed the second decade in a row as the decade of the indigenous peoples of the world (Resolution adopted by the General Assembly on December 20, 2004 [on the report of the Third Committee (A / 59/500)], 59/174 The Second International Decade of the World's Indigenous People.

1.96% of the total area of YNAO (1 million 509 thousand hectares) is occupied by 2 PAs of federal significance: the Verkhne-Tazovsky State Reserve (Krasnoselkupsky District, with an area of 631.308 hectares; it is the standard of typical northern taiga) and the State Reserve Gydansky (Tazovsky Region, with an area 878,174 thousand hectares; standard of the Arctic and subarctic tundra). The share of regional protected areas is 8.86% (6 million 815 thousand hectares) of the district's area and includes 10 state nature reserves: Nizhneobsky (142,203.92 hectares), Nadym (562,995.51 hectares), Kunovatsky (2,52860.57 hectares), Poluysky (63,196.00 hectares), Verkhne-Poluysky (195,322.00 hectares), Sobty-Yugansky (358,429.00 hectares), Yamalsky (4,113,685.70 hectares), Pyakolsky (438,560.00 hectares), Messo-Yakhinsky (86,033.00 ha), Syn-Vojkarsky (292,049.00 ha); 1 natural park Polyarno-Uralsky (310,069.67 hectares); 1 geological monument of nature Harbei (563.62 hectares). All PAs come under the jurisdiction of the Department of Natural Resources Regulation, Forest Relations and the Development of the Oil and Gas Complex in the Yamal-Nenets

Autonomous District. The listed territories of the natural reserve fund of the YNAO are completely or partially withdrawn from economic use for the preservation of unique natural systems and certain types of organisms.

The territory of the region is located in the extreme natural-climatic zone of the Extreme North and the Arctic Circle, which directly complicates the livelihoods of people and influences the development of production and social infrastructure. The increasing variability of climatic conditions is manifested in recent decades in the increase of climate extremes and the frequency of occurrence of dangerous meteorological and adverse hydrometeorological phenomena, initiating natural and man-made disasters. In about 60% of cases, natural and technological accidents are related to climatic factors. One of the reasons is the degradation of permafrost, leading to the deformation of boreholes, oil and gas pipelines, transport and social infrastructure of the region. (https://bookucheba.com/meteorologiya-klimatologiya-bu/vliyanie-izmeneniy-klimata-razvitie-otdelnyih-7045.html).

As a consequence of the above features of the Yamal-Nenets Autonomous District, another issue arises: the interest of the world community and the world scientific community in the current situation in the region and the prospects for its sustainable development.

Ensuring the sustainable development of the Arctic zone of the Russian Federation (AZRF), dictates the need to study the fundamental laws of evolution and the current state of the structures and processes of spatial development of the Russian Arctic, the development and scientific substantiation of the methodology, methods and practical approaches to the formation of alternative (innovative) models of spatial ecological and economic development of coastal continental territories, including urban development, adaptation of existing management methods resource and environmental activities (Detter, 2017).

Sustainable development can only be realized in practice when its phased plan is presented. The difficulties of drawing up a plan for the sustainable development of the YNAO, taking into account all the above facts, especially in the context of climate change and the lack of adaptation to them of infrastructures and the population of the YNAO, ensuring normal functioning and a comfortable lifestyle are obvious.

Realizing this, the Russian government decided to supportthe development of the scientific sector in the region. The need for the presence of science in the Arctic is caused not only by the need to solve numerous local problems but also by the need to integrate Russian science into the world professional community. No country is able to solve the problem of developing the Arctic alone. Russia is one of the leading players. By virtue of at least geographical, territorial proximity and, as a result, the influence of the Arctic on our country – from climate to security. In addition, practice and accumulated experience inevitably lead to the conclusion that it is necessary to implement an integrated interdisciplinary approach in this work. (http://www.meteorf.ru/upload/iblock/fc8/Izmenenie_klimata_N76_DecJan_2018_19.pdf). (http://www.mid.ru/vnesneekonomiceskie-svazi-sub-ektov-rossijskoj-federacii/-/asset_publisher/ykggrK2nCl8c/content/id/128534) (http://gfi89.uralfo.ru/soch-ekon/).

To provide scientific support for the development of such geographically isolated regions as the Yamal-Nenets Autonomous District, regional scientific centres were periodically created, whose work, as a rule, was aimed at solving narrow-local problems. So, for example, in 1932, the Yamal Zonal Vegetable Experimental Station was established at the V.I. Lenin All-Union Academy of Agricultural Sciences (since 1938 renamed the Salekhard complex zone experimental station of the Institute of Polar Agriculture, Livestock and Commercial Farming). Experimental works in the field of crop production were carried out at the station:

the development of agrotechnical measures, selection, breeding of new valuable varieties of grain, vegetable and fodder crops until 2007, when scientific work was gradually curtailed. In 2019, in the framework of the subprogram "Development of potato breeding and seed production in the Russian Federation" of the Federal Scientific and Technical Program for the Development of Agriculture for 2017–2025, the Federal Research Center "Tyumen Research Center" of the Siberian Branch of the Russian Academy of Sciences plans to revive the research work on the production of virus-free seed potatoes for the purpose of further reducing the dependence of agricultural producers of potatoes in the Russian Federation on foreign seed material and increasing the crop yield.

Currently, GKU YNAO "Scientific Center for the Study of the Arctic" (Salekhard, Nadym) is one of the most important regional scientific institutions. Researchers in the natural sciences (environmentalists, geologists, soil scientists, zoologists, botanists, physicians, landscape scientists) comprehensively study the possibility of maintaining a favourable environmental situation in the YNAO in the conditions of industrial development of the territories.

A number of joint research projects are being carried out with specialized academic institutions aimed at studying: the transformation of the cryolite zone and the spatial and temporal variability of the glaciers of the Polar Urals in a changing climate, such as studying the transformation of the cryolitezone under the influence of climate change and anthropogenic load on the basis of the monitoring network of the Gydan peninsula; creation of an integrated system for automated geotechnical monitoring of temperatures of permafrost soils in the city of Salekhard; monitoring the depth of the seasonally thawed layer and landscape research at the CALM site; modern climate change, glaciation and monitoring of the glaciers of the Polar Urals.

Within the framework of the integrated environmental monitoring of the living territory of the small indigenous peoples of the North and environmental assessment, research is being carried out on chemical, physical and biological measurements (observations) in various components of the YNAO ecosystems. Based on the results, the “Program of Integrated Scientific Environmental Monitoring in the Yamal-Nenets Autonomous Area” was developed (authors: E.N. Morgun, R.A. Kolesnikov, E.V. Agbalian, edited by A. Titovskiy).

Research is being conducted to study the health condition of representatives of the indigenous peoples of the Northleading a traditional lifestyle, studying the health of the indigenous people of the Arctic, nutrition and food safety of the inhabitants of the Arctic. Recommendations were developed to improve the health and active longevity of the indigenous people of the Arctic, enhance food security, taking into account the advantages of traditional nutrition of the inhabitants of the Arctic zone (Bogdanova, Lobanov, Popov, 2017; Bogdanova et al., 2018).

Together with scientists from academic institutions and leading universities, researchers from the GKU YNAO "Scientific Center for the Study of the Arctic" continued work aimed at obtaining a geobotanical assessment of reindeer pastures in several districts of the YNAO (Yamal, Ural regions). The state of plant and soil covers of the territory has been studied. A geobotanical map of pasture capacity has been compiled.

On the basis of the initial information obtained, work is underway to create an up-to-date database for the formation of a holistic picture of ideas about the features, factors and mechanisms for the development of negative processes in natural environments.

Thus, the GKU YNAO "Scientific Center for the Study of the Arctic" (Salekhard, Nadym), along with district educational organizations, enterprises, authorities and local governments helped to solve the problems of integrated development of the YNAO territory.

Currently, in order to implement a unified state policy in the field of scientific and technological activities, the Government of the Yamal-Nenets Autonomous District has concluded a number of cooperation agreements with the Russian Academy of Sciences, the Ural Branch of the Russian Academy of Sciences, and the Federal Agency of Scientific Organizations. The Department of External Relations of the Yamal-Nenets Autonomous District interacts with scientists from academic institutions, leading universities and other scientific organizations in the country and the world.

For example, in the scientific stations of the Yamal-Nenets Autonomous Okrug (Bely, Sabetta, Zeleny Yar, Parisento, Yerkuta, Laborova, Nadymsky, etc.) only in the field season of 2018 about 80 scientists from Russia, Norway, France, Canada, Poland, Bulgaria, Germany and New Zealand in total spent about 1000 working days there. The expedition was supported by the Department for Science and Innovation of the Yamal-Nenets Autonomous Okrug, the Scientific Center for the Study of the Arctic, the Russian Center for the Development of the Arctic, Gazpromtrans, and Yamal LNG. The research is carried out in the framework of grants from the Russian Foundation for Basic Research-Yamal, the Russian Foundation for Basic Research-Arctic, SPbSU, as well as international projects with the participation of the University of Tromso (Norway), the University of Giessen (Germany) and the INTERACT project.

Scientific environmental studies in the YNAO are conducted by specialized research institutes, Moscow State University, Tomsk State University, Tyumen State University, St. Petersburg State University, Institute of Plant and Animal Ecology, Ural Branch of the Russian Academy of Sciences, Institute of Earth Cryosphere of the Siberian Branch of the Russian Academy of Sciences, the A.N. Severtsev Institute of Ecology and Evolution RAS and others. The range of observations is planned to be continued and expanded.

Due to the specificity of the research problems to be solved, ensuring access to representative scientific stations of the YNAO of an even wider range of scientists is of particular importance. In this connection, there is an urgent need to create a network of supporting stations in the YNAO in order to coordinate the activities of research teams from Russian and foreign research institutes and universities in the framework of integrated scientific environmental monitoring in the following areas: climatology and climate change; environmental science, pollution; geology and sedimentology; geophysics; permafrost and geomorphology; soil science; terrestrial biology and ecology; anthropology, sociology and archeology.

The creation of a network should be focused on solving such priority scientific tasks as organizing collective use for conducting scientific research at environmental scientific stations in order to coordinate, increase the efficiency of their use and develop science in the Yamal-Nenets Autonomous Okrug and the Russian Arctic; research work and the provision of services aimed at external customers, conducting joint research in cooperation with Russian and foreign organizations, the creation of conditions for securing talented young people in the field of science, participation in the training of specialists and highly qualified personnel (undergraduate and graduate students).

A significant increase in network infrastructure is planned, which is expected to make it a unique platform for international and circumpolar cooperation. The network of basic stations of the YNAO must form a new model – a unified research network of well-equipped basic scientific ecological stations that optimally represent the ecosystems of the Arctic, subarctic and northern taiga zones of the West Siberian Plain and are convenient for conducting research on various programs.

Formation of the regional policy of the Yamal-Nenets Autonomous District for the management of anthropogenic pollution (for example, acidification of precipitation, climate change, pollution of soil and water with heavy metals, etc.), including reduction of industrial emissions, solid household waste, requires reliabledata on the results of integrated scientific environmental monitoring. In this connection, the conducting of scientifically based system observations should be carried out at basic and additional scientific stations, followed by modelling the results of the study of ecosystem processes.

Optimization of the network of supporting scientific ecological stations is necessary to obtain scientifically based, stable time series of various variables characterizing natural and natural-anthropogenic ecosystems of the YNAO, therefore the search for means to establish effective interaction between different research networks among independent and state bodies in the YNAO is additionally of urgency.

Due to the fact that a number of factors (poor logistics, low representativeness of sampling points, extreme climatic conditions, the presence of ice phenomena for 8-10 months a year, focal nature of the industrial and economic development of the YNAO territory) is the most limiting for the optimal choice of research sites by visiting research teams, their placement, as a rule, has been historically predetermined by localization in settlements with a more developed infrastructure. In addition, the actions of various scientific institutions among themselves were often uncoordinated. In the same territory, scientists from different institutes may conduct identical research, while other territories remain “blank spots.” The need to coordinate research work on the study of the ecosystems of the Yamal-Nenets Autonomous Area additionally makes it necessary to create a network of scientific stations for monitoring in landscapes with different qualitative and quantitative anthropogenic loads. Concerted actions will significantly reduce labor and financial costs.

Ecosystem monitoring in the Yamal-Nenets Autonomous District should be carried out through annual integrated monitoring of the floristic and faunistic composition, periodic inventory of flora, mycobiota, plant communities, fauna and animal populations, soil and landscape observations. The required inventory work as background and adjacent territories. In this connection, the territories of the reference (background) and anthropogenic landscapes of the Yamal-Nenets Autonomous District should cover the network of reference monitoring sites (scientific stations).

Each scientific station must necessarily include a catchment area sufficient to cover all the interacting components: the atmosphere, vegetation, soil, bedrock, groundwater, stream, river or lake, i.e., terrestrial ecosystem and reservoir biocenosis. At the same time, the ecosystem is traditionally regarded as a complex of living organisms (including humans) interacting with each other and with their habitat: air, soil, and water.

The general network implies the identification of base scientific stations where intensive environmental monitoring should be carried out continuously, the results of which will predict changes in the state of ecosystems of the district. There should be no more than 10-15 such representative stations.

In addition, the network also offers additional research stations (polygons), where sampling can be carried out rarely (short-term). They represent an additional source of information for comparing assessments with regard to climatic, ecological, and geological zoning. These data are necessary for model verification and hypothesis testing, but at the same time, short-term monitoring alone does not provide all the information sufficient to detect anthropogenic effects and to distinguish them from natural trends. Differentiation of basic and additional scientific stations increases the output of scientific products in relation to the costs of maintaining scientific stations.

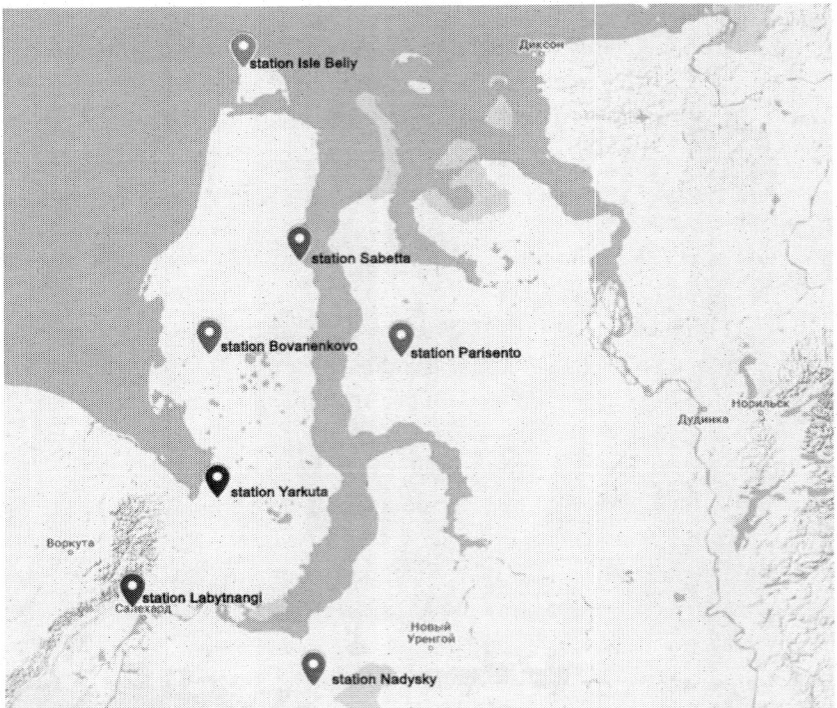

Figure 5. Existing regional research stations in YNAO.

In the structure of integrated scientific environmental monitoring, it is necessary to allocate several scientific subprograms. The most important task of such monitoring is to collect information for a comparative analysis of nutrient cycling in natural and anthropogenically transformed landscapes. To this end, it is proposed to identify a range of parameters (and indices) characterizing abiotic and biotic properties and directly affecting their ability to accumulate nutrients and stability at all stations. As a result, the list of abiotic properties includes the following characteristics: infiltration rate, runoff value, erosion hazard, evaporation soil moisture loss, soil colloids, leaching loss, soil temperature, soil density, etc. The internal circulation carried out by plants, synchronization of activity of plants and microorganisms, the diversity of biological activity over time, the diversity of plant populations, genetic diversity, potential reproduction, biological productivity.

Scientific subprograms should contain a unified methodology for chemical, physical and biological measurements (observations) for the same basic parameters in different components of ecosystems at the same time and at the same scientific stations for a long period of time. Quantitative assessment of the basic parameters, as well as monitoring the rate of their changes, is necessary for the development of the environmentally oriented policy of the YNAO.

Nowadays, on the basis of archival and cartographic materials, the natural and economic matrix, taking into account the landscape, topography, vegetation and soil cover on the territory, logistics features in the YNAO, scientific stations have already been allocated, which partly successfully represent the main parts of the natural-anthropogenic series.

The main scientific stations are Bely Island Station, a station near the Sabetta camp, a station in the Bovanenkovo area "Vaskina Dachi", a station in the Parisento area, the Erkuta station, a station near the city of Labytnangi, and the Nadym geocryological station. The base scientific stations have historically established names, many years of research have been and are being conducted on them, which makes it possible to consider them to be basic (Figure 5).

Unfortunately, until recently, not all studies at these stations were coordinated, and not all are subject to the ecosystem principle. Often, research topics are duplicated by different scientific institutions, and a number of the results obtained do not fall into the open-access of the scientific community at all.

Additional scientific stations are: "Kharp", "Laborovaya", "Purovsky-1", "Purovsky-2", "Tazovsky", Natural Park "Polar-Ural", State Nature Reserve of Regional Value "Nadym." These stations were allocated for environmental monitoring by the scientists of the GKU YNAO "Scientific Center for the Study of the Arctic" and will provide primary information about the state of ecosystems.

It is planned to organize a new research station in the Seyakha settlement to conduct long-term integrated environmental and biogeochemical monitoring of tundra landscapes of Yamal as a scientific basis for the sustainable development of the region jointly with the National Research Tomsk State University, the Siberian Environmental Change Network (SECNet) and the State Scientific Center for the Study of the Arctic.

The Seyakha station will allow organizing a world-class platform for conducting complex atmospheric and ground-based research, including the coastal zone, in close cooperation with the largest pan-arctic programs and networks, such as T-MOSAiC, PEEX, INTERACT, IPA, NEON, SecNet and by others.

It is obvious that the network integrated scientific ecological work declared here in the form of technological breakthrough and qualitative growth of the human environment for the future will be problematic without both state support of environmental research and joint scientific research of various scientific teams (including international ones). It is also obvious that such work should be carried out taking into account the interests of local residents, including small indigenous peoples. The interests of the fragile and particularly valuable natural environment of the Arctic region should also be taken into account, moreover, they are priorities. We have just identified a stakeholder circle. And this is the first necessity when using the transdisciplinary approach in the implementation of the above objectives.

The infrastructure basis for carrying out comprehensive environmental studies and year-round environmental monitoring will be unparalleled in the world, developed at the Center "Transsiberian Scientific Path" at NSTU will be the Mega-transect, 2500 km long (Kirpotin et al., 2018), which is the axis of the organization of network interactions in Siberia. In 2018, the mega-transect was designed and registered as a "unique scientific installation (USI) of the Russian Federation." The university "mega-facility" unites TSU research stations located in Western Siberia, mobile research laboratories based on automobiles and water transport, monitoring sites and regular sampling points while ensuring comparability of information, as well as infrastructure and highly qualified scientific support, collecting, storing and data processing. "USI is registered on the portal Scientific and technological infrastructure of the Russian Federation." http://ckp-rf.ru/usu/586718/.

The "mega-transect approach" allows researchers to implement simultaneously several interrelated and complementary large-scale international projects with optimal logistics, the combination of tasks and interests of which leads to synergy and a better understanding of the behavior of complex natural systems. Such an approach leads to real integration, cooperation and interaction within the framework of modern science without borders.

The integrated research program proposed at the mega-transect stations will focus on the need to understand, predict and control the effects of climatic and anthropogenic impacts on vulnerable Arctic ecosystems and the population due to the balance and coordination between field research, including paleo-, historical and environmental data collection, monitoring, carrying out field experiments, modelling and forecasting, as well as sociological research.

The program aims to identify, understand, and predict climate and human-induced changes in vulnerable ecosystems and landscapes of the Arctic at a level that could have a significant impact on mitigating the negative effects of these changes, and ideally improving the quality of life for people in the region and beyond.

The program is based on a combined, multidisciplinary approach to understanding the modern evolution of the main components of the landscape of Western Siberia. A comprehensive program, unified for all stations, is proposed, including remote sensing methods for environmental studies, landscape-ecological monitoring studies, meteorological observations, geochemical and microbiological studies of natural environments, conducting field experiments and experimental modelling, as well as methods of socio-humanitarian sciences. When conducting research at all stations, the common conceptual methodology "4M" (monitoring, manipulation, modelling, management), which was developed by Prof. Terry Callaghan and published in AMBIO magazine in 2012 (Johansson et al., 2012; https: //www.ncbi.nlm.nih), will be applied.

Data collection methods will be standardized for various research stations and mobile laboratories. Comprehensive data, spatial extent and remote sensing technologies will allow the user community to solve sensitive issues on a scale inaccessible to previous generations of environmentalists.

When implementing the Program, we consider two groups of tasks.

Scientific:

1) identification of drivers of changes in vulnerable ecosystems: past, current and potential, through a combination of monitoring methods and experimental manipulations (field experiments);
2) assessment of climatic and anthropogenic impacts on vulnerable ecosystems and identification of feedbacks of ecosystem changes with the main drivers, through changes in biodiversity, albedo, greenhouse gas emissions, dynamics of biogeochemical processes and other indicators;
3) assessment of the effects of current climatic and anthropogenic impacts on the well-being of the population of the YNAO and the Arctic zone of the Russian Federation;
4) identification and assessment of the effects of changes in the environmental management system on the state of vulnerable ecosystems and the identification of

potential flexible ways to adapt ecosystems and local people to regional and global challenges.

Strategic:

1) combining the scientific potential of different regions of Russia to align the initially different conditions of the state of science in them and create favorable opportunities for the breakthrough development of science in the actively developing "resource"regions of the country, with prospects for their release to the world level;
2) the creation of additional conditions for the diversification of the economic development of the "resource" regions of Russia through the improvement and development of the scientific and educational potential of these regions;
3) combining interdisciplinary and transdisciplinary approaches in research and education;
4) providing a "single point of entry" to obtain information about the Region and the Russian Arctic as a whole;
5) dissemination of acquired knowledge in the educational and scientific community, among the population, and in the public and political environment;
6) improving the perception of the region in the world arena in order to strengthen the geopolitical status of Russia.

The main areas of work.

In order to concentrate efforts, obtain relevant and synergistic results from research in different climatic zones by specialists with different scientific specializations and different technical skills, it is proposed to focus on a minimum monitoring program that includes the following main areas:

1) Meteorological monitoring;
2) Monitoring of atmospheric chemistry, including greenhouse gases;
3) Biogeochemical monitoring;
4) Hydrological and hydrochemical monitoring;
5) Landscape monitoring, including flora and fauna monitoring and microbiological monitoring;
6) Social monitoring.

Let us write in more detail on each of these areas.

1. Meteorological monitoring.

Most of the landscape and ecosystem functions and processes are to some extent related to climate/weather. The only way to obtain climate data from a specific area is to operate the climate station for a long period. Climate stations operated by national weather services usually only record climate parameters relevant to weather forecasting. As a result, it is desirable for the research station to manage its own climate complex, which has the most relevant sensor configuration. Therefore, the climate station should have at least the following sensors, which measure the following parameters:

- Air temperature;
- Air humidity;
- Wind speed and direction;
- Air pressure;
- Precipitation;
- Shortwave incoming radiation;
- Shortwave outgoing radiation;
- Clean radiation;
- Ultraviolet radiation;
- The power of snow cover;
- Soil temperature (at different depths);
- Soil moisture (at different depths);
- Levels of surface, permafrost and underground waters

It is necessary to install autonomous modifications of the IMCES meteorological complex of the SB RAS AMK-03 (or equivalent) with a remote data transmission and assimilation system at the research stations of the Yamal-Nenets Autonomous Area.

2. Monitoring of atmospheric chemistry, including greenhouse gases. A program has been formed with the following priorities in measuring air quality and snow cover:

- Black carbon;
- Heavy metals and nutrient salts, including those present in the snow;
- Long-term persistent organic pollutants, including those present in the snow;
- Persistent organic pollution present in the snow;
- Carbon dioxide;
- Nitrogen oxides.

3. Biogeochemical monitoring

Biogeochemical monitoring requires specialized tools, which, as a rule, are not trivial, constant laboratory work and modern data processing.

Priority should be given to monitoring:

- The overall level of nutrients in the soil;
- Carbon balance/its change.

There are many different ways to measure the exchange of greenhouse gases between soils and the atmosphere, both on the scale of a single site and on a landscape scale. For a minimum monitoring program, it is recommended to measure the landscape scale CO_2 balance on the basis of the so-called vortex covariance method.

The toolbox consists of a climate system equipped with several sensors, including an acoustic anemometer for measuring turbulence, and a so-called Li-COR device (or its equivalents) for measuring CO_2 concentrations.

In addition, the project involves the study of the dynamics of ecosystem productivity – the definition of the stock of phytomass, mortars and net primary production. To determine the intensity of destruction of plant residues in the experimental plots, the method of plant vegetation in peat will be used.

4. Hydrological and hydrochemical monitoring

Most of the studies devoted to the problems of "landscape-scale" use catchment basins as studied landscape units. This is due to the fact that the catchment areas are relatively limited systems that allow you to establish a "balance" of what goes into it and what comes out, for example, water balance and carbon balance. If the research station is located near the river, then an appropriate drainage basin is selected as the object for the study of landscape scale, and the research station should install the appropriate instruments necessary for long-term monitoring of the catchment basin.

At a minimum, the catchment basin will be equipped with both automatic sensors and equipment for manual measurements:

- atmospheric precipitation;
- water levels;
- the level of pH (pH);
- electrical conductivity;
- dissolved oxygen and carbon dioxide;
- CO_2 emissions;
- elements-nutrients.

Some of the basic measurements mentioned above can be performed automatically, although manual measurements of river water discharge are necessary to establish the so-called Q/h relationship between the water level in the river and the flow of river water ('Q'). The thickness of the snow cover will also be determined by the installed sensors.

In addition to the proposed hydrological monitoring, the following will be observed:

- Migration of suspended precipitation;
- Migration of organic substances;
- Migration of the solute.

When conducting hydrological studies, we will adhere to the recommendations of the "Manuals on Hydrological Stations and Posts." In the course of the work, hydrodynamic grids of the runoff lines will be constructed and the components of the water balance of the territories will be calculated.

5. Landscape monitoring and monitoring of biodiversity of flora, fauna and microbiological monitoring

Monitoring the fauna usually requires qualified staff with zoological competencies. In addition, conducting observations, for example, species phenology and species distribution, as a rule, takes a long time and requires many hours of fieldwork. However, some observations can be made quite easily, even if the staff does not have sufficiently specialized skills. It is assumed that the personnel of research stations should be able, at a minimum, to carry out the following monitoring of vegetation:

- Phenology of characteristic species in the corresponding characteristic vegetation zones.
- Geobotanical survey of vegetation, both at the scale of small key areas and at the scale of the landscape, using either a conventional digital camera, or a special camera with “false color,” or using satellite technologies.
- Display of vegetation zones at equal intervals of time either on maps of vegetation or along high-altitude cuts.

Many research stations also keep lists of random observations of the local fauna made by staff and invited scientists during their stay in the field. The following minimum set of observations is proposed:

- Lists of mammals, birds and other taxonomic groups that can be identified by station personnel or invited scientists.
- The first emergence of various species of birds, regularly migrating into / or through the area during the season.
- The first appearance during the season of regularly occurring breeding species (if any).

If the permanent station staff includes zoologists, this minimum set can be monitored:

- Imprints of traces on the trails, to count the species of mammals and birds characteristic of the area.
- Catching arthropods.

During the implementation of the program, it is possible to identify the main patterns of changes in microbial coenoses in the studied landscapes. Studying the diversity of microbial communities should involve using both traditional methods (identification of ecological-trophic groups) and methods of molecular biology.

A scientific group of radiophysicists at TSU is developing methods, equipment and software for assessing the ecological state of the water are based on the study of particles, dimensional, species information and behavioral activity of zooplankton. This will allow the organization of annual (seasonal) monitoring of plankton and settling particles in the study area in order to assess bio-resources and early predict the ecological status of the water area. To solve these problems, it is proposed to use a line of sensors based on DHC technology (Digital Holographic Camera), a measurement technology that provides registration, recognition and classification of particles suspended in water of various origins, including biological ones.

6. Social monitoring.

The main methods of sociological research, in the framework of expeditions using the infrastructure of Mega-transect-TSSW, will be a sociological survey, non-formalized interview and observation, and the document analysis method will be used as an auxiliary method. A non-formal interview is a free conversation with the most competent informants in a particular territory. This technique is the most flexible in the conditions of a small amount of preliminary information about the object under study. As a result of the social monitoring, new data will be obtained on the life of the indigenous peoples of Siberia, taking into account the current socio-ecological state; new information about the traditional knowledge of the indigenous population on environmental changes to correct current problems of fundamental research and search for new promising areas of research; new data on local practices and new knowledge arising regardomg local communities' response to environmental changes; new information about the response of society to environmental change as a long-term trend, including not only the scientific description of changes based on the natural sciences, not only the design and introduction of new technologies and infrastructures, but also changes in lifestyle and behavioral patterns of both "indigenous" and industrial communities. A system of indicators will be created to track the effects of environmental change on the lives of human communities.

The design of a complete landscape monitoring system is a large task, which will be developed further, in the case of obtaining adequate funding necessary for the implementation of the Program.

Based on the landscape monitoring, thematic maps or other spatial information will be compiled that are absolutely necessary for most scientists visiting research stations.

In order to establish a more accurate timeline of the changes that have occurred, the method of decoding multi-temporal space images obtained on these territories in the period 1990-2019 will be used. Operational surveys of the studied landscapes in the annual dynamics, including thermal imaging and NDVI, will be enabled with the means of unmanned aerial vehicles.

At the same time, long-term landscape-ecological studies of the area will be the basis for the implementation of diverse and diverse scientific tasks of the Program.

The Program also assumes that the stations of the mega-transect will carry out experimental manipulations with ecosystems and their components, which are powerful alternative tools for elucidating causal relationships and predicting ecosystem responses to global climate change. It is planned to createinfrastructure for carrying out multifactorial full-scale experiments, for example:

- Heating, cooling and fertilization of the soil and changes in the hydrological regime in test areas in areas of seasonal, continuous and intermittent permafrost;
- Changes in the thickness of the snow cover in different natural areas;
- Fertilization of water of the lakes with decomposition of vegetation, ashes after fires in different natural areas;
- Testing the effect of fire on productivity, renewal of vegetation, biodiversity, intensity of gas exchange with greenhouse gases with the atmosphere, emission of volatile components into the atmosphere, and runoff into the river network from the catchment of organic carbon, macro components and elements-biogenesis;

- Removal of the upper organogenic horizons and individual dominants in the vegetation cover;
- Forced cryoturbation of the soil mass.

The design of experiments will also be developed further, in the case of adequate funding for the Research Program.

When processing the information received by Mega-transect-TSSW we will use the supercomputer information and computing infrastructure of the TSU, which includes: terrestrial satellite station networks; b) supercomputer SKIF Cyberia cluster with a capacity of 100 teraflops; c) a data centre with a storage system of 300 TB; d) Aerospace Earth Monitoring Center – a high-performance hardware and software system for processing spatial data on the state of the environment, predicting the dynamics of natural and man-made processes. The project assumes the creation and filling of a single geo-analytical database of monitoring studies and field experiments, including geodata, metadata (geo-referencing, height above sea level, etc.), local data (characteristics of each experimental area, initial physical and chemical properties of soils and permafrost soils, the status of vegetation and soil biota), operational data (date, meteorological data, specific data from field experiments, etc.). Based on the data obtained, it is planned to develop a model of the processes occurring in Siberia under the influence of climatic and anthropogenic changes.

In addition, the system will include an open ethnic-demographic database on ethnic groups in Siberia; electronic database of sacred places of small ethnic groups of Siberia (identification, mapping, fixation of objects using 3D technology, recording of folklore materials); a database on the effects of industrial facilities on the cultural patterns of indigenous ethnic groups in Siberia, the fixation and content analysis of narratives that characterize indigenous ideas about the impact of industrial development and climate change on their lifestyle; a "Map of Communities" of Western Siberia, including descriptions of the most effective teams and social practices for the development of the study areas, the bases of identity and available resources.

In the process of implementing such ambitious goals, the optimal solution may be to use a transdisciplinary approach. All stakeholders feel an urgent need for dialogue, building optimal relations, however, this process does not bring the desired results –an improving intervention. The methods of the transdisciplinary approach, contributing to the establishment of strong non-hierarchical relations between all groups of stakeholders, fully meet the needs of the region.

An important milestone in the development of a real dialog between science and society was the scientific seminar "Human well-being in the Arctic in a changing climate," held in April 2019 in the city of Nadym, YNAO. The participants were reindeer herders of the Nadym, Yamal, Taz and Krasnoselkupsky districts, representatives of the government of the Yamal-Nenets Autonomous District and the administration of the Nadym region. Specialists from the enterprises of PJSC Gazprom, who conduct production activities in the region, as well as representatives of universities: Tomsk State University, Moscow State University, Tyumen State University, Lomonosov Northern Federal Arctic University(Arkhangelsk).

The workshop announced the start of a completely new project on climate change monitoring for the country. For this purpose, it is planned to create a large-scale monitoring network in Yamal and to introduce representatives of the indigenous population to data collection. Particularly interesting are the factors that affect the health, economic activity, economic and social well-being of the Arctic population. It is necessary to learn how to predict and level them.

Figure 6. Prof. Terry Callaghan (in national clothes in the centre) among participants of the Nadym seminar (photo by Olga Shaduiko).

Seminar participant Prof. Terry Callaghan (Figure 6), an outstanding Arctic explorer, noted that the climate is seriously changing not only in Russia but throughout the entire Arctic zone of the Earth. "The participation of ordinary people in our project," said Prof. Callaghan, "whose daily life is associated with observations of climatic phenomena and weather changes, could significantly complement the picture of observations made by meteorological centres and stations."

If we talk about the technical features of the project, the scientists plan to create a portal accessible to public observers, where people can transmit data in real-time using, for example, mobile phones. Further, this information will already be processed by scientists. They will analyze it, systematize and compare it with the data of satellites and meteorological stations http://nadym-worker.ru/articles/nobelevskiy-laureat-provyol-v-nadyme-chumovoy-seminar.

CONCLUSION

The extensive material cited in the Chapter, based on the analysis, reflection and synthesis of some practices of scientific stations and networks, as well as a truly unique situation in terms of opportunities for developing large-scale and diverse interaction of science and society in the Yamal-Nenets Autonomous District convincingly demonstrated that, without building a professional interest dialogue between science, government and society (different groups of stakeholders), development of network research, even in the case of the most modern and sophisticated methodologists of their agreed programs and scientific protocols will not be successful, efficient and fully beneficial to the population. In today's world, large networked research projects should ultimately be aimed at improving the quality of life of people in vulnerable regions of the planet, such as the Arctic, and far beyond.

The Yamal-Nenets Autonomous District is the current most convenient platform for testing and developing the principles of transdisciplinarity, with subsequent replication of this experience, if successful, both in other regions of the Arctic zone of the Russian Federation and in other Arctic countries.

Acknowledgments

This research was supported by the RF Federal Target Program, project RFMEFI58717X0036. We also express gratitude to large-scale research facilities "System of experimental bases located along the latitudinal gradient" for fruitful cooperation (http://ckp-rf.ru/usu/586718/).

References

Ackoff, R. *The Art of Problem Solving*. Ripol Classic, 1982: 220. (in Russian).

Ala-aho, P.; Soulsby, C.; Pokrovsky, O. S.; Kirpotin, S. N.; Karlsson, J.; Serikova, S.; Manasypov, R.; Lim, A.; Krickov, I.; Kolesnichenko, L. G.; Laudon, H.; Tetzlaff, D. 2018. Permafrost and lakes control river isotope composition across a boreal Arctic transect in the Western Siberian lowlands. *Environmental Research Letters*, Vol. 13, Paper № 034028, DOI: 10.1088/1748-9326/aaa4fe; Q1, Impact factor WoS 4,404.

Alekseychik, P., Lappalainen, H. K., Petäjä, T., Zaitseva, N., Heimann, M., Laurila, T., Lihavainen, H., Asmi, E., Arshinov, M., Shevchenko, V., Makshtas, A., Dubtsov, S., Mikhailov, E., Lapshina, E., Kirpotin, S., Kurbatova, Y., Ding, A., Guo, H., Park, S., Lavric, J. V., Reum, F., Panov, A., Prokushkin, A., Kulmala, M. 2016. Ground-based station network in Arctic and Subarctic Eurasia: an overview. *Geography Environment Sustainability*, 2 (09): 75-88.

Allison, I., Be´land, M., Alverson, K., Bell, R., Carlson, D., Danell, K., Ellis-Evans, C., Fahrbach, E., Fanta, E., Fujii, Y., Glaser, G., Goldfarb, L., Hovelsrud, G., Huber, J., Kotlyakov, V., Krupnik, I., Lopez-Martinez, J., Mohr, T., Qin, D., Rachold, V., Rapley, C., Rogne, O., Sarukhanian, E., Summerhayes, C., & Xiao, C. 2007. The scope of science for the International Polar Year 2007–2008, WMO/TD, No.1364: 1–79.

Asmolov, A. G., Chernorizov, A., Schechter, E. D. 2013. The Historical-Evolutionary Synthesis: Mutual Aid as a Factor of Evolution. *Voprosy psikhologii*, https://www.researchgate.net/publication/289465516_Historical_and_evolutionary_synthesis_Mutual_assistance_as_a_factor_of_evolution (in Russian).

Barr W. 2008. The expeditions of the first International Polar Year 1882–83. Calgary: Arctic Institute of North America: 417.

Barzun, Jacques. 1958. Darwin, Marx, Wagner: Critique of a heritage - 2. rev. ed.: Garden City (N.Y.): Doubleday, - XX: 373.

Beck U. 1992. *Risk Society: Towards a New Modernity*. London.

Bogdanova, E. N., Lobanov A. A., Popov A. I. 2017. Socio-economic policy of the Russian Federation in the field of health protection: calling out forecasts. *Economics and Entrepreneurship*. 9–3 (86–3): 862–868.

Bogdanova, E. N., Zadorin M. Yu., Lobanov A. A., Popov A. I., Andronov S. V., Kochkin R. A. 2018. *The right to the protection of the health of the indigenous peoples of the Russian Arctic: the "collision field" and the search for new solutions*. Bulletin of Tomsk State University. 432: 210-217. (in Russian).

Brown, L. R. 1981. *The Death of Nature: Women, Ecology, and the Scientific Revolution.* Carolyn Merchant Building a Sustainable Society. New York and London: Norton: 433.

Brown, J., O. J. Ferrians, Jr., J. A. Heginbottom, and E. S. Melnikov, eds. 1997. Circum-Arctic map of permafrost and ground-ice conditions. Washington, DC: U.S. Geological Survey in Cooperation with the Circum-Pacific Council for Energy and Mineral Resources. Circum-Pacific Map Series CP-45, scale 1:10,000,000, 1 sheet.

Brysina, T. N. 2006. *The network principle of the organization of society, science, education. Modern problems of science and education.* 1: 37-38 URL: http://www.science-education.ru/en/article/view?id=38 (in Russian).

Callaghan, T. V., Johansson, M., Heal, O. W., Sælthun, N. R., Barkved, L. J., Bayfield, N., Brandt, O., Brooker, R., Christiansen, H. H., Forchhammer, M., Høye, T. T., Humlum, O., Järvinen, A., Jonasson, C., Kohler, J., Magnusson, B., Meltofte, H., Mortensen, L., Neuvonen, S., Pearce, I., Rasch, M., Turner, L., Hasholt, B., Huhta, E., Leskinen, E., Nielsen, N., Siikamäki, P. 2004. SCANNET Environmental changes in the North Atlantic Region: SCANNET as a collaborative approach for documenting, understanding and predicting changes *Ambio*. 33 (Spec. Iss. 13): 39-50.

Callaghan, T. V., Johansson, M., Pchelintseva, Y., Kirpotin, S. N. 2015. Scientific cooperation throughout the Arctic: The INTERACT experience. *The New Arctic.* Springer International Publishing: 269-289.

Callaghan, T. V., Kulikova, O., Rakhmanova, L., Topp Jörgensen, E., Labba, N., Lars-Kuhmanen, A., Kirpotin, S., Shaduyko, O., Burgess, H., Rautio, A., Hindshaw, R. S., Golubyatnikov, L. L., Marshall, G. J., Lobanov, A., Soromotin, A., Sokolov, A., Sokolova, N., Filant, P. 2019. Improving Dialogue Among Researchers, Local and Indigenous Peoples and Decision Makers to Address Issues of Climate Change in the North. *Ambio*. (in press).

Detter, G. F. 2017. Models of development of resources and territories of the Yamal-Nenets Autonomous District. *Arctic and North.* 26. DOI: 10.17238 / issn2221-2698.2017.26.98 (in Russian).

Capra, F. 1983. *The Web of Life.* New York: Anchor Doubleday.

Capra, F. 1996. *The web of life: A new scientific understanding of living systems.* New York: Anchor Books.

Capra F. 2003. *The web of life: A new scientific understanding of living systems.* Transl. from English by ed. W. G. Trilis. M.: Publishing House "Sofia": 336. (in Russian).

Corby, G. A. 1982. The first International Polar Year (1882/83), *World Meteorological Organization Bulletin*, 31 (3): 197-214.

Gerson, N. C. 1958. From polar years to IGY. In: H. E. Landsberg, & J. VanMieghem (Eds). *Advances in Geophysics*. Academic Press, New York, 5: 1–52.

Gorchitskaya, E. A. 2011. Intellectual elite from the standpoint of philosophy. *Bulletin Omsk Univ.* 4: 207–210. (in Russian).

Gorokhov, V. G. 2006. Interdisciplinary research of scientific and technological development and innovation policy. *Questions of Philosophy*, 4: 80-96. (in Russian).

Gorokhov V. G. 2010. Introductory article "The duality of modern society - risk and information (transition to a knowledge society)." To the book: Behmann, Gotthard. Modern society: risk society, information society, knowledge society. - M.: Logos: 6-72 (in Russian).

Gorokhov V. G., Scherz K. 2011. Socio-ecological consequences of the development of technology (Comparative analysis of socio-cultural features of the development of nuclear technology in Russia and Germany). *Philosophical Sciences*, 6: 49-62 (in Russian).

Interact. Stories of Arctic science. 2015. Editors: Terry V. Callaghan, Hannele Savela: 186.

Interact. 2015. Research and monitoring. Editors: Elmer Topp-Jørgensen, Zhanna Tairova, Morten Rasch, Jannik Hansen: 156.

International congresses on transdisciplinarity: their importance for the emergence of a transdisciplinary méthodology. http://ciret-transdisciplinarity.org/ARTICLES/Nicolescu_fichiers/InterviewAlvarengaENG.pdf.

Jantsch E. 1972. *Technological planning and social futures,* Wiley: 256.

Jantsch E. *The Universe of Self-Organization: The Scientific and Human Values of the Emerging Development Paradigm,* New York: Pergamon Press, 1980.

Johansson, M., Jonasson, C., Sonesson, M., Christensen, T. R. 2012. The man, the myth, the legend: Professor Terry V. Callaghan and his 3m concept. *Ambio.* 41(SUPPL.3): 175-177.

Kirpotin S. N. 2005. Living self-organizing cognitive systems and the role of biots in the evolution of life //Evolution of life on Earth: Proceedings of the III International Symposium, November 1-3, 2005, Tomsk /rev. Editor V. M. Like. - Tomsk: *Tomsk State University*: 26-28.

Kirpotin S. N. 2007. *On the nature of human consciousness // Dialogue of cultures - dialogue culture: materials of international scientific and practical.* Conf. - Kostroma, September 3-7, / Res. Ed. L.N. Vaulina. - Kostroma KSU named after ON. Nekrasova: 144-152.

Kirpotin, S., Berezin, A., Bazanov, V., Polishchuk, Y., Vorobiov, S., Mironycheva-Tokoreva, N., Kosykh, N., Volkova, I., Dupre, B., Pokrovsky, O., Kouraev, A., Zakharova, E., Shirokova, L., Mognard, N., Biancamaria, S., Viers, J., Kolmakova, M. 2009. *Western Siberia wetlands as indicator and regulator of climate change on the global scale.* 66, (4): 409 – 421.

Kirpotin, S., Polishcuk, Y., Bryksina, N., Sugaipova, A., Kouraev, A., Zakharova, E., Pokrovsky, O., Shirokova, L., Kolmakova, M., Manassypov, R. & Dupre, B. 2011.West Siberian palsa peatlands: distribution, typology, cyclic development, present-day climate-driven changes, seasonal hydrology and impact on CO_2 cycle. *International Journal of Environmental Studies*. 68, (5): 1–20.

Kirpotin S. N. 2013. Organizational patterns of ecosystems and landscapes: a view through the prism of a synergistic approach and the theory of symmetry. *Lambert Academic Publishing.*

Kirpotin S. N., Vorobev S. N., Kolesnichenko L. G., Cazzolla Gatti R., Volkov I. V., Volkova I. I., Loyko S. V., Manasypov R. M., Pokrovsky O. 2015. Siberia as a natural mega-installation - the basis for the development of scientific mega-projects.*Biodiversity and development of the gene pool of flora, fauna and population of the Central Asian region*:

Materials of the IVth international scientific and practical conference (October 1-4, 2015, Kyzyl, Russia) / Ans. ed. D. Philos. n., prof. Hamster. - Kyzyl: TuvGU RIO: 4-7.

Kirpotin S. N., Callaghan T. V. 2016. Project "Trans-Siberian Scientific Way": the world level in response to global challenges. *Actual problems of researching the ethnoecological and ethnocultural traditions of the peoples of Sayano-Altai*, Materials of the IV international scientific and practical conference of young scientists, graduate students and students, dedicated to the Year of Hospitality in the Republic of Tyva. Tuvan State University: 9-11.

Kirpotin, S. N., Callaghan, T. V., Pokrovsky, O. S., Karlsson, J., Vorobiov, S. N., Kolesnichenko, L. G., Popravko, I. G., Kolesnikova, T. S. and Audry, S. 2018. Russian-EU collaboration via the mega-transect approach for large-scale projects: cases of RF Federal target Programme and SIWA JPI Climate EU Programme. *International Journal of Environmental Studies*. 75(3):385-394.

Kirpotin, S. N. 2018. *TSU open lecture "The role of living matter URL"* https://www.youtube.com/watch?v=SKA2we-QEZQ (appeal date 07.09.2018). (in Russian).

Kiyaschenko, L. Ya., Moiseev, V. I. 2009. *Transdisciplinary Philosophy*. Grew up Acad. Sciences, Institute of Philosophy. M .: IFRAN: 205.

Knyazeva, E. N. 2011. Transdisciplinary research strategies. *Bulletin of TSPU*. 10 (112: 193 - 201.

Kotlyakov, V. M. 2011. From the International Polar Year to the International Polar Decade. Arctic. *Ecology and Economics*. 1 (in Russian).

Korsmo, F. L. 2007. The genesis of the international geophysical year. *Phys.Today*, 60 (7): 38–43.

Kropotkin, P. A. 1907. *Mutual Aid as a Factor of Evolution* / transl. from English by B. Baturinskiy. – SPb.: Znaniye, – 351 p. (in Russian).

Kropotkin P. A. 2011. *Mutual Aid as a Factor of Evolution*. - Moscow: Direct-Media. (In Russian).

Kulmala, M., Lappalainen, H. K., Petäjä, T., Kurten, T., Kerminen V. M., Viisanen, Y., Hari, P., Sorvari, S., Bäck, J., Bondur, V., Kasimov, N., Kotlyakov, V., Matvienko, G., Baklanov, A.,Guo, H., Ding A., Hansson, H. C. andZilitinkevich, S. 2015. Introduction: The Pan-Eurasian Experiment (PEEX) – multi-disciplinary, multi- scale and multi-component research and capacity building initiative, *Atmos. Chem. Phys.*, 15: 13085–13096.

Kulmala, M., Lappalainen, H. K., Petäjä, T., Kerminen, V. M., Viisanen, Y., Matvienko, G., Melnikov, V., Baklanov, A., Bondur, V., Kasimov, N. and Zilitinkevich, S. 2016. Pan-Eurasian Experiment (PEEX) Program: Grant Challenges in the Arctic-boreal context. *Geography, Environment, Sustainability*. 9(2): 5-18.

Lappalainen, H. K., Petäjä, T., Kujansuu, J., and Kerminen, V. M., Shvidenko, A., Bäck, J., Vesala, T., Vihma, T., de Leeuw, G., Lauri, A., Ruuskanen, T., Flint, M., Zaitseva, N., Arshinov, M., Spracklen, D., Arnold, S., Juhola, S., Lihavainen, H., Viisanen, Y., Chubarova, N., Filatov, N., Skorokhod, A., Elansky, N., Dyukarev, E., Hari, P., Kotlyakov, V., Kasimov, N., Bondur, V., Matvienko, G., Baklanov, A., Guo H., Zilitinkevich, S. and Kulmala, M. 2014. Pan-Eurasian Experiment (PEEX) – a research initiative meeting the grand challenges of the changing environment of the northern Pan-Eurasian arctic-boreal areas, *Geography, Environment, Sustainability*, 2, 13–48.

Lappalainen, H. K., Kerminen, V. M., Petäjä, T., Kurten, T., Baklanov, A., Shvidenko, A., Bäck, J., Vihma, T., Alekseychik, P., Arnold, S., Arshinov, M., Asmi, E., Belan, B., Bobylev, L., Chalov, S., Cheng, Y., Chubarova, N., de Leeuw, G., Ding, A., Dobrolyubov, S., Dubtsov, S., Dyukarev, E., Elansky, N., Eleftheriadis, K., Esau, I., Filatov, N., Flint M., Fu, C., Glezer, O., Gliko, A., Heimann, M., Holtslag, B., Janhunen, J., Juhola, S., Järvi, L., Järvinen, H., Kanukhina, A., Konstantinov, P., Kotlyakov, V., Kieloaho, A. J., Komarov, A., Kujansuu, J., Kukkonen, I., Laaksonen, A., Laurila, T., Lihavainen, H., Lisitzin, A., Mahura, A., Makshtas, A., Mareev, E., Matishov, D., Mazon, S., Melnikov, V., Mikhailov, E., Moisseev, D., Nigmatulin, R., Noe, S. M., Ojala, A., Pihlatie M., Popovicheva, O., Pumpanen, J., Regerand, T., Repina, I., Shcherbinin, A., Shevchenko, V., Sipilä, M., Skorokhod, A., Spracklen, D. V., Su, H., Subetto, D., Sun, J., Terzhevik, A., Timofeyev, Y., Troitskaya, Y., Tynkkynen, V. P., Kharuk, V. I., Zaytseva, N., Zhang, J., Viisanen, Y., Vesala, T., Hari, P., Hansson, H. C., Matvienko, G., Kasimov, N., Guo, H., Bondur, V., Zilitinkevich, S. S. and Kulmala, M. 2016. Pan-Eurasian Experiment (PEEX): Towards holistic understanding of the feedbacks and interactions in the land – atmosphere – ocean-society continuum in the Northern Eurasian region, Atmos. *Chem. Phys. Discuss.*, doi:10.5194/acp-2016-186, in review.

Loiko, S. V., Pokrovsky, O. S., Raudina, T. V., Lim, A., Kolesnichenko, L. G., Shirokova, L. S., Vorobyev, S. N., Kirpotin, S. N. 2017. Abrupt permafrost collapse enhances organic carbon, CO_2, nutrient and metal release into surface waters. *Chemical Geology*, 471: 153-165. DOI: 10.1016/j.chemgeo.2017.10.002.

Magidovich, I. P., Magidovich, V.I. 1984. *Essays on the history of geographical discoveries*. T. 3. M.: 287. (In Russian).

Makarova, E. E. 2013. Popularization of science on the Internet: content, forms, development trends. *Moscow University Bulletin*, Series 10, Journalist, (in Russian).

Malthus, T. 2008. *Experience of the law of population*. M.: Directmedia Publishing: 465. (in Russian).

Malitskaya, E. A. 2014. The concept of an infrastructure project and its life cycle management. *Economy of the regions*. 3. – introduction (in Russian).

Manasypov R. M., Vorobyev S. N., Loiko S. V., Kritzkov I. V., Shirokova L. S., Shevchenko V. P., Kirpotin S. N., Kulizhsky S. P., Kolesnichenko L. G., Zemtzov V. A., Sinkinov V. V., and Pokrovsky O. S. 2015. Seasonal dynamics of organic carbon and metals in thermokarst lakes from the discontinuous permafrost zone of western Siberia. *Biogeosciences*, 12, (10): 3009-3028, doi: 10.5194/bg-12-3009-2015.

Mazzoli, E. 2016. *Carl Weyprecht and the International Polar Years*. Trieste, Italy: Luglioprint: 460, ISBN 978-88-6803-147-3.

Melyukhin, I. S. 1999. *Information society: origins, problems, development trend*. M .: MSU Publishing House: 208.

Merchant, C. 2006. *The Scientific Revolution and The Death of Nature*. The University of Chicago Press. 97 (3): 513-533.

Morgalev, Y. N., Lushchaeva, I. V., Morgaleva, T. G., Kolesnichenko, L. G., Loiko, S. V., Krickov, I. V., Lim, A., Raudina, T. V., Volkova, I. I., Shirokova, L. S., Morgalev, S. Y., Vorobyev, S. N., Kirpotin, S. N., Pokrovsky, O. S. 2017. Bacteria primarily metabolize at the active layer/permafrost border in the peat core from a permafrost region in western Siberia. *Polar Biology*, 40, (8): 1645-1659. DOI: 10.1007/s00300-017-2088-1.

Morgun, E. N., Kolesnikov, R. A., Agbalian, E. V. *Program of integrated scientific environmental monitoring in the Yamal-Nenets Autonomous Okrug* (edited by A.L. Titovsky). Salekhard, St. Petersburg: Geo Graph, 2019. – 148, 8 (in Russian).

Nicolescu, B. 2006. *Transdisciplinarity – past, present and future. Published in Moving Worldviews - Reshapingsciences, policies and practices for endogenous sustainable development.* COMPAS Editions. Holland, edited by Bertus Haverkort and Coen Reijntjes: 142-166.

Pokrovsky, O. S., Manasypov, R. M., Loiko, S., Shirokova, L. S., Krickov, I. A., Pokrovsky, B. G., Kolesnichenko, L. G., Kopysov, S. G., Zemtzov, V. A., Kulizhsky, S. P., Vorobyev, S. N., Kirpotin, S. N. 2015. Permafrost coverage, watershed area and season control of dissolved carbon and major elements in western Siberian rivers. *Biogeosciences*, 12 (21): 6301-6320. doi: 10.5194/bg-12-6301-2015.

Pokrovsky, O. S., Manasypov, R. M., Loiko, S. V., Krickov, I. A., Kopysov, S. G., Kolesnichenko, L. G., Vorobyev, S. N., Kirpotin, S. N. 2016. Trace element transport in western Siberian rivers across a permafrost gradient. *Biogeosciences*, 13, (6): 1877-1900.

Polishchuk, Y. M., Bogdanov, A. N., Polishchuk, V. Y., Manasypov, R. M., Shirokova, L. S., Kirpotin, S. N., Pokrovsky, O. S. 2017. Size distribution, surface coverage, water, carbon, and metal storage of thermokarst lakes in the permafrost zone of the western Siberia lowland. *Water*, 9, (3), Article number 228, doi: 10.3390/w9030228.

Popov, M. S. 2015. *The concept of infrastructure projects and the specifics of their legal nature.* https://cyberleninka.ru/article/v/ponyatie-infrastrukturnyh-proektov-i-spetsifika-ih-pravovoy-prirody (in Russian).

Rueschemeyer, D., Stephens, J. 1992. *Capitalist Development and Democracy.* Chicago: Chicago University Press.

Serikova S., Pokrovsky O. S., Ala-Aho P., Kazantsev V., Kirpotin S. N., Kopysov S. G., Krickov I. V., Laudon H., Manasypov R. M., Shirokova L. S., Soulsby C., Tetzlaff D. and Karlsson J. 2018. High riverine CO_2 emissions at the permafrost boundary of Western Siberia. *Nature geoscience.* 11: 825–829.

Serikova, S; Pokrovsky, O. S.; Laudon, H.; Krickov, I., V; Lim, A. G.; Manasypov, R. M.; Karlsson, J. High carbon emissions from thermokarst lakes of Western Siberia. *Nature communications.* 10.Paper number: 1552. DOI: 10.1038/s41467-019-09592-1.

Shevchenko Vladimir P., Pokrovsky Oleg S., Vorobyev Sergey N., Krickov Ivan V., Manasypov Rinat M., Politova Nadezhda V., Kopysov Sergey G., Dara Olga M., Auda Yves, Shirokova Liudmila S., Kolesnichenko Larisa G., Zemtsov Valery A., and Kirpotin Sergey N. 2017. Impact of snow deposition on major and trace element concentrations and elementary fluxes in surface waters of the Western Siberian Lowland across a 1700'km latitudinal gradient. *Hydrology and Earth System Sciences*, 21, 5725–5746, https://doi.org/10.5194/hess-21-5725-2017.

Shirokova L. S., Pokrovsky O. S., Kirpotin S. N., Desmukh C., Pokrovsky B. G., Audry S., Viers J. 2013. Biogeochemistry of organic carbon, CO_2, CH4, and trace elements in thermokarst water bodies in discontinuous permafrost zones of Western Siberia. *Biogeochemistry.* 113:573–593. DOI 10.1007/s10533-012-9790-4.

Skocpol, T. 1984. *Vision and Method in Historical Sociology.* Cambridge: Cambridge University Press.

Stepin, V. S., Gorokhov V. G., Rozov M. A. 1996. *Philosophy of science and technology.* M.: Gardariki. (in Russian).

Stepin, V. S. *Philosophy of science. 2006. Common problems: a textbook for graduate students and applicants for the degree of candidate of sciences*. M.: Gardariki: 384. (in Russian).

Strezhneva M. 2012. Integration and involvement as tools of global management. *International processes*. 7 (1). URL: http://www.untertrends.ru/seventh/002.htm (in Russian).

Tarasenko, F. P. 2013. *Applied system analysis*. Prospectus: 216. (in Russian).

Topp-Jørgensen, E. 2014. INTERACT Management planning for arctic and northern alpine research stations – Examples of good practices DCE – Danish Centre for Environment and Energy, *Aarhus University*: 328.

Vasil'chuk Yu. K., Shevchenko V. P., Lisitzin A. P., Budantseva N. A., Vorobiov S. N., Kirpotin S. N., Krizkov I. V., Manasypov R. M., Pokrovsky O. S., Chizhova Ju. N. 2016. Oxygen isotope and deuterium composition of snow cover on the profile of Western Siberia from Tomsk to the Ob Bay. *Doklady Earth Sciences*. 471, Part 2: 1284–1287.

Vorobyev, S. N., Pokrovsky, O. S., Kirpotin, S. N., Kolesnichenko, L. G., Shirokova, L. S., Manasypov, R. M. 2015.Flood zone biogeochemistry of the Ob River middle course. *Applied Geochemistry*, 63: 133-145.

Vorobyev Sergey N., Pokrovsky Oleg S., Serikova Svetlana, ManasypovRinat M., Krickov Ivan V., ShirokovaLiudmila S., Lim Artem, Kolesnichenko Larisa G., Kirpotin Sergey N.andKarlsson Jan. 2017. Permafrost Boundary Shift in Western Siberia May Not Modify Dissolved Nutrient Concentrations in Rivers. *Water* 9 (12), 985; doi: 10.3390/w9120985, Q2, IF 2.069.

Vorobyev S. N., Pokrovsky O. S., Kolesnichenko L. G., Manasypov R. M., Shirokova L. S., Karlsson J., Kirpotin S. N. 2019. Biogeochemistry of dissolved carbon, major and trace elements during spring flood periods on the Ob River. *Hydrological Processes*, 33(11), pp. 1579-1594.

Zakharova E. A., Kouraev A. V., Rémy F., Zemtsov V. A., Kirpotin S. N. 2014. Seasonal variability of the Western Siberia wetlands from satellite radar altimetry. *Journal of Hydrology*. 512. 366–378. doi: 10.1016/j.jhydrol.2014.03.002.

In: The Arctic: Current Issues and Challenges
Editors: Oleg S. Pokrovsky, Sergey N. Kirpotin et al.
ISBN: 978-1-53617-306-2

Chapter 6

VEGETATION OF THE SUBARCTIC AND ARCTIC OF SIBERIA: SOME TYPES AND APPROACHES TO THE STUDY

I. V. Volkov[1,*], I. I. Volkova[1], S. N. Kirpotin[1] and A. I. Volkova[2]
[1]Bio-Clim-Land Center, Tomsk State University, Tomsk, RF
[2]Institute of Archaeology and Ethnography SB RAS, Novosibirsk, RF

ABSTRACT

Vulnerability assessment is the main tool to understand where climate change and anthropogenic impacts will have severe consequences for nature and society, and which ecosystems are more susceptible to these changes. Threats related to the possible consequences for ecosystems as a result of climate change have been growing in recent years. These threats are associated with shifts in the ranges of various plant and animal species, threats to biodiversity and adverse consequences for the structure and composition of communities. According to research undertaken by the Economic Commission for Europe, the most important ecosystems continue to be at risk: forests, wetlands, biologically rich agricultural habitats, a number of arid and desert regions, and some water areas. In our opinion, the most vulnerable ecosystems, such as Arctic tundra and wetlands, are in a special risk zone. Changes in climate, about which there is a broad international scientific consensus regarding the size of anthropogenic impacts, are an extraordinarily important part of global environmental change and a significant cause of threat to global biodiversity.

Vegetation is the basis and the key component of natural ecosystems; it is the vegetation cover that determines the landscape's appearance and is the main indicator when interpreting satellite images. The vegetation of the polar tundra, by virtue of its boundary position in the biosphere, has a relatively low stability in the conditions of global climate dynamics. Therefore, research into the plant component of polar ecosystems is the most urgent task. The Russian sector of the Arctic, due to its greatest length and typical character of the manifestation of zonality of tundra vegetation, is of particular interest for monitoring climate-dependent changes.

* Corresponding Author's Email: volkovhome@yandex.ru.

Based on the analysis of Russian publications that are not familiar to the English-speaking reader, and contemporary English-written literature as well as materials from our own research, an overview of the main types of vegetation of the Subarctic and Arctic of Siberia, as well as the main approaches to its study, is given.

Keywords: arctic, vulnerability, vegetation, biodiversity

1. INTRODUCTION

The concept of the Arctic is characterized by incredible uncertainty; there are many interpretations and a very different understanding of this term. As the well-known Scottish scientist Robert Crawford (Crawford, 2013) rightly points out in his monograph "Tundra-Taiga Biology," the Arctic is not so much a definite well-defined territory, it is a concept invented by people (Aristotle) in ancient times about "a northerly Frigid Zone where human life would be impossible." Right up to the present day, the definition of the Arctic has been a big problem.

Etymologically, the word Arctic comes from the Greek "ἄρκτος," "arktikos" – "northern," "arctos" – "bear," and is associated with the constellation Ursa Major, or the "Great Bear." The Arctic has a unique geographical location. This is a single physical-geographical region of the Earth, adjacent to the North Pole, the northern circumpolar space, which includes the waters of the seas of the Arctic Ocean, many islands and archipelagoes located in it, as well as the coastal areas of the three continents around it – Europe, Asia and North America. This is why the region is sometimes called the "polar Mediterranean" (Lukin, 2013).

From a global perspective, eight Arctic countries are directly related to the Arctic: Denmark (Greenland), Iceland, Canada, Norway, Russia, the USA, Sweden, Finland, possessing northern territories beyond the Arctic Circle. Five of them (Denmark, Canada, Norway, Russia, and the USA) – on the basis of UNCLOS (1982) – own territorial waters of the Arctic Ocean in addition to the land (Figure 1). At the same time, there is not a single northern country in the world which, by its entire territory and water area, would be fully incorporated into the circumpolar zone. Calling these eight northern countries "semi-arctic" and not "arctic," we thus emphasize this important geopolitical feature. The geopolitical space "Arctic – XXI" is more than 26% of the land area; 7.4% of the population of global society (2012) and 29.2% of world GDP in 77.8 trillion American dollars (2014) according to the World Bank database (Lukin, 2013).

The simplest geographic or astronomical approach to defining the Arctic includes all territories north of the Arctic Circle (Ives and Barry, 1974; Ørbæk et al., 2007; Crawford, 2013). However, if we take into account that the average January temperature on Svalbard (78 degrees North) is about -15^{0}C, and in the boreal zone of Yakutia it can fall below -60^{0}C, the question arises: which area fits more to the Arctic mental concept – the mild climate of the Svalbard warming up by the Gulf Stream and located far beyond the Arctic Circle, or the extremely harsh climate of the forest zone of Yakutia, located substantially to the south of the Arctic Circle and underlain by almost a half-kilometer-thick permafrost?

Figure 1. Map of the Arctic region, with the Arctic Circle in blue and the July 10°C mean isotherm in red (from https://en.wikipedia.org/wiki/Arctic_Circle#/media/File:Arctic_circle.svg).

From a climatic point of view, the Arctic is often defined as a territory underlain by permafrost (Ives and Barry, 1974; Ørbæk et al., 2007; Crawford, 2013). But this principle does not work well either, since, for example, in some places even far north of the Arctic Circle (Norway, Sweden), the permafrost may be poorly represented and has an island character, while in Russia it occupies 60% of the territory and has almost a continuous distribution to the east of the Yenisei, penetrating far to the south from the Arctic Circle. One more climatic definition of the tundra zone is when the July isotherm at 10 °C is taken as the southern border (Ives and Barry, 1974; Ørbæk et al., 2007).

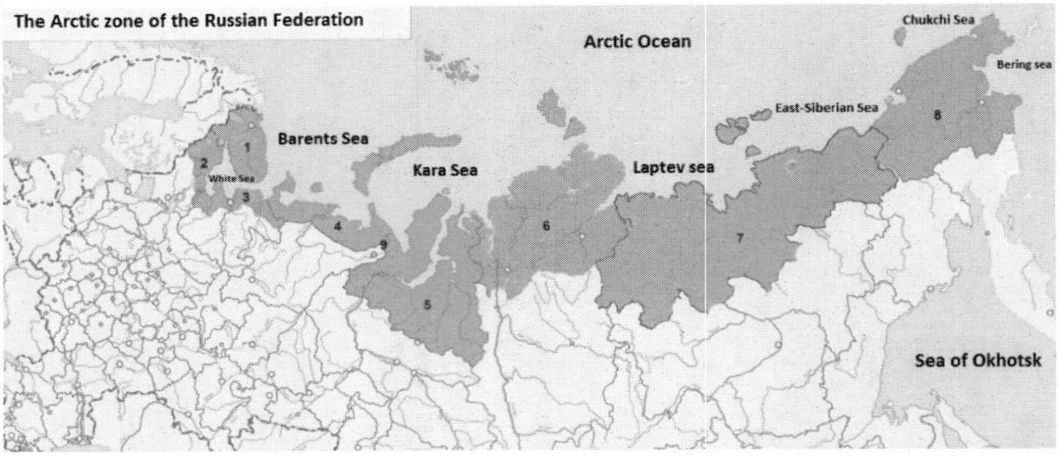

Figure 2. Map of the the Arctic zone of the Russian Federation: 1– Murmansk Region (Oblast), 2 – Republic of Karelia, 3 – Arkhangelsk Region (Oblast), 4 – Nenets Autonomous District (Okrug), 5 – Yamalo-Nenets Autonomous District (Okrug), 6 – Krasnoyarsk Territory (Krai), 7 – Republic of Sakha (Yakutia), 8 – Chukotka Autonomous District (Okrug), 9 – Komi Republic, (Translated into English from the Map of the Russian Arctic / © LukinYu. F., Eremin A. E.).

Another approach to the definition of the Arctic, which can be designated as botanical and geographical, defines the Arctic as a treeless space, where woody plant life forms are almost wholly absent (Shumilova, 1962; Ives and Barry, 1974; Ørbæk et al., 2007; Crawford, 2013). However, this approach is not clear enough, especially if we consider that in Siberia, the forest-tundra subzone can reach a width of 600 km.

Since the authors of the chapter took up describing the vegetation of the Siberian Arctic, it would be useful to provide data on how the Arctic zone is understood and interpreted in the Russian Federation. The concept of the "Arctic zone of the Russian Federation" was established by a decision of the State Commission of the Council of Ministers of the USSR on the Affairs of the Arctic on April 22, 1989 (Figure 2). The Arctic zone then included: 1) the northern territories of eight administrative subjects: the Republic of Sakha (Yakutia), Krasnoyarsk Territory, Murmansk Region, Arkhangelsk Region, Nenets Autonomous District, Yamalo-Nenets Autonomous District, Taimyr (Dolgan-Nenets) Autonomous District, Chukotka Autonomous District; 2) lands and islands specified in the Decree of the Presidium of the CEC of the USSR of April 15, 1926 "On declaring the territory and islands of the USSR located in the Arctic Ocean"; 3) inland waters, territorial sea, exclusive economic zone and continental shelf (http://lawinrussia.ru/content/arkticheskaya-zona-rossiyskoy-federacii-azrf).

Features of the Arctic zone of the Russian Federation influencing the formation of the state policy in the Arctic in 2008 were determined: 1) extreme climatic conditions, including permanent ice cover or drifting ice in the Arctic seas; 2) focal character of industrial and economic development of territories and low population density; 3) remoteness from major industrial centers, high resource intensity, and dependence of economic activities and the sustenance of the population on the supply of fuel, food and essential goods from other regions of Russia; 4) low stability of ecological systems that determine the biological balance and climate of the Earth, and their dependence even on minor anthropogenic influences (http://lawinrussia.ru/content/arkticheskaya-zona-rossiyskoy-federacii-azrf). These features not only characterize the Arctic zone from a legal point of view, but also have a scientific, primarily geographic, basis.

2. General Characteristics of the Tundra Zone and Its Vegetation

Tundra, which comes from a Sami word meaning "barren land," refers to a treeless arctic region characterized by permafrost (Terasmae, Reeves, 2009). Even Encyclopedia Britannica recognized that although the Finns called their treeless northern reaches the tunturi, the concept of a vast frozen plain as a special ecological realm called tundra was developed by the Russians (https://www.britannica.com/science/tundra). That is why it is so important to get acquainted with the fundamental works of Russian scientists on this topic.

The Russian sector of the Arctic occupies a significant part of this biome. Depending on the delineation method, the Russian Federation occupies either 33% or 43% of the World Arctic; the area of the Russian Arctic sector is about 9 million km^2, of which about 6.8 million km^2 is in the water area (Avdonina, Dolgoborodova, 2018). Therefore, the study of the Arctic without taking into account the research of Soviet scientists provides a very incomplete picture of the research of this extremely important region for humanity. However, the political contradictions during "cold war" between the Soviet Union and Western countries created certain difficulties in the exchange of scientific experience, so the Arctic territories of the USSR turned out to be a kind of "terra incognita" for English-speaking researchers. Only a small part of Soviet scientific works and previous works of scientists of Tsarist Russia became known in the West. A rare example is the monograph of Yury Chernov, "The Living Tundra (Studies in Polar Research)" (Chernov, 1985), one of the few works translated into English that attracted the attention of the scientific community and was highly regarded amongst its readership. From the 1990s onwards, as Russia became more open to the world, the situation changed for the better, but nevertheless, many scientific papers in Russian remained unknown in the West. At the same time, scientific research conducted in the Russian Arctic on the basis of long-term observations, is of great interest to world science. There are at least two reasons for this (Matveyeva, 1998):

1. For synecologists, tundra and polar-desert communities – with their extremely low biodiversity and simplified structure – are objects that can be compared with the hydrogen atom in physics, i.e., the simplest model of ecosystems to study the principal features and mechanisms of general ecological phenomena that can be studied in the Arctic with the least amount of time and effort.
2. The Russian sector of the Arctic is the only place – not only in Eurasia, but also on the globe – where all sub-zones of tundras are developed on a large part of the continental territory (Figure 3).

Studies of the vegetation of arctic landscapes touch upon the general biological problem of life in extreme conditions, where biological phenomena are exacerbated and appear more pronounced than in other types of zonal landscapes (Matveyeva, 1998). In the zonal tundras of the North, which are located in the frontier living conditions and limited by severe climatic conditions from above and from below by permafrost that thaws in summer for a short period, even a small change in conditions caused by a change in exposure on a small permafrost mound creates a significant ecological contrast, which is reflected in a dramatic change in vegetation.

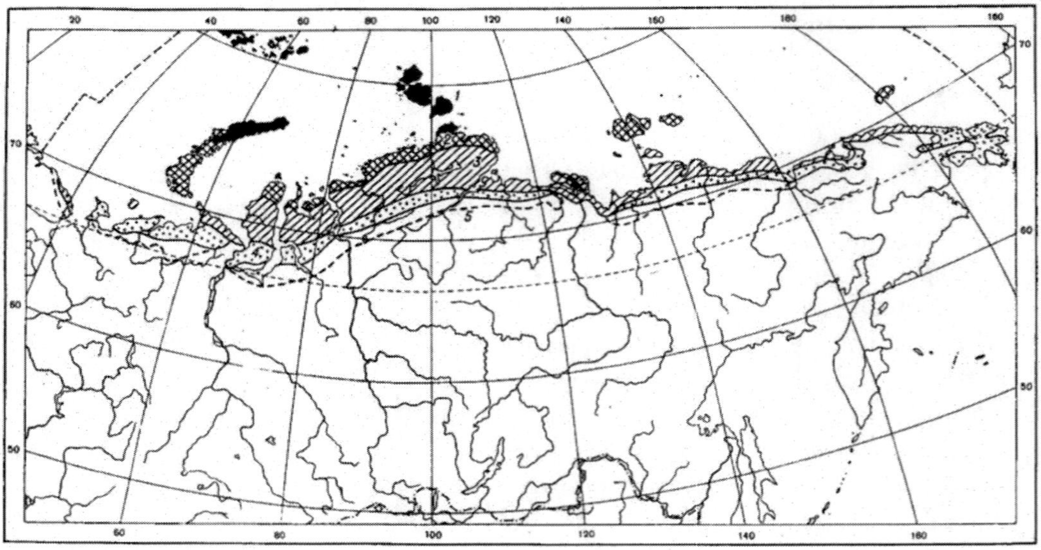

Figure 3. The main zonal landscapes of the Arctic in Eurasia: 1 – polar deserts, 2 – arctic tundras, 3 – typical tundras, 4 – southern tundras, 5 – southern border of forest-tundra (according to: Matveyeva, 1998).

The geographical generality of high-mountain and high-latitude natural environments, noted by the famouse German biogeographer A. von Humboldt, determined the assignment of these territories to a special planetary zone – the periglaciosphere. A fundamental factor explaining this generality is the low radiation balance, not exceeding 20 kcal/cm^2 per year (Golubchikov, 1996), which determines the low annual temperatures of this world of perpetual cold balancing on the verge of freezing only during a short summer. Organisms and their communities seem to be in conditions of an "acute" experiment with the prevailing influence of one factor – lack of heat – which limits other environmental factors, creating the specific conditions of tundra plants (nitrogen deficiency, physiological dryness with abundant moisture, low primary productivity, etc.) (Matveyeva, 1998). Therefore, research into the vegetation of the tundra zone implies knowledge of the specifics of the environment in which this vegetation develops. One of the first complex works devoted to a detailed study of the specifics of the physical and geographical conditions of the tundra zone belongs to academician A. A. Grigoriev (1946). In this paper, the climatic, permafrost processes and their influence on the formation of microrelief, features of soil formation, vegetation cover, and fauna of the Eastern European sector of the tundra zone are analyzed in detail. Moreover, environmental factors are not considered separately, but in combination and interaction with different seasons of the year, in various sectors of the Russian Arctic. This allowed the author to substantiate typical variants of the physical-geographical process in the zonal tundras and the laws of altitudinal zonation in the Arctic mountains.

With common periglacial processes, there is a significant differentiation in space, which is expressed in the formation of the phenomenon of zonality. The first, most reasonable, zonal division of the northern territories of Russia belongs to the well-known Russian tundra researcher B. N. Gorodkov (1935). The discussion on the principles of zonal division of the Arctic gave rise to many schemes, showing not only how contradictory judgments on this issue are, but ultimately creating the most objective system of latitudinal differentiation of natural environments in the continental Arctic based on the analysis of various points of view. Among them (according to Matveyeva, 1998):

1. Zonal division of the territory is carried out primarily on the basis of studying the structure of plakor[1] communities.
2. The leading role in identifying zones or subzones of the Arctic belongs to the vegetation cover, which accumulates the combined effects of climatic factors and has the greatest physiognomic importance compared to other components of the landscape.
3. The main criteria for zonal typology should be characteristics that are determined primarily by the features of the modern landscape shell of the Earth, and primarily the quantitative indicators of the structure of the plant cover and animal population.
4. In landscape typology, it is necessary to use categories having a landscape-physiognomic meaning, taking into account that there are different attitudes and concepts that reflect different approaches to the study of the zonal-geographical process. Thus, the concepts of "Subarctic" and "tundra zone" do not coincide, since they reflect different aspects of differentiation of the landscape envelope (Grigoriev, 1970). The former corresponds to the concept of physical-geographical environment, while the latter has primarily a landscape-physiognomic meaning.
5. Zonal categories are determined on the basis of their similarity rather than their territorial proximity, unlike regional-individual zoning, which is based on genetic unity and the spatial contiguity of territories.

The phenomenon of zonality in the circumpolar distribution of tundra vegetation prevails over sectorial differences, which are largely determined by the history of biota development and geographical location. To understand these differences, it is necessary to have a certain standard in which, based on the concept of an "ideal continent," it is possible to study patterns of the geographical zonality of tundra vegetation. In understanding the patterns of distribution of tundra vegetation, the Taimyr Peninsula is just such an "ideal continent," where the zonality of the tundras typical of the Eurasian sector of the Arctic is most fully expressed, and the unique continental section of cold deserts is located. Therefore, Taimyr can serve – like no other territory – as a model for the organization of tundra communities (Matveyeva, 1998).

---

[1] *Plakors* (from the Greek *plax* – plane, plain) are watershed drained plains (flat or slightly inclined watershed space) or plateau with the most stable (climax) communities characteristic of a specific zone. Plakors are characterized by deep groundwater, and a lack of significant erosion or accumulation, as a result of which vegetation and soil cover most closely correspond to the landscapes of this zone (for example, coniferous forests on podzolic soils in the taiga, grass communities on the black soil in the steppe zone, etc. (Kirpotin, 2018). Plakor is equivalent to the mound or hilltop in the catena concept (Milne, 1936, 1947). The term "soil catena" is used to describe the lateral variation in soils over a hillslope.

Figure 4. Forest tundra on the Taimyr Peninsula (Photo by Dr. Igor Volkov).

Figure 5. Southern tundra on the Taz Peninsula (Photo by Dr. Igor Volkov).

Figure 6. Shrub-moss tundra on the Gydan Peninsula (Photo by Dr. Igor Volkov).

Figure 7. Meadow vegetation in the drained thaw lake basins (local name: *hasyrei*) on the Taz Peninsula (Photo by Dr. Igor Volkov).

The phenomenon of zonality is reflected in the various features of plakor vegetation of polar tudras and polar deserts – species composition, structural and ecological organization. The following types of tundra are successively replaced here, from south to north: multi-tiered woodlands of forest-tundra (Figure 4), shrubby southern tundras, 1-2 meters in height (Figure 5), low-level shrub-moss or shrub-lichen-moss typical tundra, where the height of the shrub tier usually does not exceed 30-40 cm, and geofitized communities poorly differentiated into tiers, i.e., shrubs and elders immersed in the lichen-moss layer in the Arctic tundra (Figure 6). In polar deserts, individual small flowering plants are scattered around protected shelters (they may be absent altogether), forming dwarf curtains with mosses and lichens 10-40 cm in diameter (Golubchikov, 1996). Flowering plants and mosses often take cushion-like forms (Matveyeva, 1998), but single-cell algae are the maior producers of organic matter in polar deserts (Matveyeva, Chernov, 1976).

The main producers of biomass are cryptogram plants – mosses and lichens in the tundra zone; the second most important group are woody plants – shrubs in the southern tundra and shrubs, geophytic and trellis shrubs in typical and arctic tundras. Herbaceous plants are usually less abundant and have less biomass than woody plants on plakors, although they sometimes play a significant role in the formation of tundra communities and may prevail in non-zonal habitat types. For example, in the southern tundra, in drained *hasyreis*, meadows grow from fairly tall grassy plants (Figure 7). This suggests that along with the climate, tundra soils' deficiency in mineral nutrients plays some role in the formation of zonal tundra vegetation (Loiko et al., 2016).

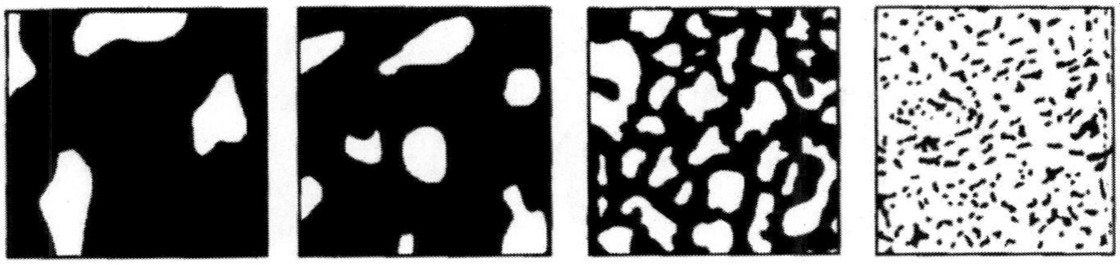

Figure 8. Vegetation distribution in zonal communities on Taimyr (Chernov, 1985): from left to right: southern tundras, typical tundras, arctic tundras, polar deserts (side of a square – 5 meters).

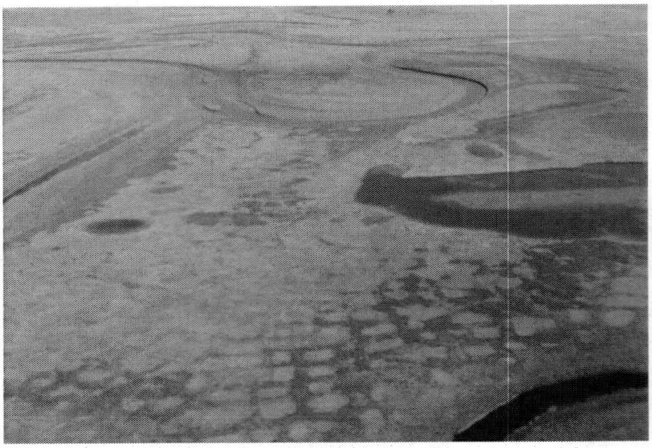

Figure 9. Permafrost polygons in typical tundra of the Gydan Peninsula (Photo by Dr. Igor Volkov).

When moving to the North, the role of the climatic factor increases. This is manifested in the compression of the vertical layer of vegetation in which life can exist. The simplified vertical tiering is being replaced by the complicated horizontal structure of plant communities. Simultaneously, phytomass reserves and the biological productivity of communities are reduced by several orders of magnitude. If the amount of phytomass of the polar cold desert is taken as a unit, then its relation to the zonal communities of the arctic tundra, typical dwarf-shrub and shrub tundra, will be approximately 1: 50: 100: 500 (Aleksandrova, 1970). Along with the growth of biomass, the projective cover of vegetation increases and the mosaic structure of the tundra decreases (Figure 8).

Although the vegetation of the tundra is stunted and outwardly uniform, it is a very well-defined discrete mosaic of communities well distinguished by geobotanists (Aleksandrova, 1971b). The predominance of the influence of abiotic factors on the formation of horizontal mosaic of tundra vegetation has its own laws. In the polar-tundra zone, permafrost is the most important factor affecting the horizontal structure of vegetation, the significance of which, in conditions of increasing environmental extremity from the southern to the polar tundra, is reflected in certain patterns of the structural organization of phytocenoses.

Permafrost is manifested in the formation of the horizontal structure of the tundra surface, which is reflected in the formation of polygons (Figure 9), medallions and other structural units. A significant role of permafrost is manifested in the formation of tundra microrelief (permafrost mounds and depressions of various sizes), whose significance in the differentiation of vegetation (especially on the flattened tundra plains of Western Siberia) increases as the climate becomes more extreme the further north the location (Polishchuk et al., 2017). Permafrost processes play a significant role in the dynamics of ecosystems of the tundra zone. Cyclic succession of thermokarst lakes and drained lake basins – *hasyreis* – is one of the best examples (Kirpotin et al., 2008, 2009).

The role of permafrost processes in the formation of the horizontal structure of tundra vegetation is highly significant. Moreover, from southern tundras, where permafrost processes are relatively weakly manifested on the soil surface, having the features of sporadic cryogenic spots and polygons, to typical and Arctic tundras, where permafrost phenomena are the main factor in the organization of the soil surface, permafrost is one of the major factors in the organization of the horizontal structure of tundra vegetation, which is essentially superimposed on the stencil formed by this phenomenon.

According to N. V. Matveyeva (1998), the Taimyr southern tundra is characterized by an irregular-mosaic type of structure (Figure 10) which is also found in typical tundra. It is then replaced by a 3-member regular-cyclic type, which prevails in the zone of typical tundra. Moreover, the closeness of vegetation during the transition from the southern to typical tundra decreases from 90 to 60%; the number of modules of the structure increases, and therefore the communities become more and more mosaic. In the subzone of the arctic tundra, a 2-member regular-cyclic type prevails, while the proportion of bare soil (55-70%) is already higher than the projective cover of vegetation. In polar deserts, this type of structure is preserved (with an even greater increase in the proportion of bare soil – 80-90%). According to N. V. Matveyeva (1998), the formation of a regular-cyclic type of structure is especially specific for the polar-tundra zone and, in part, for polar deserts, although some of its analogues can be observed under extreme conditions in the mountains.

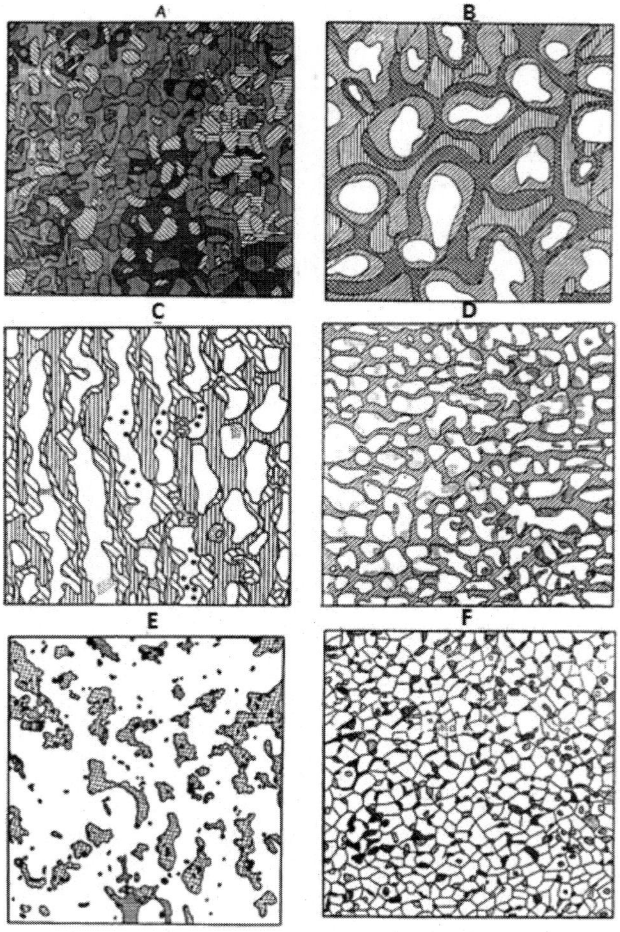

Figure 10. Types of horizontal structure of plant communities in the Arctic (by: Matveyeva, 1998): A – irregular-mosaic; B – 3-member regular-cyclic, on a horizontal surface; C – on the slope; D – 2-member regular-cyclic; E – sporadically spotted in the subzone of the Arctic tundra and F – in polar deserts. Size of the site: AE – 100 m^2, F – 25 m^2.

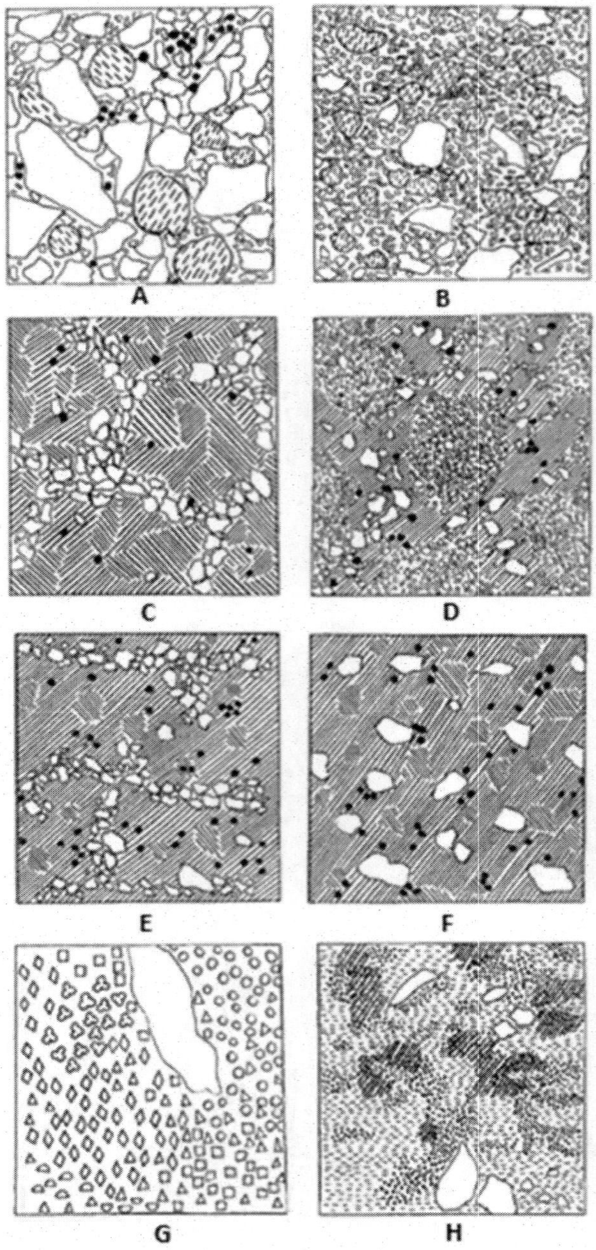

Figure 11. Some variants of the horizontal structure of plant communities in the mountains. A – sporadic-spotted homotonic type of horizontal structure of cryophyte cushions; B – sporadically-spotted homogeneous type of horizontal structure; C – 2-member regular-cyclic reversed type of horizontal structure of lichen-cereal-cobresium-dryas tundra; D – 2-member regular-cyclic type of horizontal structure of Dryas tundra; E – 2-member banded irregular-cyclic type of horizontal structure of Dryas tundra; F – regular-continual type of Dryas tundra with elements of mosaic structure; G – poly-dominant grass Alpine meadow; H – dwarf-shrub-lichen tundra. Stony debris of various sizes in the schemes is indicated by light outlines.

Another characteristic type of vegetation structure in polar deserts is the sporadically-spotted type, where the bare soil occupies up to 95%. Communities with this type of horizontal structure are permanently pioneer. The internal development of plants here is extremely hindered or cut off under the influence of abiotic environmental factors, although along with abiotic factors, biotic factors participate in the formation of a sporadically spotted structure as a consequence of the development of compact plant forms at the initial stage of succession with an unclosed cover (Matveyeva, 1998).

In the highlands of the arctic mountains and more southern climatic zones, the influence of permafrost on the horizontal structure of phytocenoses is transformed by the predominance of coarsely clastic stony underlying rock over the fine-grained substrate. Therefore, there are fewer permafrost polygons here; and permafrost medallions and clay stains are more scarce. When cryogenic sorting of stony soils, a larger rocky fraction is distributed along the edges of the medallions, and their central part occupied by a smaller fraction of rock fragments occupies vegetation. Cryogenic sorting of stony soils results in a larger rocky fraction being distributed along the edges of medallions, with their central part – occupied by a smaller fraction of rock fragments – covered by vegetation (Figure 11. C).

We called this type of horizontal structure of vegetation a 2-member regular-cyclic reversed type, which differs from the 2-member regular cyclic type of structure described by N. V. Matveyeva (1998), in that the vegetation in the first case is grouped inside the permafrost medallion, and in the second, along its edges (Figure 11. D).

Under the influence of slope processes, a 2-member regular-cyclic type of structure can be transformed into a 2-member stepped banded regular-cyclic type of horizontal structure, with fine earth and plants concentrating on the outer part of the step and large fragments of stones accumulating on top. Moreover, vegetation plays an important role in the formation of the microrelief. For example, the Dryas curtain – thanks to an extensive system of underground shoots and roots – fixes slopes, resisting the development of slope processes, which is a factor in the formation of a stepped microrelief of slopes (Figure 12) (Volkov, 2018).

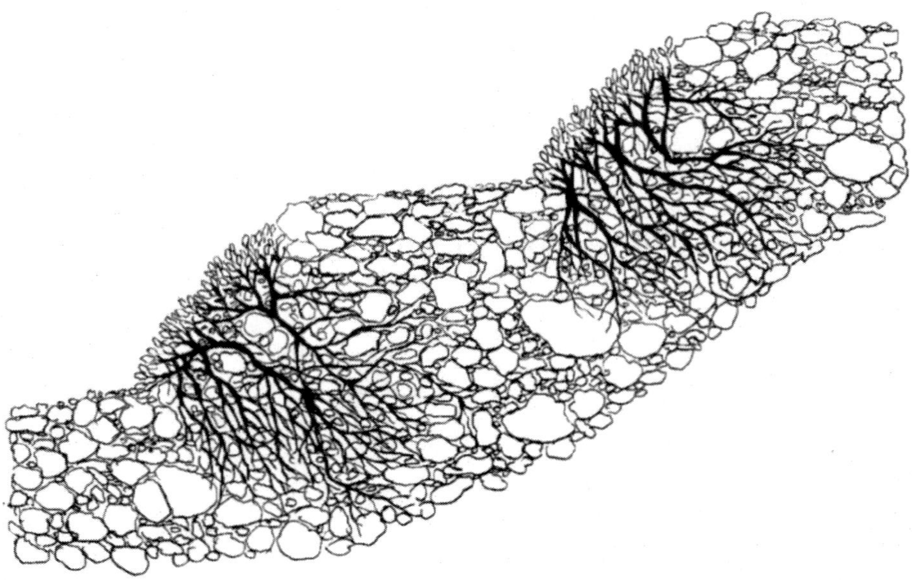

Figure 12. The formation of a 2-member stepped banded regular-cyclic type of horizontal structure is associated with the resistance of geofitized small bushes of Dryas to sloping processes.

Among Russian scientists, there were extensive discussions about the origin of the tundra biome. A. Krishtofovich (1932) considers that zonal tundras of the northern hemisphere are geologically relatively young, since for a considerable part of the Tertiary Period the climatic conditions in this territory (at least in the subzone of the present-day southern tundras) corresponded to those of the temperate zone. Accordingly, there were forests with deciduous and coniferous trees, which, according to A. I. Tolmachyov (1944), could exist in a humid coastal temperate climate with average temperatures of 15-20°C in the warmest month. Widespread podzolic soils in the present-day southern tundra, which could have been formed under the canopy of taiga-type forests, serve as evidence.

Moreover, the vegetation cover of the Arctic as part of the tundra landscape began to form at the end of the Tertiary Period and beginning of the Quaternary Period due to global cooling of the climate and the onset of the glacial epoch, and is characterized by its relatively young age (Tolmachyov, 1932).

Most plant communities of the tundra include various florocenotic elements – forest (boreal), golets (arctic-alpine), arctic, hypoarctic, which indicates their participation in the formation of modern tundra of different types of vegetation. The harsh climatic conditions of the Arctic and the widespread development of permafrost contributed to the formation of a special arctic floristic complex of species, with an original spectrum of life forms and ecological groups of species. Peculiar tundra cenotypes different from other (non-tundra) types of vegetation were formed. The heterogeneity of vegetation in the Arctic – the combination of floristic elements of different origin in the tundras during the warm and cold periods of the Pleistocene and Holocene – contributes to the different reactions of these elements to global climate change.

When cooling, heat-loving species disappear whilst cold-loving species thrive, and vice versa. The most vivid "trace" about the past was left by the recent global warming expressed in the expansion of forest vegetation in the tundra (Telyatnikov, 2005).

The most complete and fairly recent information on vegetation of the Siberian Arctic is given in Mikhail Telyatnikov's doctoral dissertation, "The Peculiarities of the Distribution of the Tundra Vegetation of the Siberian Sector of the Arctic" (Telyatnikov, 2005). It also provides a historical overview of studies of the Siberian sector of the Arctic. It is shown that a sufficiently large number of scientific works are devoted to research into the plant cover of the Siberian Arctic; there are generalized reports on both the individual regions of Yamal, Taimyr, Yakutia, and in general for the Far North. Initially, information on flora and vegetation was obtained from reports of 19th-century expeditions on the development of the North Sea paths, and included both herbarium collections and general information on the vegetation cover. In the pre-revolutionary period, studies of the Far North were geographically oriented; the material was collected from the flora of the tundra zone (Zhitkov, 1913; Gorodkov, 1916, 1929, Komarov, 1926; and others). From 1930, land surveying began throughout the Soviet Arctic.

Zoning issues in the vegetation cover of the Arctic, as well as the flora and vegetation of Taimyr, are considered in the papers and monograph written by N. V. Matveyeva (1998). Vegetation of typical Yamal tundra is considered in the papers and doctoral dissertation by M. Yu. Telyatnikov (2003, 2005). The history of the Arctic flora and its floristic zoning are discussed in the articles by B. A. Yurtsev (1981, 1982, 1987a, b, c; 1988; Yurtsev, 1994; Yurtsev, Katenin, Koroleva et al., 2001; Yurtsev, Zverev, Katenin, 2002).

3. Vegetation of Southern Tundras of the Taz Peninsula

The Taz and Gydan peninsulas form the eastern sector of the Arctic territory of Western Siberia. In terms of natural science research, and in particular, studies on vegetation, this territory seemed to have been in the shadow of the more well-known neighboring territories – the Yamal and Taimyr peninsulas. This fully applies to research into vegetation, which is reflected in the appearance of monographs summarizing the detailed study of the vegetation of Taimyr and Yamal (Matveyeva, 1998; Telyatnikov, 2003, etc.). Against this background, the knowledge of the tundra vegetation of the Taz Peninsula is clearly insufficient. There are relatively few articles (Kornienko, Jacobson, 2011; Kobeleva, 2012, 2014, 2018) devoted to studies and mapping of vegetation on the Taz Peninsula, which were conducted locally (Kobeleva, 2014). Therefore, research into the vegetation of the southern tundra, which we carried out at the interval between the villages of Tazovsky and Gassale (Figure 13) is of particular interest in terms of the primary study of vegetation and comparison with the neighboring territories.

The object of study was the tundra vegetation on plakors. Structural and physiological features of climax vegetation reflect the optimal strategy for land exploration in these climatic conditions. As noted by Y. Odum (1975), species largely replace each other in time and space. Thus, when studying the landscape level of the organization of vegetation, it is logical to rather use structural and physiognomic characteristics of vegetation than floristic, since the former are largely associated with the functioning of ecosystems, the specificity and intensity of biological cycles, and the productivity of biota, i.e., basic functional features of elementary ecosystems.

Climax vegetation of the Arctic zone is represented by tundras. Polydominant cryophytic plant communities that developed on automorphic soils and zonal habitats of the tundra zone in watershed areas, and having low height and complex horizontal structure, refer to tundras (Rebristaya, 2013).

The climate of the Taz Peninsula is characterized by the severity of the thermal regime: the average annual temperatures in the polar part are –8, –11° C. In warm time, the radiation balance is positive with the largest values in June and July. An average of 394 mm of precipitation falls (Masalkin et al., 1989).

The Taz Peninsula is located in the zone of tundra-gley and tundra illuvial-humus soils of the Subarctic, in the North-Siberian province (Vasilevskaya et al., 1986). The tundra as a whole is characterized by a low ash content of plants and an extremely low rate of its decomposition, resulting in poorly decomposed peaty bedding horizons. Gley-forming process is widespread, including the upper-permafrost. The tundra's gley soils have some very specific features such as cryogenic ironification, which is caused by the migration of restored iron compounds to the freezing fronts and deposition on the oxidation barrier, and manifested by the formation of an ocher rim on the upper boundary of the gley horizon. Cryogenic processes have a significant effect on soil morphology and properties: frost cracking, frost heaving, thermokarst (Kobeleva, 2014).

The main reason for the difficulty in identifying elementary units in the tundra vegetation cover is a high level of its spatial heterogeneity. In the tundra, along with extensive structurally and floristically homogeneous patches of vegetation, in a small area fragments of communities coexist, clearly differing in composition, longline structure and ecological requirements. Moreover, in terms of heterogeneity, there may be no correspondence between the vegetation cover and other components of the tundra landscape, such as soils, relief, and underlying rocks. The alternation of fragments of the complex may be due to the alternation of microrelief forms (especially cryogenic), the action of permafrost processes in soils, and frequent changes of rocks and surface sediments. Elements of the complex can build a succession row, being in dynamic equilibrium with each other (Koroleva, 2008).

Moreover, the weak edificator properties of tundra vegetation, which determine the predominance of the influence of abiotic environmental factors on its formation, make the tundra – like other periglacial types of vegetation – a good indicator of ecosystem change under the influence of global climate change and anthropogenic influence. Based on the high correlation of the structural features of tundra vegetation with environmental factors, it is advisable to study its structural and physiognomic features that are associated with the specificity and abundance of the main life forms of higher plants and the main cenosis-generators – mosses and lichens. Depending on the dominant groups of plants and lichens, sedge-shrub-moss-lichen tundra, shrub-lichen tundra, shrub-lichen-moss tundra and shrub-moss tundras were identified in the area of research.

In essence, these are not units of vegetation classification, but rather, types of basic landscape elements of the facies level, the distribution of which reflects the landscape structure of vegetation well and can be clearly seen on aerial and satellite images. Then it can serve as a basis for further monitoring of dynamic processes in tundra ecosystems where vegetation is the ecological framework.

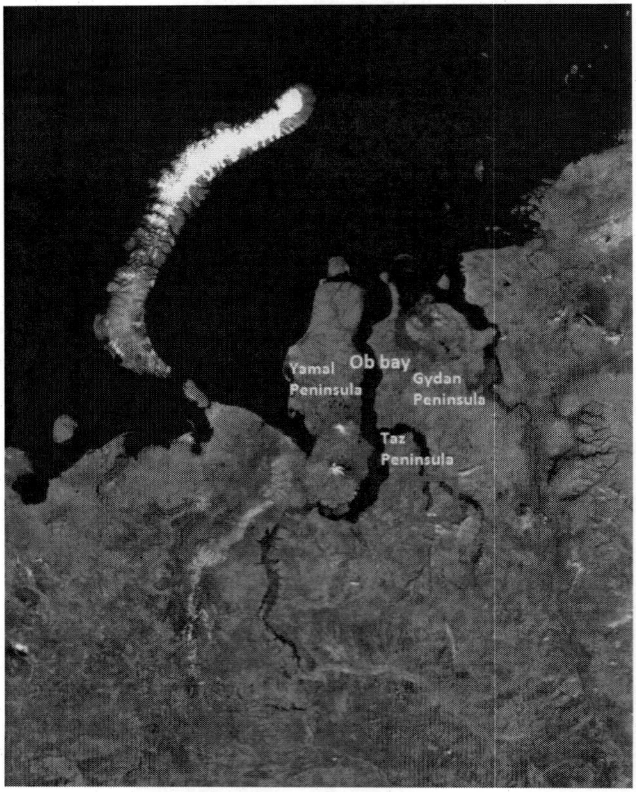

Figure 13. Research area.

Figure 14. Sedge mire in depression among the tundras occupying higher areas of mesorelief (Photo by Dr. Igor Volkov).

It should be borne in mind that in addition to tundra vegetation, which occupies the main territory of flattened tops and slopes of mounds (i.e., plakors' habitats), azonal types of vegetation are present in the area of research, whose existence is connected with the presence of specific habitats – humidified depressions, drained thaw lake basins (local name: *hasyreis*), permafrost cracks of giant frozen landmasses. The largest territories in the area of research are occupied by the cotton-sedge hydromorphic communities (Figure 14), common in depressions in the lower parts of mounds and near lakes. Meadows formed at the place of lowered *hasyreis* are of particular interest in terms of understanding the formation of vegetation (Figure 15). *Hasyreis* are lakes that drained into other lakes or hydrological nets located at a lower level (Kirpotin, 2005) due to a slow uplift of individual sections of Western Siberia (Zemtsov, 1976) and the activation of thermokarst processes as a result of present-day climate change (Kirpotin et al., 2009).

The vegetation growing on the bottom of drained lakes is represented by highly productive grass and cotton-grass meadows, where the height of the grass layer reaches one meter. The existence of such vegetation in the tundra zone in areas enriched with lake sediments suggests that the formation of moss- and lichen-dominated phytocoenoses at plakors is not only induced by climate but also connected with the low soil productivity.

Figure 15. Highly productive meadows on the site of a drained lake (Photo by Dr. Igor Volkov).

Undoubtedly, low temperatures play a major role in the formation of soils saturated with the poorly decomposed organic remains, which determines the decrease in the intensity of the biological cycle and, as a result, the poverty of the soil with mineral nutrients available to plants. In other words, there is an imbalance between the climate potential and the real productivity of vegetation in the tundra zone, which is especially noticeable in the formation of more fertile soils and more productive ecosystems of the mammoth fauna in even more severe conditions of the Late Pleistocene ice age (Lojko et al., 2016). At that time there were many species of large animals. The enzymatic activity of their digestive systems ensured the rapid decomposition of organic matter and contributed to the intensification of the biological cycle and, accordingly, the formation of soil with nutrients available to plants. In fact, this solved the problem of delaying the decomposition of plant residues under conditions of low soil temperatures, and the waste products of large mammals entering the soil contained organic matter easily accessible to plants and soil reducers, which determined the prevalence of highly productive vegetation in the tundra-steppe. With the disappearance of the mammoth fauna, modern tundras have been formed. Their vegetation corresponds to the biological circulation of the matter formed in modern conditions rather than to the climate (Zimov et al., 2006; Zimov et al., 2009; Loiko et al., 2014).

Figure 16. Shrub-lichen tundra on plakors (Photo by Dr. Igor Volkov).

Figure 17. Vertical profile of the shrub-lichen tundra (Photo by Dr. Igor Volkov).

Figure 18. Lichen cover in the shrub-lichen tundra (*Cladonia arbuscula, C. digitata*) (Photo by Dr. Igor Volkov).

Figure 19. A frost crack, a suffusional depression and a lake in the shrub-lichen tundra (Photo by Dr. Igor Volkov).

In describing the types of tundra of the study area, the characteristics of the species composition of various groups of plants and lichens and their role in the formation of tundra longlines are given. Structurally similar communities can be attributed to one type, for example, shrub-lichen tundra dominated by wild rosemary (*Ledum decumbens*) in the shrub layer and *Empetrum subholarcticum*. Despite the general structural and physiological features, these communities are considered as different subtypes of the tundra, since they have floristic and structural differences, and occupy different types of habitats.

The most common type of tundra vegetation in the study area is shrub-lichen tundra (Figure 16), occupying about 40% of the surface in the study area. This vegetation occupies flattened tops and upper parts of the slopes of mounds. *Ledum decumbens* is the main dominant in the shrub layer (Figure 17), which in the driest variants of such tundra becomes the absolute dominant in the shrub layer. Dwarf-shrub birch (*Betula nana*) usually plays a relatively minor role. Its significance increases in depressions and breaks of slopes. There are constantly, but with a small abundance, shrubs – lingonberry (*Vaccinium vitis-idea subsp. minus*), blueberries (*Vaccinium uliginosum subsp. microphyllum*), black crowberry (*Empetrum subholarcticum*) and the herb cloudberry (*Rubus chamaemorus*). In such variants of the tundra, mosses do not play a special role in communities, yielding absolute dominance in the ground cover to lichens having a projective cover of up to 90%. *Cladonia rangiferina, C. arbuscula, C. macroceras, C. deformis, Flavocetraria cuculata, Flavocetraria cucullata, Gowardia nigricans* and a number of other species are present in such tundras (Figure 18).

In their typical form, shrub-lichen tundras with wild rosemary dominance are very poor communities in the species of flowering and higher spore plants. The factors contributing to an increase in the species and biomorphological diversity of such communities include: permafrost phenomena, such as the formation of frost-cracking permafrost polygons, depressions, and lakes, all of which increase the spatial heterogeneity of habitats (Figure 19).

Figure 20. Transitional phytocenosis from the shrub-lichen tundra to the shrub-dwarf-shrub-moss tundra (Photo by Dr. Igor Volkov).

Figure 21. Cotton grass (*Eriophorum vaginatum*) in the shrub-lichen tundra (Photo by Dr. Igor Volkov).

Sedge hydrophilic communities usually develop in suffusional depressions, thermokarst subsidences, and frost cracks, introducing a variety of spatial organization of the upland tundras. Near such habitats, the shrub-lichen tundra is noticeably transformed – closeness and the height of the shrub layer increases. Dwarf birch (*Betula nana*) can act as a co-dominant (Figure 20), and in wetter depressions it can prevail over the *Ledum*, which nevertheless is always quite well represented in such phytocenoses. The significance of lichens in communities decreases, and the significance of mosses increases. Sedge (*Carex arctisibirica*) and *Arctous alpina*, and cotton grass (*Eriophorum vaginatum*) penetrate the community (Figure 21). Such communities are transitional variations to the shrub-dwarf-shrub-moss tundra, which will be described below.

Figure 22. The shrub-lichen tundra (foreground) and the sedge-shrub-lichen tundra (in the middleground) (Photo by Dr. Igor Volkov).

Figure 23. Sedge-shrub-lichen tundra (in the middleground) (Photo by Dr. Igor Volkov).

The sedge-shrub-lichen tundra can intersperse with the shrub-lichen tundra from which it is well distinguished by the reddish color that sedges (*Carex arctisibirica*) display at the end of the growing season (Figure 22). The surface of the sedge-shrub-lichen tundra is mainly covered with flat mounds 30–40 cm high and about one meter in diameter (Figure 23). The shrub tier of such communities, 10-15 cm in height, does not usually exceed 20% of the projective cover dominated by wild rosemary (*Ledum decumbens*) and dwarf birch (*Betula nana*), which is distributed in small spots, with a total cover of over 5%. Cowberry (*Vaccinium vitis-idea subsp. Minus*) and blueberries (*V. uliginosum subsp. Microphyllum*) mixed with black crowberry (*Empetrum subholarcticum*) are less abundant. A grass tier with a projective cover of about 10% is mainly represented by sedge (*Carex arctisibirica*), cloudberry (*Rubus chamaemorus*) and *Arctagrostis latifolia*.

The lichen layer in such communities varies from 90 to 100% of the projective cover. Lichens of the genus Cladonia (*Cladonia rangiferina, C. arbuscula*), *Flavocetraria cuculata* prevail. There are lichens *Cladonia stellaris, Cetraria laevigata, Flavocetraria cucullata, Peltigera aphthosa* and *Nephroma arctica*. Mosses are poorly represented in the community (about 5% of the projective cover). Lichens, spreading over the surface, suppress mosses. *Polytrichum juniperinum* is most common. Apparently, it has a greater growth rate than other mosses, and penetrates between lichen thalluses (Figure 24).

Figure 24. Moss (*Polytrichum juniperinum*) penetrates between lichen thalluses (*Cladonia arbuscula*) in the sedge-shrub-lichen tundra (Photo by Dr. Igor Volkov).

Figure 25. Permafrost mound with *Sphagnum* turf in the sedge-shrub-lichen tundra (Photo by Dr. Igor Volkov).

Fragments of the surface with the prevalence of *Sphagnum aongstrinum* (Figure 25) are preserved in some places on the hillocks. Although the growth of mosses as a result of competition with lichens largely slowed down, the peaty layer of living and decomposed mosses probably fulfills a water-retaining function, which allows the more moisture-loving and bryophilic species *Carex arctisibirica* to grow.

Shrub-moss tundras (Figure 26) are formed on more convex parts of the mounds. Their nano-relief is usually formed by permafrost mounds of 40-50 cm and depressions between them. According to the species composition of higher plants, this type of tundra resembles the shrub-lichen tundra, with the exception of the fact that in the ground cover mosses prevail over lichens, covering 60-80% of the soil surface, while lichen spots occupy only 7-10%.

Figure 26. Permafrost mound with *Sphagnum* turf in the sedge-shrub-lichen tundra (Photo by Dr. Igor Volkov).

In the dwarf-shrub layer of the dwarf-shrub-moss tundra, about 10 cm high, wild rosemary dominates, but almost always there is an admixture of dwarf birch, blueberry, lingonberry, and crowberry.

The mosaic of the shrub-moss tundra is associated with the heterogeneity of the moss cover and the activity of different species of moss. The main elements of the mossy mosaic are the spots of green mosses (*Aulacomnium turgidum, Polytrishum juniperinum, P. hyperboreum, Dicranum elongatum.*) (Figure 27) and the spots of Sphagnum moss (*Sphagnum aongstrinum*) (Figure 27).

In such communities, there is an increase in *Sphagnum aongstrinum* that actively "crawls" onto the surfaces covered with green mosses (Figure 28).

Figure 27. Moss cover interspersed with lichens in the shrub-moss tundra (Photo by Dr. Igor Volkov).

Figure 28. Mossy tier in the shrub-moss tundra with *Sphagnum aongstrinum* prevailing (Photo by Dr. Igor Volkov).

Figure 29. Shrub-grass-moss tundra on watercourses (Photo by Dr. Igor Volkov).

Figure 30. The ground cover and grass layer in shrub-grass-moss tundras (Photo by Dr. Igor Volkov).

Figure 31. Shrub-dwarf-shrub-moss-tundras (Photo by Dr. Igor Volkov).

This process can be supported by local watercourse and the presence of permafrost acting as an impermeable horizon and moisture donor, which is manifested in the activity of Sphagnum mosses that retain moisture well, due to the presence of hyaline cells.

The shrub-grass-moss (local – "yernik") tundras (Figure 29) usually stretch along the most humidified central depressions on the slopes of mounds along which streams flow. Usually, the upper shrub layer with a projective cover of about 30% is formed by dwarf birch of 50-60 cm in height, with the co-dominating *Salix glauka* and *S. lanata* that both reach 80 cm in height (Figure 29).

These are the tallest tundra communities in the study area. It is explained by their growth in the largest depressions of the relief. As a result, in winter the thickest snow layer forms here, which protects bushes from snow corrasion.

The shrub cover in such communities is relatively rarefied (Figure 30) and consists of blueberry, wild rosemary, and crowberry. The grass cover with a projective cover of over 30% is more developed. It mainly consists of *Equisetum boreale*, cloudberries, *Carex arctisibirica*, *Poa alpigena*, *Eriophorum angustifolium* and *Pedicularis lapponica*.

The moss cover has about 70% of the projective cover. *Sphagnum girgenshonii* prevails; *Hylocomium splendens*, *Polytrichum commune* are quite abundant.

The shrub-shrub-moss (willow) tundras (Figure 31) is the second most common type of vegetation after shrub-lichen tundras, occupying about 20% of the surface of the study area. This type of tundra, along with shrub-moss tundras, usually occupies slopes of mounds, giving preference to lower areas, where more snow facilitates the growth of a higher shrub layer.

In the shrub layer, from 30 to 50 (60) cm in height, there are dwarf birches and willows (*Salix glauka*), to a lesser extent *S. lanata*, occasionally *Duschekia fruticosa*. The projective cover of the shrub layer amounts to 25-40%. On more stretched/flattened areas, the height of shrubs decreases. In various types of depression and ravine that get covered with snow in winter, the height of shrubs increases.

Usually in such communities the shrub layer consisting of *Vaccinium vitis-idea subsp. minus*, *V. uliginosum subsp. microphyllum*, *Empetrum subholarcticum*, *Ledum decumbens*, *Arctous alpina*, and *Salex pulchra* is rather well developed.

Among herbaceous plants, *Carex arctisibirica*, *Calamagrostis laponica*, *Rubus chamaemorus*, *Pedicularis lapponica*, *Festuca ovina*, *Tofieldia coccinea* (on rare spots of bare ground covered with algae and black lichens), *Eryophorum vaginatum*, *E. angustifium*, and *Eryophorum vaginatum* are common.

Figure 32. Medallion tundras (Photo by Dr. Igor Volkov).

The projective cover of the lichen-moss layer exceeds 90%. The dominant species are mainly mosses from the genus *Polytrichum*, *Hylokomium splendens*. Lichens occupy no more than 20% of the projective cover, usually dominated by *Peltigera aphthosa*, *Cladonia rangiferina*, *C. arbuscula*, and *C. stellaris*, which cover the tops of frozen mounds or are found as an impurity on the surface of the mossy layer.

The formation of complexes with shrub-moss tundra dominated by *Ledum decumbens* is a specific feature of the shrub-shrub-moss tundra. The horizontal mosaic of such phytocenotic complexes is supplemented with spots dominated by *Empetrum subholarcticum*, *Arctous alpina*, mosses, and tussocks of cotton grass (*Eriophorum vaginatum*).

Despite the widespread use of cryogenic micro- and mesorelief in the form of permafrost hillocks and depressions between them, soils with signs of surface reclamation in the form of clay medallions, or frost heaving zones of soils, occupy a relatively small part of the study area (about 5%). On even, relatively flat surfaces of mounds, in places where the wind blows snow away in winter, there are medallion tundras (Figure 32). The vegetation around clay medallions resembles dwarf-lichen tundra dominated by wild rosemary; however, its phytocenotic significance decreases here. *Empetrum subholarcticum* and *Arctous alpina* are usually used as co-dominants of wild rosemary. The role of shrub birch is quite significant, being in some instances a part of the dominants. There are *V. uliginosum subsp. microphyllum, V. vitis-idea subsp. Minus, Salix polaris, Salix glauka*, which do not rise more than 7 cm above the soil surface. Among herbaceous plants *Eryophorum angustifolium*, *Carex arctisibirica*, *Festuca ovina*, *Tofieldia coccinea* (on clay medallions) are found.

On high hill tops, shrub-lichen tundras with crowberry dominant in the shrub layer – crowberry tundras – are common (Figure 33). This type of tundra is characterized by the significant role performed by mosses and lichens in the formation of phytocenosis (50-60% of the projective cover), which, together with algae, develop on the surface of the soil affected by cryogenic processes. Such communities form in conditions of minimum snow cover or its absence during the winter period, which is reflected in a decrease in the significance of wild rosemary and dwarf birch. Therefore, the dominance in the shrub layer moves to crowberry and *Arctous* (Figure 34). And if wild rosemary is still found in almost all the descriptions of this type of vegetation, dwarf birch is practically absent. The projective cover of *Equisetum boreale* sometimes reaches 10%. Less abundant, but practically always present, are lingonberries, blueberries, *Tofieldia coccinea*, *Festuca ovina*, and *Carex arctisibirica. Poa alpina*, *Pyrola grandiflora*, and *Salix glauka* are rarely present; *Minuartia biflora*, growing in a cushion-shaped life form, was found only once.

Mosses and lichens in this tundra cover a significant part of the surface – anything from 20 to 80%. In some phytocenoses, most of the surface can be occupied by bare ground.

The characteristic feature of the crowberry tundras is the dark-colored bare substrate patches covered with a film of algae and mosses from the *Polytrichum* genus (Figure 35) and a rare cover of lichens (mainly *Flavocetraria cuculata* and species from the genus *Cladonia*).

Finally, on the surface areas most exposed to cryogenic processes, the most marginal phytocenoses are formed, where the permafrost forms of the microrelief are covered with a rare cover of lichens, and the higher plants retain the turf Festuca ovina (Figure 36).

Figure 33. Crowberry-lichen tundra dominated by *Empetrum subholarcticum* (Photo by Dr. Igor Volkov).

Figure 34. Fragment of the crowberry tundra with dominance of Arctous alpina (Photo by Dr. Igor Volkov).

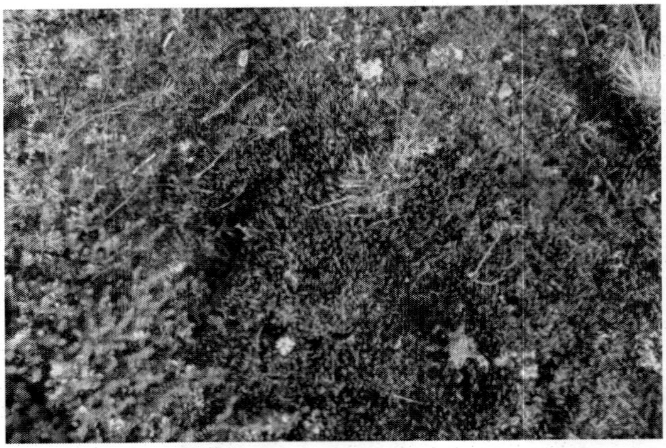

Figure 35. Spots of bare ground covered with mosses from the genus *Polytrichum* in the crowberry tundra (Photo by Dr. Igor Volkov).

Figure 36. Permafrost heaving areas (Photo by Dr. Igor Volkov).

4. Discussion

In the course of the research, different types of tundra vegetation on the transect from the village of Tazovsky to the village of Gazsale in the subzone of the southern tundra of the Taz Peninsula were studied. The study of tundra vegetation in terms of the structure of its features (types of life forms and their role in the formation of phytocenoses) made it possible to identify the shrub-lichen tundra with the dominance of wild rosemary (*Ledum*), dwarf-shrub-lichen (crowberry) tundra, sedge-shrub-lichen tundra, shrub-moss tundra with the dominance of wild rosemary in the dwarf-shrub layer, shrub-grass-moss (willow) tundra, shrub-dwarf-shrub-moss (willow) tundra and medallion tundra, which form conjugate ecological range corresponding to the diversity of habitats of tundra vegetation. Ledum rosemary and dwarf birch are the main coenogenizers of woody plants in the tundra types described. It is characteristic that dwarf birch absolutely dominates in shrub communities about 40 cm high, and with a height below 30 cm, domination passes to *Ledum*, regardless of the characteristics of the ground layer. At intermediate values of the height of the tiers of woody plants, mixed communities are formed — with co-domination of *Ledum* and dwarf-birch.

The predominant development of dwarf-shrub tundras on watershed drained plains – plakors – is associated with a small snow cover during the winter period, which does not contribute to the development of shrubs, but serves as a shield for shrubs from the action of winds. This is also noted in the Yamal Peninsula (Telyatnikov, 2003). A characteristic feature of the study area is the wide distribution of wild rosemary tundra. Ledum (*Ledum decumbens*) in this territory is the most active coenogeous species of both lichen and moss variants of tundra on plakors. This feature is the most characteristic feature of the vegetation of the subzone of the tundra of the Taz Peninsula, which seems to be a zone of ecological optimum for this species. Characteristically, the wild rosemary tundra in the south of the Yamal occupies small areas and is found on peat mounds and flat-polygonal tundra-mire complexes (Rebristaya, 2013). In Taimyr, wild rosemary is a member of the codominants in the cotton-grass-sphagnum communities (association Sphagno-Eriophoretum vaginati subass. Tipicum Waalrtr et al., 1994), which themselves represent the western “outposts” of tussock tundras, widespread in Eastern Siberia tundra (Matveyeva, 1998), i.e., in habitats not typical for the area of our study. Similar shrub-lichen tundras are widespread in the Bolshezemelskaya tundra in the European part of Russia (Larenko, 2014), which allows us to consider the southern tundras of the Taz Peninsula as an eastern outpost of this type of vegetation.

The sedge-dwarf-shrub-lichen tundra is a transitional variant to the transformation of the dwarf-shrub-moss tundra in the direction of the formation of climax for this territory – dwarf-shrub-lichen tundra. The development of the lichen cover is clearly secondary in nature and is likely due to climate change, as evidenced by a fairly thick layer of moss (5–10 cm) underlying the lichen layer prevailing on the surface. This trend is similar to the degradation of sphagnum moss in raised bogs which, at a certain stage of development of the accumulation of peat deposits, are covered with a layer of lichens blocking the development of mosses. Unlike such processes in bogs, which may reflect their stage of development, the process by which the tundra with moss cover is replaced by tundra with a predominance of lichens is clearly largely related to climate and can be used as the basis for monitoring the quantitative changes of tundra vegetation, taking into account the different color of these communities – darker in the moss cover and lighter in lichens. The sedge-dwarf-shrub-lichen tundra, as a kind of transitional variant, is well distinguished by its reddish coloration, which acquires sedge in the autumn, and can serve as another element for monitoring.

The horizontal type of the medallion tundra structure can be attributed to a 2-member regular-cyclic type consisting of two elements – a clay medallion resulting from an increase in pressure between the layers of the soil that freezes from the surface during the autumn period, and a layer of permafrost not thawed over the summer. N. V. Matveyeva (1998) considers this type of structure as characteristic of the Arctic tundras and found in typical tundras. For the spotted southern tundra, the 3-member regular-cyclic type of structure is more common. However, we observed a wide distribution of the 2-member regular-cyclic type in tundra communities in the forest-tundra zone of Taimyr (northern suburbs of Lake Pyasino), which, along with the observation of this type of structure in the southern tundras of the Taz Peninsula, makes us question the “dependence” of this type of structure on the zone as its formation is most likely influenced by the peculiarities of the behavior of the soil during the manifestation of surface frozen processes.

The high projective cover of the dwarf-shrub layer around the permafrost medallions is probably due to the increased availability of nutrient elements to the roots of plants in such habitats. S. Jonasson and S. E. Sköld (1983) noted basophilic vascular plants and cryptogams on polygonal soil, which were absent in the surrounding nutrient-poor dwarf-shrub wastelands. They were able to explain this phenomenon by a higher nutrient content in the heaving zone, as well as by the mechanical instability of polygonal soils (Körner, 1999). On the other hand, the very low height of the dwarf-shrub layer does not allow mosses and lichens to develop, which thrive in other types of dwarf-shrub and shrub tundras, whose conditions allow woody plants to rise above the soil surface. In fact, it is in this type of community that conditions are created that allow higher plants to prevail over the almost universally dominant tundra mosses and lichens.

As noted by B. A. Tikhomirov and G. I. Galazy (1952), many shrubs and dwarf shrubs in their northern and vertical limits almost completely lose their ability for generative reproduction, and completely switch to vegetative growth by forming young vegetative shoots with continuous dieback of the trunk in its basal part. Usually, this is accompanied by a small height in the shoots, barely rising above the surface of the soil (prostrate geophitized shrubs).

Figure 37. Hemiprotic life form of dwarf birch bushes in the southern tundras of the Taz Peninsula (Photo by Dr. Igor Volkov).

Figure 38. Sounding shoots of dwarf birch shown by arrows (Photo by Dr. Igor Volkov).

Such a life form of birch, in the study area, is characteristic of the shrub-dwarf-shrub-moss tundra, where its above-ground shoots rise to a height of 30-40 cm, forming hemiprostrate geophitized shrubs (Figure 37). More favorable conditions in the southern tundra zone allow the birch not only to form a greater height of the shrub layer (which is actually limited by the thickness of the snow cover, but also contributes to the preservation of skeletal shoots. Therefore, individual birch bushes often turn out to be tied with rhizomes more than 2.5 meters long. These cord-like rhizomes are formed from thin shoots (Figure 38), which in the ground and the layer of moss depart from the dwarf birch.

When the apical meristem dies off, so called "probing" shoots form vertical shoots and roots – forming a partial bush – and gradually thicken. The preferential distribution of the shoots in the uppermost layer of the soil and on its surface, under the layer of moss, probably makes it possible to avoid mechanical damage as a result of cryotube and other permafrost processes in the soil. Whilst aging, the connection with the parent plant may be interrupted, but such clonal groups of birch trees play a certain role in the formation of the complexity characteristic of shrubs and moss tundras. In addition, more favorable growing conditions contribute to seed reproduction of birch trees, although seed productivity in the shrub-dwarf-shrub-moss, according to our observations, is not very high.

In the medallion tundras, dwarf birch forms prostrate geophitized shrubs which, although only a few centimeters, can rise above the surrounding shrubs of *Empetrum subholarcticum* and *Arctous alpine*; but this summer increase is probably destroyed in the winter period as a result of snow corrosion. The vegetative growth of dwarf birch in the medallion tundra is clearly suppressed as a result of cryogenic processes in the soil, as a result of which it grows in the form of small compact shrubs. Under these conditions, seed reproduction is probably the main source of distribution of dwarf birch in communities. At the same time, the most probable source of seed distribution is birch, which grows in places with a deeper snow cover – in shrub-grass-moss tundras.

In shrub-grass-moss tundra, under the conditions of the deepest snow cover, dwarf birch forms orotropic shrubs. It is likely that the vertical growth of birch shoots in these habitats to some extent blocks the vegetative mobility of plants. As a result, resources are spent on the formation of a high shrub layer and seed productivity, which is nevertheless not very high.

Thus, dwarf birch in the tundra zone has at least three strategies corresponding to the growing conditions. The high ecological plasticity of this species in the tundra zone determines its wide distribution in various types of tundra vegetation.

In more extreme conditions at the elevations of the relief which are open to winds, dominance passes to more hardy shrubs – *Empetrum subholarcticum* and *Arctous alpine* – which form crowberry tundras.

The harsh habitat conditions of such communities reduce the competitive pressure of shrubs, which contributes to the wide penetration of mosses and lichens, as more enduring organisms.

Another characteristic botanical and geographical feature of the tundra vegetation of the study area is the complete absence of Dryas tundras. This type of vegetation, according to M.Yu. Telyatnikova (2005), is found in the southern tundras of all sectors of the Siberian Arctic. Willow-Dryas-grassy tundras with *Dryas punctata* and, less often, with *D. octopetala subsp. subincisa* are characteristic of the southern tundra of Yamal (Rebristaya, 2013). We observed a wide distribution of communities with *Dryas punctata* on the shores of Lake Pyasino and in the southern tundras, where the *Dryas* is quite active in the permafrost zone.

The absence of species from this genus in the study area is probably the floristic feature of this area, associated with the history of the origin of the flora, and to some extent contradicts the version of the wide, often circumpolar, distribution of species typical of tundra vegetation. In general, the specificity of the vegetation of tundra in the investigated area suggests a certain uniqueness of the southern tundra of this sector of the Arctic.

CONCLUSION

The research conducted and an analysis of publications allowed us to reliably reveal some unique features of the vegetation of the southern tundras of the Taz sector of the Siberian Arctic. These features can be summarized in several compact conclusions.

1. The types of tundras in the study area form an ecological series determined by abiotic environmental factors.
2. For various types of tundra vegetation, specific structural and physiognomic features are characteristic, reflecting the peculiarities of functioning of the plants laying out their life forms in certain habitats.
3. For dwarf birch, which is one of the main tundra cenocedent plants, three phytocenotic strategies have been identified, related to its biomorphological features, manifested in various habitats.
4. The specific features of this southern tundra sector on the Taz Peninsula include the wide representation in the vegetation cover of the lichen-wild rosemary tundra (characteristic of the Bolshezemelsky sector of the tundra in the European Arctic sector of Russia) and the absence of *Dryas* tundras.

ACKNOWLEDGMENTS

This study was partially supported by funds for basic projects of the Russian Academy of Sciences and by grants № 18-05-60264 and 17-55-16008 from the Russian Foundation of Basic Research. We also express gratitude to the large-scale research facility «System of experimental bases located along the latitudinal gradient» for fruitful cooperation (http://ckp-rf.ru/usu/586718/).

REFERENCES

Aleksandrova, V. D. 1971. Experience in the analysis of the structure of vegetation on the border of phytocenoses of spotted and tubercular tundra in Western Taimyr. *Biocenoses of the Taimyr tundra and their productivity*. 185–197. (in Russian).

Aleksandrova, V. D. 1970. The mosaic dynamics of plant communities of spotted tundra in Arctic Yakutia. *Mosaic plant communities and their dynamics: Reports at a scientific symposium*. 3–5. (in Russian).

Avdonina, N. S., Dolgoborodova, S. O. 2018. Covering geopolitical problems in the context of the Arctic exploration in the American media discourse (based on The New York Times content analysis). *Arktika i Sever* [*Arctic and North*], 33, pp. 178–191.

Bliss, L. C., Feng Sheng Hu. Tundra. *Encyclopedia Britannica*. https://www.britannica.com/science/tundra.

Chernov, Yu. 1985. The Living Tundra. D. Love, translator. *New York: Cambridge University Press.*

Gorodkov, B. N. 1916. The experience of dividing the West Siberian lowland into botanical and geographical areas. *Ezheg. Tobol, lips. Museum XXVII* 56 p. (in Russian).

Gorodkov, B. N. 1929. *Tundraless woodland (botanical and geographical observations in Western Siberia).* Priroda. 3, 219–240. (in Russian).

Golubchikov, Yu. N. 1996. Geography of mountain and polar countries. *Publ. Moscow State University.* 304 p. (in Russian).

Grigoriev, A. A. 1946. Subarctic. Experience characteristics of the main types of physiographic environment. *Publisher Academy of Sciences of the USSR. Moscow-Leningrad.* 171 p. (in Russian).

Ives, J. D. and Barry, R. G. 1974. Arctic and Alpine Environments. *London: Methuen.*

Jonasson, P. L., Sköld, S. E. 1983. Influences of frost-heaving on vegetation and nutrient regime of polygon-patterned ground. *Vegetatio*. 53, 97–112.

Josephson P. R. 2014. The conquest of the Russian Arctic. *Cambridge: Harvard University Press*, 441 p.: ill., maps.

Kirpotin, S., Polishchuk, Yu., Zakharova, E., Shirokova, L., Pokrovsky, O., Kolmakova, M., Dupre, B. 2008. One of Possible Mechanisms of Thermokarst Lakes Drainage in West-Siberian North. *International Journal of Environmental Studies.* 65 (5), 631–635.

Kirpotin, S., Berezin, A., Bazanov, V., Polishchuk, Y., Vorobiov, S., Mironycheva-Tokoreva, N., Kosykh, N., Volkova, I., Dupre, B., Pokrovsky, O., Kouraev, A., Zakharova, E., Shirokova, L., Mognard, N., Biancamaria, S., Viers, J., Kolmakova, M. 2009. Western Siberia wetlands as indicator and regulator of climate change on the global scale. *International Journal of Environmental Studies.* 66 (4), 409–421.

Kirpotin, S., Polishchuk, Yu., Bryksina, N. 2009. Abrupt Changes of Thermokarst Lakes in Western Siberia: Impacts of Climatic Warming on Permafrost Melting. *International Journal of Environmental Studies. 1029-0400*, 66 (4): 423–431.

Kirpotin, S. N. 2018. Landscape ecology with the basics of environmental management: a training manual. *3rd edition, Tomsk: Publishing House of Tomsk State University.* 288 p. ISBN 978-5-94621-701-1. (in Russian).

Kobeleva, N. V., Chumakova, A. V., Cherepanov, S. V. 2018. The structure of the land cover of typical tundras of the Taz Peninsula. *International research journal*. 8 (74), 67–73. (in Russian).

Kobeleva, N. V. 2012. Large-scale ecological-phytocenotic mapping on the basis of aerial photographs and gis-technologies (on the example of the central part of the Taz Peninsula). *News of the Samara Scientific Center of the Russian Academy of Sciences*. 14, 1 (6), 1607–1617. (in Russian).

Kobeleva, N. V. 2014. The indicator role of tundra vegetation in the compilation of large-scale soil maps (using the example of the Tazov Peninsula). *Theoretical and Applied Ecology*. 1. 87–92. (in Russian).

Komarov, V. L. 1926. Introduction to the study of vegetation of Yakutia. *USSR Academy of Sciences.* 183 p. (in Russian).

Körner Ch. (1999). Alpine Plant Life. *Berlin.* 338 p.

Kornienko, S. G., Yakubson, K. I. 2011. Study of the transformation of vegetation in the areas of the Taz Peninsula according to satellite imagery. *Arctic: Ecology and Economics*. 4, 46–57. (in Russian).

Krishtofovich, A. 1932. *Climates of the past of the North Polar Region*. Nature, № 4. (in Russian).

Koroleva, N. E. 2008. The main biotopes of mountain and zonal tundras of the Murmansk region. *Vestnik MGTU*. 11, 3, 533–543. (in Russian).

Laruelle, M. 2014. Russia's Arctic Strategies and the Future of the Far North. *Armonk,: M. E. Sharpe, Inc.*. 251 p.

Loyko, S. V., Bobrovsky, M. V., Istigechev, G. I., Kritskov, I. V. 2016. Non-climatic causes of low productivity of tundra ecosystems. *Modern concepts of ecology of biosystems and their role in solving problems of nature conservation and nature management: materials of the All-Russian school-conference devoted to the 115th anniversary of A. A. Uranov*. 253–256. (in Russian).

Loyko, S. V., Kulizhskiy, S. P., Mironycheva-Tokareva, N. P., Istigechev, G. I., Kritskov, I. V. 2014. Non-climatic causes for low productivity of Siberian tundra ecosystem. *International Journal of Environmental Studies*. 71(5). 605–610. https://doi.org/10.1080/00207233.2014.942544.

Lukin, Y. F. 2013. Russian Arctic in a changing world: *a monograph ... (Arctic) Federal University Lomonosov*. 284 p. (in Russian).

Masalkin, S. D., Rebriskaya OV, Kobeleva N. V., Ilina I. S. and others. 1998. Characteristics of geological and soil-plant features of the territory of the gas condensate Location of the North of the Tyumen region. *Moscow.* 98 p. (in Russian).

Matveyeva, N. V., Chernov Yu. I. 1978. Polar deserts of the Taimyr Peninsula. *Botanical Journal, Vol. 61, No. 3*. 313–325. (in Russian).

Matveyeva, N. V. 1998. Zonality in the vegetation cover of the Arctic. *SNb.: BNN RAS*. 220 p (in Russian).

Milne, G. 1936. Normal erosion as a factor in soil profile development. *Nature.* 138. 548–549. doi:10.1038/138548c0.

Milne, G. 1947. A soil reconnaissance journey through part of Tanganyika Territory December 1935 to February 1936. *Journal of Ecology.* 35. 192–265. doi:10.2307/2256508.

Odum, Y. 1975. Fundamentals of Ecology. Moscow: *Mir.* 740 p. (in Russian).

Ørbæk, J. B. et al., 2007. Arctic alpine ecosystems and people in a changing environment, Meppel. *Springer-Verlag Berlin Hindenburg*.

Polishchuk, Y.M., Bogdanov, A. N., Polishchuk, V. Y., Manasypov, R. M., Shirokova, L. S., Kirpotin, S. N., Pokrovsky, O. S. 2017. Size distribution, surface coverage, water, carbon, and metal storage of thermokarst lakes in the permafrost zone of the western siberialowland. *Water, Volume 9, Issue 3, Article number.* 228, doi: 10.3390/w9030228.

Rebristaya, O. V. 2013. Flora of the Yamal Peninsula. *Modern state and conditions of formation*. St. Petersburg. 312 p. (in Russian).

Shumilova, L. V. 1962. Botanical geography of Siberia. *Tomsk State University Press*. 440 p. (in Russian).

Telyatnikov, M. Yu. 2001a. Activity and species richness of latitudinal geographic groups of species (on the example of shrub-green moss tundra of the Yamal Peninsula). *Botan. journals*. 86, # 3, 86–96 (in Russian).

Telyatnikov, M. Yu. 1998. Analysis of tsenoflor of typical tundras of the Yamal Islands on the island. *Study of biological diversity using comparative floristics. SNb: SRIH SNbGU.* 201–208 (in Russian).

Telyatnikov M. Yu. 2016. Comparison of latitudinal geographic groups of coenoflora sedge-chilocomium (moss) tundras of the Yamal Peninsula, Taimyr Peninsula and Yakutia. *Flora and Vegetation of Siberia and the Far East. Materials of the III Russian conference. Krasnoyarsk: Kyrgyz State National University.* 187-188 (in Russian).

Telyatnikov, M. Yu. 1993. On some peculiarities of vegetation cover in typical tundras of the Yamal Peninsula. *Sib. biol. Journals*. 42–46.

Telyatnikov, M. Yu. 2000a. On the main regularities of the distribution of latitudinal and altitudinal zones of the coenofloras of the subarctic forest-tundra and high mountains. *Modern Problems of Botanical Geography, Cartography, Geobotany, Ecology. SNB: BIN RAS*. 91–92/ (in Russian).

Telyatnikov, M. Yu. 2006. On the main regularities of the distribution of the latitudinal elements of the tsenoflor of the forest-tundra in the West Siberian Subarctic. *Problems of the study of the plant Nokrov of Siberia. Tez. report II Russian scientific. conference Tomsk. TSU*. 141–142. (in Russian).

Telyatnikov, M. Yu. 1996. Tundra communities. *Green Book of Siberia. Rare and protected plant communities. Novosibirsk: Science*. p. 185, 190-197, 199-207, 213-214, 219-222, 226-229, 231-232. (in Russian).

Telyatnikov, M. Yu. 2003a. Comparison of the latitudinal elements of the coenoflora of sedge-heilocomium tundras of Yamal and Taimyr. *Botan. Journal*. 88. №1. 69–76. (in Russian).

Telyatnikov, M. Yu. 2006. Vegetation typical of the Yamal tundra peninsula. *Novosibirsk: Science*. 123 p. (in Russian).

Terasmae J., Reeves A. 2009. Tundra. The Canadian Encyclopedia. https://www.thecanadian encyclopedia.ca/en/article/tundra.

Tikhomirov, B. A., Galaziy, G. I. 1952. Determination of the age of the siversii ice (*Sieversia glacialis* R. Br.) and some questions about the life span of a plant. *Botanical Journal*. 32, № 3. (in Russian).

Tolmachev, A. I. 1932. Flora of the central part of eastern Taimyr. *Part 1 Proceedings of the Polar Commission. L. USSR Academy of Sciences*. 8, 126 p. (in Russian).

Tolmachev, A. I. 1944. On the Conditions for the Existence of Tertiary Tundras of the Arctic. *USSR Botanical Journal*. 29, No. 1. (in Russian).

Vasilevskaya, V. D., Ivanov V. V., Bogatyrev L. G. 1986. Soils of the north of Western Siberia. *Publishing House of Moscow. Univ.* 227 p. (in Russian).

Volkov, I. V. 2018. Study of the spatial organization of high-mountain phytosystems // Problems of Botany of Southern Siberia and Mongolia. *A collection of scientific articles on maleials of the XVII International Scientific and Practical Conference. Barnaul.* 33–37. (in Russian).

Yurtsev, B. A. 1959. Alpine flora of Mount Sokuydah and its place among the mountain floras of the Arctic Yakutia. *Botan. Journals*. 44, No 8, 1171–1177.

Yurtsev, B. A. 1982. Flora as a natural system. *Bul. MONN Separate biol.* 87, No. 4, 3–22. (in Russian).

Yurtsev, B. A. 1987. Flora as a basic concept of floristics: concept content, approaches to the study. *Theoretical and methodological problems of comparative floristics. D: BINRAN.* 13–28. (in Russian).

Yurtsev, B. A. 1982. Flora as a natural system. *Bul. MOIP Separate biol.* 87, No. 4, 3–22. (in Russian).

Yurtsev, B. A. 1968. Flora Suntar-Khayat: The Problems of the History of the High-Mountain Landscapes of the North-East of Siberia. D. *Nauka,* 235 p. (in Russian).

Yurtsev, B. A., Kamelin, R. V. 1991. Basic concepts and terms of floristics. *Perm: vocational school.* 80 p. (in Russian).

Yurtsev, B. A., Katenin, A. E., Koroleva, T. M., et al. 2001. The Experience of Creating a Network of Biodiversity Monitoring Points in the Asian Arctic at the Level of Local Floras: Zonal Trends. *Botan. journals.* 86, № 9, 1–27. (in Russian).

Yurtsev, B. A. 1966. *Komarovskie readings, Hypoarctic botanical-geographical belt and the origin of its flora.* M.; D .: Nauka, 19. 94 p. (in Russian).

Yurtsev, B. A. 1991. *Problems of allocation of the tundra vegetation type.* Botan. journals. 76, No. 1. P. 30–41. (in Russian).

Yurtsev B. A., Netrovsky V. V., Korobkov A. A. and others. 1979a. Overview of the geographical distribution of vascular plants of the Chukchi tundra. *Message 3121. Bul. MOIP Separate biol.* 84, issue. 5, 111–122. (in Russian).

Yurtsev, B. A. 1970. On the position of the polar coast of northeastern Siberia in the Pleistocene. *Northern Arctic Oceans and its coast in the Cenozoic. D.: Gidrometizdat.* 494–499. (in Russian).

Yurtsev, B. A. 1977a. On the ratio of arctic and alpine subarctic floras. *Problems of ecology, geobotany, botanical geography and floristics. D.: Nauka,* Pp. 125-138. (in Russian).

Yurtsev, B. A., Petrovsky, V. V., Korobkov, A. A. and others. 1979b. Overview of the geographical distribution of vascular plants of the Chukchi tundra. Post2. *Bul. MOIP Separate biol.* 84, vol. 6, 74–83. (in Russian).

Yurtsev, B. A. 1987a. Plant populations as an object of geobotany, floristics, botanical geography. *Botan. Journals.* 72, №5, 581–588. (in Russian).

Yurtsev, B. A. 1981. Relict steppe complexes of Northeast Asia. *Novosibirsk: Science.* 168 p. (in Russian).

Yurtsev, B. A. 1977b. Some issues of botanical geography of Northeast Asia. *Botan. journals.* 62, № 6, 832–845. (in Russian).

Yurtsev, B. A. 1988. *The main directions of modern science of vegetation.* Botan. Journals. 73, №10, 1380–1396. (in Russian).

Yurtsev, B. A. 1987b. The role of the historical factor in the development by plants of extreme conditions of the subzone of the arctic tundra (on the example of Wrangel Island*). Botan. journals.* 72, №11, 1649–1661. (in Russian).

Yurtsev, B. A., Tolmachev, A. I., Ribristaya, O. V. (1978) Floristic restriction and separation of the Arctic. *Arctic floristic region. D.: Nauka.* 9–104. (in Russian).

Yurtsev, B. A., Zverev, A. A., Katenin, A. E., et al. 2002. Gradients of taxonomic parameters of the local and regional floras of the Asian Arctic (in the network of biodiversity monitoring points). *Botan. Journals.* 87, № 6, 1–28. (in Russian).

Zhitkov, B. M. 1913. Yamal Peninsula. *Notes of the Russian Geographical Society on general geography*. 49. 384 p.

Zimov, A., Schuur Edward, A. G., Chapin, F. Stuart III. 2006. Permafrost and the Global Carbon Budget. *Science*. 312, 1612–1613. www.sciencemag.org.

Zimov, N. S., Zimov, S. A., Zimova, A. E., Zimova, G. M., Chuprynin, V. I. and Chapin F. S.III. 2009. Carbon storage in permafrost and soils of the mammoth tundra-steppe biome: Role in the global carbon budget. *Geophysical Research Letters*. 36, L02502, doi:10.1029/2008GL036332.

In: The Arctic: Current Issues and Challenges
Editors: Oleg S. Pokrovsky, Sergey N. Kirpotin et al.
ISBN: 978-1-53617-306-2

Chapter 7

CONSERVATION ISSUES OF MIGRATORY ANSERIFORMES IN THE ARCTIC: THE EXPERIENCE OF THE YAMAL-NENETS AUTONOMOUS OKRUG

S. B. Rosenfeld[1,2,*], G. V. Kirtaev[1], M. Yu. Soloviev[3], N. V. Rogova[4], S. N. Kirpotin[5] and D. O. Zamyatin[6]

[1]The A.N. Severtsov Institute of Ecology and Evolution of the RAS, Moscow, Russia
[2]The Gydanskiy State Nature Reserve, Russia
[3]Lomonosov Moscow State University, Moscow, Russia
[4]The Goose, Swan and Duck Study Group of Northern Eurasia
[5]Bio-Clim-Land Centre of excellence, Tomsk State University, Russia
[6]The Department of External Relations of the Yamal-Nenets Region, Russia

ABSTRACT

The principles of nature conservation and hunting management in Russia at the present stage are far from the principles of sustainable use of natural resources. Hunting legislation is not capable of acting as a regulatory mechanism. There is no monitoring of species, subspecies and populations. There is no record of breeding success, the procedure for obtaining a hunting license has become much simpler. In addition, the Okhotminimum exam has been abolished.

The reduction in the number of waterfowl resources is also caused by an additional set of factors: spring hunting of geese continues; imperfection of the rules of hunting; imperfection of Article 11 (Paragraph 2) of the Federal Law No. 209 dated July 24, 2009; lack of international cooperation on wise use of waterfowl; lack of monitoring of the status of populations and their habitats; lack of mechanisms for assessing production, wetland's conservation, andpoaching.

Migratory Anseriformes especially suffer from this, since Russia has not yet adopted standard approaches to managing their resources and conservation, which have proven themselves in all civilized countries.

* Corresponding Author's Email: rozenfeldbro@mail.ru.

These approaches are as follows: The territorial unit of management is the "flyway"; the elementary unit of management is the "migratory population".

Since 2010 our work has been carried out in the Yamal-Nenets Autonomous Okrug to improve the methods of monitoring and conservation of migratory waterfowl. At present, a method of aerial counts from ultralight airplanes and statistical data processing (spring and autumn migration, breeding and moulting sites) have been created.

These innovations are the quintessence of the best foreign experience and are unique to Russia. To assess the status of populations we use: landsat snapshots to determine extrapolation sites; GSM tagging to determine the timing and location of counts; identification of key areas for the creation of protected areas; estimation of the abundance and breeding success; evaluation of population dynamics; evaluation of anthropogenic impact, including hunting.

The result of this approach has been a modern method of calculating the estimated number and density of waterfowl. It is based on the following action algorithm: determination of extrapolation area; determining the type of habitat for each survey point for each species; determination of the area of all identified habitat types surveyed in the course of aerial surveys; determination of the total area of all identified habitat types in the work area; determination of the density of each species in each habitat; calculation of the number of each species in each type of habitat.

According to data from counts in the YANAO, temporary hunting free zones during the spring hunt period have been created in key locations identified during the monitoring. This protection method has no analogues in Russia, has proven its effectiveness and deserves to be extended to other regions.

Keywords: conservation, migratory Anseriformes, Arctic, Yamal

INTRODUCTION

The impact of groups of people – numbering many millions, and rapidly increasing – on the Earth's natural environment causes reduction in the numbers of most animal species, along with a decrease in the habitats available for them. The group of birds known as the Anseriformes (goose-family waterfowl) are no exception to this impact. Even in a country such as Russia, where one of the last great wilderness environments has survived, the Anseriformes continue to face many problems, arising primarily from breaches in environmental legislation, and from poor solutions during the implementation of programs for the preservation of the environment. Of all kinds of birds, the Anseriformes in particular have played, and continue to play an important role for human life – most especially as a valuable food source. For thousands of years, ducks, geese and brants have been sought after by hunters and trappers. Seasonal hunting of the Anseriformes has occasionally determined the survival of entire tribes and peoples. For many of the indigenous peoples of the North, the Anseriformes are, even today, a major source of food, with local populations carrying out large-scale collection of their eggs or hunting moulting birds.

Sports hunting of the Anseriformes is also a hobby pursued by huge numbers of people – and today, such hunting is one of the main factors affecting the Anseriformes population.

Regulating the hunting of the Anseriformes is not simple, because issues of hunting and preservation of the Anseriformes go hand in hand. Firstly, hunted and endangered species of these birds frequently migrate together, stopping to rest at the same locations. Secondly, regulating hunting and conservation of these birds must be carried out across their entire range, including their nesting sites, migratory and wintering spots. If protection in even just one of these locations becomes ineffective, and hunting is extensive, then the benefit of control measures at the other locations will be of little value. For example, if hunting of a species of which a major or even the entire world population goes on at migratory stops which are concentrated in a limited area, their total number can be severely reduced in just one season (Rosenfeld, 2010-2012). Deaths among species which migrate along sea coasts, and/or across few countries are significantly lower than among species making land-bound flights of several thousand kilometres across the land of countries where hunting them is permitted. This is the reason why, in civilised countries, hunting of the Anseriformes is restricted by international agreements and conventions governing migratory species. The most conspicuous success in resolving these problems, as well as in the coordination and effectiveness of hunting management, has been achieved in North America (Rosenfeld, 2014).

A rather worse situation applies in Europe, arising from the lack of coordination between different countries (Rosenfeld, 2014). Even so, the introduction of good practice from North America, and the establishment of a so-called Unified Anseriformes Protection Platform is currently under active discussion, and in December 2016 the first meeting of the European Goose Management International Working Group took place in Sweden. The participants of the meeting discussed rolling out an Action Plan for the Spitzbergen population of the Pink-Footed Goose, (Madsen, Williams, 2012) and the Taiga Bean Goose (Marjakangas et al., 2015).

Hunting has been well regulated in Kazakhstan, which operates very strict restrictions, with tough controls fulfilled by three different government services – where illegal hunting of protected species of birds results not only in huge fines, but also in criminal liability (Timoshenko & Rosenfeld, 2013). The legislation of Kazakhstan demands that if even a single case of poaching is established on the lands of any particular hunting farm, then the lease agreement for the farm is peremptorily terminated. From 2017 onwards, a full ban on spring hunting has been implemented within Kazakhstan.

Sustainable management of the Anseriformes is carried out on the basis of the following principles:

1. The unit of management will be “the migratory population”.
2. The territorial unit of management will be “the flyway”.
3. Constructive regulation of cull numbers will be based on the regular annual monitoring of the status of the populations.
4. The interests of the population (including the interests of indigenous peoples of the Far North) will be fully considered, since hunting of the Anseriformes is the basis of their livelihood, as well as that of sports hunters.

The breeding grounds for many Anseriformes species are located in the Russian Arctic. The welfare of the global population depends largely on a controlled system for migratory bird populations in the Arctic regions of Russia.

Management of the hunting of the Anseriformes in Russia is delegated officially to regional administrations. This particular factor is the largest problem and obstacle to implementing effective management – because it is impossible to implement efficient management of migratory species at a regional level. This situation is made worse by the fact that Russia has not signed up to any of the international conventions on migratory species, nor to their framework agreements, such as the Bonn Convention, or the Agreement on the Conservation of African-Eurasian Migratory Water Birds.

In order to regulate hunting, adequate assessment of the “resource volume” is required. This involves knowledge of the numbers of waterfowl and their breeding success; how many birds, and of what age and sex were culled; and how to calculate the proportion which could be safely culled. This would require annual monitoring (Rosenfeld, 2015). Yet how could this be achieved at a regional level? Birds fly through the areas of many territorial regions of the Russian Federation without stopping – while in some regions they may make brief stops, and in yet other regions they may rest for longer periods, and the third group of regions may be used as breeding grounds. If there is no federal, in other words, nationwide, state-organised monitoring, how can calculations be made of the number of birds culled in a particular region? In most regions, monitoring is either not carried out at all, or is carried out on such an ad-hoc basis that it is incapable of providing objective evidence on the state of the population, or on the state of a species as a whole. In most regions, the survey data on estimation of the Anseriformes numbers is missing altogether, while the objective data on multi-year counts is only available for single regions of Russia. At present, state monitoring of population figures for the Anseriformes in Russia is not compiled whatsoever. The lack of annually compiled and accurate data on flight patterns or population dynamics renders any objective approach to managing the hunting of the Anseriformes or competent managing their populations impossible. Population data trends and monitoring data exist only for specific species, involved in special projects. The information available to estimate global numbers has been compiled within the framework of international counting programs in the countries within the wintering range. Even so, the data permitting an assessment of the dynamics of the numbers in the Russian populations is very scarce. Individual areas – most particular those in the north and east of Russia – have not ever been studied. The boundaries of habitats in these areas require clarification.

We should additionally mention assessments of culls. At the federal level in Russia, no objective assessments of culls are possible, since according to Article 11 (Paragraph 2) of the Federal Law No. 209-FZ dated July 24, 2009, operations are carried out without regard to species, subspecies or populations, but on a generalised category covering all kinds of geese, brants and ducks. Formally, calculation of the cull numbers in Russia is in the hands of the Federal State Budget Institution “Centre Okhot Konrol”. However, since data about the culls of geese and ducks (listed as “geese” and “ducks”, but without any mention of which species) is received by this federal institution from the Russian regions, these figures have no connection with actual reality. Instead the information is deduced from the analysis of the numbers of hunting licences handed in, in which hunters have indicated how many ducks or geese were culled. Furthermore, this information is impossible to verify, and the level of discipline among hunters with regard to returning completed hunting licences is extremely low.

A further blow to proper management of the hunt was the abolition of the mandatory State Hunting Examination (Okhotminimum Exam). The procedures for obtaining a hunting licence have been so far simplified that all hunters must do is to pay the fee and sign a declaration that they are familiar with the minimum hunt requirements. The title of “a hunter” can be given exclusively to those who know and observe the hunting laws, the rules and ethics of the hunt. Most people who come to the hunting grounds with a gun and sign a paper that they are familiar with the rules, actually have little clue about the rules and traditions of the hunt, or of the species who make up the wildfowl population, or of the biology of the primary hunting species. In the majority of cases, these so-called "hunters" have no idea which birds are listed in the Red Data Book of the Russian Federation – nor can they even tell a male from a female. There are groups of wealthy and high-positioned "hunters", very often belonging to the governing ranks of the regions. These, too, can hunt with impunity within the (theoretically) hunt-free areas of the Nature Reserves – since inspectors have little chance and even less motivation to challenge such people. Furthermore, the technical equipment in the hands of such wealthy "hunters" has moved on significantly.

Competent hunt management requires the provision of the so-called "rest zones" for the game birds and the preservation of hunting resources. This can ensure the long-term presence of game birds in the hunting grounds. Currently, the role of such zones is fulfilled solely by the so-called Specially Protected Natural Territories (SPNTs). The existing system for the Specially Protected Natural Territories of federal, regional and local significance in the Russian Federation does not fully meet the needs of conservation of the Anseriformes – primarily because the possibility to establish new SPNTs is limited, and comes into continuous conflict with the hunting lobby. Moreover, existing SPNTs sometimes disappear under pressure from hunt lobbyists.

Up to now, spring hunts have been officially permitted in the Russian Federation. The history of the struggle against spring hunts has now been going on for more than three hundred years – in the Russian Empire, in the USSR and in modern Russia. It's a separate problem that requires discussion in greater detail. Hunts during the spring migration, when many birds at reproductive age die, cause the greatest damage. Their stressed condition prevents them from building up reserves of fat that are needed for further migratory journeys (Blokhin, 2004, 2009). Monogamous species suffer particularly during spring hunts – swans, geese, and brants (Klaassen et al., 2006). Sustainable use of these susceptible target species is the use of the populations in ways and numbers that prevent their further depletion. This means that spring hunts should only be carried out on the males of species which do not form permanent couples, and who take no part in the brooding and raising of the young. There are no species of geese or brants which meet such conditions. From a visible point of view, male and female geese are indistinguishable. Geese do not only form couples, but remain in those couples for many years, often lifelong. The involvement of the male in guarding the nest and raising the goslings is an essential situation if the young are to survive. Geese fly in spring from their wintering grounds in already-established couples. The death of a partner means that no nesting will take place for at least a further breeding season. In spring, hunters kill birds which have successfully survived the autumn migration, wintering, and are ready to begin breeding. It has been shown that spring hunts are one of the major factors which most negatively affects the state of the goose and brant populations (Rosenfeld, Sheremetyev, 2016). Currently, the management of waterfowl hunting, and its timing, are decisions taken by the regions with no concern for the flight patterns or the biotopical distribution of protected species. No environmental conditions need be met before regional governors sign the orders to open a hunt. Decisions about quotas are taken on the basis of information which is far from any scientific studies. The locations of stops along the spring migration routes are well-known to the locals, and also to Russian hunters who go there each spring for goose-hunting. This means that every year geese are under pressure of the hunt, which inflicts significant damage to the flock. It's not only the pressure of hunting that causes such negative impact during spring hunts, but also the high level of anxiety that occurs in significant key locations. In spite of legal restrictions, the use of decoy ducks has practically disappeared in spring hunts – modern hunts presuppose killing birds while they are in flight and take no regard of either the gender or the species of the game birds. The destruction of flocks during the autumn migration has a lesser effect on population numbers. This is primarily due to the fact that statistically in autumn, as a rule, the abundance of the populations is higher and the damage to the reproductive members of the flock is less.

Up to now, Russia has permitted the use of lead buckshot, whilst its use has been long since banned in countries of Europe and North America. Lead poisoning of waterfowl happens when the birds swallow the lead pellets, mistaking them for food, or harmless stomach stones. In the vast majority of cases lead poisoning of birds is caused by the ingestion of lead shot pellets (Newth et al., 2013; Falk et al., 2006).

Tons of such shot pellets are scattered around hunting grounds – particularly those inhabited by feeding water birds. The toxic lead salts quickly enter their bloodstream, and lead poisoning occurs, with the birds dying of exhaustion resulting from muscular paralysis of their intestines.

This all means that at present, there is simply no scientifically-based or competent management of the hunting of waterfowl in Russia. This fact was specifically noted in the conclusions of the last two international conferences on the Anseriformes in Northern Eurasia, held in Elista (2011) and in Salekhard (2015). The legal framework for controlling hunting is wrongly-framed, while corruption among officials, and the low level of education and awareness among the population, allied to a lack of proper control, produce persistent and all-encompassing poaching. Hunting of waterfowl in Russia is an endemic problem, since it is carried out on non-scientific grounding, in conditions of an absence of control or effective legislation. As a result, the present numbers of the Anseriformes species which nest in the Russian Federation are steadily decreasing. The steady decline in the number of many flocks over recent decades has mounted to between 30% and 50% of their known numbers in the mid-twentieth century. There is an ongoing fall, which is at catastrophic proportions in Asia, in the range and numbers of the Anseriformes populations (Syroechkovskiy, Jr., 2006; Fox, Leaflor, 2018).

The responsibilities for protecting waterfowl are not limited to hunting alone, as it is often thought. Hunting, regulated by rationally-devised time periods and quotas, does not cause appreciable loss to the hunting species of birds.

Therefore, the most important aspects of waterfowl conservation are as follows:

1. Preventing the impact of all kinds of trauma (arising from the actions of people, vehicles or equipment, domesticated animals – especially in heavily-populated areas) on birds during their critical periods. By being located in the nesting areas of ducks or in the geese moulting sites, or simply by walking along the shores of water pools, it's possible to cause serious damage to these birds.
2. Control of habitat changes (river floodplains, lake shores, marshes, meadows, and so on) in respect to building works, land reclamation, and so forth. During the development of an area, it is vital to retain at least some part of its wetlands, for the conservation of the Anseriformes.

3. Compliance with legislative stipulations for environmental management, with particular reference to tackling environmental pollution, especially pollution of water areas by industrial or domestic waste. A principal cause of environmental damage that severely affects the Anseriformes is low ecological awareness among people of the negative effects of pollution, resulting from low public perception of environmental issues.

During their migrations, birds cross the borders of different administrative regions and countries. Of course, birds are entirely unaware of the existence of such man-made borders. However, national policy in different countries and regions can differ widely, and thus crossing these borders may be a crucial element in the annual cycle of the Anseriformes, with significant impact on their overall chances for survival. The Anseriformes need specific environments for their stops during migration, yet the quality of such sites is being degraded by human activity and worsening environmental conditions. In addition to this, the Anseriformes can become victims of naturally-occurring phenomena – such as climate change, epidemics or bird flu. Furthermore, the Anseriformes population is particularly susceptible to intensification and decline in the agricultural development, and also to the economic use of wetland locations.

To solve at least some of these problems and to work out a scientifically-based scheme for the management of the Anseriformes in the targeted pilot region – the Yamal-Nenets Autonomous Okrug – there has been devised a checklist of the following objectives:

- setting up aerial counting methods with the use of an ultralight aircraft and statistical data processing (during spring and autumn migrations, control of breeding and moulting sites);
- using Landsat imaging to clarify zones of extrapolation;
- using GSM tagging to determine the timing and location of head-counts;
- evaluation of the numbers and breeding success of several species, and assessment of the flock dynamics.

Related to the planned and current intensive development of hydrocarbon products in the Yamal-Nenets Autonomous Okrug, an urgent need has arisen to identify the key locations for the Anseriformes population. Locations of these places must be taken into consideration and used as one of the major criteria when establishing a network of Specially Protected Natural Territories.

1. The Use of an Ultralight Aircraft for Monitoring Migrating Anseriformes

When talking about modern methods of studying the Anseriformes, we do not only mean the effectiveness and credibility of the data compiled, but also the speed and the coverage area. This means that we are interested in acquiring large volumes of data over the briefest possible time-period. Doing this involves the use of an ultralight aircraft. Despite the obvious benefits available, no ornithological studies in Russia currently make use of an ultralight aircraft. Processing data in order to extrapolate the figures and determine key bird locations compiled using aerial counts, will require modification of the existing foreign methods with regard to their adaptation to the conditions of the Russian Arctic (Rosenfeld et al., 2016; Rosenfeld et al., 2017).

An aerial survey was carried out using a STERKH S-1 ultralight two-seat seaplane, Superstol category, equipped with Full Lotus inflatable floats (Figure 1). This kind of plane is ideal for count surveys, since it possesses the following characteristics: 180 degree view: upper wing, glazed cabin with opening side windows; cruising speed approximately 120 km/h, thus offering a small turning radius (200–250 m per manoeuvre); the minimum speed of controlled flight is less than 70 km/h, thus making it possible to fly at the same speed as the birds; inflatable floats, which make it possible to land on virtually any water body, as well as on ice and snow; the empty aircraft weighs roughly 350 kg; load capacity is 300 kg; gasoline consumption is less than 20 litres per hour; flight duration without refuelling is up to 6 hours.

Figure 1. The STERKH S-1 ultralight survey aircraft.

Over the entire period of work in the Yamal-Nenets Autonomous Okrug, the total mileage of aerial surveys has exceeded 150,000 km.

Carrying Out the Aerial Count

Aerial bird-counting flights were undertaken over pre-selected routes, using topographical maps showing potentially viable habitats. The length of these routes and intervals between them varied according to the locations and the areas of sites chosen for the survey. This means that in the potentially richest areas for waterfowl, such as Dvuobye, the interval was 0.5-2 km, while in the basins of the Pur, Taz and Nadym rivers, where the distance is not intensive, the interval was around 20 km. Every 10 seconds, by using an Iridium tracker, the coordinates of the plane were detected and recorded automatically. This meant that the entire flight trajectory was recorded. The flightpath was recorded with a Garmin GPS. The speed of the plane was between 80 and 120 km/h, and the altitude between 30 and 50 metres. The count was made over a strip of 800 m, 400 m on either side of the aircraft. The birds encountered were mapped using a Garmin GPS Dacota navigator, and photographed from a height of 10 to 20 metres (the camera was a Canon D 700, a 100-400 mm lens). Isolated birds or groups of birds numbering 10 or less, were similarly noted and photographed, if it proved impossible to identify their species using binoculars (Swarovski 10 x 42). For geo-referencing of the photographs, a GPS Receiver GP-12 Canon was used, and additionally, the time on the camera and GPS navigator was synchronized. Next, the photos were fixed to track points in the GEOSETTER freely distributed program. All information about the encounters of individual birds or clusters was recorded on a dictaphone.

Identifying the Birds

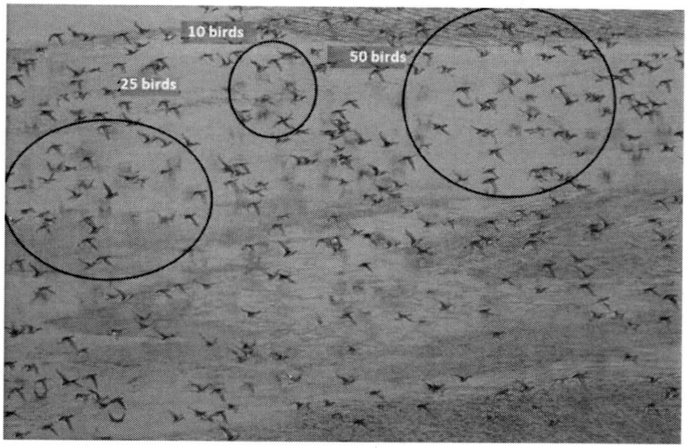

A

B

Figure 2. Example of photo-identification of numbers and species in mixed flocks.
A – Counting the numbers of birds while in flight.
B – Different species

To identify the make-up and number of the flocks of the Anseriformes, they were photographed using a Canon camera with a 100-400 mm lens – to be able to count the number of birds and determine the species composition of the flock later, in office conditions (Figure 2). Where the birds had formed large clusters (> 100 birds) we took a series of photographs at different parts of the flock, determined the percentage ratio of species and the proportions of young versus adult birds. This data was then extrapolated to the entire cluster.

2. Processing the Data

The areas for the aerial counts were calculated using a buffer in a GIS environment. Analysis of Landsat8 satellite images, and allocation of classes was handled using methods of visual and automatic interpretation, using neural networks with special training undergone in the ScanEx IMAGE Processor program. Further data processing and area calculation was handled in the Quantum GIS program. Clarification of the characteristics of vegetation cover was achieved using the Circumpolar Arctic Vegetation map (http://www.geobotany.uaf.edu/cavm/download.php).

2.1. GIS Data

The GIS data was created in digital form, using shapefiles in the international WGS 84 format. The GIS data includes the following thematic layers: aerial survey routes; the boundaries of the key territories of the Yamal-Nenets Autonomous Okrug most valuable and important for the conservation and restoration of waterfowl; main migration routes in the work area; waterfowl concentration sites; places of detection of the species listed in the Red Data Book of the Russian Federation and in the Red Data Book of the Yamal-Nenets Autonomous Okrug. These thematic layers were then converted to formats compatible with the ArcGIS Geographic Information System (Esri). The layers were obtained from http://gis-lab.info/qa/osm-adm.html, where a detailed description may also be found. During the construction of the thematic maps for visualisation, we have used water flow system layers (for line objects) and water bodies (for area objects) taken from the OpenStreetMap layers for the Yamal-Nenets Autonomous Okrug (http://beryllium.gis-lab.info/project/osmshp/region/RU-YAN). Thematic formatting for all layers was created using the Graduated Symbols method, with a width of 200 in the range of 0-1,000, a width of 1,000 in the range of 1,000-4,000, and a width of 2,000 in the range of 4,000-12,000.

2.2. Evaluating the Overall Population Numbers

We rate all the count data compiled during the population survey to be the minimum numbers. The figures do not include: birds which were located on the ground, on the water surface, or in flight. We accounted for the numbers of birds counted on each survey flight as a single sample, and thus the population number for the surveyed area was estimated from a total number of samples, weighted according to the distance of the survey flights. To reach a proper estimate of the undercount, with consideration of the extrapolation methodology, we should take into account the mosaic landscape patterns in the area being studied; the area of biotopes conducive to waterfowl; and similarly the population density of each species in its respective biotope. Selection of these biotopes was by considering the use made of the biotopes by waterfowl, and the most significant (the broad river floodplains, complexes of lakes and marshes) were mapped in greater detail (Figure 3). Satellite image analysis and class allocation were made using a neural network method of automatic classification, with special training undergone in the ScanEx IMAGE Processor program with further processing and area calculation being done in the Quantum GIS program. An obtained landscape map of the area was used to pinpoint registration locations of waterfowl. The extrapolation area was only marked with the locations of biotopes which were used by waterfowl.

2.3. Methodology for Calculating the Estimated Numbers and Densities

Estimated numbers represent an extrapolation of bird numbers counted during the aerial survey throughout the entire territory, including those areas not surveyed within the terms of the aerial counts. Statistical modelling makes it possible to extrapolate bird numbers to areas which were not surveyed, by using both qualitative and quantitative data as independent variables. This modelling makes it possible to put aside the method popularly used in aerial surveys which is predicated on a preliminary number of transects.

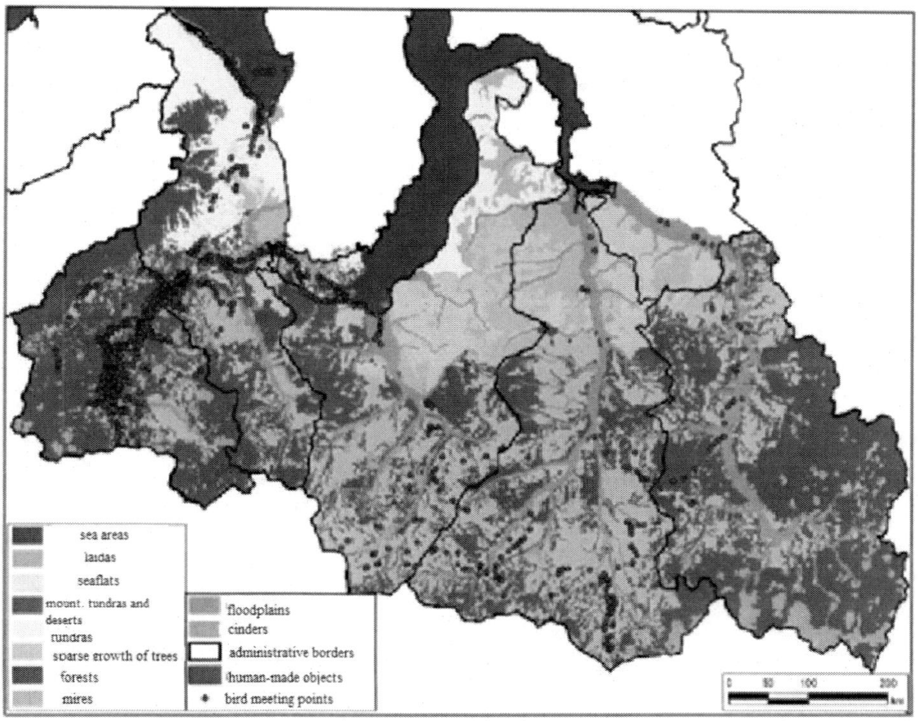

Figure 3. Example of defining the boundaries of waterfowl population areas in the Yamal-Nenets Autonomous Okrug during the autumn migration.

The areas of aerial counts were assessed by creating a buffer within a GIS environment. Analysis of Landsat8 satellite images and class allocations were made using visual and automatic interpretation methods, using neural networks with special training undergone in the ScanEx IMAGE Processor program. The Circumpolar Arctic Vegetation map (http://www.geobotany.uaf.edu/cavm/download.php) was used to clarify the habitat characteristics. The density calculation (individuals per km^2) and distribution in different biotopes were handled using GIS layers which contained the boundaries of biotopes identified by interpretation, survey points comprising the geographical coordinates of the interaction point with the specific number of animals, and the surveyed area within the range of aerial study of the territory (Figure 3). Calculations which required the use of geographic operators were handled in the Quantum GIS program. Additional data processing was handled in the Paradox 9.0 Database Management System (DBMS).

Calculation of the distribution was managed using a density modelling approach (Miller et al., 2013). GAM theory deployed for calculation of the degree of uncertainty was used to calculate a standard error and reliability intervals for estimating the distribution (Figure 4). The calculations were made using the dsm 2.2.9 package (Miller et al., 2016; Solovjev et al., 2017) for the statistical language R (R Core Team, 2016). The number of birds is modelled as the total of the smoothing function of independent variables (coordinates, habitat, and the height of the terrain above sea level).

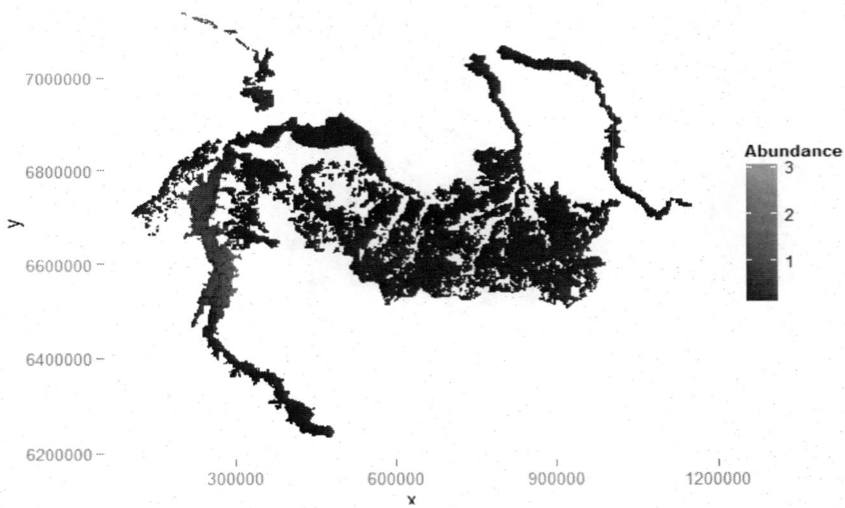

Figure 4. Illustration of the surface modelling of density to establish the numbers in the case example of the White-Fronted Goose (Anser albifrons):

Biotope extrapolation
39917.71134
> sum(mod2.pred)
[1] 38728.57
Summary of uncertainty in a density surface model calculated
analytically for GAM, with delta method
Approximate asymptotic confidence interval:
5% Mean 95%
28957.27 38728.57 51797.09
(Using delta method)
Point estimate : 38728.57
Standard error : 5776.946

Thus, the scheme for determining estimated numbers comprises the following stages:

1. Determining the type of habitat for each survey point of birds.
2. Determining the area of all identified habitat types examined in the course of aerial surveys.
3. Determining the total area of all identified habitat types in the survey zone.
4. Determining the overall number of recorded birds of each species for each identified habitat type.
5. Calculation of typical distribution of each species for all identified habitat types examined in the course of aerial surveys.
6. Calculation of estimated numbers of each species for all identified habitat types (multiplication of the typical habitat density per species by the area of habitat within the survey zone).
7. Calculation of the extrapolation coefficient by dividing the extrapolation number by the number of birds counted.

2.4. Deciding on the Timings for the Work in Accordance with Remote Data Tracking

To decide the optimal timings for the project, so that the aerial count coincided with the migration peak, we used data analysis of satellite and GSM tagging.

Figure 5. The Bewick's Swan and the Red-Breasted Goose with loggers, which allow to track the beginning of migration.

To determine the migration peak, we used data from late arriving and early departing species from the tundra, such as the Red-Breasted Goose (Branta ruficollis). At the same time, data from large species that arrive early to the nesting sites and leave them last (Syroechkovsky, 2013), such as the Bewick's Swan (Cygnus bewickii) (Figure 5), was used. With data on the beginning of migration of large and small species, it is possible to isolate the peak of the span and to coincide with holding aerial surveys exactly to the most intensive migration period.

2.5. Determining the Key Locations and the Human Impact

Key locations are determined to be those places where clusters of over 1000 birds were indicated (Figure 6, Table 1) as well as all locations of species registered as endangered in the Red Data Book of the Russian Federation and in the Red Data Book of the Yamal-Nenets Autonomous Okrug. Moreover, to estimate pressure of the human impact we kept record of hunters who made use of the hunting grounds, of the Nenets nomadic camps and also of herds of domesticated reindeer.

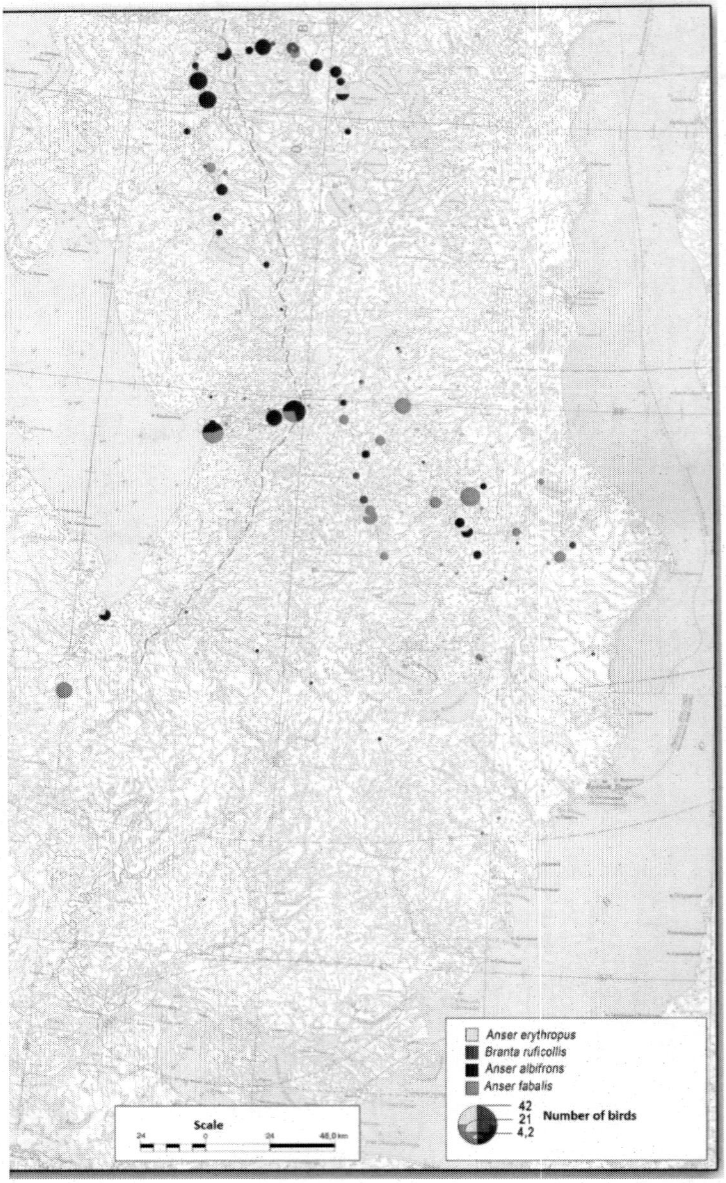

Figure 6. Examples of maps of registered species of birds, used to establish the key locations.

Table 1. Estimated numbers of the Anseriformes in the Yamal-Nenets Autonomous Okrug in autumn 2014 (rounded up to 10 individuals)

Species	Individuals counted	Overall estimated numbers	Average density (individuals per km^2)
Cygnus cygnus	11637	36960	0,39
Cygnus bewickii	1440	3180	0,08
Anser fabalis rossicus	9552	45200	1,8
Anser albifrons	18511	141190	3,3
Anser erythropus	3905	24900	0,6
Branta ruficollis	5604	13900	0,8
Branta bernicla	5934	13810	9,1
Anas penelope	32534	68300	4,4
Anas platyrhyncos	11847	23630	0,5
Anas acuta	17423	56320	0,6
Anas clypeata	18822	26080	6,9
Anas crecca	19772	44740	1,9
Anas querquedula	795	1200	0,1
Aythya fuligula	12813	67765	1,0
Aythya marila	6803	18045	1,1
Melanitta fusca	230	1405	0,3
Melanitta nigra	1171	3855	0,3
Bucephala clangula	2766	9990	0,4
Mergellus albellus	5295	44300	0,6
Mergus serrator	5106	32035	0,8
Mergus merganser	2644	20480	0,2

3. The Impact of Hunting

Giving an objective assessment of the impact of hunting on the state of the populations of all species of waterfowl in the Yamal-Nenets Autonomous Okrug currently is made more complicated due to the fact that we do not have access to statistics on the overall numbers of hunters in the Yamal-Nenets Autonomous Okrug; nor of how many hunters (including both the locals, and outsiders who come for hunting) hunt in spring at the wetlands, or of how many in autumn. Similarly, we lack overall numbers of waterfowl culled by all hunters each season, nor do we have species-specific listings of the seasonal "takes" of hunters, and their discipline in reporting their takes is very low. Despite this, we have succeeded in compiling some statistics, which enable some estimates of the impact hunting has on populations of the rare Anseriformes species in the Yamal-Nenets Autonomous Okrug, including those whose culling is banned by both federal and local legislation.

In fact, data from the return of bird-rings is the only information we have for making any assessment of the pressure of hunting. There is no quantitative data since the information is lacking, and there are no cull figures. Analysis of data from the Bird Ringing Centre shows that in the period of work relating to the intensity of the spring hunting season, hunting activity is approximately seven times higher than in autumn. This means that of the 774 returned bird rings from waterfowl which had been shot over the year, only 128 were returns from the autumn season. The impact of autumn hunting is not so significant. In spite of its more extended season, it is less intense, and the large proportion of young individuals in clusters offers more chance of surviving to adult individuals, which are valuable in terms of their experience and ability to go on breeding. The main pressure of hunting occurs at Dvuobye – a region which sees the most intensive migrations of waterfowl. The pressure of hunting is also high at the Obskay Bay, the mouth of the Nadym River, the area around Aksarka Village and the area between the Ob and Poluy Rivers near the city of Salekhard.

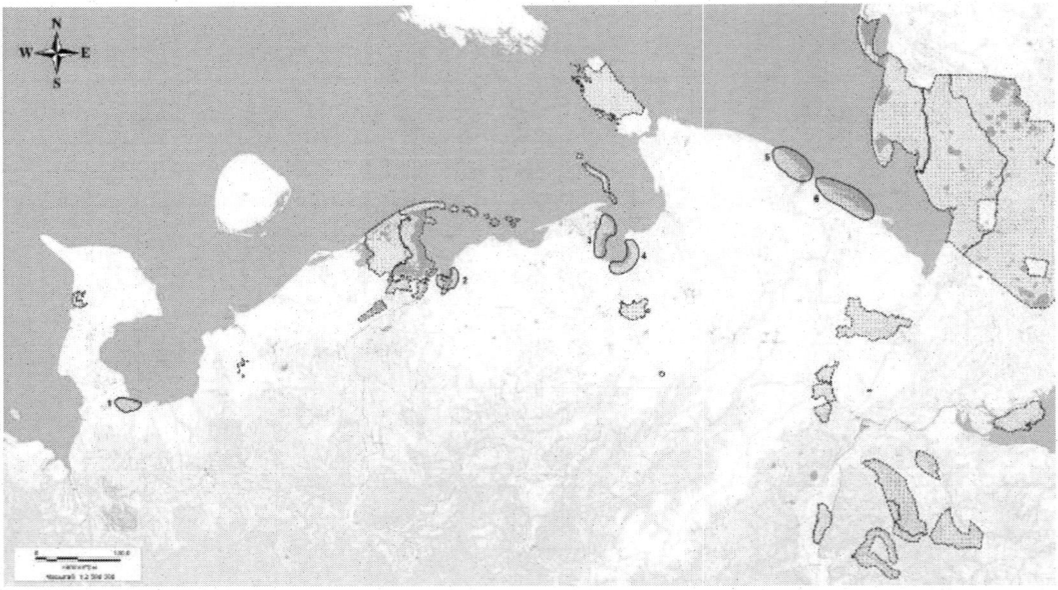

Figure 7. Locations of the autumn concentrations of the Lesser White-Fronted Goose and the regional SPNTs.

3.1. The Negligible Effectiveness of Existing Specially Protected Natural Territories (SPNTs) for the Conservation of the Rare Species of Anseriformes

Not a single SPNT was set up for the preservation of waterfowl for the entire period of their annual cycle. The result of this lax approach has been that existing protected areas are unable to protect endangered species outside the nesting period or migratory stops. We have proven the negligible effectiveness of SPNTs, located on the territory of the Yamal-Nenets Autonomous Okrug, regarding the Bewick's Swan and the Lesser White-Fronted Goose (Figure 7). The concept of setting up protected areas needs to be revised, and they should be sited along the migratory flyways. In concentrated locations of endangered waterfowl species, there is a need to set up SEASONAL (based on the times when endangered species settle there) protected reserves at which a complete ban on hunting, and all other actions which result in environmental degradation applies.

3.2. Key Locations for the Anseriformes in Migration Periods

Summarizing the information compiled for all seasons of the aerial survey, we have identified the following key locations for the Anseriformes during their migrations (Figure 8)

1. Levdiev Island and the adjacent mainland tundras of the south coast of the Baydaratskaya Bay (the Priuralsky District).
2. The delta of the Baydarata River with the adjacent waters of the Baydaratskaya Bay (the Priuralsky District).
3. The delta of the Ob River (the Priuralsky District), partially included in the Lower Dvuobye Ramsar Site (http://russia.wetlands.org).
4. The left bank floodplain of the Igorskaya Ob River at the mouth of the Sobtyegan River (the Priuralsky District).
5. The left bank floodplain of the Greater Ob River to the south of Pitlyar Village (the Shuryshkarsky District).
6. The floodplain of the Smaller Ob River to the south of Azovy Village (the Shuryshkarsky District).
7. The delta of the Taz River (the Tazovsky District), included in the shadow list of the Ramsar Sites "Lower Reaches of the Taz River" (http://russia.wetlands.org).
8. The lake and marsh complexes of the sources of the Nadym and Left Khetta Rivers (the Purovsky District).
9. The wetlands between the Khadyteypur and Kharampur Rivers (the Purovsky District), partially included in the shadow list of the Ramsar Sites "The Group of Lakes between the Chaselka and Kharampur Rivers" (http://russia.wetlands.org).
10. Lakes Chaselskoye and Nuito, including the wetlands between them (the Krasnoselkupsky District) partially included in the shadow list of the Ramsar Sites "Devil's Lake System" (http://russia.wetlands.org).

In eight key locations of the ten we selected in 2013-2016, resting sites were introduced during the spring hunt by the Governor's decree. Such experience is unique to Russia, since it establishes the scheme for conservation of waterfowl without any interference with the rights of hunters.

3.3. Comparison of the Data Compiled with the Data from Previous Surveys Carried out with an AN-2 Aircraft and with a Mi-8 Helicopter

The only source to provide overall data on the numbers of waterfowl in the region where these surveys were undertaken is the monograph by Krivenko and Vinogradov (2008, Tables 30-37). We aimed to compare our data with those compiled in the 1980s, 1990s, and early 2000s (Figure 9). This comparison showed that, according to our data, the numbers for most species were much lower than those given in Krivenko and Vinogradov's monograph. This either indicates a catastrophic decline in the numbers of common species, or incorrectness in the methodology used for recording and extrapolating the earlier data. However, we were unable to find the published details of how those data was compiled. It is our belief that the sharp decline in numbers for many species listed is impossible – since in the region concerned, there are no such natural or human-caused factors which might prompt it. Using this method enabled us to clarify and estimate the numbers of the Anseriformes species in real terms. Previously, aerial surveys were carried out using Russian Mi-8 helicopters, and they did not allow for species assessment. The results of counts undertaken with this level of engine noise can never be accurate, since the birds are either frightened away by the engine noise before they even come into sight or they dive.

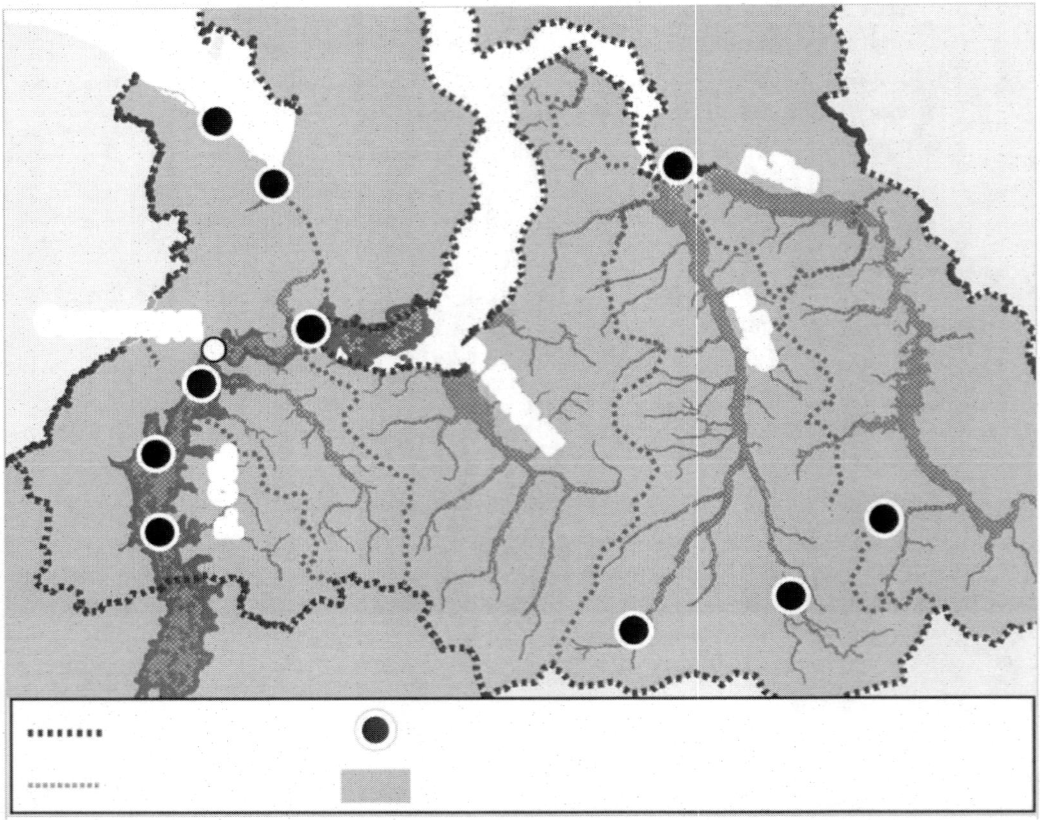

Figure 8. Key locations for the Anseriformes in the Yamal-Nenets Autonomous Okrug during the periods of migration, determined by the aerial surveys and confirmed by their 2010-2014 results.

4. FINDINGS

This was the first time in many years that in the Yamal-Nenets Autonomous Okrug of Russia, a modern methodology involving surveys with an ultralight aircraft was used to study waterfowl populations of 28 species of birds, the state of their habitats, their numbers and distribution, the key locations both during their nesting periods and the time of migration, and GIS was set up.

New data was obtained on the migratory habits of the Bewick's Swan, the Taiga Bean Goose (*Anser fabalis fabalis*), the Lesser White-Fronted Goose (*Anser erythropus*) and the Red-Breasted Goose.

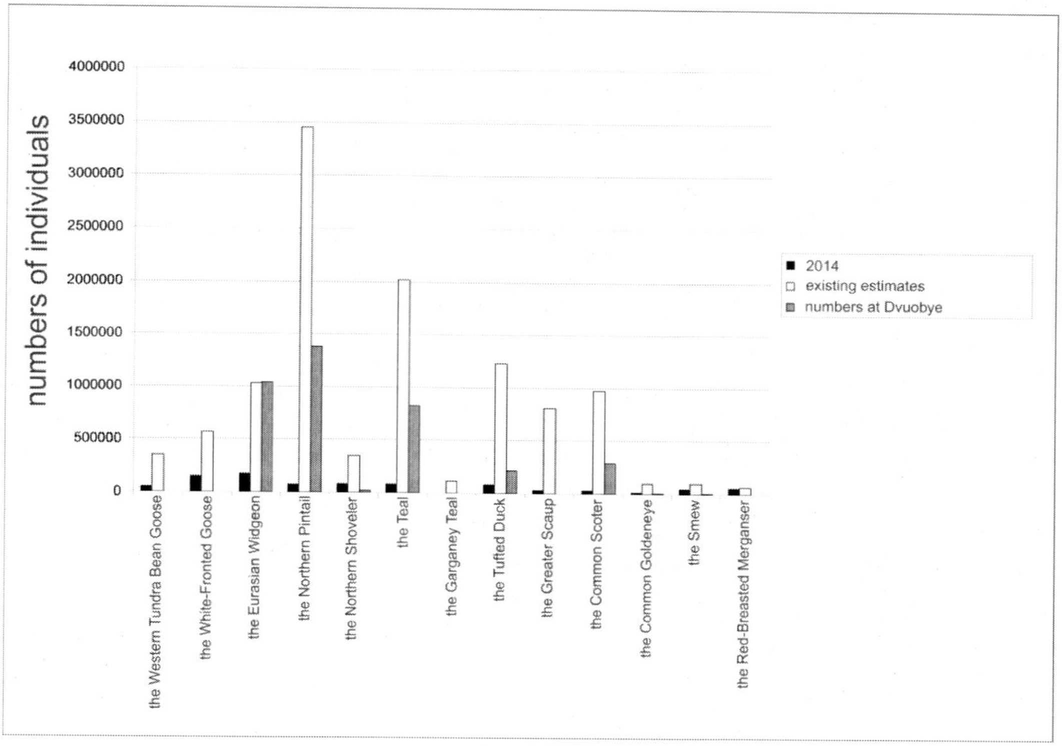

Figure 9. Statistics for numbers of different species in the survey, compared with statistics from previous surveys.

Monitoring and conservation recommendations were established for these waterfowl and their habitats.

CONCLUSION

Under modern conditions in Russia, the most effective method for implementing measures for the conservation of waterfowl will be the creation of nature reserves for rest during the periods of their spring and autumn migration. Justification for the setting up of such zones lies not only in the knowledge of the numbers of these birds and their tendencies, but also in the identification of the key zones for the Anseriformes during their migratory periods. Regrettably, today there is a severe shortage of long-term data on the numbers and breeding success of many species of the Anseriformes. Considering the enormous, remote areas involved, the only feasible way to compile such data would be to survey this territory using an ultralight aircraft. Unfortunately, the present methods for carrying out aerial counts and monitoring are extremely outdated. We have developed a system of monitoring using aerial flyovers modelled on areas in Western Siberia. This is an area through which one of the most numerous migratory flyways for geese, ducks and swans passes.

Based on the results of the research, key areas were identified of significant importance for the conservation of migratory waterfowl. In these areas waterfowl resting zones were created in 2013-2019, operational during the spring hunt, and general recommendations for regulating waterfowl hunting in the Yamal-Nenets Autonomous Okrug were made. The experience of individual regions in solving the problems of managing migratory Anseriformes is unique and valuable. The Yamal-Nenets Autonomous Okrug became a pilot region, where we were able to test and partially implement effective management schemes for the Anseriformes. The YANAO's experience is worth distributing to other Arctic regions of Russia. Since 2015, monitoring of waterfowl using aerial surveys has been adopted in the Nenets Autonomous Okrug, where in 2017 in the key areas identified, four SPNTs were created, which have a ban on hunting the Anseriformes. It will be necessary in the future to roll out a similar scheme for responsible management of hunting culls based on the monitoring figures, set up interregional and international hunting quota schemes, even if only for the populations of the main kinds of waterfowl which are sought by hunters.

Acknowledgments

The results were obtained as part of the assignment of the Ministry of Education and Science of Russia, project No 5.4004.2017/4.6. We express gratitude to Large-scale research facilities «System of experimental bases located along the latitudinal gradient» for fruitful cooperation (http://www.secnet.online/megaustanovka-ru.html).

References

Blokhin Yu.Yu., 2004. Duck- and Goose-Hunting on the Basis of Licences in the Yamal-Nenets Autonomous Okrug. *Kazarka*, 10: 119–141. (in Russian).

Blokhin Yu.Yu., 2010. The Hunt of Waterfowl Game in the Moscow Region Calculated from Data of Registered One-Time Licenses. *Kazarka*, 13: 104–122. (in Russian).

Falk K., Merkel F.R., Jamieson, S.E., 2006. The Effect of Embedded Lead Shot on Body Condition. *Journal of Wildlife Management*, 70 (6): 1644–1649.

Fox A.D., Leaflor J.O. (eds.), 2018. Hemisphere Goose Populations (Component 2: Population Accounts). *Conservation of the Arctic Flora and Fauna.* International secretariat, Acureyri, Iceland. 173 p.

Klaassen, M., Bauer, S., Madsen, J., Tombre, I., 2006. Modelling of the Spring Flightpath. *Journal of Applied Ecology*, 43 (1): 92–100.

Krivenko V.G, Vinogradov V.G, 2008. Birds of the Aquatic Environment and the Rhythms of the Climate of Northern Eurasia. M.: *Nauka*. 588. (in Russian).

Madsen J., Williams J.H. (Compilers), 2012. Goose Anser brachyrhynchus. *AEWA Technical Series*. 48. Bonn, Germany: 1–51.

Marjakangas A., Alhainen M., Fox A.D., Heinicke T., Madsen J., Nilsson L, Rozenfeld S. (Compilers), 2015. Bean Goose (*Anser fabalis fabalis*). *AEWA Technical Series*. 56. Bonn, Germany: 1–88.

Miller D.L., Rexstad E., Burt L., Bravington M.V., Hedley S., 2016. dsm: Density Surface Modeling of Sampling Data. R package version 2.2.13. URL https://CRAN.R-project.org/package=dsm.

Newth J.L., Cromie R.L., Brown M.J., Delahay R.J., Meharg A.A., Deacon C., Norton G.J., O'Brien M.F., Pain D.J., 2013. Poisoning from Lead Gunshot: *European Journal of Wildlife Research*, 59 (2): 195–204.

R Core Team, 2016). R: A Language and Environment for Statistical Computing. R Foundation for Statistical Computing, Vienna, Austria. URL https://www.R-project.org/.

Rosenfeld S.B, 2010. A Unique Key Ornithological Area on the Border of Russia and Kazakhstan Requires Protection. *Steppe Bulletin*, 30: 40–42. (in Russian).

Rosenfeld S., 2011. Is it Possible to Save Endangered Species? The Legitimacy of Spring Hunting of Geese in the South of Russia. – *Hunting,* 9 (157): 10–14. (in Russian).

Rosenfeld S., 2012. Why is this Necessary? Restriction of Hunting of Geese at Migratory Stops. *Hunting, 1* (161): 76–82. (in Russian).

Rosenfeld, S., 2014. Hunting of the Anseriformes: How it is Done in America and Europe. *Russian hunting magazine*, July 2014: 12–17. (in Russian).

Rosenfeld S., 2014a. Hunting of the Anseriformes: How it is Done in America and Europe. *Russian hunting magazine*, August 2014: 10–15. (in Russian).

Rosenfeld S., 2015. How to Bring Back Monitoring of Waterfowl? Innovative Methods of Regulating Spring Hunting According to Long-Term Monitoring Based on the Example of the Yamal-Nenets Autonomous Okrug. *Hunting*, 5 (201): 76–84. (in Russian).

Rosenfeld S.B., Kirtaev G.V., Vanzhelyuv D., Zamyatin D.O, 2016. Results of the Application of Innovative Methods of Monitoring and Studying Migratory Anseriformes. *Scientific Herald of the YANAO*, 4 (93): 51-57. (in Russian).

Rosenfeld S.B., Sheremetyev I.S., 2016. Arctic Geese (Anser) and Geese (Branta) of Eurasia: Analysis of Population Dynamics Factors and Habitats. *Journal of General Biology*, 77 (1): 17–77. (in Russian).

Rosenfeld S.B., Soloviev M.Yu., Kirtaev G.V., Rogova N.V., Ivanov M.N., 2017. New Approaches to Estimating the Number of the Anseriformes: the Experience of Aerial Surveys in Western Siberia Using Ultralight Aviation. – *Zool. Journal, 96* (2): 201–221. (in Russian).

Soloviev M.Yu., Rozenfeld S.B., Kirtaev G.V., Rogova N.V., Ivanov M.N., 2018. The Experience of Using Density Surface Modelling to Extrapolate Numbers of Birds According to Aerial Surveys – *First All-Russian Ornithological Congress* (Tver, Russia, January 29 - February 4, 2018). Abstracts of reports (ed. A.B. Popovkina, S.P. Kharitonov). Tver, 2018: 312. (in Russian).

Syroechkovsky E.V., 2013. Ways of Adaptation of Anserini Anseriformes to the Arctic habitat. M.: Partnership of Scientific Publications KMK. 297 p. (in Russian).

Timoshenko A., Rosenfeld S., 2013. Hunting of Anseriformes in Northern Kazakhstan, or the Right Approach to Hunting Management. *Hunting*, 4 (176): 64-72.

In: The Arctic: Current Issues and Challenges
Editors: Oleg S. Pokrovsky, Sergey N. Kirpotin et al.
ISBN: 978-1-53617-306-2

Chapter 8

REMOTE RESEARCH OF SPATIOTEMPORAL DYNAMICS OF THERMOKARST LAKES FIELDS IN SIBERIAN PERMAFROST

***Yury M. Polishchuk*[1,*], *Ildar N. Muratov*[1] and *Vladimir Y. Polishchuk*[2,3,†]**

[1]Ugra Research Institute of Information Technology, Khanty-Mansiysk, Russia
[2]Institute of Monitoring of Climatic and Ecological Systems, SB RAS, Tomsk, Russia
[3]Tomsk Polytechnic University, Russia

ABSTRACT

The results of a remote study of the long-term dynamics of lake areas in the permafrost zone of Western and Eastern Siberia are considered. The results of the study of little-studied changes in the number of thermokarst lakes in the permafrost zone are presented. These results are based on data on changes in the number of extinct and newly formed lakes over a 40-year study period, depending on the geocryological zoning of the study area. The results of the analysis of climatic changes according to weather stations and reanalysis data in the permafrost zone of Western Siberia are presented. The influence of climate change on the dynamics of thermokarst lakes is considered. Based on the regression analysis, an equation was found of the relationship between changes in lake areas and climatic parameters. The determining influence of the increase in average annual temperature on the reduction of lake areas has been established. Studies of the long-term dynamics of lake areas on the vast territory of the Eastern Siberia cryolithozone were carried out. These studies allowed to establish an increase in the average size of thermokarst lakes in this territory in the period under study. At that time, on the territory of the permafrost zone of Western Siberia, there is a tendency to reduce the areas of lakes. To determine the law of the distribution of lakes by size, a histogram of the distribution of lake areas in a very wide range of sizes from units of m^2 to hundreds of km^2 was constructed based on the integration of satellite images of ultrahigh and

* Corresponding Author's Email: yupolishchuk@gmail.com.
† Corresponding Author's Email: liquid_metal@mail.ru.

medium resolutions. It is shown that the areas of the lakes are distributed according to a lognormal law.

Keywords: thermokarst lakes, permafrost, remote sensing, satellite imagery, climatic changes, methane, carbon dioxide, long-term dynamics of lake areas, lakes population dynamics, the log-normal lake size-distribution, Siberia

INTRODUCTION

It is known that the majority of oil and gas fields in Western Siberia are located in the permafrost zone. In recent decades, under the conditions of climate warming, a decrease in the strength of permafrost leads to an increase in the accident rate on pipelines and other structures of the oil and gas complex, accompanied by great economic and environmental damage (Anisimov and Lavrov, 2004; Permafrost, 2002). In addition, permafrost, being a repository of conserved carbon in the vast frozen peatlands of Siberia, Scandinavia and North America, under global warming can cause even more warming when greenhouse gases are released, especially methane (Karol, 1996), as a product of the vital activity of microorganisms that process thawed organic matter (Walter et al., 2007; Krasnov et al., 2015; Pokrovsky et al., 2011).

The projected continuation of temperature increases on the globe are of concern to both the scientific community and the governments of most countries of the world. In particular, at the climate conference held in Paris in December 2015-January 2016 with the participation of leaders of the leading countries of the world, measures aimed at preventing the global temperature of the Earth by more than 2 degrees in the coming decades were discussed.

The main sources of methane emissions to the atmosphere in the northern territories are swamps and thermokarst reservoirs. Due to the inaccessibility of the territories in the permafrost zone, the study of such landscapes is impossible without the use of remote sensing methods. At present, a lot of experience has been gained in the use of satellite imagery in the study of various natural systems and processes, which has shown the effectiveness of remote sensing methods.

According to (Viktorov, 2006; Zuidhoff and Kolstrup, 2000), lake-thermokarst landscapes are most sensitive to temperature changes in the permafrost territory. For remote study of changes in thermokarst processes in permafrost areas, thermokarst lakes are usually used as indicators of permafrost changes, which are well interpreted on satellite images (Luoto and Seppala, 2003; Dneprovskaya et al., 2009; Kravtsova and Rodionova, 2016). It is known (Viktorov, 2006) that the landscapes of the lake thermokarst plains form on a slightly wavy earth surface with characteristic tundra vegetation and interspersed lakes, as a rule, of rounded shape and randomly scattered throughout the territory.

According to (Viktorov, 2006), the formation of thermokarst lakes occurs under the influence of several factors, the main of which are the presence of high ice permafrost rocks (accumulation of critical water thickness) and the flat character of the territory. Therefore, the majority of researchers of the thermokarst lake plains proceed from the assumption that in areas of high ice permafrost distribution in the cryolithozone, lakes of a thermokarst origin are prevalent. For example, in the works of V. I. Kravtsova and co-authors (Kravtsova, 2009; Kravtsova and Bystrova, 2009; Kravtsova and Tarasenko, 2010) based on the geomorphological and climatic analysis, the whole cryolithozone of Western Siberia and most of Eastern Siberia are classified as areas with widespread distribution of thermokarst lakes. Therefore, the lakes studied in this paper on the basis of satellite images are considered as thermokarst lakes.

In connection with melting the permafrost, thermokarst lakes arise and develop in a relatively short time - within a few decades and can also quickly disappear, becoming khasyreys (pits of drained lakes). However, the life span of some ancient thermokarst lakes can be from several hundred to several thousand years (Hinkel et al., 2003).

Currently, the formation of thermokarst reservoirs and depressions due to the degradation of permafrost over the past 50 years has been observed in Alaska, Canada and Europe (Luoto and Seppala, 2003; Riordan et al., 2006; Zuidhoff and Kolstrup, 2000). As shown in (Bryksina et al., 2011; Kirpotin et al., 2008), melting of permafrost under conditions of current climate warming accelerates thermokarst processes and changes in the number and area of lakes in the permafrost zone. The results of a study of the long-term dynamics of lake areas in the permafrost zone of Western Siberia are presented further in subsection 1.1.

Very important are the issues of studying the changes in the number of thermokarst lakes in the permafrost zone, which have been insufficiently considered in the literature. The results of a remote study of changes in the number of thermokarst lakes in the permafrost zone of Western Siberia, depending on the geocryological zoning of its territory, obtained (Bryksina and Polischuk, 2015; Shiklomanov et al., 2013; Polishchuk and Polishchuk, 2017) on base of the analysis of data on the change in the number of disappeared and newly formed lakes for a 40-year study period will be considered in subsection 1.2.

Studies (Luoto and Seppala, 2003; Smith et al., 2005; Riordan et al., 2006; Dneprovskaya et al., 2009; Kirpotin et al., 2009; Shiklomanov et al., 2013; Karlsson et al., 2014; Polishchuk and Polishchuk, 2017), conducted in Alaska, Scandinavia and Siberia, revealed the dependence of changes in thermokarst processes, in particular, lake sizes, from climate change. However, some researchers of the dynamics of thermokarst processes in the permafrost zone, for example (Kravtsova and Rodionova, 2016), deny the dependence of the observed changes in thermokarst lakes on climate. This can be considered as a consequence of insufficient knowledge of the relationship of thermokarst processes with climate change, which does not allow solving the problem of forming reasonable predictions of changes in thermokarst processes under conditions of climatic changes in the permafrost territory. The need to take this dependence into account in modeling climate prediction problems is indicated in (Kabanov and Lykosov, 2006; Machulskaya and Lykosov, 2002).

As noted above, the melting of permafrost under the conditions of modern climate warming in the permafrost zone of Western Siberia is manifested in a reduction in the average area of thermokarst lakes. However, the insufficiency of previously performed studies of the dynamics of thermokarst processes in the permafrost zone of Eastern Siberia did not allow to establish patterns of geocryological changes in this vast territory. Our studies of the long-term dynamics of lake areas in the zone of continuous distribution of permafrost in Eastern Siberia (Polishchuk et al., 2014; Polishchuk et al., 2016) made it possible to identify patterns of dynamics of lake areas in this territory and to determine their features in comparison with those found in the Western Siberia. These research results are described further in subsection 1.3.

In recent years, remote studies of the spatial structure of the fields of thermokarst lakes have been conducted. The laws of the distribution of areas of thermokarst lakes in Siberia and in some other northern regions were also studied (Viktorov et al., 2015, 2016; Grosse et al, 2008). Most of these studies used Landsat medium resolution (30 m) satellite images, as they provide multiple complete coverage of the globe. But in these images are not found lakes of small sizes. Therefore, the empirical distributions of lake areas obtained from these data do not take into account the contribution of small thermokarst lakes, the concentration of methane in which, according to (Pokrovsky et al, 2011), is more than an order of magnitude higher than its concentration in lakes with sizes greater than 0.1-0.5 ha.

In a number of works, for example, (Grosse et al, 2008; Polishchuk et al., 2015), ultra-high resolution images (less than 1 m) were used to study the distributions of thermokarst lakes in the permafrost zone of Siberia. However, these images due to the small coverage area provide a relatively small coverage of the study area, which does not allow to obtain histograms of the lake size-distribution, including data on the lakes of medium and large sizes.

In connection with the above, the task of constructing histograms of the lake size-distribution of according to the results of remote sensing studies, which would take into account almost all the lakes of the study territory in the permafrost in a wide range of sizes from several meters to tens of kilometers, becomes important. Such histograms of the lake size-distribution can be constructed only on the basis of the integration of data on the areas of water bodies obtained from satellite images of both high (superhigh) and medium resolutions. The need to combine data on lakes obtained from images of different spatial resolution to build integrated histograms of the lake size-distribution in a wide range of lake sizes required the development of methodological issues of integrating data obtained using medium and high / ultra-high resolution satellite images in order to build generalized the lake size-distribution histograms. Such a methodology and the results of the construction and analysis of integrated histograms of the distribution of thermokarst lakes by size on the example of the territory of the cryolithozone of Western Siberia are presented in section 3.

1. Experimental Study of the Long-Term Dynamics of the Sizes of Thermokarst Lakes in the Permafrost Zone of Western Siberia

1.1. Study of Changes in Lake Sizes

The most sensitive indicator of thermokarst processes to temperature changes are the areas of thermokarst lakes, which is illustrated by the results of remote sensing in different regions of the world, described, for example, in works (Kravtsova, Bystrova, 2009; Kravtsova, 2009; Kirpotin et al., 2009; Viktorov, 2006). To obtain data on the areas of thermokarst lakes in the permafrost of Western Siberia, a set of test sites (TS) was selected. The selection of test sites in our work was carried out taking into account the specifics of the zonal-geocryological differentiation of the territory (Atlas, 1984; Landscape, 1987). In each subzone of permafrost, some number of the test sites were selected, which makes it possible to investigate the dynamics of thermokarst lakes depending on the geocryological zoning of the territory. The map-layout of the TS in the study area is presented in Figure 1. The test sites number is selected approximately evenly in each zone of the permafrost in Western Siberia.

As can be seen from the figure, there are 7, 10, and 12 sites in the insular, discontinuous, and continuous zones of permafrost, respectively. The sizes of the test sites were chosen approximately equal in area. The average area of test sites was 2,600 km^2. The total area of the test sites studied was about 78.5 thousand km^2.

Due to the very large number of cloudy days per year in the northern territories of Siberia, for each test site, 3-6 cloudless Landsat images were collected for the period of research starting from 1973. To obtain experimental data on lake areas in the study area, a total of 134 cloudless Landsat images collected during the warm months of the study period were collected. All images were selected from the publicly available "Global Land Cover Facility" archive and relate to the 1T processing level, including radiometric and geometric correction using digital elevation models. Landsat images included in the collection are geo-referenced in the UTM projection. The used satellite images belong to the research period 1973-2013, 90% of them were obtained in the second half of the summer seasons (mostly in July-August), when the ice cover of the lakes completely disappears, which makes it difficult to deciphering lakes with automatic interpretation of images.

Processing and interpretation of satellite images were carried out using the software tools of geo-information systems (GIS) ENVI 4.7, ArcGIS 9.3 and ArcGIS 10.3. At each test site, GIS determined from several hundred to several thousand lakes. The total number of investigated lakes in the test areas of Western Siberia was more than 50 thousand.

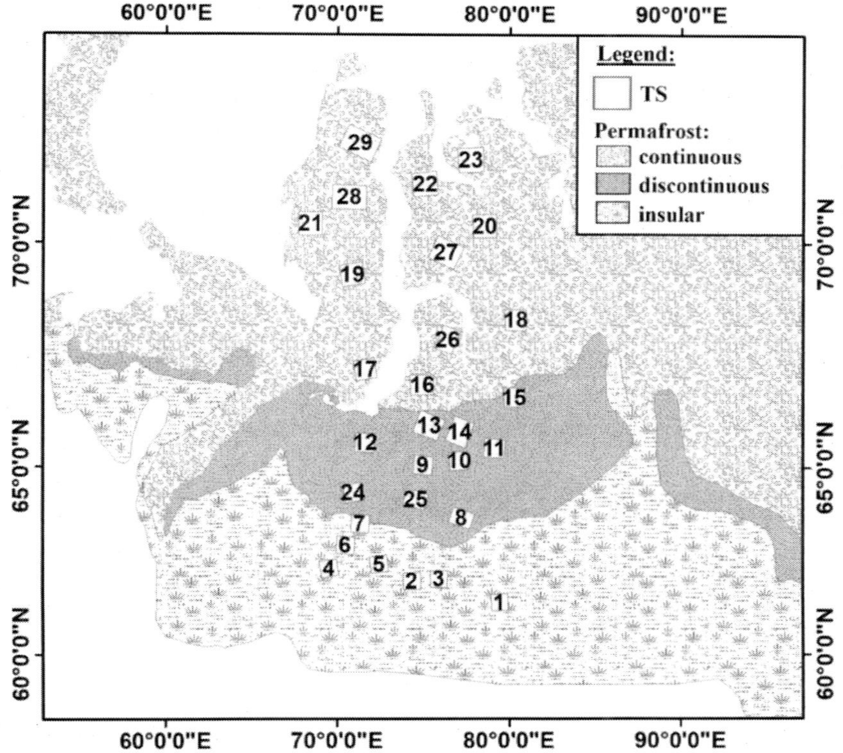

Figure 1. Map of the location of test sites in the study territory.

On the basis of the obtained data on the areas of lakes for all the TS, time series of average values of the lakes areas were formed, which allowed the approximation by a linear equation of the form:

$$y = \alpha \times x + B, \qquad (1)$$

where a is the coefficient of the linear approximation equation (linear trend); B - the free term of the approximation equation; x - time (years).

For illustration in Figure 2 shows the time dependence of the average area of thermokarst lakes at TS–13 (Polishchuk and Polishchuk, 2013), showing, on average, a decrease in the mean area of lakes with a linear trend coefficient $\alpha = -0.11$ ha/year. The positive sign of the linear trend coefficient α in equation (1) indicates that the areas of lakes in the study period increase on average, and a negative value of coefficient α shows an average decrease in the value of this indicator during the observation time period. Table 1 shows the values of the coefficient α for the time series of the average areas of thermokarst lakes at each test site in different zones of permafrost. Most (almost 90%) of the values of α in the Table 1 are negative, which means the prevalence in the change of areas of the tendency to reduce on average the areas of lakes in time. Note that on only three of the 29 TS, the value of α has a small positive value, which, however, does not violate the general trend of a decrease in the average lake areas in time.

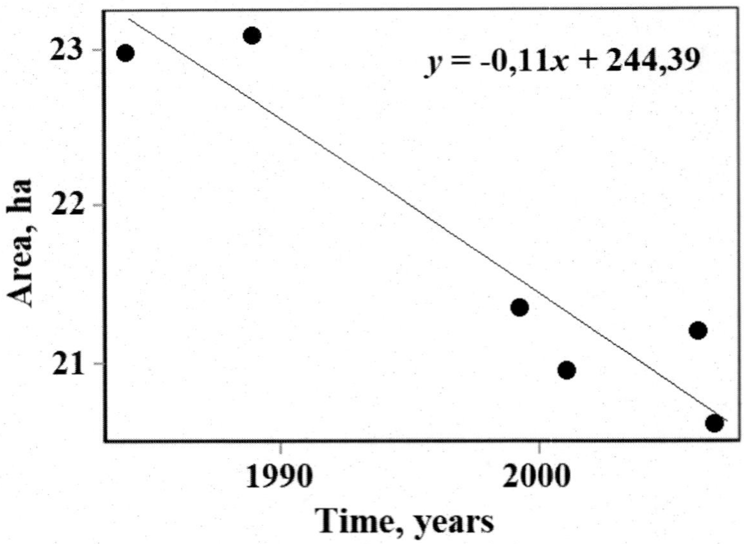

Figure 2. The time dependence of the average area of thermokarst lakes in the TS-13.

According to the Table 1 the average values of the linear trend coefficient in the three permafrost zones were calculated, which for the continuous, discontinuous and insular zones were the following values: -0.1 ha/year, - 0.08 ha/year and -0.1 ha/year, respectively. The value of the linear trend coefficient averaged over all TSs, obtained according to the Table 1 and equal to -0.09 ha/year, shows a global trend of reducing the size of lakes in the permafrost zone of Western Siberia on average over the study period.

Table 1. The coefficient of the linear trend of changes in the average areas of thermokarst lakes in different test sites

TS number	Latitude, degree N	Longitude, degree E	α, ha/year
1	79,31	61,26	-0,15
2	74,23	62,03	-0,16
3	75,59	62,05	-0,03
4	69,35	62,25	-0,13
5	72,28	62,3	-0,15
6	70,27	63,01	-0,01
7	71,2	63,33	-0,14
8	77,13	63,4	-0,01
9	74,59	65,03	-0,04
10	77,03	65,1	-0,13
11	79,12	65,27	-0,09
12	71,37	65,35	0,01
13	75,23	65,6	-0,11
14	77,8	65,51	-0,05
15	80,15	66,41	-0,05
24	70,56	64,23	-0,19
25	74,36	64,12	-0,06
16	74,54	66,57	-0,09
17	71,37	67,17	-0,05
18	80,23	68,23	-0,03
19	70,5	69,19	0,01
20	78,31	70,2	0,00
21	68,24	70,22	-0,03
22	75,04	71,08	-0,16
23	77,44	71,37	-0,28
26	76,26	67,58	-0,20
27	76,18	69,51	-0,18
28	70,42	70,54	0,03
29	71,18	71,55	-0,18

1.2. Analysis of the Change in the Number of Thermokarst Lakes in the Permafrost Zone of Western Siberia

To study the changes in the number of thermokarst lakes in the permafrost zone of Western Siberia, based on the interpretation of different-time satellite images, digital maps were obtained showing the location of the lakes at each test site for a particular year of shooting. The thermokarst lakes that had disappeared and newly formed during the study period were determined by comparing the initial and final maps of the location of lakes at each test site. To create the initial maps in (Bryksina and Polischuk, 2015), Landsat-1 (1973), Landsat-2 (1981), Landsat-4 (1983 and 1988), Landsat-5 (1984 and 1987) images were used, and maps for the final year of the study were formed from Landsat-8 images obtained in

2013. The Landsat-1 images from 1973 were used to create base maps on 16 test plots, and for the rest of the test plots Landsat images were used with shooting dates in the period 1981 - 1988, indicated above. The use of Landsat-8 images from the 2013 survey made it possible to estimate the change (decrease or increase) in the number of thermokarst lakes over a sufficiently long period of time (from 25 to 40 years at different test sites).

As follows from the above, by comparing the initial and final maps of the location of lakes at each test site, both thermokarst lakes that have disappeared during the study period and the newly formed lakes were identified. Then, using ArcGIS 9.3, we determined the coordinates of the centers of the disappeared and newly formed lakes, convenient for a cartographic display of the location of these lakes.

The map of the location of the disappeared lakes in the studied territory of Western Siberia is shown in Figure 3. As seen in Figure 3, the number of disappeared lakes in different subzones of permafrost is different, with the highest density of these lakes observed in discontinuous permafrost. As shown by remote measurements, the total number of disappeared lakes in the permafrost zone of Western Siberia over the study period was 391, with a total water surface area of these lakes of 14,826 ha. Consequently, the average area of the disappeared lakes in the study area was 37.9 ha.

Consider the change in the number of disappeared lakes depending on the geographic latitude of their location in the study territory. In Figure 4-a is a graph of the change in the number of disappeared lakes (N_1) depending on the latitude of the location and on the geocryological zoning.

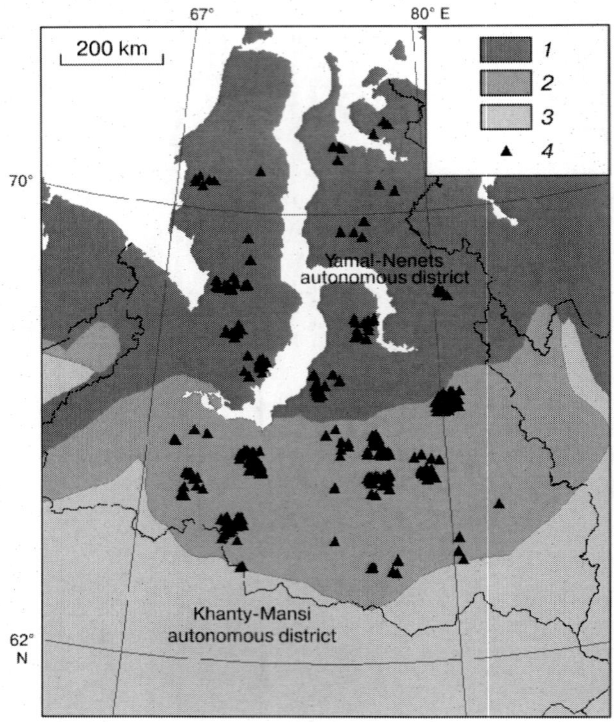

Figure 3. A schematic map of the location of disappeared lakes in different permafrost subzones in Western Siberia. Permafrost: 1–continuous, 2–discontinuous, 3–insular; 4–khasyreys (shrinking ponds).

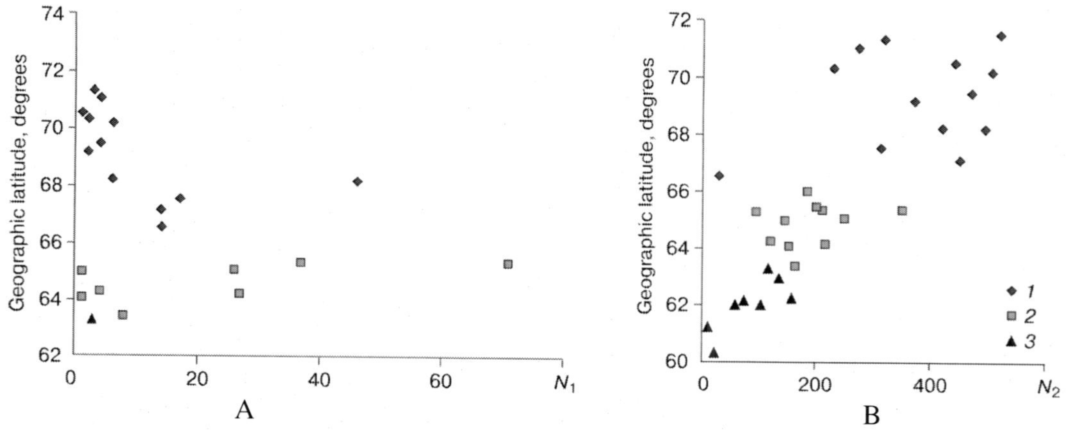

Figure 4. Dependences of the number of disappeared lakes $N_1(a)$ and of newly formed lakes $N_2(b)$ on geographic latitude. Permafrost: 1–continuous, 2–discontinuous, 3–insular.

The individual points on the graph, displayed as triangles, squares or rhombuses, depending on their location in different permafrost subzones, show the number of disappeared lakes in each test site. As can be seen in the graph (Figure 4-a), the number of disappeared lakes decreases on average with increasing latitude. At the same time, it is noticeably less in the continuous permafrost subzone than in discontinuous permafrost, i.e., the greatest reduction in the number of lakes in the cryolithozone of Western Siberia is observed mainly in the subzone of discontinuous permafrost.

Consider the results of studying the change in the number of newly formed lakes in the study area. In Figure 4-b shows a plot of the number of lakes formed during the period of research (N_2) on the geographical latitude of the location of the test sites. As in the previous graph, each point in Figure 4-b displays the number of lakes formed on a separate test site. As follows from the graph, the number of newly formed lakes on average increases significantly with increasing latitude. The predominant formation of new lakes is observed in the zone of continuous permafrost. Consequently, based on the analysis of graphs in Figure 4 we can formulate the conclusion that the formation of new lakes occurs mainly in the continuous subzone of permafrost, and the disappearance of lakes is manifested mainly in the subzone of discontinuous permafrost.

This conclusion is quantitatively confirmed by the results of a comparative analysis of changes in the number and total area of the disappeared and newly formed lakes. According to (Bryksina and Polischuk, 2015), the number of lakes formed during the period of study in the permafrost zone of Western Siberia was 7751, which is much higher than the number of disappeared lakes. And the total water surface area of the formed lakes, equal to 13,649 ha, differs little from the total area of the disappeared lakes. The data presented here give an average area of newly formed lakes of 1.76 hectares, which is 22 times less than the average area of the disappeared lakes.

We will conduct a comparative analysis of the above data on the number of disappeared and newly formed lakes and their total areas throughout the investigated territory of the cryolithozone of Western Siberia. Comparison of these data leads to the establishment of another important regularity: with almost identical values of the total areas of the disappeared (14,826 hectares) and the formed (13,649 hectares) lakes, the number of newly formed lakes is approximately 18 times the number of disappeared lakes. Consequently, the new thermokarst lakes are significantly smaller by area than disappearing lakes. Therefore, it can be assumed that the observed acceleration of thermokarst processes as a result of warming will be accompanied by a significant increase in the number of small thermokarst lakes in the permafrost of Western Siberia.

According to the results of experimental studies (Audry et al, 2011; Pokrovsky et al, 2011; Shirokova et. al., 2013; Manasypov et. al., 2014), small thermokarst lakes in Western Siberia are intensive natural sources of greenhouse gases, and in particular, carbon dioxide and methane. Therefore, the predominant increase in the number of small thermokarst lakes discovered as a result of our research will contribute to the growth of the accumulation of lake greenhouse gases.

Consequently, the acceleration of thermokarst processes in the permafrost zone of Western Siberia under the influence of climate warming causes a much more intensive formation of new lakes in the continuous permafrost subzone compared to the discontinuous permafrost subzone, and a more intensive reduction in the number of thermokarst lakes is observed in the discontinuous permafrost subzone compared to the continuous subzone.

As shown in (Dneprovskaya et al., 2009; Kravtsova and Bystrova, 2009; Polishchuk and Polishchuk, 2014), the most important feature of the discontinuous cryolithozone of Western Siberia is the reduction in the total area of thermokarst lakes, which (Dneprovskaya et al., 2009) is explained by an increase in average annual air temperature. The study of the relationship between changes in lake area and climatic characteristics in the territory of the Western Siberia permafrost zone carried out in next section of the book confirms that an increase in average annual air temperature is accompanied by a reduction in areas of thermokarst lakes.

Since the disappearance of the lake (the formation of khasyrey at the site of the thermokarst lake) is the final stage of the process of reducing the water surface of the lake, the reasons for the reduction in the number of lakes are the same as the reasons causing the reduction in the size of the lakes. Researchers of thermokarst in Alaska (Hinkel et al., 2003; Riordan et al., 2006) explained soil drainage as a result of melting permafrost under the influence of climate warming.

One of the possible mechanisms for the descent of water from a thermokarst lake due to soil drainage with increasing soil temperature is described in (Kirpotin et al, 2008). As a rule, large lakes are older and have a lower level of water surface than the surrounding relatively young lakes. This creates conditions for the descent of water from small lakes to larger ones due to soil drainage when the soil thaws. Under the conditions of warming, the depth of thawing of frozen soils increases during warm seasons, which accelerates the thermal erosion of water from the lakes and leads to a reduction in the area of the lakes and to the disappearance of their parts.

1.3. Study of the Dynamics of Lakes in Western and Eastern Siberia

In Figure 5 shows the location of test sites in the study areas of Western and Eastern Siberia. The test sites were selected in the areas of the most active thermokarst, determined by the large accumulation of thermokarst lakes, taking into account the presence of pairs of multi-temporal pairs of cloudless images for each TS relating to a fairly long time interval. As seen in Figure 5, where the subzones of permafrost are shown in accordance with (Atlas, 1984; Bryksina et al., 2016), all selected test sites in Eastern Siberia, located in the northern territories of the Krasnoyarsk Territory and Yakutia, are located in the subzone of continuous permafrost. Accounting the large (almost 75%) prevalence of continuous permafrost in Eastern Siberia, it was the subzone of continuous permafrost in the study territory that was chosen for research (Atlas, 1984).

The total number of investigated TS was 53, of which 14 are located in Western Siberia and 39 - in Eastern Siberia (Table 2). To conduct remote studies for each test site, multi-temporal pairs of cloudless Landsat images taken during the summer months in the 40-year period (1973-2013) were selected. All images are selected from the global "Global Land Cover Facility" archive, geo-referenced in the UTM projection and refer to the 1T processing level, which includes radiometric and geometric correction using digital elevation models. The total number of satellite images in the study territory of Western and Eastern Siberia was 48.

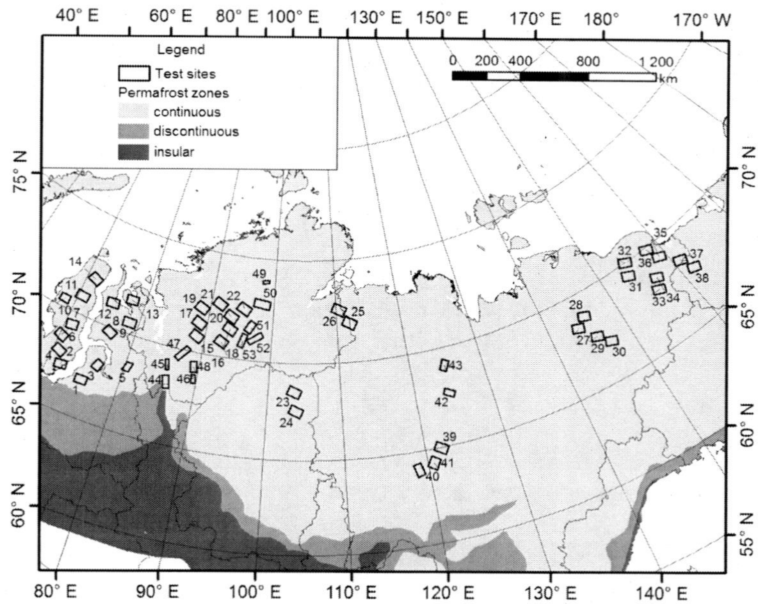

Figure 5. Map of the location of permafrost subzones in the territory of Eastern and Western Siberia with designated boundaries of test sites.

Table 2. Distribution of test sites by landscape zones

Study areas	Landscape zones	Number of TS
Eastern Siberia	Arctic tundra	0
	Subarctic tundra	16
	Forest tundra	10
	Taiga	13
Western Siberia	Arctic tundra	3
	Subarctic tundra	10
	Forest tundra	1
	Taiga	0

The processing of space images and the determination of lake areas were carried out using the ENVI 4.7 and ArcGIS 9.3 software. The water surface classification in Landsat images was carried out using the binary encoding method (Binary encoding classification algorithm in ENVI 4.7 software). On each test site, several hundred to several thousand thermokarst lakes were identified.

Based on the data obtained on the areas of lakes, a quantitative analysis of changes in the areas of thermokarst lakes with time was carried out. For each test site in the continuous permafrost zone of Western and Eastern Siberia, the total areas of lakes in different years of observations were calculated and the absolute and relative values of their changes over the observation period were determined. The value of the relative change in the total area was determined in accordance with the formula:

$$R = \frac{(S_f - S_{in})}{S_{in}},$$

where S_f and S_{in} are the total area of lakes on the test site in the final and initial years of research, respectively.

Positive values of R show an increase in the total area of lakes, and negative values of this magnitude show a reduction in the area of lakes. The zero value of R, meaning the absence of changes in lake areas over the observation period, shows the balancing of two trends in area changes, namely, trends of their growth and contraction.

For convenience, a comparative analysis of the dynamics of thermokarst processes in Western and Eastern Siberia in Figure 6 shows a combined graph of the latitudinal dependence of the magnitude of the relative changes in the areas R in the studied test sites. Hollow circles and red points show the R values for Eastern and Western Siberia, respectively.

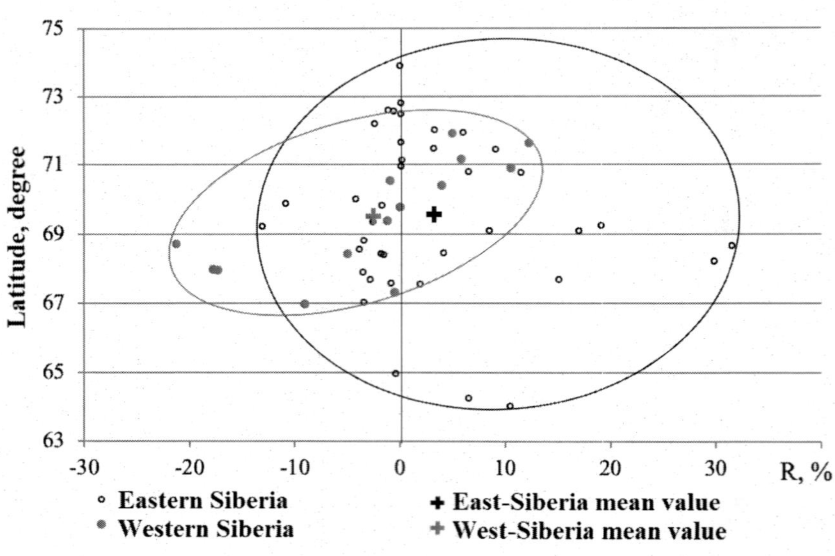

Figure 6. Relative values of changes in the area of thermokarst lakes depending on the geographical latitude in Western and Eastern Siberia.

In Figure 6 red circles show the relative magnitudes of changes in the total area of thermokarst lakes in Western Siberia, the Red Cross represents their average value equal -2.4%. This value shows that in Western Siberia the reduction in the area of thermokarst lakes prevails, which was previously confirmed by the results experimental studies in section 1.1 of this book. Hollow circles in Figure 6 shows the values of the relative change in the total area of lakes from the geographical latitude in Eastern Siberia, and the cross shows their average value for all test sites in this territory, equal to 3.2%. From this it follows that in Eastern Siberia the trend of increasing the area of thermokarst lakes prevails on average.

Note that at latitude of about 70°N in Western and Eastern Siberia, R takes on a value close to zero. Consequently, we can assume that in the continuous zone of the permafrost of Siberia, the range of latitudes in neighborhood of the value 70°N is an area of transition from the trend of reducing the average area of thermokarst lakes to the trend of their growth on average with an increase in geographic latitude. It is to this area of latitudes that the average values of R in Western and Eastern Siberia, shown by crosses in Figure 6.

A comparative analysis of the study of changes in the total areas of lakes depending on the geographic latitude in the continuous permafrost zone of Western and Eastern Siberia showed that the continuous zone of the frozen ground of Siberia can be divided into two latitudinal subzones: the northern and southern ones. In the northern subzone at latitudes greater than 70°N there is an average increase in the total area of lakes, and in the southern subzone at latitudes less than 70°N - reduction in the average area of lakes. Consequently, thermokarst processes in the southern subzone of continuous permafrost in Siberia are similar to the processes occurring in the discontinuous zone of permafrost in Western Siberia (Kirpotin et al., 2008; Polishchuk et al., 2014), which lead to a trend of reducing the areas of lakes under conditions of accelerating thermokarst processes under the influence of global warming.

It can be assumed that the observed trend of growth in the total areas of lakes in the permafrost zone of Eastern Siberia will be accompanied by an increase in the accumulation of methane and carbon dioxide in thermokarst lakes of the cryolithozone of Eastern Siberia, which may contribute to an increase in methane and carbon dioxide emissions to the atmosphere.

2. Research of Climate Changes and Their Influence on the Dynamics of the Fields of Thermokarst Lakes in the Permafrost Zone of Western Siberia

2.1. Climate Change Analysis Based on Weather Station Data

In this section, we analyzed the time series of air temperature and precipitation based on the observations of weather stations obtained in recent decades in the study area (Zheleznyak, 2005; Dneprovskaya and Polishchuk. 2007; Polishchuk and Polishchuk. 2013; Hijmans et al., 2005). The location of these meteorological stations in different zones of permafrost is shown in the map-scheme of Figure 7.

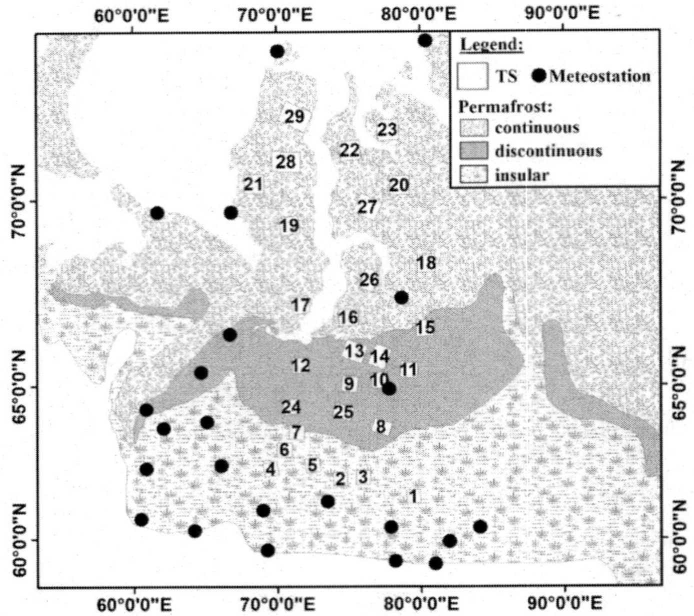

Figure 7. Map of the location of the weather stations and test sites in the study territory.

Table 3. Values of linear trend coefficients for average annual values of temperature and annual amount of precipitation at different meteorological stations

The name of the weather station	Latitude, degree N	Longitude, degree E	α_T, °C/year	α_P, mm/year
Amderma	69,7	61,7	0,02	-0,80
Maresale	69,7	66,8	-0,02	-0,60
Im.M.V.Popov	73,3	70,1	-0,02	-0,48
Tazovskoe	67,5	78,7	0,09	0,65
OstrovDikson	73,5	80,4	0,03	-0,31
Saran-paul	64,3	60,9	0,06	-0,87
Muzi	65,4	64,7	-0,02	-1,32
Salehard	66,5	66,7	0,02	-1,16
Tarko-sale	64,9	77,8	0,04	-0,76
Sosva	63,7	62,1	-0,03	-1,13
Berezovo	63,9	65,1	0,00	-1,08
Njaksimvol	62,4	60,9	0,12	-1,17
Oktjabrskoe	62,5	66,1	0,02	-1,28
Surgut	61,3	73,5	-0,03	-1,25
Ivdel	60,7	60,5	0,01	-1,02
Saim	60,3	64,2	0,00	0,38
Ko	60,4	77,9	0,01	-0,3
Vanzil-kynak	60,4	84,1	0,02	0,01
Demjanskoe	59,6	69,3	0,06	-0,59
Napas	59,9	82,0	0,02	-0,10
Vasjugan	59,2	78,2	0,02	0,19
Kargasok	59,1	81,0	0,03	-0,18
Hanty-Mansijsk	61,0	69,0	0,02	-1,40

Table 4. Average values of linear trend coefficients for temperature and precipitation in different permafrost zones

Permafrost zones	**Mean value of α_T,°C/year**	**Mean value of α_P, mm/year**
Continuous	0,02	- 0,37
Discontinuous	0,01	- 1,09
Insular	0,01	- 0,53

Table 3 shows the values of the linear trend coefficients of time series of the average annual values of temperature (α_T) and the annual total precipitation (α_p) obtained for each meteorological station. In addition, this table shows the calculated averaged values of these coefficients for certain permafrost zones.

A similar analysis of the time series of annual total precipitation with the aim of establishing linear trends and determining the values of the coefficient α_p was carried out both for individual meteorological stations of the study area and for areas of continuous, discontinuous and insular permafrost. The results of this analysis in the kind of the values of the linear trend coefficients α_T and α_p, obtained by averaging the data of weather stations in the territories of the corresponding permafrost, are given in Table 4.

As can be seen from the Tables 3 and 4, in the continuous permafrost zone, the values of α_T vary from -0.02 to + 0.09°C/year with a mean value of 0.02°C/year. In the discontinuous zone, the trend coefficient varies from -0.02 to + 0.06°C / year with a mean value of 0.01°C/year. And in the island zone, the limits of α_T change are much wider (from -0.03 to + 0.12°C/year), the mean value is 0.01°C/year. It should also be noted that the mean values of the trend coefficient at all weather stations located in each of the specified permafrost zones are positive. This means that during the research, an increase in the average air temperature in these zones is observed, which is considered as a manifestation of global warming in recent decades.

To study the relationship between climatic and geocryological changes, a coincidence of the location (or at least a close location) of weather stations and test sites is necessary. Consider a map of the location of weather stations in the study area, combined with the layout of the test sites (Figure 7). As can be seen from Figure 7, weather stations and test sites, as a rule, with rare exceptions do not coincide in location. Only TS-2 and TS-10 are located near weather stations. This discrepancy between the locations of test sites and weather stations can undoubtedly be the cause of significant errors in the analysis of the above-mentioned relationship of climatic and geocryological changes in the study area. It is known that in hard-to-reach permafrost areas, weather stations are usually located along river banks or in populated areas. Therefore, both in the continuous and in the discontinuous zones the stations are significantly removed from the test sites. In this regard, a study aimed at determining the relationship between changes in the areas of thermokarst lakes and climatic indicators of average annual air temperature and precipitation required obtaining temperature and precipitation data using the reanalysis method of meteorological data which allows to determine the values of climatic indicators at the TS locations, which the following subsection.

2.2. Study of Climate Changes in the Permafrost Zone Based on Reanalysis Data

As noted above, the discrepancy between the locations of meteorological stations and test sites leads to the need to involve reanalysis (repeated meteorological analysis) methodology. Reanalysis (Meteorological, 2019) is a method of obtaining meteorological information at specified points in a study area, based on the "data assimilation" procedure. The purpose of reanalysis is to "assimilate" the retrospective observational data accumulated over a period of several decades in various archives of meteorological observations. The advantage of this reanalysis is uniform coverage of the territory. There are various reanalysis systems. In the present work, data from the reanalysis systems ERA-40, ERA-INTERIM and APHRODITE JMA were used to determine the climatic characteristics in the study area.

We note the features of the use of reanalysis in our work. With the help of reanalysis, the monthly average values of air temperature and annual precipitation at locations of test sites for the time period from 1970 to 2010 were obtained. Since the reanalysis data were presented as a set of graphic files in one of the common graphic formats (*.png), a template map was created to extract the necessary climate information representing the location of the test sites studied. The resulting map template using the graphic editor MSPaint was superimposed on the reanalysis data map and a new image was obtained in the form of a combined map of the location of test plots and temperature fields (or precipitation). An example of such a combination of cards is shown in Figure 8.

Using such combined maps, the values of mean monthly (and average annual) air temperatures and annual precipitation were determined at the center of each test site. Then the average temperatures for each TS for all months and years were calculated. Statistical analysis of time series of average summer, average winter and annual average air temperatures, carried out in (Polishchuk and Polishchuk, 2013) for all test sites, showed the most pronounced correlation between changes in average winter and average annual surface air temperatures (the correlation coefficient is 0,68).

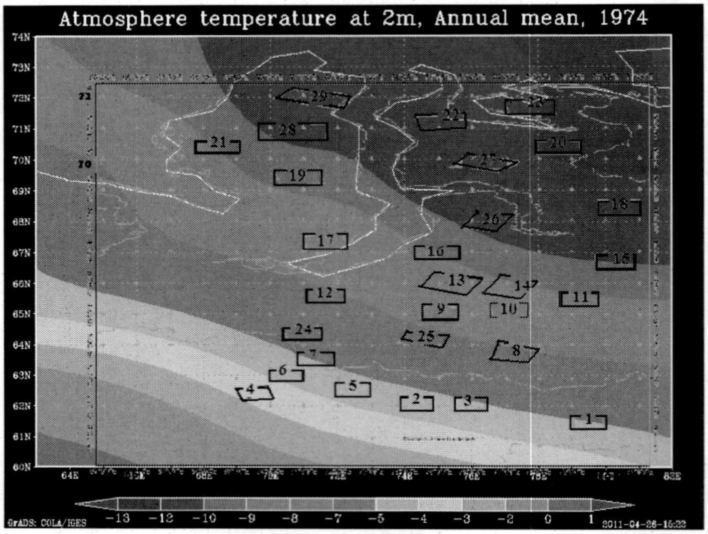

Figure 8. Combination of location maps of test sites and temperature field, Western Siberia, July 1974.

Table 5. The correlation coefficients of temperature changes in different zones of permafrost

Permafrost zones	Insular	Discontinuous	Continuous
Insular	1	0,94	0,78
Discontinuous	0,94	1	0,92
Continuous	0,78	0,92	1

Table 6. Values of the linear trend coefficient for average annual temperature on different test sites

TS number	Latitude, degree N	Longitude, degree E	α_T, °C/year
1	61,26	79,31	0,052
2	62,03	74,23	0,057
3	62,05	75,59	0,061
4	62,25	69,35	0,039
5	62,3	72,28	0,057
6	63,01	70,27	0,052
7	63,33	71,2	0,053
8	63,4	77,13	0,04
9	65,03	74,59	0,066
10	65,1	77,03	0,056
11	65,27	79,12	0,042
12	65,35	71,37	0,04
13	65,6	75,23	0,053
14	65,51	77,8	0,055
15	66,41	80,15	0,06
24	64,23	70,56	0,051
25	64,12	74,36	0,051
16	66,57	74,54	0,062
17	67,17	71,37	0,051
18	68,23	80,23	0,072
19	69,19	70,5	0,062
20	70,2	78,31	0,074
21	70,22	68,24	0,051
22	71,08	75,04	0,064
23	71,37	77,44	0,066
26	67,58	76,26	0,07
27	69,51	76,18	0,071
28	70,54	70,42	0,066
29	71,55	71,18	0,049

The statistical relationship of temperature changes in different zones of permafrost ("amicability" of temperature changes) was studied. For this, the correlation coefficients between average annual air temperature values in different permafrost zones were calculated. The results of the calculations are given in Table 5, from which it is clear that the highest correlation (0.94 and 0.92) show temperature changes in the neighboring zones (insular - discontinuous, discontinuous - continuous). Temperature changes in the far away continuous and insular zones are also correlated, but the magnitude of the correlation coefficient has a smaller value (0.78).

Table 6 shows data on the trend coefficient of average annual temperature values (α_T) according to reanalysis data on each TS, which allows studying the spatial variability of the linear trend coefficient in permafrost territory. Comparison of the average α_T values for each permafrost zone showed that in the insular and discontinuous zones a similar increase in air temperature is observed, and in the continuous one - a higher rate of temperature increase.

Figure 9 shows a graph of the time course of the average annual air temperature values according to reanalysis, which shows an increase in the average annual temperature with time as a result of warming. The points shown in the graph represent the average annual temperature values obtained by averaging the average monthly temperature values for all test areas. The solid line in the graph of Figure 9 displays a linear approximation (linear trend) of the time course of the average annual temperature with a linear trend coefficient of 0.0505 ° C/year. Consequently, the reanalysis data show a larger trend ratio of average annual temperatures compared with the data of weather stations presented in the previous section, which can be explained by the location of weather stations near the rivers and sea shores. It is the result of the warming effect of large water reservoirs affects.

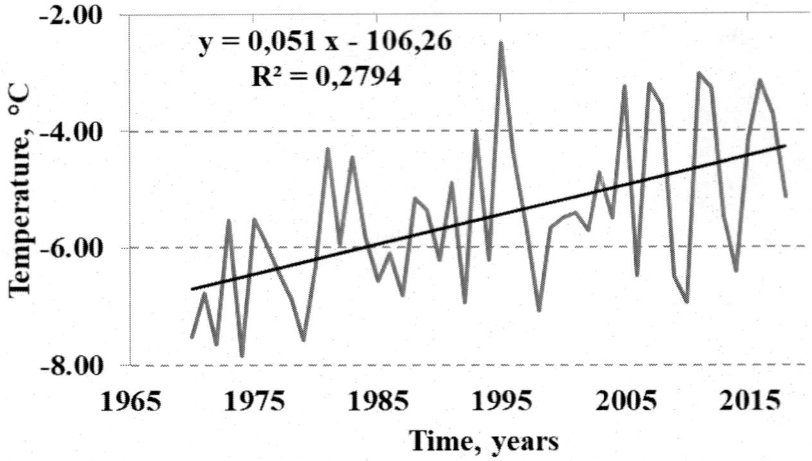

Figure 9. Time dependence of the average annual air temperature in the permafrost of western Siberia according to reanalysis data.

2.3. Analysis of the Relationship of Climatic and Geocryological Changes in the Permafrost Zone

The study of the relationship between climate change and the dynamics of areas of thermokarst lakes in the cryolithozone of Western Siberia is carried out here based on a multifactor regression analysis of data from joint studies of temporary changes in areas of thermokarst lakes, air temperature and precipitation (Polishchuk, 2017). The research data for more than 40-year period was used at 29 test sites of the cryolithozone of Western Siberia located in different landscape and geocryological conditions. The time series of averaged over all test plots of average values of lake areas, average annual temperature and annual total precipitation for the regression analysis were formed. These data were combined into a 3-dimensional dataset (lakes area - temperature - precipitation).

Further analysis of the resulting data set was carried out by the method of multidimensional linear regression analysis, the results of which can be represented by the equation of the multiple linear regression in the form:

$$\overline{S} = c_0 + c_1 \cdot x_1 + c_2 \cdot x_2 + c_3 \cdot x_3 , \tag{2}$$

where $\overline{S}$ is the area of lakes averaged over all investigated test sites for a certain year (dependent variable of the regression equation);

c_i – regression coefficients, i = 0, ..., 3;
x_1 – average annual air temperature;
x_2 – annual total precipitation;
x_3 – time (years).

As a result of the regression analysis of a multidimensional data set using MSExcel, the following values of the coefficients of the regression equation (2) were obtained:

$c_0 = 315{,}15$ ha;

$c_1 = -\,0{,}129$ ha/ºC;

$c_2 = -\,0{,}013$ ha/mm;

$c_3 = -\,0{,}145$ ha/year.

It is known that the magnitude of the regression coefficients in (2) determines the relative contribution of the corresponding independent variable of the regression equation to the change in magnitude of study variable. Comparing the values of the regression coefficients c_1 and c_2 (for temperature and precipitation, respectively), we see that the absolute value of the coefficient c_1 is approximately 10 times the value of c_2. Therefore, as can be seen from the above, the effect of changes in average annual temperature is an order of magnitude greater than the effect of changes in precipitation on the average area of thermokarst lakes.

Therefore, it can be concluded that the reduction in the size of lakes in the permafrost zone established by remote studies is determined mainly by the increase in average annual temperature, and the effect of changes in precipitation can be neglected. The observed pattern can be explained by the influence on permafrost of the climate warming observed in the last decades, accelerating thermokarst processes, which is accompanied by a decrease in the average area of thermokarst lakes in the permafrost zone of Western Siberia. Our conclusion confirms the assumptions of a number of researchers, for example, in Alaska (Riordan et al., 2006), based on qualitative considerations on the impact of climate warming on the reduction of areas of thermokarst lakes in the permafrost zone.

3. Study of the Spatial Structure of the Fields of Thermokarst Lakes

3.1. Background

The tasks of estimating and predicting methane accumulation in lakes on a planetary scale typically use a power law for the distribution of lakes by region based on the results of the analysis of global water basin inventory data using mainly cartographic materials (Lehner and Doll, 2004; Downing and Prairie, 2006). As noted above, the warming of the recent decades has led to a significant acceleration of thermokarst processes in the permafrost zone, causing an increase in methane concentration in thermokarst lakes. Previously conducted distance studies of the distribution of the territories of thermokarst lakes in Siberia and other northern regions (Viktorov et al., 2015; Polishchuk et al., 2015; Polishchuk et al., 2016; Muster et al., 2017) confirm the power law of the area distribution of the law. Most of these studies used medium-resolution Landsat satellite images (30 m), small lakes cannot be detected on such images. Therefore, the empirical size distributions of lakes obtained from Landsat images do not take into account the contribution of small thermokarst lakes with a high concentration of methane, which significantly exceeds its concentration in larger lakes (Pokrovsky et al., 2011; Holgerson and Raymond, 2016).

Analysis of data on estimated methane emissions from small lakes obtained in (Holgerson and Raymond, 2016) based on mathematical modeling using the power law of the distribution of lakes by size, showed a very significant increase in methane emissions from small lakes. In this regard, there is an urgent need to find the theoretical law of the distribution of lakes in size, which more adequately corresponds to the experimental data and takes into account the totality of small lakes. For the study and registration of small lakes it is proposed to use high-resolution and ultra-high-resolution satellite images. However, these images, due to the small coverage of the study area, do not allow to obtain histograms of the distribution of lakes in size, taking into account larger lake. Consequently, high and ultra-high resolution images must be used in combination with medium resolution images.

In connection with the above, the purpose of this section is to construct histograms of the lake size-distribution based on the results of remote sensing studies that take into account almost all the lakes of the study area in a wide range of sizes, from a few meters to tens kilometers. Such histograms of the lake size-distribution can be constructed only on the basis of the integration of data on the areas of lakes obtained from satellite images of both medium and high and ultra-high resolution. The methodology for integrating data on lake areas from satellite images with high and medium spatial resolution, intended for constructing an integrated histogram for the lake size-distribution, is considered in (Polishchuk et al., 2016). Using this methodology (Polishchuk et al., 2018), a synthesized histogram of the lake size-distribution was built based on medium and high resolution satellite images, which made it possible to include lakes of 20 m^2 in size. The methodology and results of its applying to construct a synthesized histogram of the lake size-distribution based on medium and ultra-high resolution satellite images to account for small lakes of 5 m^2 in size are discussed in more detail.

3.2. Data and Methods

Thermokarst lakes are characterized by an extremely wide range of changes in their sizes - from units of m^2 to hundreds of thousands of hectares. For convenience of consideration, we conditionally divide the size of the lake into three groups: large (more than 2 hectares), small (from 500 m^2 to 2 hectares) and extra small (less than 500 m^2). The expediency of separating a separate group of extra small thermokarst lakes is due to the fact that the concentration of methane in them, according to (Holgerson and Raymond, 2016; Pokrovsky et al., 2011), is more than an order of magnitude greater than its concentration in lakes larger than 500 m^2.

The study of the distribution of lakes in this work was carried out in the cryolithozone of Western Siberia using ultra-high resolution Resurs-P (resolution 0.7 m), QuickBird and GeoEye-1 (resolution 0.6 m) in combination with Landsat-8 images with an average resolution (30 m). All images were selected in a rather short period of the summer season (end of June - August) in order to minimize the effect of seasonal fluctuations in the water level in the lakes. During this period, the ice cover which prevents their detection by interpreting the images disappears completely on the lakes. Processing of satellite images using the standard tools of the ArcGIS 10.3 geographic information system was aimed at obtaining data on the number and areas of lakes in order to construct a histogram of the distribution of lakes in a very wide range of their sizes.

The issues of constructing histograms for the lake size-distribution in an extremely wide range of sizes (from units of m^2 to hundreds of km^2) based on the integration of images with different resolutions were developed in (Polishchuk et al., 2016). In connection with the need to build a single histogram for the lake size-distribution based on the synthesis of two initial histograms of the lake size-distribution in this work, it was proposed to build partial intervals with an uneven pitch for building histograms (according to the logarithmic law): 5-10 m^2, 10-20 m^2, 20-50 m^2, 50-100 m^2, etc., up to 20,000 ha. This made it possible to present data of the lake size-distribution by intervals of their areas rather compactly over a very wide range of lake sizes. There were selected only 23 intervals.

To determine the type of the theoretical law of the lake size-distribution for building a synthesized histogram of the distribution of lakes based on the integration of ultra-high and medium-resolution images, a methodology has been proposed that includes a list of the stages (Polishchuk 2017; Polishchuk et al., 2018):

Stage 1. Construction of the first initial histogram of the lake size-distribution from medium-resolution images, taking into account lakes of predominantly large sizes.

Stage 2. Construction of the second initial histogram of the lake size-distribution according to ultra-high-resolution images in order to take into account the lakes of small and ultra-small sizes according to data obtained on a limited set of test sites.

Stage 3. Construction a synthesized histogram of the lake size-distribution in a wide range of sizes (5 m2 - 20,000 ha) based on the synthesis of two initial histograms of the distribution of lakes.

Stage 4. Determination of the theoretical form of the law of lake size-distribution based on the analysis of the synthesized empirical histogram using the procedure of testing statistical hypotheses according to the form of the laws of distribution (Kremer, 2004; Wentzel and Ovcharov, 2000)

3.3. The Construction of a Single Histogram of the Lake Size-Distribution in a Wide Range of Their Sizes

At the first stage, to obtain the first initial histogram of the lake size-distribution of relatively large size, a mosaic of images of medium resolution Landsat-8 for the period 2013-2014 was compiled in the cryolithozone of Western Siberia. The minimum size of the lakes in the images was chosen to be 0.5 hectares, which corresponds to an area of approximately 6 pixels and allows you to reliably determine the water bodies in the image.

As a result of processing the mosaic of Landsat-8 satellite images using ArcGIS 10.3, in the territory of the cryolithozone of Western Siberia, vector layers of thermokarst lakes were obtained, which made it possible to determine the number and area of all lakes exceeding 0.5 ha. The total number of identified lakes in the study area in the period 2013-2014 was 727.7 thousand with a total area of about 6 million hectares. The obtained data on the number and area of lakes in the medium resolution images in the study area, distributed over the above partial area intervals on a logarithmic scale, are presented as the first initial histogram of the size distribution of lake areas. From 0.5 to 20,000 ha in Figure 10 (black columns). Note that the data on the lakes in the specified size range, on the basis of which the histogram of the lake size-distribution of large lakes in Figure 10, are given in the tabular form (Polishchuk et al., 2016) as the dependence of the number of lakes from sizes.

To obtain a second initial histogram on stage 2, representing lakes of predominantly small and ultra-small sizes, ultra-high resolution images (Resource-P, QuickBird, and GeoEye-1) with a spatial resolution of 0.6 m were used. Ultra-high resolution images were processed using ArcGIS 10.3 on stages within 37 test sites (TS), evenly distributed over the territories of the insular, discontinuous and continuous permafrost of Western Siberia. The location of test sites in studying territory is shown in Figure 11, and a brief description of these test sites and ultra-high resolution images used in the work are given in Table. 7. Selected test sites were about the same size. The total area of all TS was about 0.2 million hectares, which corresponds to 0.002% of the total area of studying territory of permafrost in Western Siberia.

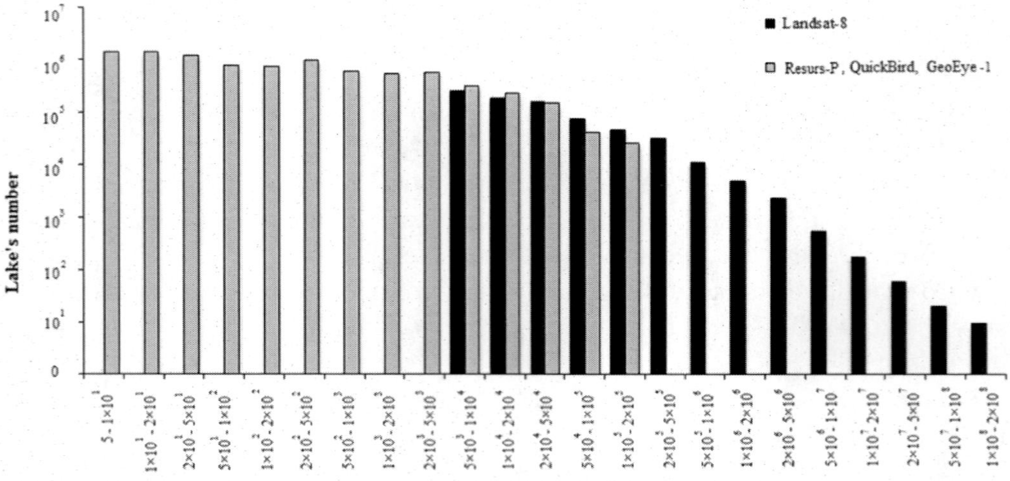

Figure 10. Illustration of a combination of two source histograms.

Table 7. Characteristics of the test sites and ultra-high resolution satellite imagery

Permafrost zone	Camera	Number of TS	Average area of TS
Insular	Resource-P	10	372
	GeoEye-1	4	369
	QuickBird	1	353
Discontinuous	GeoEye-1	5	382
	QuickBird	7	351
Continuous	GeoEye-1	4	293
	QuickBird	8	353

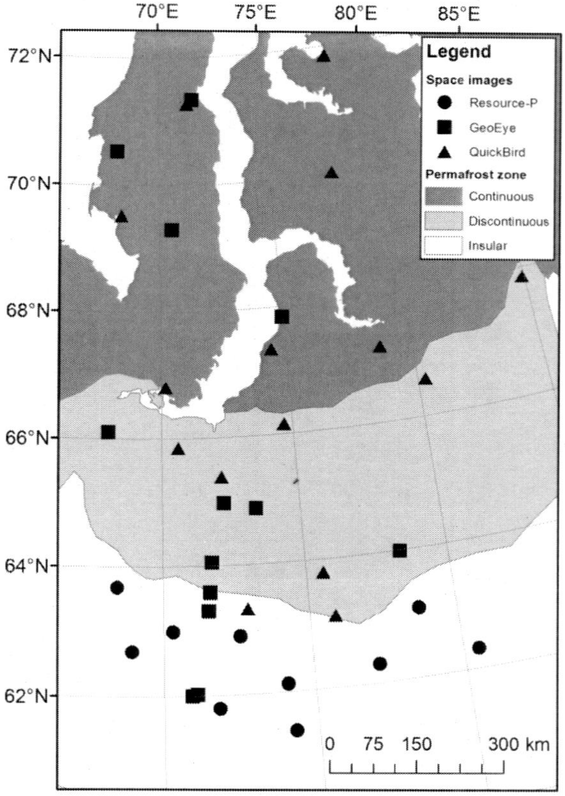

Figure 11. Location of polygons in the permafrost territory of Western Siberia.

The test sites were chosen by us, as is customary in similar studies, in areas of accumulation of thermokarst lakes, i.e., in areas where there are conditions for the formation of a center of thermokarst processes (Viktorov, 2006), named in (Polishchuk and Bogdanov, 2015) zones of active thermokarst (ZAT). Therefore, the ZAT map in the territory of the Western Siberia permafrost zone, constructed from Landsat-8 images using ArcGIS 10.3, was used to select test sites in accordance with the method of forming the cartographic layer of the boundary of the active thermokarst zone (Polishchuk, Bogdanov, 2015).

When constructing the 2nd initial histogram of the distribution of the areas of small and ultra-small lakes, an important methodological question arises: how to extrapolate the data on the number and area of lakes obtained in individual test sites to the entire studying territory? Questions of extrapolating data on the number and areas of lakes obtained in test plots for the entire research area are discussed in sufficient detail in (Polishchuk et al., 2016; Polishchuk et al., 2018). The second initial histogram of the lake size-distribution, constructed from ultra-high-resolution images using the extrapolation technique, is represented by gray bars in Figure 10 in the range of lakes from 5 to 2 × 10^5 m^2.

Table 8. The lake size-distribution intervals

Interval of lake areas, ha	Number of lakes	Total area of lakes, ha	Relative total area, %	Relative number of lakes, %
0,0005-0,001	1 373 457	935,7	0,01	15,83
0,001-0,002	1 316 592	1 845,2	0,03	15,18
0,002-0,005	1 143 321	3 562,0	0,06	13,18
0,005-0,01	738 576	5 339,5	0,08	8,52
0,01-0,02	718 506	10 428,7	0,16	8,29
0,02-0,05	899 805	28 864,5	0,45	10,38
0,05-0,1	580 692	41 433,0	0,65	6,69
0,1-0,2	511 785	72 413,0	1,13	5,90
0,2-0,5	551 925	178 059,0	2,78	6,36
0,5-1	307 740	214 681,0	3,35	3,55
1-2	213 411	297 020,1	4,64	2,46
2-5	152 240	481 650,6	7,53	1,76
5-10	72 091	507 487,8	7,93	0,83
10-20	43 443	609 712,6	9,53	0,50
20-50	30 081	926 364,3	14,48	0,35
50-100	10 354	717 571,4	11,21	0,12
100-200	4 636	638 174,6	9,97	0,05
200-500	2 227	666 245,2	10,41	0,02
500-1000	511	352 498,5	5,51	0,0059
1000-2000	169	233 803,1	3,65	0,006
2000-5000	57	162 679,6	2,54	0,0007
5000-10000	19	125 671,2	1,96	0,0002
10000-20000	9	123 066,3	1,92	0,0001
Total	8 671 647	6 399 506,9	100	100

Let us consider the creation of a single (synthesized) histogram of the distribution of lake areas in a wide range of their sizes based on the combination (synthesis) of two initial histograms (stage 3). In Figure 10 initial histograms have an overlapping area (overlapping intervals) with a width of 5 intervals in the lakes sizes range from 5 × 10^3 to 2 × 10^5 m^2. According to (Bryksina and Polishchuk, 2013), a practically acceptable error of remote measurement of the lake area from medium-resolution images is achieved when the lake is 2 ha (2 × 10^4 m^2) or more. Consequently, a synthesized histogram of the distribution of lake areas can be obtained by "stitching" two initial histograms at a point corresponding to an area of 2 × 10^4 m^2. The synthesized histogram obtained as a result of this "stitching" represents the distribution of lakes in a wide range of lakes sizes from 5 to 2×10^8 m^2. At the same time, in the histogram intervals of sizes less than 2×10^4 m^2, data were obtained from ultra-high resolution images, and in the intervals of more than 2×10^4 m^2 - from images of medium resolution. According to the Table. 8, the total number of lakes in the cryolithozone of Western Siberia is 8.7 million lakes, the total area of which is about 6.4 million hectares. Of these, about half of the number of lakes (44.2%) are occupied by extra small lakes, the total area of which is extremely small and does not exceed 0.1%. However, large lakes (more than 2×10^4 m^2), which constitute only 3.6% of the total number, occupy 86.6% of the total area of the lakes.

3.4. Determination of the Theoretical Form of the Law of Lake Size-Distribution in a Wide Range of Their Sizes

The main aim of this subsection is the implementation of stage 4. This stage includes the determining theoretic law type by the histogram of the lake size-distribution based on the integration of data from satellite images of different resolutions. This histogram takes into account both large and small and extra small lakes. This allowed for the first time to study the lake size-distribution in a very wide range of their sizes (Polishchuk et al., 2018). The graph of the synthesized histogram of the empirical lake size-distribution over a wide range of lake sizes (from 5 to 2×10^8 m^2) is shown in red in Figure 12, where empirical histograms of the lakes distribution in Sweden and the planet generally (Cael and Seekell, 2016) shown for comparison. However, unlike the data (Verpoorter et al., 2014; Cael and Seekell, 2016) for Sweden and the planet as a whole, taking into account the lakes larger than 1 hectare, the synthesized histogram for Western Siberia obtained in our work reflects the distribution of lakes in a much wider range (from 5 m^2), including small and super small lakes.

In Figure 12 in the distribution of lakes both at the global and regional levels (Western Siberia, Sweden) with a lake size of less than 2×10^5 m^2, a significant deviation from the power law is observed, which manifests itself as an excess of the number of lakes. Of course, estimates of methane and carbon dioxide reserves in small lakes (and especially in extra small ones) with the power law of their distribution, which is usually used for these purposes, will be significantly overestimated, as can be seen from the calculations in (Holgerson and Raymond, 2016). Thus, the histogram obtained in our work can be used to determine the theoretical form of the law of the distribution of lakes by their size. It is expected this law describes a studying distribution better than power-law.

According to (Kremer, 2004; Wentzel and Ovcharov, 2000), in mathematical statistics there are two approaches to solving the problem of determining the form of the distribution law of a random variable from empirical data: either using a theoretical distribution law based on the theory of the object under study or searching for the distribution laws studied in statistics. To determine the most appropriate to experimental data the distribution law using a mathematical apparatus for testing statistical hypotheses, for example, using software Statistica, Excel, etc.

Remaining within the frameworks of the first approach, let us turn to the results of a theoretical study of the mechanisms of formation and development of thermokarst foci in the lake – thermokarst plains of the cryolithozone, made by Viktorov (Viktorov, 2006; Viktorov et al., 2016), who theoretically substantiated the lognormal distribution of thermokarst lakes in size. Experimentally, this law was confirmed by him and his colleagues (Viktorov et al., 2015) in various parts of the cryolithozone located in different geocryological and geographic conditions. Plots for research, a brief description of which was given in (Viktorov et al., 2015), were selected in the permafrost zones of different regions of the world (Alaska, Canada, Western and Eastern Siberia).

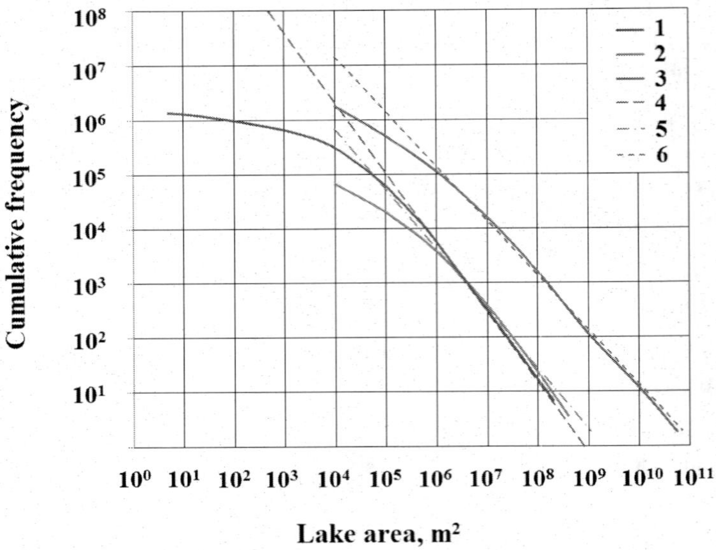

Figure 12. Graph of the synthesized histogram of the distribution of lakes by size in the permafrost zone of Western Siberia (1) in comparison with similar histograms for Sweden (2) and for the Earth as a whole (3). Dotted and dashed-dotted lines denote of the histograms approximation by power-law: 4 - for Western Siberia, 5 - for Sweden, 6 - for the entire planet.

According to (Wentzel and Ovcharov, 2000), the probability density for the lognormal distribution law of the lakes area (*s*) is determined by the equation:

$$f(s) = 1/(2\pi s\sigma)^{1/2}\exp(-(\ln s-\mu)^2/2\sigma^2),$$

where s is the lakes area, μ is the mathematical expectation, σ is the standard deviation.

Estimates of the parameters μ and σ of the theoretical lognormal distribution law of lake areas can be obtained from sample data. In our case, estimates of the mathematical expectation (M) and variance (D) for the lakes of Western Siberia, according to the data of Table 8 were calculated by the formulas:

$$M = \sum^{m}_{i=1}(\ln s_i \times w_i), \quad (3)$$

$$D = \sum^{m}_{i=1}((\ln s_i)^2 \times w_i) - M^2, \quad (4)$$

where w_i is the relative number of lakes (Table 8), m is the number of partial intervals, i is the number of the interval.

Using the above data on the number and area of lakes in a wide range of their sizes (Table 8) and given that $m = 23$, we find using formulas (3) and (4): $M = 4.49$ and $D = 6.08$.

The empirically obtained (synthesized from images of different spatial resolution) the theoretical lognormal law of the lake size-distribution was confirmed using the standard procedure for testing statistical hypotheses about the form of the distribution law (Kremer, 2004; Wentzel and Ovcharov 2000). A study of the correspondence between the empirical and theoretical distributions using the Pearson test, performed using the Excel software package, showed that the calculated χ^2 value is 0.23 and, therefore, a synthesized histogram of the lake size-distribution obtained in a wide the range of their sizes can be accepted corresponding to a lognormal law with a high probability of 0.99.

CONCLUSION

The section is devoted to the presentation of the results of remote sensing studies of the spatial and temporal dynamics of thermokarst lakes in the permafrost zone of Western and Eastern Siberia. The studies used medium-resolution images of Landsat. The temporal dynamics was studied at 29 test sites in Western Siberia and at 39 test sites in Eastern Siberia. Research in Eastern Siberia, there is a tendency to increase the total area of lakes, while in Western Siberia - a tendency to reduce the areas of lakes.

An analysis of climate changes was carried out. The data on average annual air temperature and annual precipitation data from weather stations and reanalysis data obtained from the ERA-INTERIM and APHRODITE JMA archives of temperature and precipitation data, respectively, were used. Due to the fact that the weather stations are remote from the test sites, reanalysis data were used, since they allow obtaining climate information directly at the locations of the lakes. The influence of climate changes on the dynamics of areas of thermokarst lakes is considered. Based on the regression analysis, an equation was found for the relationship between changes in lake areas and climatic parameters. It is established that the effect of increasing the average annual temperature on the reduction of lake areas exceeds the effect of changes in the annual amount of precipitation.

The spatial structure of random fields of thermokarst lakes is investigated. To determine the law of the lake size-distribution, a histogram of the lake size-distribution in a very wide range of sizes from units of m^2 to hundreds of km^2 was constructed based on the integration of data on areas of lakes obtained from satellite images of both ultra-high and medium resolutions. The studies used medium resolution images of Landsat-8 and ultra-high resolution Resource-P, GeoEye-1, QuickBird. It is shown that the areas of lakes are distributed according to a lognormal law, in contrast to the Pareto law widely used in modern literature, which is leading to significantly overestimated volumes of methane and carbon dioxide emissions from thermokarst lakes in the permafrost zones.

ACKNOWLEDGMENT

This work was financially supported by the Russian Foundation for Basic Research (grants no.19-07-00282, 18-45-860002, 18-47-700001 and 18-45-703001).

REFERENCES

Anisimov, O., and Lavrov, S. 2004. Global warming and melting of permafrost: risk assessment for industrial objects of the fuel and energy complex of the Russian Federation. *Technologies of the Fuel and Energy Complex*. 3, 78-83 (in Russian).

Atlas of the Tyumen Region. 1971. Edited by Galkin L. A. *Chief Department for Survey and Mapping under the Council of Ministers of USSR*. Moscow, Tyumen, 1, 216. (in Russian).

Atlas of the USSR. 1984. Edited by Sidorenkov T. P. *Chief Department for Survey and Mapping under the Council of Ministers of USSR*, Moscow, 260. (in Russian).

Audry, S., Pokrovsky, O. S., Shirokova, L. S., Kirpotin, S. N., and Dupre, B. 2011. Early diagenesis processes and their effect on organic matter mineralization and trace element post-depositional redistribution in Western Siberia thermokarst lakes. *Biogeosciences Discuss*. 8, 8845-8894.

Bryksina, N. A., and Polishchuk, Y. M. 2015. Analysis of changes in the number of thermokarst lakes in the permafrost zone of Western Siberia based on satellite images. *Earth's Cryosphere*. 19, 2, 114–120 (in Russian).

Bryksina, N. A., Polishchuk, Y. M., and Kupriyanov M. A. 2016. Comparative analysis of changes in the size of lakes in the permafrost zone of Western and Eastern Siberia using satellite images. *Geoinformatics*. 1, 64-67 (in Russian).

Bryksina, N. A., Polishchuk, V. Y., and Polishchuk, Y. M. 2011. The satellite images database for the thermokarst lakes of Western Siberia and the possibilities of its practical use. *Modern Problems of Remote Sensing of the Earth from Space*. 8, 3, 175–181. (in Russian).

Bryksina, NA, and Polischuk, Y. M. 2013. Investigation of the accuracy of remote measurement of lake areas using space images. *Geoinformatics*. 1, 64-68. (in Russian).

Cael, B. B., and Seekell, D. A. 2016. *The size-distribution of Earth's lakes. Sci. Rep*. 6, 29633, doi: 10.1038/srep29633.

Dneprovskaya, V. P., Bryksina, N. A., and Polishchuk, Y. M. 2009. A satellite image based study of thermokarst changes in the discontinuous permafrost zone of Western Siberia. *Earth study from space*. 4, 88–96. (in Russian).

Dneprovskaya, V. P., and Polishchuk, Y. M. 2007. A study of interaction between the climatic condition and the spatial structure of vegetation in Western Siberia. Bulletin of Yugra State University. 6, 31–38. (in Russian).

Downing, J. A., and Prairie, Y. T. 2006. The global abundance and size of distribution of lakes, ponds, and impoundments. *Limnology and Oceanography*. 51, 2388-2397.

Grosse, G., Romanovsky, V., Walter, K., Morgenstern, A., Lantuit, H. and Zimov, S. 2008. Distribution of thermokarst lakes and ponds at three yedoma sites in Siberia. *Proc. 9th Intern. Conf. Permafrost* (June 29 - July 3, 2008), Fairbanks, Alaska, 551-556.

Hijmans, R. J., Cameron, S., and Parra, *J. World Clim - Global Climate Data - 2005*. [electronic resource]. – Mode access: http://www.worldclim.org/.

Hinkel, K. M., Eisner, W. R., Bockheim, J. G., Nelson, F. E., Peterson, K. M., and Dai, X. 2003. Spatial Extent, Age, and Carbon Stoks in Drained Thaw Lake Basins on the Barrow Peninsula, Alaska. *Arctic, Antarctic, and Alpine Research*. 35, 3, 291-300.

Holgerson, M. A., and Raymond, P. A. 2016. Large contribution to inland water CO2 and CH4 emissions from very small ponds. *Nature Geoscience Letters*. 9, 222-226.

Kabanov, V. M., and Lykosov, V. N. 2006. Monitoring and modeling of climatic changes in Siberia. *Atmospheric and Oceanic Optics*. 19, 9, 753-765 (in Russian).

Karol, I. L. 1996. Estimates of the characteristics of the relative contribution of greenhouse gases to global climate warming. *Meteorology and Hydrology*. 11, 5-12 (in Russian).

Karlsson, J. M., Lyon, S. W. and Destouni, G. 2014. Temporal behavior of lake size-distribution in a thawing permafrost landscape in Northwestern Siberia. Remote sensing. 6, 621-636, doi: 10.3390/rs6010621.

Kirpotin, S., Polishchuk, Y., and Bryksina, N. 2009. Abrupt changes of thermokarst lakes in Western Siberia: impacts of climatic warming on permafrost melting. *International Journal of Environmental Studies*. 66, 4, 423-431.

Kirpotin, S., Polishchuk, Y., Zakharova, E., Shirokova, L., Pokrovsky, O., Kolmakova, M., and Dupre, B. 2008. One of the possible mechanisms of thermokarst lake drainage in West-Siberian North. *International Journal of Environmental Studies*. 65, 5, 631-635.

Krasnov, OA, Maksutov, S. S., Davydov, D. K., Fofonov, A. V., Glagolev, M. V., and Inoue, G. 2015. Monitoring of methane and carbon dioxide emissions from soil to atmosphere and soil parameters. Bakcharskoe swamp of the Tomsk region. *Atmospheric and oceanic optics*. 28, 7, 644-654 (in Russian).

Kravtsova, V. I. 2009. Distribution of thermokarst lakes in Russia. *Bulletin of MGU*. 5, 3, 33–42. (in Russian).

Kravtsova, V. I., and Bystrova, A. G. 2009. Changes in the size of thermokarst lakes in various regions of Russia over the past 30 years. *Cryosphere of the Earth*. 13, 2, 16-26. (in Russian).

Kravtsova, V. I., and Rodionova, T. V. 2016. Investigation of the dynamics of the area and number of thermokarst lakes in various regions of the Russian cryolithozone from satellite images. *Earth's Cryosphere*. 20, 1, 81-89 (in Russian).

Kravtsova, V. I., and Tarasenko, T. V. 2010. A study and mapping of change patterns of the thermokarst lakes in Western Siberia based on multitemporal satellite images. *Environmental dynamics and global climate change*. 1, 82–87.

Kremer, N. S. 2004. *Theory of probability and mathematical statistics*. Textbook for high schools. Moscow: UNITY-DANA. 573. (in Russian).

Landscape map of the USSR. 1987. Edited by I. S. Goodilina. Moscow: MGU. [electronic resource]. – Mode access: http://webarchive.iiasa.ac.at/Research/FOR/forest_cdrom/data/landscapes/ (in Russian).

Lehner, B., and Doll P. 2004. Development and validation of a global database of lakes, reservoirs and wetlands. *Journal of Hydrology*. 296, 1-22.

Luoto, M., and Seppala M. 2003. Thermokarst ponds as indicator of the former distribution of palsas inFinnish Lapland. *Permafrost and Periglacial Processes*. 14, 19-27.

Machulskaya, E. E., Lykosov V. N. 2002. Simulation of the thermodynamic reaction of permafrost on seasonal and interannual variations of atmospheric parameters//Izvestiya RAN. *Physics of the Atmosphere and the Ocea*n. 38, 1, 20-33 (in Russian).

Manasypov, R. M., Pokrovsky, O. S., Kirpotin, S. N. and Shirokova, L. S. 2014. Thermokarst lake waters across permafrost zones of Western Siberia. *Cryosphere*. 8, 1177-1193 doi:10.5194/tc-8-1177-2014.

Meteorological reanalysis. 2019. [electronic resource]. – Mode access: http://en.wikipedia.org/wiki/Meteorological_reanalysis.

Muster, S., Roth, K., Langer, M., Lange, S., CrestoAleina, F., Bartsch, A., Morgenstern, A., Grosse, G., Jones, B., Sannel, A. B. K., Sjoberg, Y., Gunter, F., Andresen, C., Veremeeva, A., Lindgren, P. R., Bouchard, F., Lara, M. J., Fortier, D., Charbonneau, S., Virtanen, T. A., Hugelius, G., Palmtag, J, Siewert, M. B., Riley, W. J., Koven, C. D. and Boike, J. 2017. PeRL: A circum-Arctic permafrost region pond and lake database. *Earth Syst. Sci. Data*. 9, 317–348.

Permafrost and development of oil and gas regions. 2002. Edited by E. S. Melnikov and S. E. Grechishcheva. Moscow: GEOS. 402 (in Russian).

Pokrovsky, O. S., Shirokova, L. S., Kirpotin, S. N., Audry, S., Viers, J., and Dupre, B. 2011. Effect of permafrost thawing on the organic carbon and metal speciation in thermokarst lakes of Western Siberia. *Biogeosciences*. 8, 565–583.

Polishchuk, V. Y. 2017. Analysis of the relationship between climate changes and thermokarst processes in the permafrost of Western Siberia. *Atmos. and the ocean. optics*. 30, 237–242.

Polishchuk, V. Y., and Polishchuk,Y. M. 2013. *Geosimulation of the Thermokarst Lake Fields in the Permafrost Zones*. KhantyMansiysk: UPI YuGU. 129 (inRussian).

Polishchuk, V. Y., and Polishchuk, Y. M. 2014. Modeling of thermokarst lake dynamics in West Siberian permafrost. Chapter 6. In *Permafrost: Distribution, Composition and Impacts on Infrastructure and Ecosystems*, edited by O. Pokrovsky. NY: Nova Science Publishers. 205–234.

Polishchuk, Y. M., and Bogdanov A. N. 2015. Zones of active thermokarst in the territory of permafrost and their detection by space images. *Bulletin of Tomsk Polytechnic University*. 326, 12, 104-114. (in Russian).

Polishchuk, Y. M., Bogdanov A. N., and Muratov I. N. 2016. Methodological issues of constructing generalized histograms of the lake areas distribution in the permafrost zone on the basis of space images of medium and high resolution. *Modern problems of remote sensing of the Earth from Space*. 13, 6, 224-232. (in Russian).

Polishchuk, Y. M., Bogdanov, A. N., Muratov, I. N., Polishchuk, V. Y., Lim, A., Manasypov, R. M., Shirokova, L. S., and Pokrovsky, O. S. 2018. Minor contribution of small thaw ponds to the pools of carbon and methane in the inland waters of the permafrost - affected part of the Western Siberian lowland. *Environmental Research Letters*. 13, 045002, 1-16.

Polishchuk, Y. M., Bogdanov, A. N., Polischuk, V. Y., Manasypov, R, M., Shirokova, L. S., Kirpotin, S. N. and Pokrovsky, O. S. 2017. Size-distribution, surface coverage, water, carbon and metal storage of thermokarst lakes (>0.5 ha) in permafrost zone of the Western Siberia Lowland. *Water*. 9, 228, doi: 10.3390/w9030228.

Polishchuk, Y. M., Bryksina, N. A., and Polishchuk, V. Y. 2015. Remote analysis of changes in the number and distribution of small thermokarst lakes by sizes in Cryolithozone of Western Siberia. Izvestia. *Atmospheric and Oceanic Physics*. 19, 2, 100-105.

Polishchuk, Y. M., Kirpotin, S. N., and Bryksina, N. A. 2014. Remote study of thermokarst lakes dynamics in West–Siberian permafrost. Chapter 5. In *Permafrost: Distribution, Composition and Impacts on Infrastructure and Ecosystems,* edited by O. Pokrovsky. NY: Nova Science Publishers. 173–204.

Polishchuk, Y. M., and Polishchuk, V. Y. 2017. Forecast of thermokarst lakes dynamics in permafrost based on geo-simulation modeling and remote sensing data. *CEUR Workshop Proceedings*. 1839, 393-405.

Riordan, B., Verbyla D., and McGuire A. D. 2006. Shrinking ponds in subarctic Alaska based on 1950-2002 remotely sensed images. *Journal of Geophysical Research*. 111, G04002.

Shiklomanov, A. I., Lammers, R. B., Lettermaier, D. P., Polishchuk, Y. M., Savichev, O. G., and Smith, L. C. 2013. Regional Environmental Changes in Siberia and Their Global Consequences. Chapter 4. In *Hydrological Changes: Historical Analysis, Contemporary Status, and Future Projections,* edited by P. Y. Groisman and G. Gutman. Dordrecht – Heidelberg – New–York – London: Springer. 111–154.

Shirokova, L. S., Pokrovsky O. S., Kirpotin S. N., Desmukh C., Pokrovsky B. G., Audry S., and Viers J. 2013. Biogeochemistry of organic carbon, CO_2, CH_4, and trace elements in thermokarst water bodies in discontinuous permafrost zones of Western Siberia. *Biogeochemistry*. 113, 573–593.

Smith, L., Sheng, Y., Macdonald, G., and Hinzman L. 2005. *Disappearing Arctic Lakes*. Science. 308, 3, 14.

Verpoorter, C., Kutser, T., Seekel, D. A., and Tranvik, L. J. 2014. A global inventory of lakes based on high resolution satellite imagery. *Geophys. Res. Lett*. 41, 1-7. doi:10.1002/201 4GL060641.

Viktorov, A. S. 2006. *The main problems of the mathematical morphology of the landscape*. Moscow*: Nauka*. 252. (in Russian).

Viktorov, A. S., Kapralova, V. N., and Trapeznikova, O. N. 2015. A mathematical model of the morphological structure of lake-thermokarst plains in changing climatic conditions. *Cryosphere of the Earth*. 19, 2, 26-34. (in Russian).

Viktorov, A. S., Orlov, T. V., Kapralova, V. N., Trapeznikova, O. N., Arkhipova, M. B., Sadkov, S. A., Zverev, A. V., Panchenko, E. G., and Bererzin, P. V. 2016. *A mathematical morphology of the landscapes of cryolithozone*. Moscow: RUDN. 232. (in Russian).

Walter, K. M., Edwards, M. E., Grosse, G., Zimov, S. A., and Chapin, F. S. 2007. Thermokarst Lakes as a Source of Atmospheric CH4 During the Last Deglaciation. *Science*. 318. 633-636.

Wentzel E. S., Ovcharov L. A. Theory of Probability and its engineering applications. - M.: *Higher School,* 2000. - 480 p. (in Russian).

Zheleznyak, M. N. 2005. *The geotemperature field and the cryolithozone of the southeast of the Siberian platform*. Novosibirsk: Publishing House "Nauka" of the SB RAS. 227 (in Russian).

Zuidhoff, F. S., and Kolstrup, E. 2000. Changes in palsa distribution in relation to climate change in Laivadalen, Northern Sweden, especially 1960–1997. *Permafrost and Periglacial Processes*. 11, 55–69.

In: The Arctic: Current Issues and Challenges
Editors: Oleg S. Pokrovsky, Sergey N. Kirpotin et al.
ISBN: 978-1-53617-306-2

Chapter 9

GROUNDWATER RESOURCES OF THE WHITE SEA BASIN (NW RUSSIA)

Alexander I. Malov*
N. Laverov Federal Center for Integrated Arctic Research of Russian Academy of Sciences, Arkhangel, Russia

ABSTRACT

According to the conditions of formation of the subterranean inflow into the rivers, the territory of the White Sea basin belongs to three basins of underground runoff: the Northern Seas; Baltic Shield; Timan basin. Underground runoff occurs from zones of active and relatively active water exchange in the ratio of 99:1. The zone of active water exchange is characterized by the presence of fresh infiltration waters. Brackish waters appear here on the areas of access to the surface of gypsum-bearing rocks of the Permian age, as well as on local sites of the inflow of saline waters from the zone of relatively active water exchange. In the zone of relatively active water exchange, brackish and saline waters with salinity up to 35 g/L are present. In most of the territory of the Northern Seas basin, the underground runoff is formed due to the drainage of aquifers from the Permian, Carboniferous and Quaternary sediments. In some small areas, water from the Jurassic, Triassic, Devonian and Vendian sediments takes part in its formation. On the territory of the Baltic Shield, underground runoff into rivers occurs mainly from zones of fracturing of sandstones and crystalline rocks of the Proterozoic and Archean. On the territory of the Timan basin, it comes from fractured zones on the raised elongated blocks of the earth's crust lying between two faults (horsts of the basement) or from aquifers of the Upper-Middle Paleozoic sediments, located in graben-like depressions. Quaternary sediment groundwater also participates in the underground runoff. The share of underground runoff in the total river flow in most of the territory is 10 - 20%. Maximum values (30 - 40%, and sometimes more than 50%) are characteristic of intensely karst areas. The total mass of chemical runoff into the White Sea basin from the territory of the Northern Seas basin is 6 million tons per year, from the territory of the fractured waters basins of the Baltic Shield, it is 0.25 million tons per year. The elements are carried out mainly in the ionic form; the main ions are Ca^{2+}, Mg^{2+}, Na^{+}, K^{+}, HCO^{3-},

* Corresponding Author's Email: malovai@yandex.ru.

SO_4^{2-}, Cl^-. Natural groundwater resources of 95% certainty defined in the amount of more than 40 million m^3/day.

Keywords: aquifers, underground runoff, chemical runoff, basin of the white sea, NW Russia

INTRODUCTION

The United Nations has published 17 Sustainable Development Goals. Goal 6 is Clean water and sanitation for all (United Nations, 2017). Groundwater plays a key role in water supplies worldwide because they are better protected from pollution than surface water. More than two billion people depend on groundwater for their daily water supply, and more than half of the global population depends on it for drinking purposes (United Nations, 2015, Bui et al., 2018). In the same time, water scarcity affects more than 40 percent of people around the world, a figure that is projected to increase with the rise of global temperatures as a result of climate change (United Nations, 2017). The greatest problems with water supply of the population are characteristic of dry regions (Morsy et al., 2018). The arctic territories are located in a humid climate. However, their location near the coastline causes another problem: the salinity of groundwater. In many instances, the area contaminated by saltwater is limited to small parts of an aquifer; in other instances, saltwater contamination is of regional extent. In addition, coastal aquifers are subject to salinity problems as a result of rising deep saline water and brines, and residual salinity in aquitards, associated with ancient transgressions. This chapter discusses the conditions for the formation of underground runoff and chemical runoff in the White Sea Basin (NW of Russia), a typical region of the Arctic coast.

GENERAL CHARACTERISTICS OF THE CONDITIONS OF FORMATION OF UNDERGROUND RUNOFF

According to the conditions of formation of the subterranean inflow into the rivers, the territory of the White Sea basin can be attributed to three basins of underground runoff: the Northern Seas basin; Baltic Shield; Timan basin. The Northern Seas basin belongs to the Russian Plate (Eastern European Province); the Timan basin belongs to the Timan-Ural province (mountain folded region) (Figure 1) (Gromova et al., 1980).

According to the hydrogeological classification, these basins correspond to the Northern Dvina artesian basin; Baltic system of the fissure groundwater basins; Canino-Timan system of the fissure and artesian groundwater basins.

The Northern Dvina artesian basin is a basin of the second order and it is separated from the Central Russian artesian basin of the first order by the features of its underground runoff, which is completely discharged into the White Sea. In turn, the Baltic fissure basins system and the Middle Russian artesian basin are included in the hydrogeological area of the East European (Russian) platform. The Canino-Timan system of basin of fissure and artesian waters belongs to the Timan-Ural hydrogeological folded area (Rogovskaya, 1976).

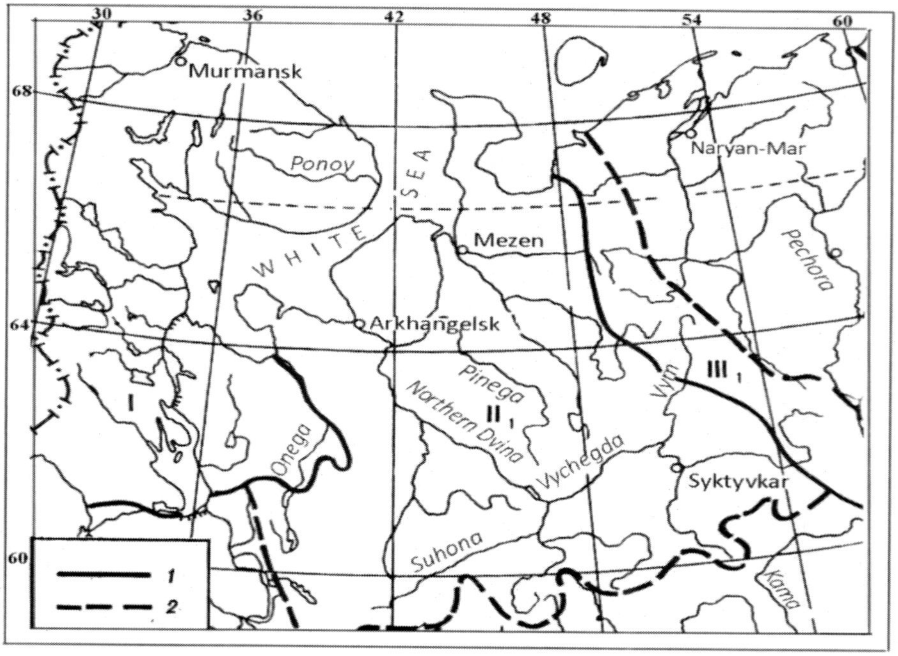

Figure 1. The zoning of the White Sea basin under the conditions of the formation of underground runoff of the zone of intensive water exchange (modified after Gromova et al., 1980). The boundaries of the provinces (1), basins (2). I - Baltic Shield; II - Russian Plate (Eastern European Province): basin - II_1 Northern Seas; III - Timan-Ural province: basin - III_1 Timan.

In the Northern Dvina artesian basin, four different hydrodynamic and hydrochemical zones of groundwater are determined depending on depth, as shown in Figure 2 (Malov, 2003, $2004_{1,2}$). These zones are determined to a first approximation by the filtration properties of water-bearing rocks, their depth of occurrence (intensity of the draining effect of surface watercourses and reservoirs), the openness of the recharges and discharges areas due to the presence or absence of regional aquicludes.

The highest filtration properties (hydraulic conductivity, or the coefficient of permeability - $k_f \geq 0.5$ - 1 m/day) have sediments lying in close proximity to the surface of the earth: above -100 m.a.s.l. This is the active (intensive) water exchange zone. It is characterized by filtration rates (with hydraulic gradients of 0.01 - 0.001) not lower than 10^{-1} - 10^{-2} m/day. The time of complete water exchange at such speeds and active porosity (n_0) 0.25 (with a filtration path from recharge to discharge area of the order of 10 km) is no more than $n \cdot 10^2$ - $n \cdot 10^3$ years. In the immediate vicinity of rivers, especially those flowing in fractured and karst rocks, complete water exchange occurs much faster.

For the Vendian deposits (V in Figure 2), located in the northwestern part of the Northern Seas basin in the interval from - 100 to -600 m.a.s.l., k_f = 0.03 - 0.01 m/day. Filtration rates here are already reduced to 10^{-3} - 10^{-4} m/day, and the time of complete water exchange for a filtration path of about 100 km increases to $n \cdot 10^4$ - $n \cdot 10^5$ years. This is the relatively active water exchange zone. In the southeastern part of the Northern Seas basin, this zone is restricted in depth by the regional aquiclude layer of gypsum and anhydrite of the Lower Permian (P1 in Figure 2) with $k_f = 10^{-5}$ m/day and includes sediments of the Upper Permian (P2 in Figure 2).

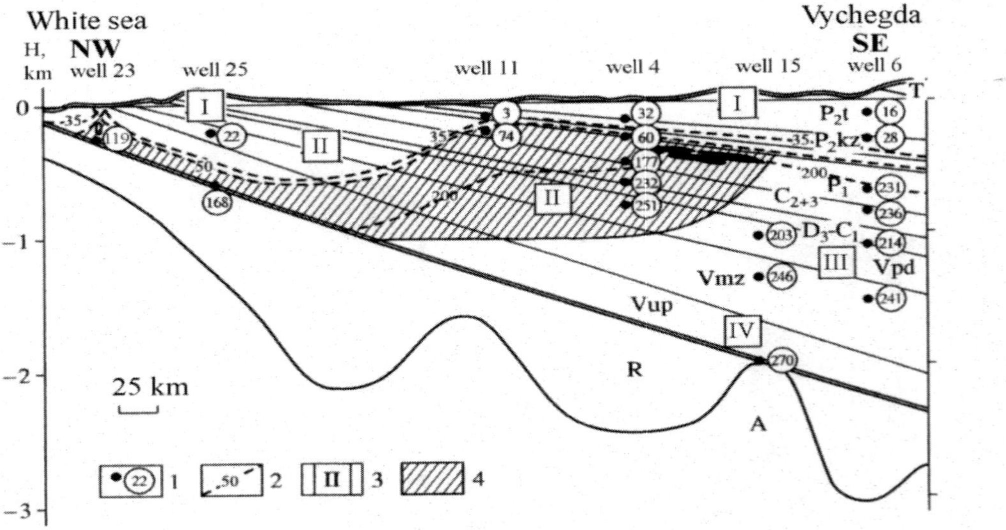

Figure 2. Hydrochemical cross-section across the Northern Seas Basin. 1 - sampling point and groundwater total dissolved solids (TDS), g/L; 2 - isolines of groundwater TDS, g/L; 3 - numbers of water exchange zones (I - active, II - relatively active, III - difficult, IV - very difficult); 4 - zone localization of the brine associated with dissolution of rock salt (unmetamorphosed brines). Geological legend – in Figure 3.

The presence of fresh infiltration water is typical of the active water exchange zone. Brackish waters appear here on the areas with outcrops of Permian gypsum-bearing rocks, as well as on local sites of the rising of saline waters and brines from the relatively active water exchange zone.

The relatively active water exchange zone is marked by the presence of brackish and salt waters with TDS content less than 35 g/L (Figure 2); i.e., this zone is characterized by the penetration of fresh infiltration waters and the dilution of marine (sedimentogenic) waters and brines. This zone also includes the domain of unmetamorphosed brines (Figure 2) that can be traced down to a depth of approximately 1 km. These brines are related to the dissolution of Permian rock salts by infiltration waters (Malov, 2001). They are found in areas where the Lower Permian aquiclude is eroded or perforated.

The remaining part of the sedimentary cover of the Northern Seas basin is occupied by medium and high metamorphosed brines (Figure 2). The Permian age of medium metamorphosed brines has been determined by the paleo-hydrogeological, kinetic-geochemical, and helium-argon methods (Korotkov, 1983; Popov, 1994). These brines were preserved owing to the screening role of the Lower Permian aquiclude. Hence, their domain is characterized by a longer time of complete water exchange (more than 250 Ma) at filtration rates ~ 10^{-5} - 10^{-6} m/day. This is identified as the difficult water exchange zone. Vendian rocks in this zone lie to a depths more than 1 km and have significantly lower filtration and capacity properties.

High metamorphosed brine, especially those in sandstones of the bottom layers of the Vendian and the top layers of the Riphean (R in Figure 2), which are screened from above by the regional aquiclude of mudstones of the Vendian Ust-Pinega Formation (Vup in Figure 2), differ from medium metamorphosed brines by a significantly higher metamorphism grade

(Malov, 2001), suggesting their older (Late Devonian) age. The time of their complete water exchange is more than 360 Ma. This is possible at filtration rates ranging from 10^{-7} to - 10^{-8} m/day and the predominance of diffusion mass exchange (Zverev, 2006). This is identified as the very difficult water exchange zone. Convective mass transfer prevails in all of the remaining hydrodynamic zones.

The formation of hydrogeological zonality in the Mezen Syneclise is related to its geodynamic evolution characterized by the predominance of uplifts in the Cambrian–Devonian when groundwater in Vendian rocks were freshened.

The Upper Devonian is marked by accumulation of the Seregovo saliferous formation in the southern Timan region. It should be noted that the Moscow, Pripyat, and Dnepr–Donets basins include Upper Devonian rocks with potash salt deposits, testifying to a higher degree of evaporation concentration of halogenic basins at that time. These basins could be the sources of high metamorphosed brines confined to underlying aquifers in the eastern part of the Northern Seas basin, because they could percolate below the Vendian aquiclude during the Late Devonian tectonomagmatic activation.

Accumulation of salt and formation of gypsum–anhydrite sequences also took place in the Northern Seas basin in the Early Permian. The downward migration of brines from the bottom of Early Permian halogenic basins produced medium metamorphosed brines that initially filled the entire underlying sedimentary sequence down to the Vendian aquiclude. The high metamorphosed brines are preserved in the pure form only below the Vendian aquiclude.

The Mesozoic–Cenozoic time is marked by maximum difference of geodynamic regimes in the western and eastern parts of the Northern Seas basin (Figure 2). In the west, uplifts of the territory and hiatuses promoted the denudation of carbonate and gypsum sequences, exhumation of Vendian rocks, intense draining and freshening of groundwater, and removal of soluble components from rocks. The composition of groundwater and rocks was conserved in the eastern part of the Northern Seas basin.

In the upper Pleistocene, the western part of the Northern Seas basin was marked by the Boreal transgression of the Mikulino Interglaciation, leading to the accumulation of thick clayey–silty sequences on the coast of the White Sea and in deep paleo-valleys, in particular, within the mouth of the Severnaya Dvina River, and salinization of groundwater in the Vendian Padun Formation. The retreat of the Valdai Glacier 10,000 yr ago and formation of recent river valleys again created hydrodynamic conditions favorable for the freshening of groundwater. However, the salinization was partly retained because of a short duration of the favorable hydrodynamic episode.

Underground runoff into rivers occurs from the first two zones: active and relatively active water exchange, with the active water exchange zone being of primary importance.

Underground discharge from the zone of relatively active water exchange averages about 1% of the total groundwater runoff. These are mainly brackish and saline waters with mineralization up to 35 g/L. Their discharge, as a rule, goes through "hydrogeological windows" in aquicludes. An example of such an inflow with groundwater flow of 3,800 m^3/day with a salinity of 22 g/L from the Vendian sandstone complex to the Northern Dvina alluvial deposits is given in the work of (Malov, 2003). Of the latter, salt water is discharged directly into the river.

Full discharge of brines in the basins of the Onega, Northern Dvina and Mezen rivers is estimated at 30.5 L/s (Korotkov, 1983), that is 2,600 m^3/day, which is about 0.007% of the total groundwater runoff. In the work of (Malov, 2003) it was shown that almost all the brine discharge occurs from the zone of relatively active water exchange in a hidden form. Brines are mixed with fresh infiltration waters, increasing their mineralization. However, one area of brine discharge is set explicitly. In the area of the village Nenoksa on the Onega Peninsula (well 23 in Figure 2) currently has brines discharged in the deep paleo-valley, which includes an almost complete section of the Pleistocene with fragments of Neogene sediments in the bottom. This paleovalley almost exposes the bottom layers of the Vendian and the top layers of the Riphean. Its draining effect was maximal in the Upper Pliocene. At present, it is filled with Quaternary sandyclay sediments, which are more permeable than the Ust-Pinega host rocks. Because of this, brines continue to discharge into the paleovalley after its filling with the Quaternary sediments. A zone of maximal permeability of the Quaternary rocks is observed near the Nenoksa settlement, as a result of which the groundwater salinity is maximal here at minimal depths. The investigation indicated that the water salinity is increased to 5.05 g/L over a distance of about 1.5 km in the Solonyi Creek crossing Nenoksa. The creek water salinity upstream of this zone is 0.68 g/L. The same TDS is characteristic of the Verkhovka River, into which this creek flows. That is, the intensity of the concentrated discharge in the Soloniy Creek of brines with a salinity of about 100 g/L, at a flow rate of 10 L/s, may be about 0.5 L/s. Therefore, during about 600 years, brines have been extracted here for salt production from up to 10-m-deep boreholes with an average amount of water inflow of 2 L/s. The salinity of water pumped out from the boreholes was 120 g/L at a constant operation and decreased to 60 g/L and lower during interruptions.

The waters of the third and fourth zones of difficult and very difficult water exchanges can be unloaded onto the day surface only "in special periods" when paths open to them as a result of tectonic or other activity. Natural events of this kind in the territory of the Northern Seas Basin in historical time were not observed. Anthropogenic activity - the drilling of deep wells with pumping brines - took place.

In the Baltic Fissure Basins System, groundwater inflows into rivers mainly come from zones of fissure sandstone and crystalline rock of the Proterozoic and Archean, located to a depth of 30–50 m, and also from local areas of distribution of sandstone deposits of Quaternary age of great thickness.

The Canino-Timan system of fissure and artesian water basins is a transitional form from platform basins to fissure basins in folded areas. Here, in many graben-like tectonic block-depressions, typical platform conditions of water-bearing conditions, similar to those occurring in the Northern Dvina artesian basin, have been preserved. They contain several storied aquifers confined to terrigenous and carbonate sediments of the Middle Paleozoic, containing pressurized waters. Many boreholes gushes. The upper aquifers are more water-abundant in conditions of intensive water exchange with the surface (active water exchange zone). In carbonate rocks, well flow rates are quite diverse - from 1 - 2 to 20 - 50 L/s, and the specific productivity lies within the range of 0.1 - 0.5 L/(s·m). The groundwater are fresh, with TDS from 0.2 to 0.8 g/L, calcium-carbonate composition. With increasing depth up to 500 m, wells flow rates are reduced to 0.2 - 0.5 L/s (relatively active water exchange zone), and even deeper, in terrigenous rocks of the middle and upper Devonian, do not exceed 0.1 L/s (with reducing the hydraulic head of 100 - 150 m) (difficult water exchange zone). The chemical composition of the water also changes accordingly: at depths of 350 - 500 m, they acquire sodium-chloride composition and TDS up to 35 - 50 g/L; and at 500 - 2000 m the TDS increases to 200 - 215 g/L.

In the system of block uplifts (of the type of horst protrusions), hydrogeological conditions are typical for folded areas. Here, ancient, sometimes Proterozoic, sedimentary-volcanogenic and metamorphic formations, very diverse in composition, were brought to the surface. In the exogenous zone, they contain one water-bearing horizon of fissure and fissure-vein groundwater, recharged due to infiltration of precipitation in the area of aquifer distribution. Discharging takes place in the nearest river network, where numerous sources are formed at a flow rate usually not exceeding 1 L/s. Very rarely, the flow rates of sources reaches 2 - 9 L/s. Relatively short groundwater movement paths from recharge areas to places of discharge caused their intensive water exchange, as a result of which ultrafresh and fresh waters are formed with TDS from 0.1 to 0.3 g/L, hydrocarbonate calcium composition (Rogovskaya, 1976).

AQIFERS OF THE DRAINAGE ZONE

A schematic description of the filtration properties of the main aquifers located in the zones of active and relatively active water exchange on the territory of the Northern Dvina artesian basin (see Figure 3) is given in Table 1. Explanation should be given regarding some values.

In the section of Quaternary sediments with a thickness of 20 to 100 m (Q), clays and loams dominate in the Northern Seas basin, defining average values of hydraulic conductivity (k_f) at the level of 10^{-4} m/day, transmissivity of the aquifer (T) - 0.002 - 0.01 m^2/day. Sandy-clay deposits of river valleys, hollows of glacier runoff, sandy fields, etc. (Q), have a k_f of up to 5 m/day, T - up to 50 - 250 m^2/day. Average values of k_f up to 1 m/day, T - up to 250 m^2/day are observed in sandy-argillaceous Neogene-Quaternary sediments of deep paleo-valleys (Q – N).

Table 1. Filtration properties of sediments of zones of active and relatively active water exchange on the territory of the Northern Seas basin

Stratigraphic Index	Thickness, m	Hydraulic conductivity (k_f), m/day	Water conduction (km), m^2/day
Q Q - N Q	20 -100 250 10 - 50	10^{-4} 1 5	0.002 - 0.01 250 50 - 250
J	50	0.5	25
T	100	0.5	50
P_2t	200	0.5	100
P_2kz	150	7	1050
P_2u	50	1	50
P_1ar-kg	70	10^{-4}	0.01
P_1a-s - C_{2+3}	50 150	50 7	2500 1050
C_2ur-vrc - C_1 - D_3	30	3.5	105
Vpd	200	1	200
Vmz	250	0.03	7.5
Vup (zm)	100	0.01	1

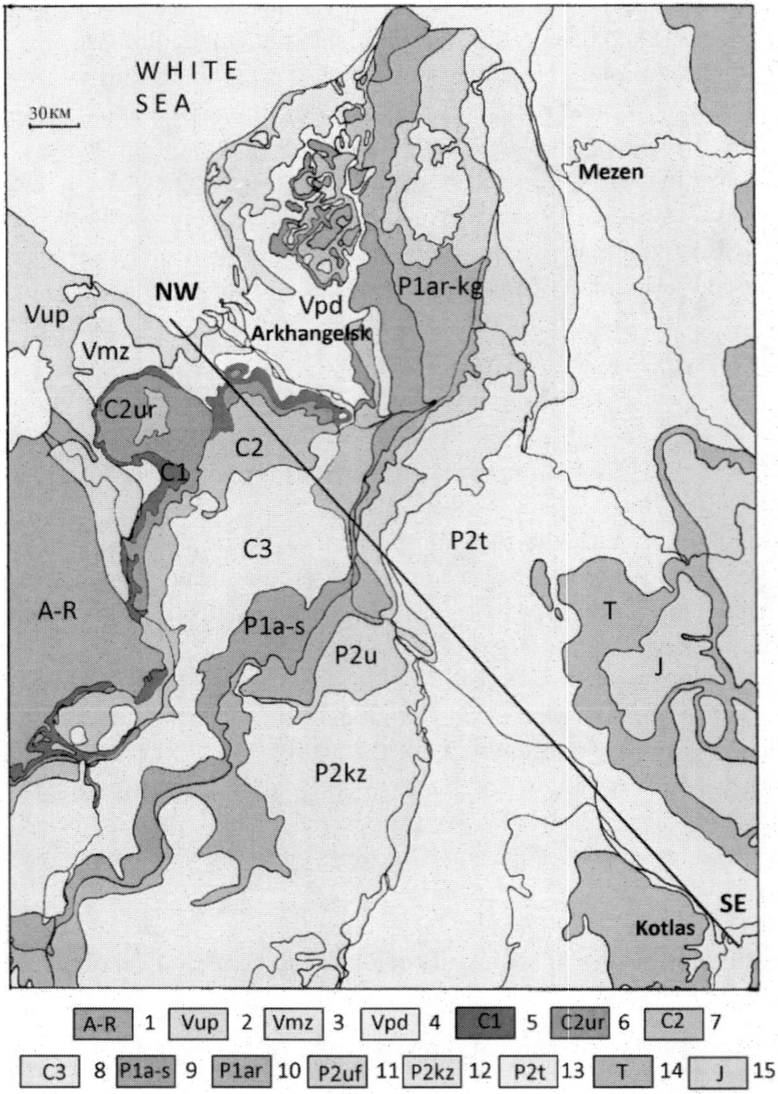

Figure 3. Geological map of the Northern Seas Basin. 1 – Archean-Riphean, 2 – 4 - Ust Pinega, Mezen and Padun suites of the Vend, 5 - Lower Carboniferous, 6 - Ursuga-Voerechensk suite of the Middle Carboniferous, 7 - Middle Carboniferous, 8 - Upper Carboniferous, 9 – 10 - Assel-Sakmarian and Artinsky-Kungur stages of the Lower Permian, 11 – 13 – Ufa, Kazan and Tatar stages of the Upper Permian, 14 - Trias, 15 – Jura. NW-SE - Hydrochemical cross-section across the Northern Seas Basin in Figure 2.

For the predominantly sandy-argillaceous deposits of the Jurassic (J) and Triassic (T) periods, Tatar (P_2t) and Ufa (P_2u) stages of the Upper Permian, the Ursuga-Voerechensk suite of the Middle Carboniferous (C_2ur-vrc), the Lower Carboniferous (C_1) and the Upper Devonian (D_3), the Padun, Mezen and Ust Pinega (Zimnegorskie layers, zm) suites of the Vend is characteristic the dependence of filtration properties on the ratio of clayey and sand components in the section.

The maximum values of filtration parameters have carbonate deposits of the Kazan (P_2kz) and Middle Carboniferous-Lower Permian ages. In turn, in the sediments of the Middle Carboniferous-Lower Permian age (P_1a-s - C_{2+3}), a more permeable part with a thickness of about 50 m lies near the surface of the earth and is less permeable at depths below 50 m.

The increased permeability of the upper part of the section at the places where carbonate sediments reach the surface is associated with a high content of open cracks and karst cavities in it, "healed" by sandy-clay formations and gypsum on deeper horizons. For the lower part of the Kazan deposits, in addition, it is characterized by high clay content, which is partially preserved in the upper part of the section.

Among the cracks in the upper part of the carbonate sequence, three systems are distinguished:

- steeply falling (80 - 900) with azimuths of 280 – 320° - 40 - 45%;
- steeply falling (80 - 900) with azimuths of 40 – 50° - 15 - 20%;
- gently falling (0 - 100) with azimuths of 10 – 50° - 40 - 45%.

The cracks of the first two systems are apparently related to the periodic revitalization of the tectonic structures of the northwestern directions (Murmansk, Kola, White Sea and Karelian Archean megablocks; the Riphean graben and horst system) and the northeastern directions (the Baltic-Mezen and Velsko-Tsilmensky Paleozoic trans-block tectonic zones) (Yudahin, 2001). Gently-falling cracks of the third system inherit the primary formation cracks. The number of cracks is from 10 - 15 to 1 - 4 per 1 m^2. The width of sub-vertical cracks is 1-10 mm, up to 2-3 cm; subhorizontal - 0.1 - 1 mm.

All cracks, especially vertical ones, are affected by karst. Karst processes in the thickness of carbonate rocks are manifested on the one hand in increasing their emptiness due to the expansion of cracks and the formation of numerous pores and cavities, and on the other in the violation of their continuity, that is, the destruction of carbonate rocks to a powdery, sandy or gravelly state. Separate intervals of intensely cavernous rocks stand out against the background of a fairly uniform porosity and karst section. Such intervals often associated with organogenic detrital limestone. The coefficient of volume karst in them reaches 14 - 15%. They are also associated with the intersection of sub-horizontal and sub-vertical cracks. The dimensions of individual caverns reach 1.5 m.

The deposits of the Artinsky and Kungur stages (P_1ar-kg) (the Sotkin Formation of the Tastubi horizon of the Sakmarian stage on the territory of the White Sea-Kuloi Plateau), are represented mainly by gypsum and anhydrite, and have minimal filtration properties, determined by sandstone and dolomite interlayers and lenses.

In terms of area, the Permian sediments (Figure 2) are the most widespread within the territory, occupying about 60%.

According to (Makkaveev, 1969), the drainage area of predominantly aquifers of the Permian sediments is extended by a wide strip from the coast of the Mezen Bay in the north to the Sukhonsky shaft in the south. The surface of the area is a hollow hilly plain with abs. marks 10 - 50 m in the valleys and 160 - 200 m in watershed spaces. The main drains of the region are: the Northern Dvina with the tributaries Vaga, Pinega; the Mezen with the tributary Vashka and others. The depth of the erosion channels of the large river valleys is 30-50 m. In the western and central parts of the region, Permian sediments are deposited directly under the stratum of Quaternary rocks, in the east they are overlapped by Mesozoic sediments.

The aquifers of the Permian sediments in this area take the main part in the underground recharging of the rivers. The limestone and dolomites of the Kazan stage are characterized by the greatest water abundance; sands and sandstones of the Tatar stage are less abundant (Table 1). The flow rates of the springs reach 2 - 8 L/s, and on karst sites 5 - 10 L/s.

In most parts this territory, the groundwater of the Permian are concentrated in the first near-surface aquifer, which is widely drained by rivers. In the northern part of the region, Permian sediments are located above the level of local rivers, and water is discharged into river valleys in the form of descending sources through alluvium.

In the southern half of the region (the basin of the Vaga River), the Permian aquifers descends below the level of local rivers, and the discharge of the headwaters of the Permian sediments occurs along tectonic cracks in the form of ascending sources. Thus, over the most part of the described territory, the waters of these horizons have a close hydraulic connection with the rivers and take the main part in the formation of the underground flow of the upper zone. Ground and low pressure waters of the Quaternary sediments have a subordinate significance in the formation of groundwater runoff.

Drainage area of the predominantly aquifers of Carboniferous deposits lies within the eastern slope of the Baltic crystalline massif. This area is extended by a wide strip from the White Sea coast in the north to the latitude of the Suhona river in the south.

The terrain relief is characterized by alternating flat inter-river plateaus with wide valleys of large rivers: the Onego-Dvina interfluve; the Onega and its tributaries (Churoks, Udrega, Telza, Voya, Ochenga, Kodina rivers, etc.); basins of the left bank tributaries of the Northern Dvina (the Yemtsa, Sheleksa, Vaimuga, Kyama rivers, etc.); White Sea-Kuloi Plateau; basins of the Kuloi and its tributaries (the Kelda, Polta, Kepina, Kotuga, Soyana rivers, etc.). Absolute elevations vary from 10 - 15 m in depressions to 200 - 260 m in watershed areas. The main drains have wide valleys and a relatively small erosion incision, reaching 60 - 80 m.

Carboniferous deposits are represented by limestones and dolomite of the Upper and Middle Carboniferous and terrigenous-carbonate deposits of the Ursuga-Voerechensk suite of the Middle Carboniferous. The carbonate complex of Carboniferous-age sediments containing karst and fissure-karst beds is characterized by considerable water abundance (Table 1). The thickness of the karst stratum of Carboniferous rocks reaches 80 - 100 m, approximately to the depth of the erosive incision of the ancient preglacial troughs (Makkaveev, 1969).

The groundwater of the carbonate complex is discharged without reaching the surface, directly into the thickness of alluvial deposits or in the form of numerous springs along the valleys of large rivers. The debit of springs, depending on the degree of karst water-bearing rocks, varies from 0.5 - 1.0 to 100 L/s and more. Karst springs are widespread along the shores of Yemtsa, Onega and other smaller rivers that do not freeze in winter in areas of karst development. The Quaternary water complex of the described area has a low water abundance and takes a subordinate part in the underground feeding of the rivers.

Fractured and pore reservoir, mainly, pressure waters of Triassic sediments take part in the formation of underground recharging of rivers in certain parts of the basins of the rivers Mezen, Vashka, Uftyuga, Viled, and others. The water abundance of Triassic deposits is on average low (Table 1); groundwater is confined to the interlayers and lenses of coarse clastic material in the clayey stratum.

The pore waters in the Jurassic sediments, mainly associated with sandy interlayers in the clays, take the main part in the underground recharging of rivers in the basin of tributaries of the Mezen (rivers Kyma, Sulla, Pyssa, etc.).

The areas with peculiar drainage conditions include the Lower Dvina, Lower Onega, and Mezen depressions. Aquifer systems here are confined mainly to river valleys. The most significant are the valley complexes of the Northern Dvina, Onega, Laya, Lodma, Mudyuga, Kuya, Zolotitsa, Kuloy, Mezen rivers. The main recharge of them is due to the infiltration of atmospheric precipitation. However, there is also an intrusion of the sea waters of the White Sea in the form of a "wedge", and the rising artesian water from the lower-lying horizons of Vendianary sediments. Noteworthy is the combined presence of fresh and salt waters in sedimentary complexes. As an example, consider the valley of the Northern Dvina.

The Lower Dvina Basin represents an onshore continuation of the Dvina Bay (White Sea). The basin is 120 km long. Maximal dimension of its transverse section (along the coast of the Dvina Bay) is as follows: width of the top zone 100 - 110 km, width of the bottom 30 - 40 km. The bottom is located at - 60 m.a.s.l and slopes are inclined at 2 - 3°. The basin is largely filled with clayey sediments of the Mikulinian interglacial Boreal sea (mIIImk at Figure 4) that existed 130 - 117 ka ago. The clays are underlain by a 10 to 15-m-thick sequence of loams of the Moskovian moraine (gIIms at Figure 4) estimated at > 130 - 140 ka. The underlying Vendian Padun Formation is composed of siltstones and subordinate sandstones with intercalations of mudstones.

The valley at the lower reaches of the Nothern Dvina River passes approximately along the thalweg of its basin. Width of the valley varies from 15 km at the estuary to 2 km in the narrowest part. The erosion valley was formed by meltwater flows from the Valdaian glacier. The valley cuts the Mikulinian interglacial clay sequence almost completely in some places, and partly cuts loams of the Moskovian moraine (Figure 4). The river valley is filled with Upper Quaternary and Recent (QIII–IV) sandy-clayey fluvioglacial, marine, alluvial, and lacustrine boggy sediments. Their total thickness amounts to 40–50 m.

The upper part of the sandy-clayey stratum in the valley of the Northern Dvina River is represented by a 3 - 8 meter layer of loams, clays and sandy loams, which makes infiltration of the atmosferic precipitation to the aquifer complex difficult. With a layer of precipitation of 500 mm per year, about 20% is spent on infiltration. The rest of the water is spent on surface runoff, evaporation and transpiration.

Groundwater is dicharged to the Northern Dvina River. The river is imperfect in the degree of opening of the aquiferous complex: its depth is 5 - 15 meters, and at the sleeves and canals it is even smaller. The width of the river reaches 5 km. For individual branches and channels, a width of 0.05 - 1.0 km is typical.

The hydrochemical situation in the aquiferous complex of the Quaternary sandy-clay sediments of the Northern Dvina river valley is rather complex. A general idea of the distribution of salinity in the groundwater of the aquiferous complex is given in Figure 4a, which shows a cross-section along the thalweg of the Northern Dvina River valley.

Saline waters are located in the lower part of the sedimentary stratum in the river valley. The degree of filling with the salt waters of the river valley decreases in the direction from the Dvina Bay up the valley, and at a distance of 63 km from the bay the salt waters practically disappear.

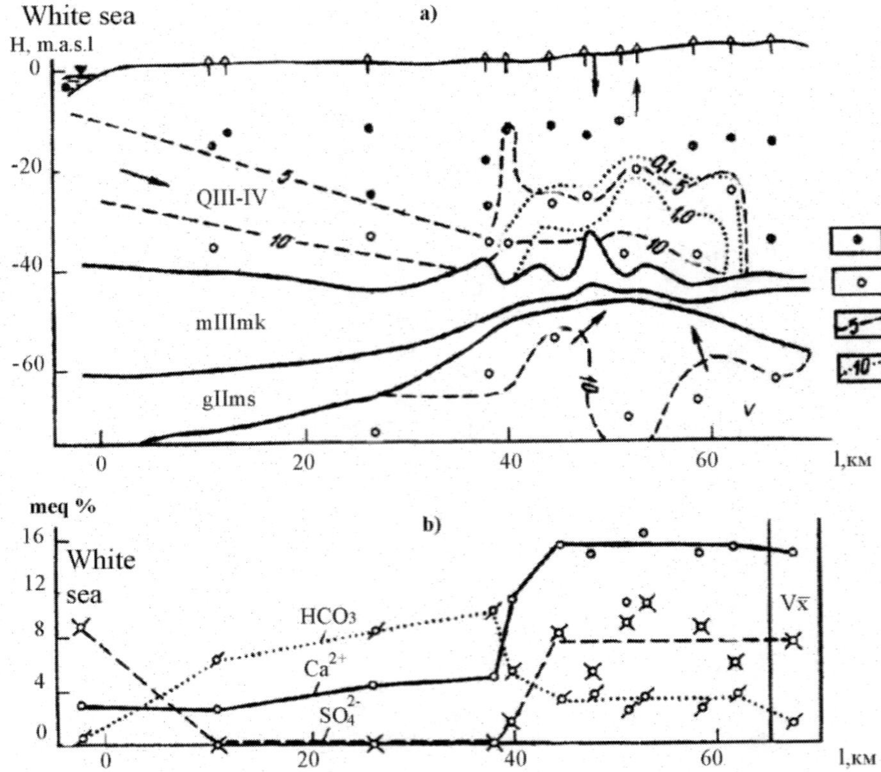

Figure 4. Hydrochemical profile along the thalweg of the Northern Dvina river valley (a) and the C = f (l) graph, where l is the distance of wells from the sea (b): 1 is the middle of the sampling interval in the well; 2 - points used for plotting graph (b); 3 - isolines of chlorine ion content, g/l; 4 - sulfate ion content isolines, g/l. "$V\overline{x}$" on the right at the graph (b) indicates the arithmetic average content of components in the aquifer of the Padun suite of the Vendian.

The composition of salt water in different parts of the river valley is not the same. There are both "sea-looking" waters and waters that coincide in composition with the salt waters of the Vendian sediments located below.

In order to separate zones occupied by waters of various origins, the hydrochemical profile was built along the thalweg of the Northern Dvina valley from Dvina Bay of the White Sea to the border of saline water distribution (Figure 4b) and changes in the chemical composition of saline waters along this profile were analyzed.

It was established that in the section from Dvina Bay up to 38 km from it up the river valley, the salt waters have salinity in the lower part of the valley of 20-21 g/L, which corresponds to the salinity of water in Dvina Bay. The ratio of cations (meq. %) typical for seawater is preserved: Na77Mg18Ca3K2, and the anionic composition differs from the composition of seawater: $Cl93HCO_37SO_40$, while in seawater it is: $Cl91SO_49\ HCO_30$.

Such a change in the anionic composition of seawater is characteristic of sulfate reduction, which may occur as a result of chemical interaction groundwater with organic matter of the sediment of the river valley. Relative stagnation of groundwater in the 38-km stretch of the river valley near Dvina Bay contribute to the reduction conditions. It is caused by the increased clay content of deposits in this area compared to the more sandy deposits developed higher along the river valley and a very small slope of the ground flow towards the sea: $6 \cdot 10^{-6}$ (almost no groundwater movement towards the sea). As a result of sulfate reduction, the concentration of sulfate ion falls from 2.1 g/L in seawater to analytical zero, the concentration of bicarbonate ion increases from 0.1 g/L to 1.1 - 1.3 g/L; up to 6.8 mg/L of iodine appears in water.

The "tongue" of the reducted sea waters of such a composition is thus traced 38 km up the valley from Dvina Bay. These sea waters have, apparently, the Late Valdai-modern age. The form of salt water occurrence in the river valley is caused by differences in the specific gravities of sea and underflow waters: the former are heavier. Due to the greater weight, they lie in the form of a "wedge", directed with the tip upwards along the bottom of the river valley - to the section, where their pressure is balanced by the pressure of the groundwater.

At a distance of 38 - 63 km from the Dvina Bay, in the lower part of the valley complex, there is also salt waters, which, like the reduced sea water, has sodium chloride composition and close mineralization: 20 - 23 g/L. However, their hydrochemical features are different. In composition, they completely coincide with the groundwater of the aquifer of the terrigenous sediments of the Padun suite of the Vendian. They differ from sea unchanged and reduced waters both by the ratio of anions: $Cl88SO_49HCO_33$ and by the ratio of cations: Na70Ca15Mg15. Their distinctive features are: high content of sulfates (1.2 - 1.7 g/L), a significant increase in the proportion of calcium in the composition of cations (about 16 eq.% against 3 to 4 eq.% in seawater), low content of bicarbonate-ion, compared to reduced water. These waters are enriched with iodine coming from Mikulinsky clays, up to 7.6 - 8.4 mg/L.

Due to the lack of an aquicludes over the Quaternary aquiferous complex of the Northern Dvina River valley, atmospheric precipitation infiltrates over the entire area of its distribution. As a result, in the upper part of the complex there is a zone of fresh water of bicarbonate magnesium-calcium composition with mineralization of 0.1 - 0.9 g/L. The chlorine content in these waters near the coast is 20–60 mg/L; at a distance of 50 - 70 km, it decreases to 5 mg/L.

The thickness of the freshwater zone is from zero along the bank of the Dvina Bay to 20–30 meters on distance from 30 to 63 km from it. Higher along the river valley in the aquiferous complex of the Quaternary sediments of the Northern Dvina river valley, fresh water is contained almost throughout the section.

The rising of the saline water from the aquifer of terrigenous sediments of the Padun suite of the Vendian into the aquiferous complex of sandy-clay Quaternary sediments of the valley of the Northern Dvina River is well fixed by the depression cone formed in this area. Also there are "tongues" of waters with low salinity in the aquiferous complex of the Padun suite, directed to the places of discharge (Malov, 2003). The discharge occurs in places with a minimum thickness of impermeable clay deposits along the thalweg of the valley of the Northern Dvina River. In this area, the composition of pore waters in the clays of the Mikulinsky interglacial is identical to the composition of the groundwater of the aquifer of terrigenous sediments of the Padun suite of the Vendian.

Within this area, there may also be a discontinuous disturbance in the aquiclude.

In the aquiferous complex of sandy-clay Quaternary sediments of the valley of the Northern Dvina River, the discharge is well fixed in the form of a continuous zone of waters with a high mineralization extended along the thalweg of the river valley. Salt water is mainly concentrated in the deepest parts of the river valley, and their mineralization decreases from bottom to top, due to differences in the specific gravities of waters of different salinity, to diffusion phenomena and to the discharge of salt water into the river.

Due to the latter factor, the zone of distribution of salt waters has a dome-shaped shape in the section.

According to well testing, water with increased mineralization was found only at distances of up to 400 m from the thalweg of the river valley.

The distances from the wells in which the salt waters are found to the river are 50 - 300 m, which is commensurate with the capacity of the aquiferous complex (34 - 46 m). The highest height of the rising saline water was recorded in the wells located closest to the river (50 m): in the range of –8 to -18 m.a.s.l., water salinity was 10 g/L.

Salt waters discharged from the aquifer of terrigenous sediments of the Padun suite of the Vendian into the aquiferous complex of sandy-clay Quaternary sediments of the valley of the Northern Dvina river mainly in a narrow zone confined to the thalweg of the river valley. In the aquiferous complex of the Quaternary sediments of the river valley, they, due to their large specific gravity, spread in all directions from the discharge zone, increasing the coverage area due to the diffusion of salts.

Part of the salt water is captured by the groundwater flow and is transferred to the discharge zone: to the Northern Dvina river.

According to calculations, discharge saline water with a salinity of 22 g/L into the river at a distance of 38 - 63 km from Dvina Bay is estimated at 3,830 m^3/day, underground freshwater flow with a salinity of 0.4 g/L in this section is 8,170 m^3/day. Salinity of water discharged into the river is 7.3 g/L.

Within the Maritime Lowland, in the areas of the White Sea coast outside the valley complexes the main part in feeding the rivers is taken by the groundwater of the moraine sands or fluvioglacial, lake-glacial sediments of the Quaternary age. The subordinate role is played by the discharge of pressure water from the Vendian aquifers. During the occurrence of the latter under Quaternary sediments of considerable thickness, the inflow from below of brackish and saline waters is characteristic, sometimes in the form of sources.

In the area of access to the surface of the Vendian sandstones of the Padun suite, which have comparatively high filtration properties (Table 1), the value of the fracture waters of these rocks in the underground feeding of small rivers with fresh waters increases (Malov, 2003). This area is located on the territory of the White Sea-Kuloi Plateau.

A large area with the presence of Quaternary sediments aquifers is located in the southwestern part of the Arkhangelsk region and in the north and northeast of the Vologda region (Makkaveev, 1969).

By the nature of the relief, the area is a flat-hilly plain, within which the low marshy areas of the ancient lake basins alternate with low watersheds.

Quaternary sediments have widespread development. Their thickness varies from 30 – 50 to 100 – 150 m in the ancient depressions (Mologo-Sheksninskaya and Sukhonskaya depressions) and in the areas of the final moraine ridges (Konoshsko-Nyandomskaya).

The main rivers of the region are Kubena, Sukhona (upper), Sheksna, Mologa. Rivers are young with poorly developed valleys and small incisions (up to 10 m). They do not cut through the entire Quaternary sediments.

Groundwater is confined to alluvial, fluvioglacial and lake-alluvial deposits. Fluvioglacial interglacial sands drained by the river through alluvium are characterized by the greatest water abundance. The thickness of the mid-moraine sand ranges from 2 - 5 to 40 m.

On the sites of the ancient lake hollows, rivers drain the ancient and modern lacustrine-alluvial sediments with weak water abundance.

In the Predtimansky trough zone (along the Vashka river basin, the upper reaches of the Pinega), groundwater is distributed in the subsurface fluvioglacial and lake-alluvial sediments. The depth of the groundwater of the Quaternary complex varies from 0 to 15 - 20 m. Groundwater is discharged through the thickness of modern alluvium.

Thus, the underground runoff on the most part of the North Dvina artesian basin is formed due to the drainage of aquifers of the Permian, Carboniferous and Quaternary sediments. In some small areas, the waters of the Jurassic, Triassic, Devonian and Vendian participate in the formation of underground runoff.

On the territory of the Baltic Shield, rivers mainly drain the fissure waters of the Archean-Proterozoic crystalline rocks. Groundwater in this rocks is confined to the zone of intense fracturing, usually traced to a depth of 30–50 m. Crystalline rocks lying under quaternary deposits or directly exposed on the earth's surface are usually weak aquifers due to the absence of ancient weathering crust (destroyed by glacial plowing), a small depth of penetration of exogenous fracturing and a rare network of tectonic cracks. Crushed stone in the zones of tectonic faults contains more groundwater. The specific well flow rate is mainly thousandths, rarely hundredths and rarely tenths of a liter per second.

The thin Quaternary cover of this territory is characterized by sporadic distribution of groundwater, their participation in the feeding of rivers is insignificant. The predominantly loamy composition of the Quaternary sediments prevents the intensive infiltration of atmospheric precipitation, and the shallow incision of the relatively young river network of the region causes a slow drainage of aquifers of the Quaternary sediments and crystalline rocks.

Local significant accumulations of groundwater are associated with the distribution areas of Quaternary deposits of high thickness and sandy composition in river valleys and lake basins.

In most of the territory of the Kanino-Timan system of the basins of fissure and artesian waters, the underground recharging of rivers occurs mainly due to the fissure waters of the Paleozoic and Riphean aquifer complexes. Here, the groundwater of the Quaternary sediments practically does not participate in the feeding of the rivers. The aquifers of the intensely fractured and karsted Upper-Middle Paleozoic carbonate rocks, which form the underground recharging of the Chetlas, Pizhma, Vmym, Yelva and other rivers in Middle and South Timan, are the most abundant in the area.

Regime of Underground Runoff in Rivers

In most of the territory of the White Sea basin, the groundwater runoff of the zone of intensive water exchange comes from aquifers that have a close hydraulic connection with river waters, which have a significant impact on the regime of groundwater during floods and prolonged floods. Backwater from high river waters leads to temporary accumulation of groundwater in the coastal zone. During this period, groundwater does not flow into the river bed, river water flows into the banks. This phenomenon is called coastal regulation and characterizes one of the phases of the underground flow, its minimum corresponding to the maximum value of the surface flow.

During the winter period, there is a fairly steady baseflow on the rivers, which makes it possible to judge the monotonous regime of underground runoff into the rivers during this season of the year. As for the summer-autumn period, depending on the humidity of the year, baseflow is either well pronounced or is disturbed by frequent rain floods, which also affects the groundwater flow regime for rivers that have a hydraulic connection with aquifers. Thus, the regime of groundwater is very strongly dependent on the presence or absence of a hydraulic connection with river water (Makkaveev, 1969).

The long-term variability of groundwater flow in rivers is characterized by coefficients of variation. Practically for the entire territory of the White Sea basin, their values are in the range of 0.1 - 0.2 (Gromova et al., 1980), which indicates a steady supply of rivers with groundwater over a long-term stretch. Only in the upper reaches of the Onega, Vaga and Sukhona rivers, they increase to 0.2 - 0.3, which is associated with a sharp increase in the thickness of low-permeable Quaternary sediments here, which impede the stable recharging of rivers: from 30 - 50 to 100 - 150 m in ancient depressions (Mologo-Sheksninskaya and Sukhonskaya depressions) and in the region of the final moraine ridges (Konoshsko-Nyandomskaya). In addition, it should be noted that the increase in the coefficients of variation from north to south can also be associated with geographical zonality, since the values of the coefficients of variation of 0.2 - 0.3 are generally characteristic of most of the territory of the central part of the Russian Plate, decreasing only in local areas of localization of highly permeable sediments mainly carbonate composition.

To characterize the intra-annual distribution of underground runoff into rivers, two periods can be distinguished: with the greatest and least dynamic underground runoff into the rivers. The period of the most dynamic underground runoff almost coincides with the spring - the beginning of the summer and can be called the spring season. The period of the least dynamic is observed in summer, autumn and winter. This period can be divided into two shorter seasons: summer-autumn and winter. The boundaries of the seasons in the White Sea basin are as follows: spring – April–July, summer–autumn – August-November, winter – December–March (Gromova et al., 1980).

Using the values of the average long-term intra-annual distribution of the underground runoff was determined the values for each season as a percentage of the annual underground inflow.

During the spring season, the underground inflow into the rivers in the Baltic Shield amounts to 25 - 35%; Northern Sea Basin - 30 - 35% (35 - 40% in the south, in the basins of the Sukhona and South Rivers). For the Timan basin, the inflow increases from north to south: from 25 – 30% on the Kanin Peninsula and Northern Timan to 30 - 35% on Middle Timan and to 35 - 40% on Southern Timan.

During the summer-autumn season, the underground inflow into the rivers in the territory of the Baltic Shield amounts to 35 - 45%; Northern Sea Basin - 35 - 40%. For the Timan basin, the inflow decreases from north to south: from 40–50% on the Kanin Peninsula and Northern Timan to 35–40% on the Middle and Southern Timan.

In the period of the least dynamic - the winter season - the underground inflow into the rivers makes up 20 - 40% on the territory of the Baltic Shield; the Northern Seas basin - 25 - 35% (20 - 30% in the south, in the basins of the Sukhona and South rivers); Timan basin - 20 - 35% (Gromova et al., 1980).

QUANTITATIVE INDICATORS OF UNDERGROUND RUNOFF IN RIVERS

The distribution of underground runoff on the territory of East European is covered in many papers (Kudelin, 1966; Underground runoff, 1982, Map of the groundwater runoff, 1983 and, etc.).

The quantitative characterization of the underground runoff into the rivers of the White Sea basin is based on the results of the research of the underground runoff department of the State Hydrometeorology Institute, compiled in the complex of works on the compilation of maps of the underground runoff of the USSR and certain regions (joint work of the State Hydrometeorological Institute, the Institute of Water Problems of the Academy of Sciences of the USSR, MSU and other institutions and organizations (Gromova et al., 1980).

The underground component of the river runoff was obtained by the method of genetic analysis of the hydrograph of the general runoff, developed by B. I. Kudelin. It characterizes the total value of the underground runoff of the free water exchange zone within a given river basin. This value is expressed in a module or a layer of underground runoff and characterizes numerically the average long-term value of groundwater recharge. Quantitative characteristics of the values of underground runoff into rivers are also the coefficient of underground runoff, representing the part (in percent) of the total amount of precipitation that participates in the formation of underground runoff, and the coefficient of underground recharge, expressing that share of the total river flow which is formed due to underground runoff.

General patterns of underground runoff into the rivers of the White Sea basin can be shown on the basis of an analysis of the complex maps of its main characteristics (Figure 5 - 7).

For this territory, which is in relatively homogeneous physiographic conditions, the magnitude of the modulus of underground runoff is determined primarily by the structural features of individual areas of drainage of an aquifers and the lithological composition of the sediments composing these aquifers in the drainage area of the rivers.

The range of values of mean annual modules of total underground runoff is from 1 to 6 $L/(s{\cdot}km^2)$, which corresponds to a layer of 30 - 180 mm.

Underground runoff reaches the largest values in karst areas. These are the basins of the Kepina, Pachuga, Kelda, Sotka rivers located within the White Sea-Kuloi Plateau (runoff modules 5 - 6 $L/s{\cdot}km^2$, which corresponds to a layer of 150 - 180 mm); and the Yemtsa river basin on the Onega-Dvina watershed and the interfluve of the Northern Dvina and Pinega, where the modules vary within 3 - 5 $L/s{\cdot}km^2$) (90 - 150 mm of the layer).

The coefficient of underground runoff (as a percentage of precipitation) varies over the territory from less than 5 to 40%, with the highest values (30 – 40%) confined to karst areas. On average, for most of the White Sea basin, ratios of 10 – 15% are typical. The share of underground runoff in the total river flow in most of the territory is 20 – 30%. Maximum values (30 - 40%, and sometimes more than 50%) are characteristic of intensely karst areas.

In the Northern Seas basin, the main runoff is from the aquifers of Permian, Carboniferousl and Quaternary sediments.

In the drainage area of predominantly aquifers of the Permian sediments, the average module of the groundwater flow is equal to 2.5 $L/s{\cdot}km^2$) (75 mm layer). The maximum values (3.5 $L/s{\cdot}km^2$, that is 110 mm of the layer) fall on that part of the territory where the thickness of the Quaternary sediments is minimal and the sub Quaternary collectors are carbonate rocks of the Kazan stage of the Upper Permian with high filtration properties. The coefficient of underground runoff for groundwater of the Permian sediments varies from 15 to 10% in the southern part of the region. The share of underground runoff in the total river flow is from 10 - 20 to 20 - 30%.

In the area of predominant drainage of aquifers of Carboniferous deposits, the average values modules of underground runoff increase and are in the range of 2.5 - 3 L/(s·km^2) (80 - 100 mm of layer), which corresponds to a change in the share of underground inflow in the total river flow from 20 to 30% and the coefficient of underground runoff - 15%. The highest values of the modulus of underground runoff (3 - 6 L/(s·km^2, or 100 - 180 mm of the layer), are confined to the intensively karst areas of the Onega-Dvina interfluve and the White Sea-Kuloi plateau. Here, the share of underground inflow in the total river flow increases to 40 - 50% and the coefficient of underground runoff increases to 15 - 40%.

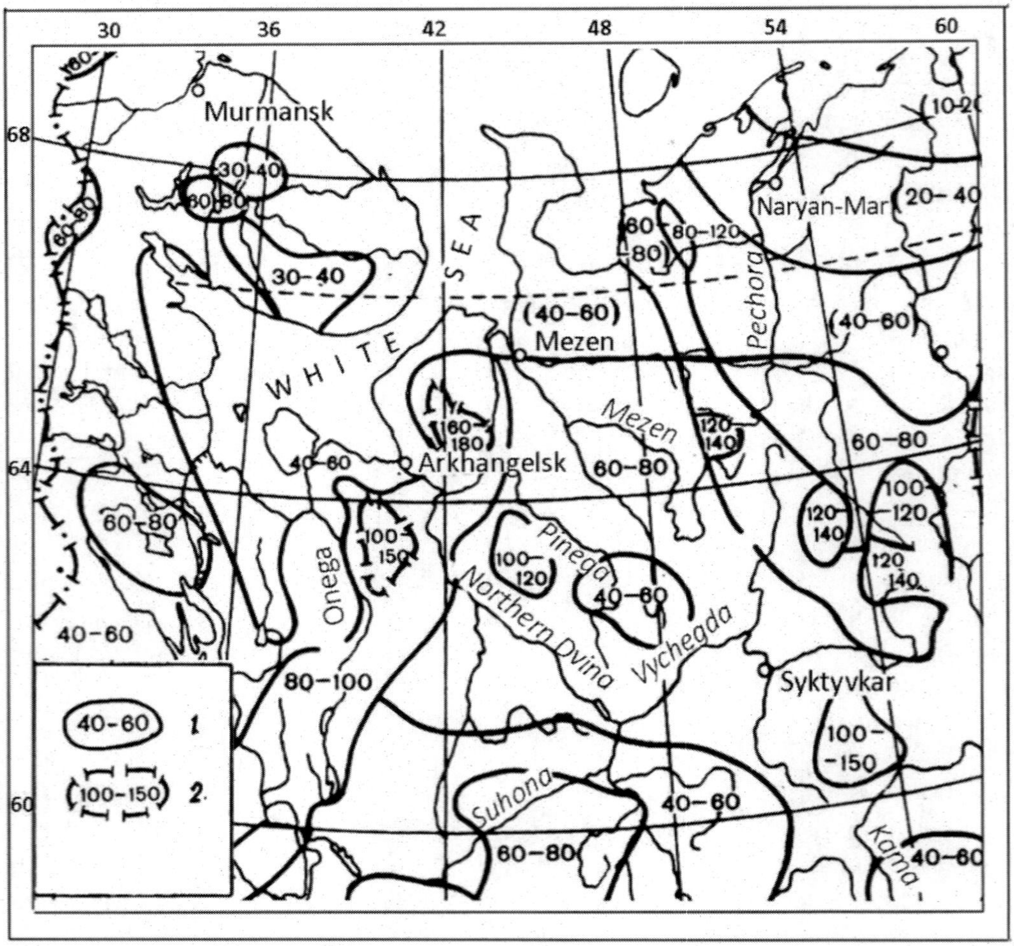

Figure 5. Underground runoff into the rivers of the White Sea basin, characterizing the underground component of surface water resources and natural groundwater resources of the zone of intensive water exchange (Gromova et al., 1980). 1 - the boundary of areas with different values of the underground runoff (numbers - the limits of its changes in millimeters of the layer, in brackets - by assumption); 2 - areas of underground runoff formation in the conditions of formation of karsting rocks (figures are the limits of changes in the underground flow in millimeters of the layer).

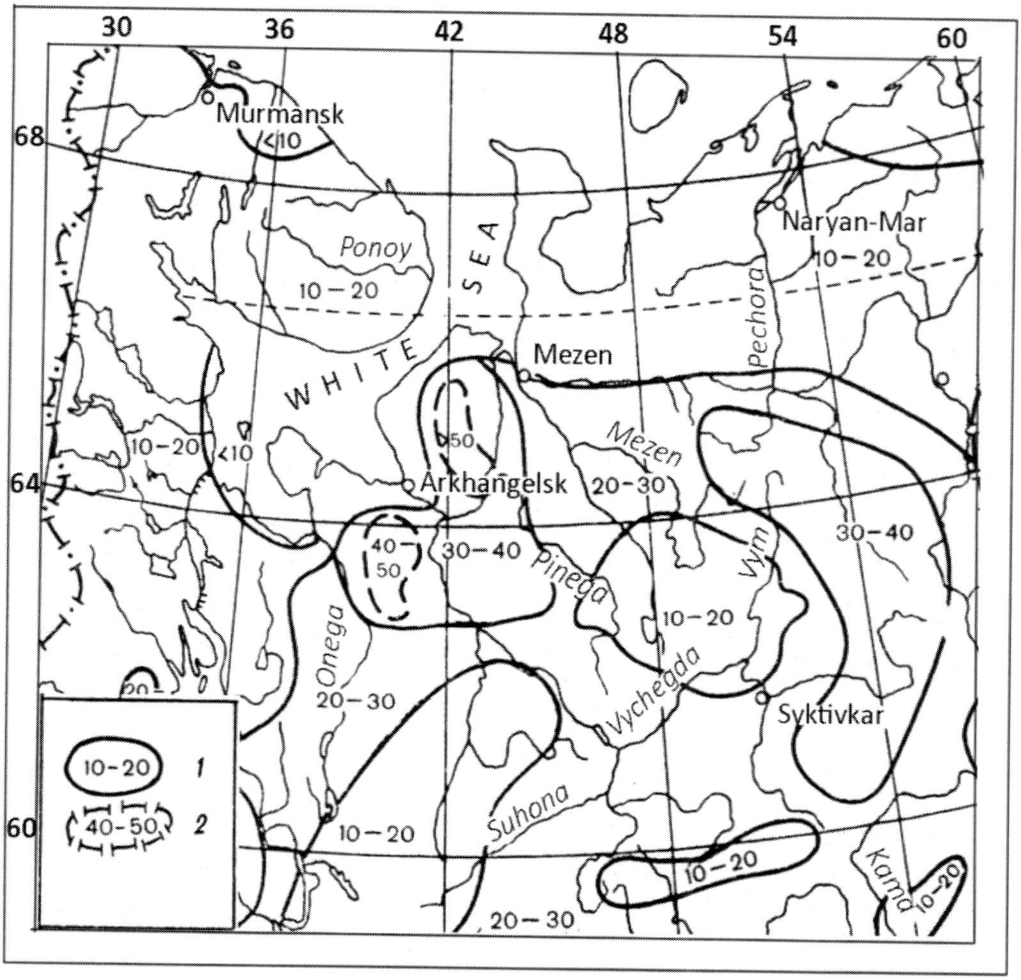

Figure 6. The coefficients of the underground recharge of the rivers of the White Sea basin (Gromova et al., 1980). 1 - the boundary of the areas with different values of the coefficients (figures - the limits of change of the underground inflow in percents of the total river flow); 2 - areas of underground inflow formation in the conditions of development of karsting rocks (figures are the limits of changes in underground inflow as a percentage of the total river flow).

Lower values of the modulus of underground runoff (2.0 – 1.0 L/(s·km^2 and lower, that is, 60 – 30 mm of layer) with a share of underground recharge of 10 - 20% of the river flow and 5 – 8% of precipitation are observed in the north-west part of the Northern Seas basin. Here under the layer of low-permeable Quaternary sediments, low-permeable sediments of the Mezen and Ust-Pinega Formations of the Vendian, and crystalline rock of the Proterozoic and Archean are exposed. The same conditions are observed and in the southern part of the region, where the thickness and value of low-permeable Quaternary sediments increase. Quaternary rocks are characterized by a change in the modules of underground runoff from 1.5 to 2.5 L/s·km^2, while minimum values (< 1.0 to 1.5 L/s·km^2) are observed in areas of widespread moraine loam and mostly clayey lacustrine-alluvial sediments (in the upper Sukhona river, on the watersheds of the Vologda and Lezha rivers). The coefficient of underground runoff here varies from 10 to 5%, the value of the share of underground inflow in the total river flow is 10 to 20%.

The underground recharge of the rivers of the Baltic Shield (Karelian-Kola Province) is formed mainly by fractured waters of crystalline rocks and, to a lesser extent, by pore waters of loose Quaternary sediments. For most of the territory, the underground recharge of the rivers of this region amounts to 10–20% with an module of underground runoff of 1.0 - 2.0 L/(s·km²) (about 50 mm layer). The highest values of underground inflow of up to 2.5 L/(s·km²) and more (60 – 80 mm layer), reaching 20 – 30% in the river flow, are observed in certain areas with the most favorable drainage conditions for aquiferous complexes. Low values of the underground recharge of rivers (1.0 – 1.3 L/(s·km², or 30 – 40 mm of the layer), less than 10% of the total river flow, are observed on the coast of the White Sea, which represents a poorly divided and poorly drained marshy accumulative marine plain. The coefficient of underground runoff in most of Karelia and the Kola Peninsula is only 5 – 10% of the average long-term amount of precipitation.

The Timan-Ural mountain-folded region (Timan-Ural province of underground runoff) is characterized by elevated values of the underground inflow into the rivers compared to the East European Plain.

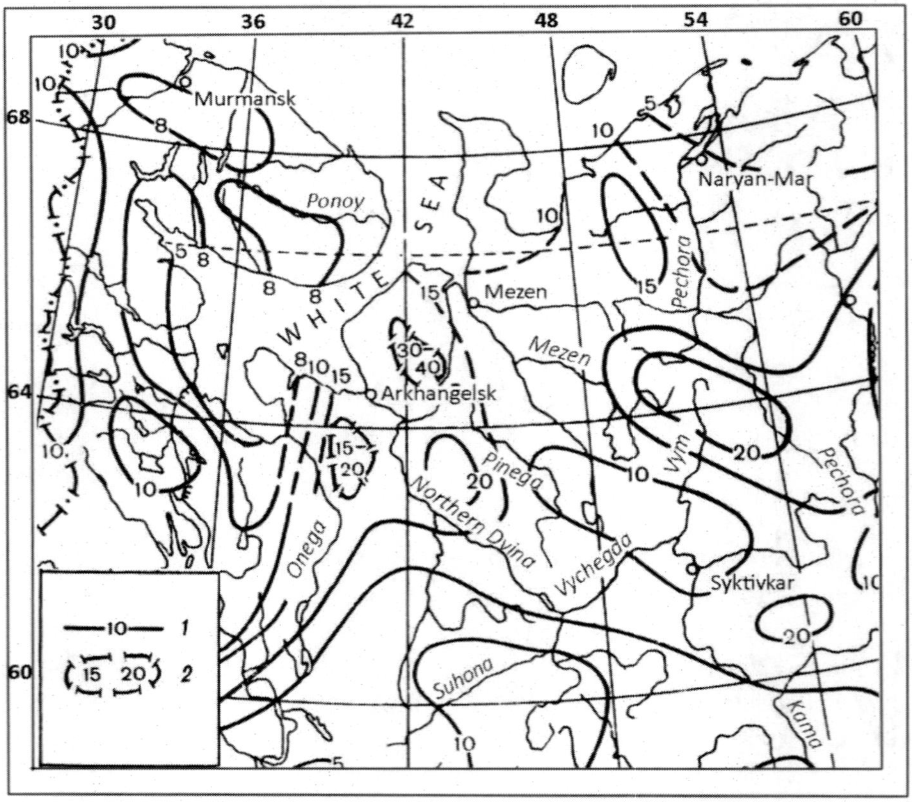

Figure 7. The coefficients of the underground runoff for the territory of the White Sea basin (Gromova et al., 1980). 1 - contours of the coefficients of the underground runoff, dotted line - by assumption (figures - values of the underground runoff as a percentage of precipitation); 2 - areas of formation of underground runoff in the conditions of development of karsting rocks (figures are the limits of changes in the coefficients of underground runoff).

On the territory of the Timan Ridge (the Timan basin of underground runoff), the underground inflow into rivers is formed mainly due to fissure and fissure-karst waters of complexly dislocated metamorphic rocks and terrigenous, terrigenous-effusive and terrigenous-carbonate sediments. The subordinate participation in the underground recharge of rivers belongs to the pore-reservoir waters of the Quaternary sediments. In accordance with the latitudinal climatic zonality (reduction of the annual amount of precipitation in the northern direction), the values of the layer and module of the groundwater runoff vary from 120 - 100 mm (4.0 - 3.0 L/(s·km^2) within the southern and middle parts of Timan to 60 mm and less (2.0 - 1.5 L/(s·km^2) in the north. The maximum values of the groundwater runoff are characteristic for the areas of distribution of karst sediments and intensely fractured rocks of the middle and southern Timan and reach 120 – 140 mm, which exceeds 4.0 L/(s·km^2). Underground inflow as a percentage of the total river flow in accordance with changes in the conditions of infiltration of precipitation and drainage of the territory varies mainly from 30 – 40% in southern and middle Timan to 10 – 20% in northern Timan, which is 15 – 10% of the annual precipitation. In the higher and karst areas of the Timan, the coefficient of underground runoff increases to 20% or more. Later, more detailed work was carried out on estimating the parameters of the underground runoff, but they do not change the schematic diagram shown in Figure 5 - 7.

In 1994, employees of the State Hydrological Institute (O.V. Popov and others) compiled a scientific report on the assessment of natural resources of groundwater for the territory of the Arkhangelsk region, which contains maps of the underground runoff of a scale of 1: 500,000. On these maps, the maximum values of the modulus of underground runoff in the areas of carbonate sediments development in the central part of the White Sea-Kuloi Plateau and in the Yemtsi River basin are even more vividly displayed: 5 - 7 L/(s·km^2). The minimum values - 0.7 - 1.0 L/(s·km^2) emphasize the low role of the fissure waters of the Baltic Shield in the formation of underground runoff. Also, low values (<1.0 – 1.5 L/(s·km^2) are characteristic of the areas of development of quaternary deposits of increased thickness: the Northern Dvina depression, the Primorsk lowlands, and the south-western regions of the Arkhangelsk region.

The authors also calculated natural groundwater resources within the Arkhangelsk region (Table 2).

CHEMICAL RUNOFF

The removal of chemical elements by groundwater on the territory of the East European Platform was estimated by V.P. Zverev (Zverev, 2006), A.I. Korotkov (Korotkov, 1983) and others.

The average values of the module for the removal of chemical elements by groundwater from the territory of the Baltic Shield are given to V.A. Zverev: 0.14 t/(month·km^2), that is - 1.7 t/(year·km^2); from the territory of the Russian Plate: 1.5 t/(month·km^2), that is - 18 t/(year·km^2). It follows from this that the specific removal of chemical elements from the territory of the Northern Dvina artesian basin is on average more than an order of magnitude higher than from the territory of the fractured waters basin of the Baltic Shield. At the same time, the White Sea catchment area within the Northern Dvina artesian basin is approximately two times larger than within the fissure basin of the Baltic Shield. In addition, in the northwest of the Russian Plate there are regions with abnormally high values of the modulus of chemical runoff in karst regions. Taking into account these factors, we can conclude that the total mass of chemical runoff into the White Sea basin from the territory of the Northern Dvina artesian basin is 20 – 30 times greater than from the territory of the fractured waters basin of the Baltic Shield. Approximate ratio: 6 and 0.25 million tons per year.

The elements are remove mainly in the ionic form; the main ions are Ca^{2+}, Mg^{2+}, Na^{+}, K^{+}, HCO_3^{-}, SO_4^{2-}, Cl^{-}.

In the structure of the chemical runoff, the reduced removal of Na^{+} and Cl^{-} from the basin of the fissure waters of the Baltic Shield is especially noticeable compared to the territory of the North Dvina artesian basin. These are elements of "marine" origin, which fell into the aquifers of the Mezen syneclise during the times of sea transgressions.

A.I. Korotkov (Korotkov, 1983) compiled maps of indicators of runoff of major ions from the territory of the North Dvina artesian basin in winter baseflow (Figure 8).

These maps show that the leaching of calcium carbonates is maximal in areas where Carboniferous-Permian carbonate deposits located at the surface.

The foci of sulphate-calcium water discharge control the position of the outcrops of the gypsum-bearing strata of the Artinsky and Kungur stages of the lower Permian. The maximum flow of sulfate ion (up to 8.7 t/(month·km²)) is characterized by the Kuloy river basin. The estimated flow of calcium sulfate here is 684 t/(year·km²) (260 m³/(year·km²)).

Table 2. Natural groundwater resources in the main basins within the territory of the Arkhangelsk region

Basin	Square, km²	Average long-term resources		
		Volume of runoff, million m³/year	Runoff module, L/(s·km²)	Runoff layer, mm
Northern Dvina artesian	274,631	20,747	2.40	76
Fissure waters of the Baltic Shield	21,117	760	1.14	36
Fissure waters of the Timan Ridge	7,252	709	3.10	98
In total within the Arkhangelsk region	303,000	22,215	2.32	73

Basin	Average long-term resources		Resources of 95% certainty	
	Resources		Resources	
	m³/s	thousand m³/day	m³/s	thousand m³/day
Northern Dvina artesian	658	56 840	414	35 801
Fissure waters of the Baltic Shield	24	2 082	9	815
Fissure waters of the Timan Ridge	22	1 942	16	1 374
In total within the Arkhangelsk region	704	**60 864**	439	**37 990**

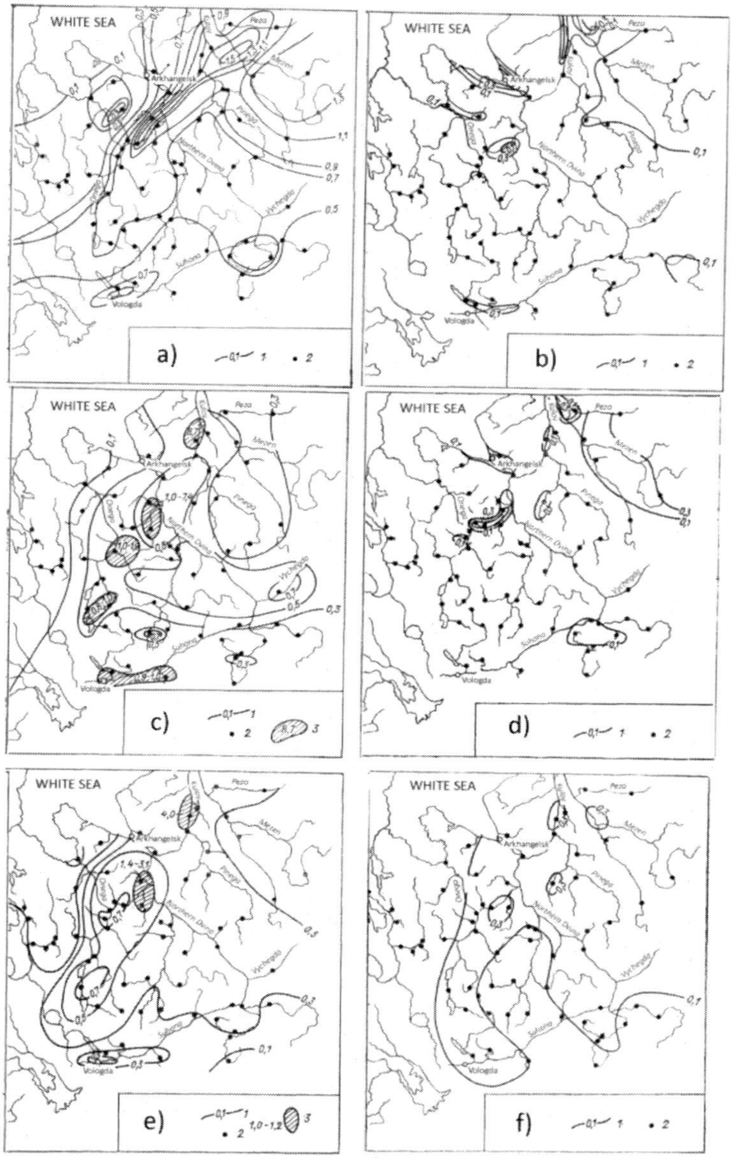

Figure 8. Maps of indicators of runoff of major ions in winter baseflow (Korotkov, 1983): a) carbonate; b) chlorine; c) sulfate; g) amounts of sodium and potassium; e) calcium; e) magnesium; 1 — isolines of runoff of major ions, t/(km^2· month); 2 - points at which the ion runoff is calculated; 3 - areas with an abnormally high ion runoff, figures - an indicator of runoff of ions in winter baseflow at the site, t/(km^2· month).

The general background of runoff of sodium and potassium chlorides in winter baseflow is 0.1 – 0.2 t/(km^2· month). Against this background, the lower reaches of the Onega, Northern Dvina, Kuloy and Mezen rivers stand out, where, as shown above, there is an intrusion of marine waters of the White Sea in the form of a "wedge" and artesian waters from the lower-lying horizons of sub-Quaternary sediments, due to which there is a joint presence in sedimentary complexes of fresh and salt waters of sodium chloride composition. Saltwater unloading from the lower-lying horizons of sub-Quaternary sediments is also observed in the upper reaches of the Sukhona River valley.

In addition to the main elements listed above, practically all other elements of the periodic system leached from the rocks by groundwater circulating in them also take part in the underground chemical runoff.

The most common are silicon and aluminum, which fall into groundwater during the hydrolysis of aluminosilicates and silicates; iron, whose sources are its hydroxides, ore minerals and many igneous rocks; manganese, the epigenetic hydrogenic formations of which are found in sediments (manganite, pyrolusite, sinter formation); nitrogen, phosphorus (mineral fertilizers, saltpeter, apatite); bromine associated with sedimentation sea waters and brines; strontium, falling into the groundwater during the dissolution of celestine and strontium.

Ferrous and iodine-bromine, industrial iodine groundwater are developed on the territory of the North Dvina depression; groundwater with high (up to 50 mg/l) strontium content are confined to the Upper Permian development area (Malov, 2004_1).

In order to study the distribution of strontium in surface waters, we conducted an expedition along the rivers Kepina-Soyana-Kuloy-Pinega. 15 water samples were taken from the mouths of the Kepina and Kotuga rivers, the mouth of the Soyana and its tributaries (the Turya, Letopala, Nyrzanga), the Kuloy river and its tributaries (Nemnyuga, Laka, Eshuga, Olma, Polta, Kelda, Sotka) and from the tributary of the Pinega River - Chuplega. According to the results of chemical analysis of samples, a map of strontium distribution in surface waters was constructed, which practically repeats its distribution pattern in groundwater. Strontium content of less than 0.5 mg/L is observed in the Kepina, Kotuga and Chuplega rivers, draining the aquiferous complex of carbonate sediments of the Middle-Upper Carboniferous and Assel stage of the Lower Permian. Values of 0.5 – 2.0 mg/L are recorded in the mouths of the Soyana River and its tributaries: Turya, Letopala, Nyrzanga, draining sandstones of the Turyanskaya suite of the Ufa stage and carbonate rocks of the Kazan stage, which are characterized by the absence of gypsum, anhydrite and celestine. Contents 2.0 – 4.0 mg/L and more were obtained along the rivers Ezhuga, Polta, Laka, Kuloy, Sotka, Kelda, Olma, draining gypsums and anhydrites of the Sakmara stage of the Lower Permian and gypsum sandstones with manifestations of celestine of the Ufa stage of the Upper Permian. The maximum contents of strontium are found in the waters of the right tributary of the Kuloy River: the Olma River - 4.36 mg/L, which is not surprising, since the sources of this river are located in the Vihtovsky ore manifestation of celestine located on the border of the Ufa and Kazan stages of the Upper Permian.

For a quantitative assessment of the possibilities of dissolution by groundwater of minerals and rocks, developed in the area of the Lomonosov diamond deposit in the central part of the White Sea-Kuloi Plateau, thermodynamic modeling was performed (Zverev et al., 2005).

The water samples from quaternary sands of the uppermost part of the section (to depth of 10 – 30 m) consisted of low-mineralized (TDS ~ 140 mg/l) bicarbonate-calcium groundwater, characterized by pH values of about 7.8 and low (less than 1 mg/L) concentrations potassium, sulphates and silicic acids. The groundwater from the sandstones of the Ursugsky suite of the Middle Carboniferous had a fairly close chemical composition (TDS ~ 150 mg/L, hydrocarbonate-calcium composition)

The lower part of the zone of active water exchange is confined to the Vend sandstones of the Padun suite. The groundwater belonging to it is characterized by a slightly higher TDS, fluctuating within 210 - 420 mg/L. Their composition varies from magnesium-calcium-bicarbonate to sodium-bicarbonate with a corresponding increase in pH to values greater than 8.

The explosion pipes break through the Vendian deposits and overlap by Carboniferous and Quaternary rocks. Groundwater of kimberlite breccias in the considered interval are characterized by a complex and variable composition. Their TDS varies in the range of 200 - 1300 mg/L. With TDS increase, the composition of groundwater changes: from bicarbonate-calcium to bicarbonate-sodium with pH > 9 and sulphate-sodium-chloride with pH 7. It is obviously connected with both the heterogeneous composition of water-bearing rocks and the significant variability of their permeability (k_f varies from 10^{-4} to 1 m/day), and by the penetration of more mineralized waters of deep horizons upwards along weakened zones. The formation of groundwater with a pH of about 7 is probably associated with a fairly high content of aggressive CO_2 (over 15 mg/L), as well as with the oxidation of iron-containing minerals (olivine) in mantle xenoliths.

The results of thermodynamic modeling led to the following conclusions.

Figure 9 shows that rock-forming minerals can be divided into three groups according to the features of equilibrium-nonequilibrium conditions between them and studied groundwater: i) equilibrium (SI is greater than 0), ii) nonequilibrium (SI is less than 0), iii) with changing of the equilibrium-nonequilibrium conditions. The first group includes microcline and muscovite stable under similar conditions, as well as chemical conversion products of silicates - clay minerals kaolinite, illite and montmorillonite. The second group includes minerals, nonequilibrium everywhere with groundwater. These are primary igneous minerals of kimberlites - diopside, forsterite and one of the products of its transformation under hydrothermal conditions - sepiolite, and also gypsum.

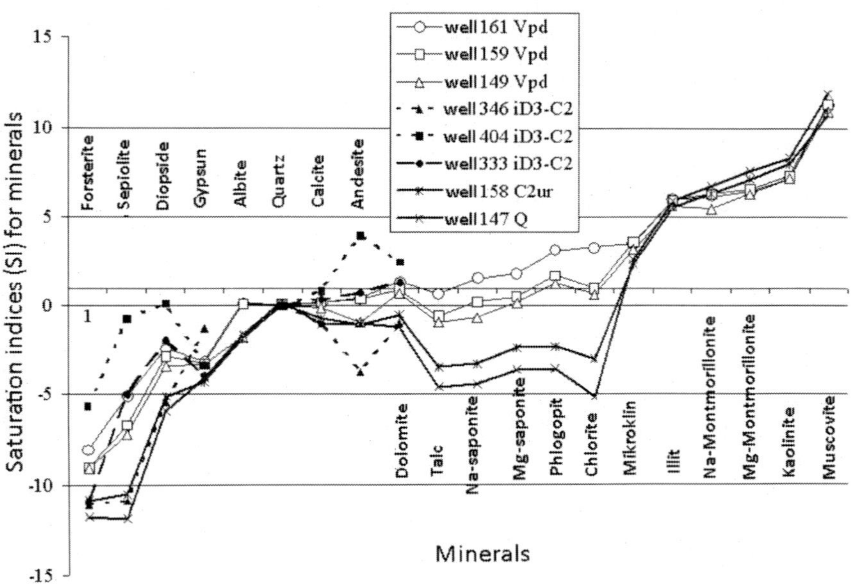

Figure 9. The change in the index of nonequilibrium (saturation indices, SI) of rock-forming aluminosilicates with groundwater in the area of the Lomonosov diamond deposit.

The third group consists of minerals whose state changes from equilibrium at depth (in the groundwater of the Vendian Padun suite) to non-equilibrium closer to the earth's surface (in the groundwater of the Ursugsky suite of the Carboniferous and Quaternary sediments). These minerals include calcite, dolomite, albite, andesite, chlorite, phlogopite and all saponite varieties. In Figure 9 it is clearly seen how for these minerals a decrease in their SI is observed, as they move from the lower to the upper parts of the section and the concentration of individual components in groundwater decreases. This creates favorable conditions for chemical weathering and the emergence of secondary products: clay minerals wich are equilibrium with groundwater; silicic acid in water-soluble forms and, possibly, quartz.

A somewhat more complicated picture is observed for groundwater of kimberlite breccias, which are characterized by considerable variability, and above all - the acid-base state of the system. For this reason, the pH of such waters varies from 7.1 to 9.1. Accordingly, for acidic conditions, the SI of carbonates and silicates has relatively low values, and for basic state, SI are greatest. Under such conditions individual minerals (for example, calcite and diopside) can be in equilibrium with the liquid phase.

CONCLUSION

According to the conditions of formation of the subterranean inflow into the rivers, the territory of the White Sea basin belongs to three basins of underground runoff: the North Seas; Baltic Shield; Timan basin.

Underground runoff occurs from zones of active and relatively active water exchange in the ratio of 99:1. For the zone of active water exchange is characterized by the presence of fresh infiltration waters. Brackish waters appear here on the areas of access to the surface of gypsum-bearing rocks of the Permian age, as well as on local sites of the inflow of saline waters and brines from the zone of relatively active water exchange. In the zone of relatively active water exchange, brackish and saline waters with salinity up to 35 g/L are present.

In most of the territory of the North Seas basin, the underground runoff is formed due to the drainage of aquifers from the Permian, Carboniferous and Quaternary sediments. In some small areas, water from the Jurassic, Triassic, Devonian and Vendian sediments takes part in its formation. On the territory of the Baltic Shield, underground runoff into rivers occurs mainly from zones of fracturing of sandstones and crystalline rocks of the Proterozoic and Archean. On the territory of the Timan basin, it comes from fractured zones on the raised elongated blocks of the earth's crust lying between two faults (horsts of the basement) or from aquifers of the Upper-Middle Paleozoic, located in graben-like depressions. Quaternary sediment groundwater also participates in the underground runoff.

The share of underground runoff in the total river flow in most of the territory is 10–20%. Maximum values (30 – 40%, and sometimes more than 50%) are characteristic of intensely karst areas.

The total mass of chemical runoff into the White Sea basin from the territory of the Northern Seas basin is 6 million tons per year, from the territory of the fractured waters basins of the Baltic Shield - 0.25 million tons per year. The elements are carried out mainly in the ionic form; the main ions are Ca^{2+}, Mg^{2+}, Na^{+}, K^{+}, HCO^{3-}, SO_4^{2-}, Cl^{-}.

Natural groundwater discharge to the White Sea basin is definied as 40 million m^3/day with 95% certainty.

ACKNOWLEDGMENTS

This work was supported by the Russian Ministry of Education and Science (project no. № AAAA-A19-119011890018-3), the UB RAS (project no. AAAA-A18-118012390242-5), the Russian Foundation for Basic Research (projects no. 18-05-60151_Arctic; no. 18-05-01041_A).

REFERENCES

Bui, NT; Kawamura, A.; Amaguch, H.; Bui, D. D.; Truong, N. T.; Nakagawa, K. Social sustainability assessment of groundwater resources: A case study of Hanoi, Vietnam// *Ecological Indicators.*, 2018, Vol. 93. p. 1034-1042.

Gromova, R. V.; Shchuranova, N. N.; Kharchenko, K. I.; et al. Vodnyye resursy Nechernozemnoy zony RSFSR (*Water resources of the Non-chernozem zone of the RSFSR*). Leningrad: Hydrometeoizdat., 1980, 214 p.

Karta podzemnogo stoka Tsentral'noy i Vostochnoy Yevropy masshtaba 1:1 500000 (*Map of the groundwater runoff of Central and Eastern Europe, scale 1: 1 500 000*). Moscow: Mingeo USSR., 1983.

Kudelin, BI. (ed.). Podzemny stok na territorii SSSR (*Underground runoff in the USSR*). Moscow: MSU., 1966, 291 p.

Korotkov, AI. Gidrogeokhimicheskiy analiz pri regional'nykh geologicheskikh i gidrogeologicheskikh issledovaniyakh (*Hydrogeochemical analysis in regional geological and hydrogeological studies*). Leningrad: Nedra., 1983, 232 p.

Makkaveev, AA. (ed.). Gidrogeologiya SSSR. T. XLIV. Arkhangel'skaya i Vologodskaya oblasti (*Hydrogeology of the USSR. T. XLIV. Arkhangelsk and Vologda regions*). Moscow: Nedra., 1969.

Malov, AI. Brine Formation in the Mezen Syneclise//*Water Resources.*, 2001, Vol. 28, p. 618–623.

Malov, AI. Podzemnyye vody Yugo-Vostochnogo Belomor'ya: formirovaniye, rol' v geologicheskikh protsessakh (*Groundwaters of the White Sea region: formation, role in geological processes*). Ekaterinburg: UB RAS., 2003, 234 p.

Malov, AI. Water-Rock Interaction in Vendian Sandy-Clayey Rocks of the Mezen Syneclise //*Lithologi and Mineral Resources.*, 2004$_1$, Vol. 39-4, p. 345 – 356.

Malov, AI. Ekologicheskiye funktsii podzemnykh vod (*Ecological functions of groundwater*). Ekaterinburg: UB RAS., 2004$_2$, 168 p.

Morsy, KM; Morsy, AM; Hassan, AE. Groundwater sustainability: Opportunity out of threat //*Groundwater for Sustainable Development.*, 2018, Vol. 7, p. 277-285.

Popov, VG. Geokhimiya i vozrast rassolov v Volgo-Ural'skom basseyne. (*Geochemistry and Age of Brines in the Volga–Ural Basin*)//*Domestic geology.*, 1994, Vol. 2, p. 62–66.

Podzemnyy stok territorii Tsentral'noy i Vostochnoy Yevropy [*Underground runoff in Central and Eastern Europe*]. 1982. Moscow: VSEGINGEO., 288 p.

Rogovskaya, NV. (ed.). *Gidrogeologiya SSSR. Svodnyy tom v pyati vypuskakh. Vyp.1. Osnovnyye zakonomernosti rasprostraneniya podzemnykh vod na territorii SSSR* [*Hydrogeology of the USSR. Consolidated volume in the five editions. Issue 1. The main patterns of distribution of groundwater in the territory of the USSR*]. Moscow: Nedra., 1976, 656 p.

United Nations. Water for a sustainable world, facts and figures. *The United Nations World Water Development Report*. 2015. 122 p.

United Nations. *Sustainable Development Goals*. 2017. http://www.undp.org/content/undp/en/home/sustainable-development-goals.html

Yudahin, FN. (ed.). *Litosfera i gidrosfera Yevropeyskogo Severa Rossii. Geoekologicheskiye problem* [*Lithosphere and hydrosphere of the European North of Russia. Geoecological problems*]. Ekaterinburg: UB RAS., 2001, 408 p.

Zverev, VP; Malov, AI; Kostikova, IA. *Geokhimicheskoye sostoyaniye podzemnykh vod zony aktivnogo vodoobmena v rayone mestorozhdeniya almazov im.M.V. Lomonosova.* [*Geochemical condition of groundwater of the zone of active water exchange in the area of the Lomonosov diamond deposit*]//*Geoecology*., 2005, Vol. 4, p. 298-303.

Zverev, VP. *Podzemnyye vody zemnoy kory i geologicheskiye protsessy* [*Groundwater of the earth's crust and geological processes*]. Moscow: Scientific World., 2006, 256 p.

In: The Arctic: Current Issues and Challenges
Editors: Oleg S. Pokrovsky, Sergey N. Kirpotin et al.
ISBN: 978-1-53617-306-2

Chapter 10

ESTIMATION OF THE CONDITION OF ARCTIC TUNDRA LANDSCAPES SOIL COVER TAKING INTO ACCOUNT PREVIOUS AND CURRENT ECONOMIC ACTIVITIES: AN EXAMPLE OF THE WESTERN COAST OF MOLLER BAY (ARCHIPELAGO NOVAYA ZEMLYA)

Vidas V. Kriauciunas, Irina A. Kuznetsova, Stanislav A. Iglovsky*, Ekaterina I. Kotova and Kirill A. Mironenko
N. Laverov FCIARctic - N. Laverov for Integrated Arctic Research,
Russian Academy of Sciences, Institute of Geodynamics and Geology, Russia

ABSTRACT

Description of the soils of the west coast of Moller Bay Southern, Yuzhny Island, archipelago Novaya Zemlya is given. Spatial distribution of natural and technogenic radionuclides in the surface layer of the soil was estimated. The presence of short-lived isotopes ^{134}Cs and ^{60}Co has been determinate in a number of soil sections, which indicates the continued influx of technogenic radionuclides into the soils of the Arctic. The area of effective transfer to the Novaya Zemlya area during the winter period covers significant mainland areas, while in the summer the main transfer comes from water areas.

Keywords: Novaya Zemlya, Moller bay, tundra landscapes, Arctic, soils, radionuclides ^{134}Cs, ^{137}Cs, ^{60}Co, ^{226}Ra, ^{232}Th, ^{40}K, long-range atmospheric transport

* Corresponding Author's Email: iglovskys@mail.ru.

INTRODUCTION

Global climate change and environmental pollution have a great impact on the state of the ecosystems of the oceans, seas and land bodies of water, especially of the Arctic territories. Soils of the Novaya Zemlya archipelago bring the imprint of past and modern impacts associated with weapon testing, the deposition of radioactive waste, secondary pollution from nuclear waste dumps and melting due to global warming of glaciers with pollutants during nuclear explosions (Aybulatov, 2000; Kiselev et al., 2006; Schwartzman et al., 2008; Usacheva et al. 2016; Miroshnikov et al., 2017; Kriauciunas et al., 2018 a, b). Also, the pollution of landscapes of the archipelago may be associated with the transboundary flow of impurities from factories processing nuclear fuel in the France and the UK (Northern Polygon, 1997; Arctic, 1997). In addition, the technogenic source of pollutants of the west coast of Novaya Zemlya may be the enterprises of the Murmansk region (Baklanov et al., 1994; Baklanov et al., 1996; Matishov et al., 1997; Matishov, 2001; Kuznetsov, 2002; Radiation, 2006; Nikitin et al., 2011).

Arctic ecosystems are very vulnerable, mostly due to the very low rate of their recovery (Arctic, 1997; The Arctic, 2000; AMAP, 2015). Soil cover stability largely determines the stability and diversity of arctic ecosystems.

Under the conditions of slow geochemical processes, including soil formation (Evseev, 1982; Pechurov, 1983), and a significant influx of anthropogenic pollutants into the territory of the insular Arctic (Kryauchuynas et al., 2014; Demeshkin, 2015; Petriga, 2016) monitoring of the soil cover and identifying factors of the formation of its elemental composition are particularly relevant in view of the construction of a mining and processing plant at the Pavlovskoye field (isl. Severny arch. Novaya Zemlya) (Demeshkin, 2015), which can lead to surface transformation of the soil (State and security 2016) in remote areas.

The need of this study is due, on the one hand, to the fragile, highly sensitive to equilibrium changes of Arctic ecosystems with a very low recovery rate, and on the other hand to the development of the lead-zinc ore deposit Pavlovskoe o. Yuzhny arch. Novaya Zemlya, which can lead to significant changes in the elemental composition of the soil (Chernova, Beketskaya, 2011). Geochemical landscapes, as unique territorial units with a certain type of migration of chemical elements, have been studied relatively poorly in the Novaya Zemlya archipelago due to the inaccessibility of the area. The information obtained will allow minimizing the harmful effects of radiation on human health, permanently or temporarily in high latitudes, where, as established (Gudkov, 2015), structural features of the magnetosphere, leads to the more permeable atmosphere to X-rays and gamma rays. The main contribution to the dose of radiation above the Earth's surface is made by top-soil radionuclides.

MATERIALS AND METHODS

The object of the study was the landscapes of the Arctic tundra of the Moller Bay coast in the area of the Malye Karmakuly meteorological station (Figure 1).

The soils sampling of the upper five-centimeter horizon was carried out in the vicinity of the village of Malye Karmakuly (Moller's Bay, Novaya Zemlya archipelago) (Figures 1, 2).

Soil sampling points were selected taking into account topography, geomorphology, plant and soil diversity. Samples were taken in places with undisturbed (sodded) soil surface, with no signs of flushing or washing of the surface layer. To be able to assess the anthropogenic impact of the sampling site, it was laid both in the territory not subject to direct human impact and in the places of economic activity (at the moment or in the past): at the dumps of fuel and lubricant barrels (FLB) soil profiles number 1, 14, 21, 31, 40; in the zone of influence of the tractor road - № 1-8, 25, pipeline fuels and lubricants - № 1-8. Surface sampling of the soil was carried out in accordance with the methodological recommendations developed at the (Gablin et al., 2012; Gablin, 2014). At sites with the same characteristics, by "envelope method" 5 soil samples to a depth of 5 cm were taken, then the soil samples were mixed, the resulting sample weighing at least 1 kg was packed in a plastic bag (Radiation environment, 2016). The selection technique provided representativeness of the sample.

Soil classification is based on morphological and genetic analysis of the structure of soil profiles. Primary sample preparation included procedures for separating undigested plant residues and foreign inclusions, drying the samples to a completely dry state (at 105^{0}C), grinding in a PM 100 ball mill at 400 rpm for 4 minutes to a fraction of about 71 microns. The content of radionuclides in soils was determined in the accredited laboratory of ecological radiology of the Federal Center for Integrated Arctic Research RAS in accordance with the «Method for measuring the activity of radionuclides in counting samples on a scintillation gamma spectrometer using the software PROGRESS».

Statistical data processing included the calculation of the arithmetic mean, standard deviation, standard error of the mean, pair correlation coefficients k (square matrix) and was carried out using the software StatSoft, Inc. (2011). STATISTICA (data analysis software system), version 10. A critical level of significance was assumed to be $p < 0.05$.

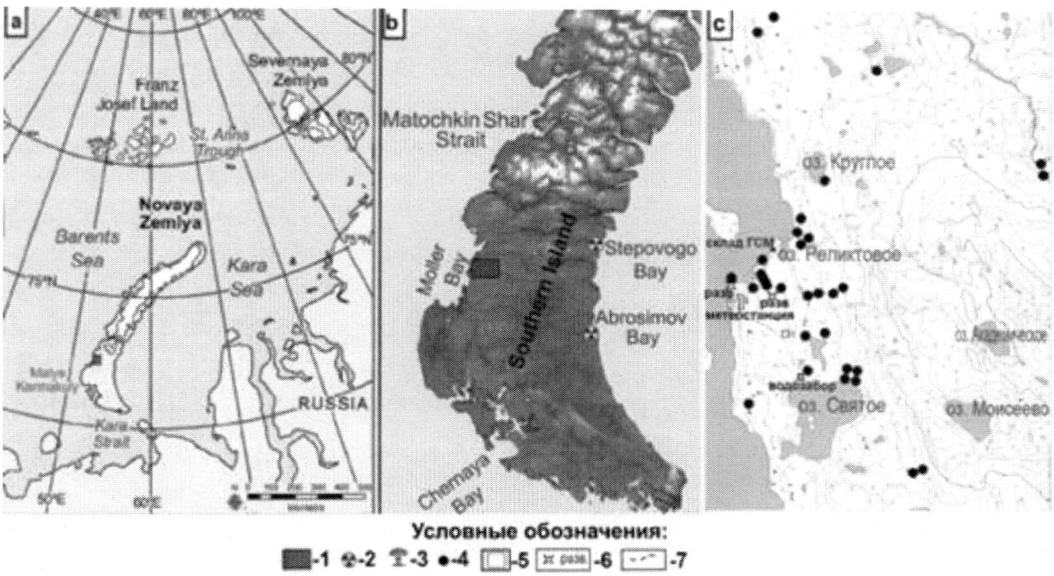

Figure 1. Investigated area of the meteorological station Malye Karmakuly. Location of the study area (1), radioactive waste disposal (2) and nuclear tests sites (3); trial plots location (4), machinery and fuel barrel location (5), buildings ruins (6), fuel pipeline (7). a – landfill, b - machinery, c - barrels of fuel and energy, d – ruins, e - pipeline fuel and lubricants, f - tractor road, g - soil sections.

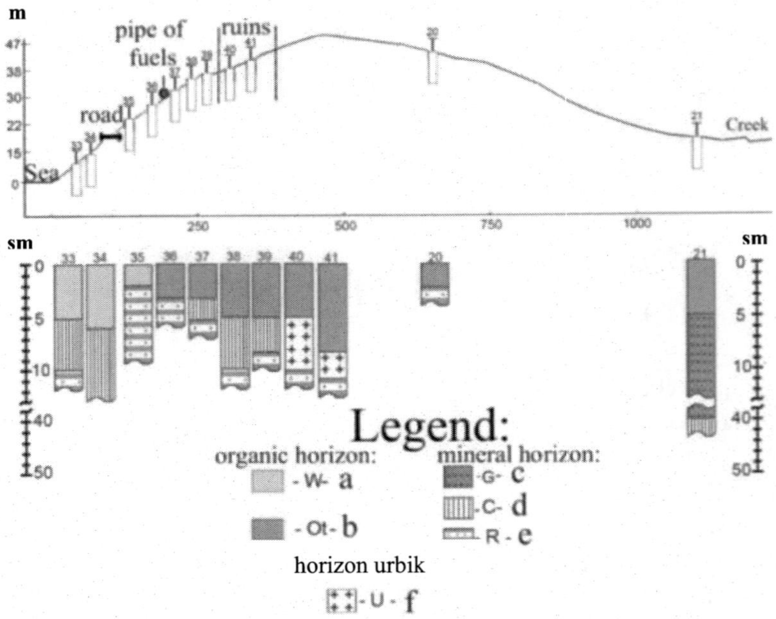

Figure 2. Scheme of location and structure of soil profiles. Legend. a – W - humus-underdeveloped horizon, b – Ot - litter-peat horizon, c – G - gley horizon, d – C - maternal horizon, e – R - dense rock on which the soil profile is formed, f – U - horizon urbik.

The construction of maps of the spatial distribution of elements in the upper five-centimeter soil horizon was carried out using the software Surfer (Golden Software, LLC).

The calculation of the long-range transport of pollution was carried out on the basis of the method of statistics of the trajectories of air masses, which is widely used for similar studies (Hirdman et al., 2010; Salvador et al., 2010; Vinogradova, Ponomareva, 2012; Vinogradova, 2014). The basis of the approach is the analysis of multiyear trajectories of air mass and impurity transfer trajectories to an arbitrary geographic location, the methodology of which was developed at the IAP RAS and improved in recent years (Vinogradova, 2014).

The model HYSPLIT4 and the NCEP/NCAR meteorological elements reanalysis data (Draxler and Rolph, 2013) were used to calculate the return trajectories. The calculation was carried out for the winter and summer period for 2010-2015. In this method, the distribution of impurities during its long-range transport in the atmosphere is modeled similarly to (Vinogradova, 2000; Halperin et al., 1995; Rovinsky et al, 1994). It is assumed that the transferred impurity is chemically passive, and the transfer of impurities in the atmosphere occurs on aerosol particles of submicron size.

Area of the Research

The combination of abiotic stress factors contributes to the formation of two types of landscapes on the archipelago: polar deserts and arctic tundra's (Novaya Zemlya, 1998).

The study area has a slope topography, which is characterized by the presence of stony and crushed stone surfaces (bedrock outcrop), outcrops, stony placers, located both on the coastal cliffs and on rafted areas. The most common elements of meso- and microrelief are bedrock outcrops, which are placers of phyllite rubble, characterized by good drainage in the area there are very dark deposits along the shores of lakes and streams.

The heterogeneous geomorphological, lithological and hydrogeological structure of the study area causes the contrast of soil formation factors and the formation of a number of soil types (Novaya Zemlya, 1998). The soil cover of the study area have local characterization and is represented by the soils of the primary soil formation soil - petrozems, as well as gray-humus and peaty-gley soils. In areas of intensive economic activity there are anthropogenic soils.

The studied petrozems are characterized by a general morphological structure, in which it is possible to distinguish the litter-peat or underdeveloped humus horizon, which lies on fine-grained-gravelly stratum, sometimes directly on solid rock. Lithozems are gray-humus, the soil profile of which is composed of organogenic litter-peat and gray-humus horizons, formed on the rubble of schilite schists, are widely spread. Dry peat-lithozems, whose diagnostic horizon is composed of residues of growing lichens and mosses, are found mainly in local spots in the middle part of the slopes, in upland areas.

Gley soils are formed in the accumulative positions of landscapes in the conditions of the subsoil water retention by permafrost (Novaya Zemlya, 1998; Krupskaya, 2017). Signs of gleing observed in the thickness of the parent rock in the form of gray and rust spots. In the study area, gray-humus gley soils, gley, peat-gley and peat-gley soils were identified. The genesis of these soils directly depends on the features of the relief regulating of the water regime of the soil: the flow regime in conditions of over-wetting contributes to the formation of gley, and the lack of runoff and stagnant moisture initiates the process of peat accumulation (peaty-gley and peat soils).

The plant community of the study area is mainly represented by moss-sedge-shrub vegetation formed in depressions of the relief and the river valleys most protected from the wind on the marshy bumpy surfaces; meadow vegetation with hummocks is fragmentarily distributed. There are also moss areas on the surface of stony placers (Figure 1, 2).

The entire study area is located in the zone of continuous distribution of permafrost. The depth of the seasonally thawed layer varies from 0.03 to 1.5 m.

RESULTS AND DISCUSSION

The radioactivity of the soil, air, water and food products is determined mainly by natural isotopes, as well as global fallout, the average specific activity of which ranges from 40 to 150 or more Bq/kg. The main share of all natural radionuclides falls on ^{226}Ra, ^{232}Th, ^{40}K of the earth's surface and reaches 98.7%. The main source of natural radionuclides (NRN) in the soil are soil-forming and underlying rocks, creating a natural radioactive background area. In the study area, parent rocks are represented by black shales and phyllites with a high content of ^{40}K (Melnik et al., 2005). The specific activity of natural radionuclides ^{232}Th and ^{40}K in shale is 45 and 700 Bq/kg, respectively (Alexandrov, 2007). Sources of direct admission of technogenic radionuclides (TRN) in the study area are not available.

According to the frequency of occurrence of radionuclides (according to the classification of V. N. Kreshtapova 1974) for the territory under consideration ^{137}Cs, ^{232}Th,

^{40}K are widespread radionuclides (occurrence from 75 to 100%), ^{226}Ra - the average degree of prevalence (50 to 75% of samples).

The specific activity of ^{137}Cs (half-life $T_{1/2}$ = 30 years) in the upper soil horizon of the study area reached 144 Bq/kg (Figure 2, b). It has been established that the main accumulation of ^{137}Cs occurs in gley, peat-gley and peat-gley soils, formed in accumulative positions of the landscape, where it is strongly associated with humic acids. The activity of this nuclide below the minimum determined was obtained in the area of water intake from the lake Svetloe, at the lake Crugloe.

The specific activity of ^{226}Ra ($T_{1/2}$ = 1590 years) in the upper five-centimeter layer of soil varies from less than 7 Bq/kg to 40 Bq/kg. In the coastal area, where thin petrozems and lithozems developed, two local anomalies with an increased concentration of ^{226}Ra in the soil were detected (Figure 2, d).

Thorium is a typically lithophilic, weakly mobile element in any geochemical setting, $T_{1/2} = 1.41 \cdot 10^{10}$ years. The average concentration of ^{232}Th in the soil profile is determined by the content of this radionuclide in the underlying and soil-forming rocks. The specific activity of ^{232}Th in the upper organic soil horizon ranged from 4 to 35 Bq/kg (Figure 3, a).

Potassium is one of the main NRNs in soils, whose half-life is $1.3 \cdot 10^{9}$ years. In the territory under consideration, isolated anomalies with a relatively high content of ^{40}K (more than 300 Bq/kg) in the soil were found, which may be due to its concentration on the sorption barrier (Figure 3, b). These soils are mainly developed on deluvial-proluvial sediments and are characterized by a thin soil profile.

An analysis of the lateral distribution of ^{134}Cs in the upper five-centimeter layer of soil showed a decrease in the concentration of this radionuclide from the coast deep into the island. Due to the absence of a direct source of the radionuclide intake in the study area and on the basis of the revealed distribution of concentrations, it can be assumed that it is delivered with marine aerosols (Kostenkov, Klyshevskaya, 2014). Taking into account the nature of sea currents and atmospheric circulation in the northwestern part of the Arctic region (Aybulatov, 2000), it can be assumed that the main source of ^{134}Cs is located in the southwest of the archipelago.

The maximum concentrations of ^{60}Co are fixed mainly in the upper organic horizons and in the mineral horizons containing clay material. Maximum concentrations of ^{60}Co were found in soil samples of the coastal area, which may indicate a general route of entry with the ^{134}Cs radionuclide.

In general, the content of ^{226}Ra from the coast inland decreases. The genesis of local anomalies and the distribution of ^{226}Ra as a whole is probably associated with the sorption concentration of ^{226}Ra manganese oxides entering the coast with marine aerosols.

The geographical location of the thorium anomalies coincides with the potassium. Thorium anomalies were found in soils formed on deluvial sediments at the foot of hills, and in soils developed on proluvial sediments in the delta of mountain streams before flowing into lakes (with seasonal hydrological fluctuations). Lateral migration of ^{232}Th, as a rule, adsorbed by suspended particles, occurs during high water in the spring period by mechanical migration in the composition of the colloidal fraction.

The average specific activity of radionuclides in the soil of the study area and some other statistical parameters are shown in Table 1. Compared with the data on the specific activity of soil radionuclides on the east coast of the Novaya Zemlya (Usacheva et al., 2016), a significantly lower level of ^{137}Cs activity should be noted (ranged from 100 to 159 Bq/kg) and

^{40}K (from 354 to 448 Bq/kg). The average values of specific activities of ^{226}Ra (14.5 Bq/kg) and ^{232}Th (18.0 Bq/kg) practically coincide with those obtained in this study.

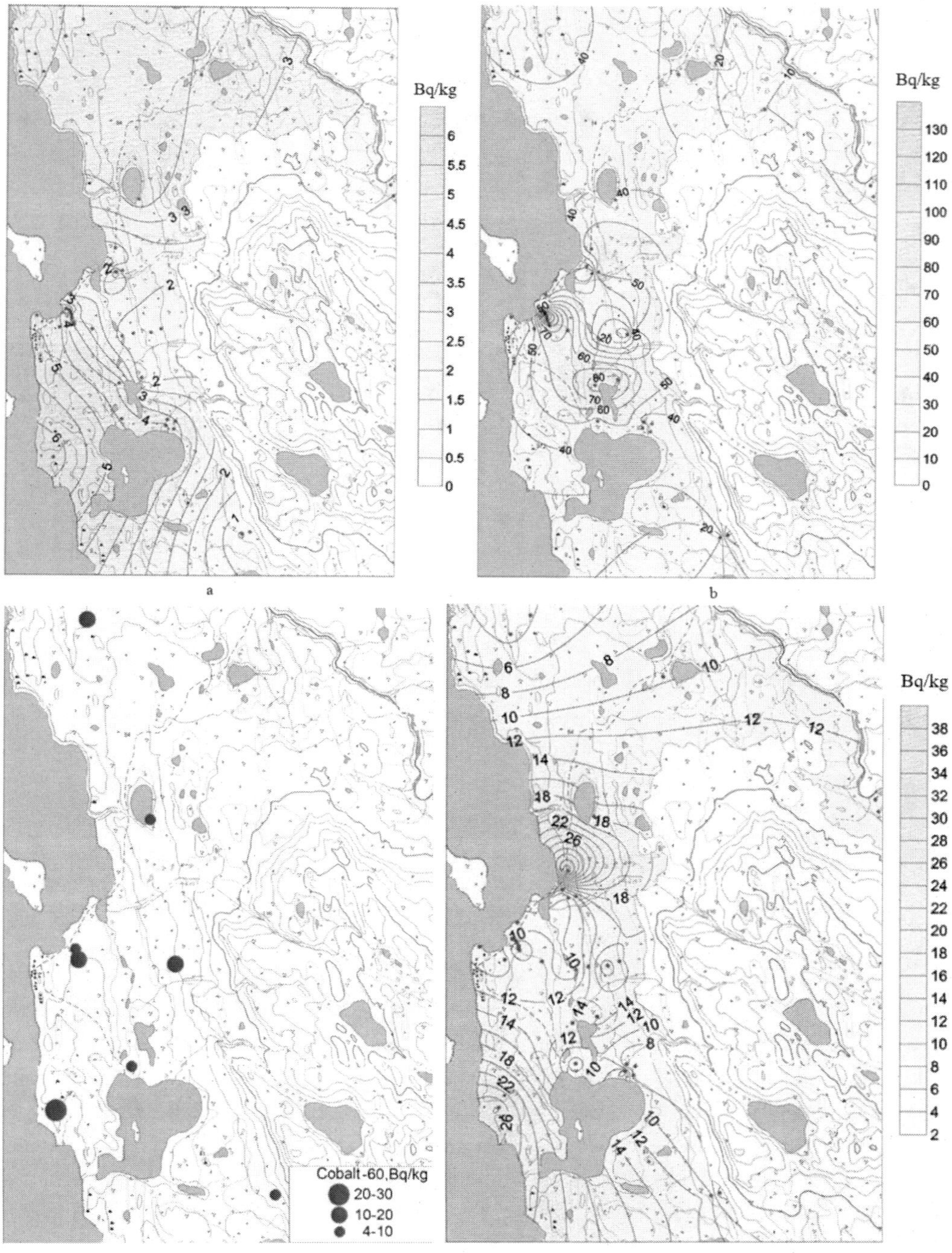

Figure 3. Specific activity of radionuclides in soils, Bq/kg: a – ^{134}Cs, b – ^{137}Cs, c – ^{60}Co, d – ^{226}Ra.

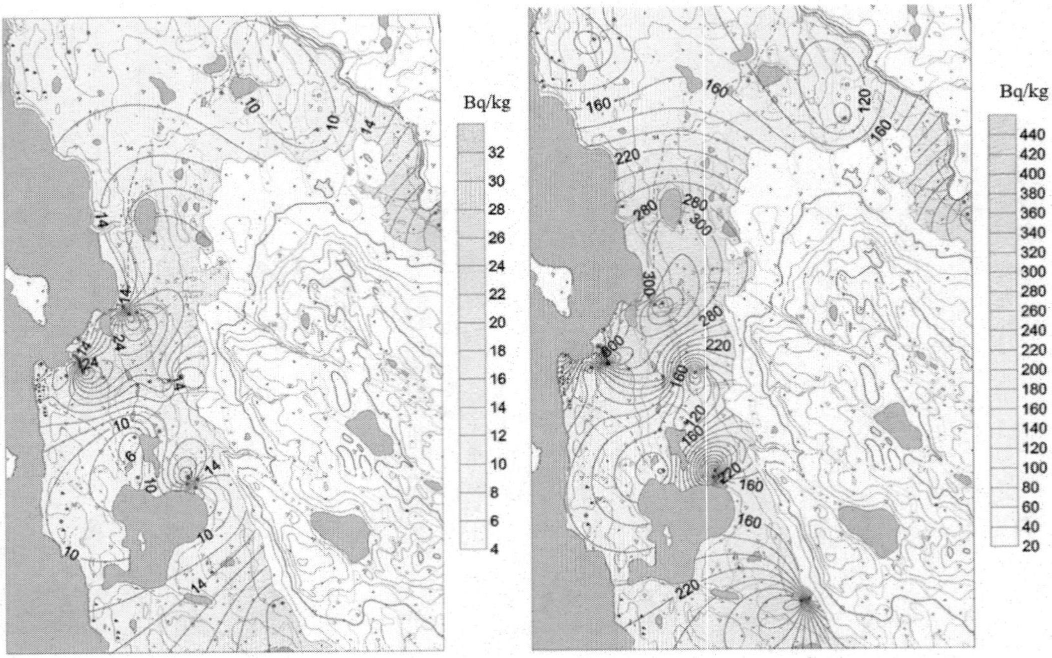

a b

Figure 4. Specific activity of radionuclides in soils, Bq/kg: a – ^{232}Th, b –^{40}K.

Table 1. The specific activity of radionuclides in soils, Bq/kg

^{134}Cs	^{137}Cs	^{60}Co
$\frac{0.00-7.80}{1.94\pm1.99(1.70)}$ *	$\frac{0,0-159}{40.0\pm38.6(34.7)}$	$\frac{0,00-29.7}{2.12\pm5.24(0.00)}$
^{40}K	^{232}Th	^{226}Ra
$\frac{0-510}{242\pm137(282)}$	$\frac{0.0-33.9}{17.1\pm10.3(20.6)}$	$\frac{0.0-52.4}{13.4\pm8.9(12.0)}$

Note: * - the minimum and maximum values are indicated in the numerator, the mean value and standard deviation in the denominator, the median value in parentheses, the number of samples n = 65.

The coefficient of pair correlation between ^{232}Th and ^{40}K was 0.55 ($p < 0.001$), which characterizes the degree of proximity of the direct connection of the specific activities of these elements as average and indicates the general pathway of accumulation and accumulation in the studied soils (Harrington, 1965). For ^{226}Ra and ^{137}Cs, a direct relationship of a low degree of closeness (k = 0.26, p = 0.035) was revealed, and a feedback of the same degree of closeness was found between ^{232}Th and ^{137}Cs (k = 0.31, p = 0.012). Revealed relationships in the future should be taken into account when analyzing the sources of radionuclide entry and migration in the soil.

To assess the aerogenic input of substances into the territory of Novaya Zemlya, the spatial distribution of the function of potential sources Z_{ij} (X) was obtained, which characterizes all atmospheric processes determining the transfer of impurity X from the source m to point n: probability, duration, range, vertical deposition to the surface and horizontal spreading.

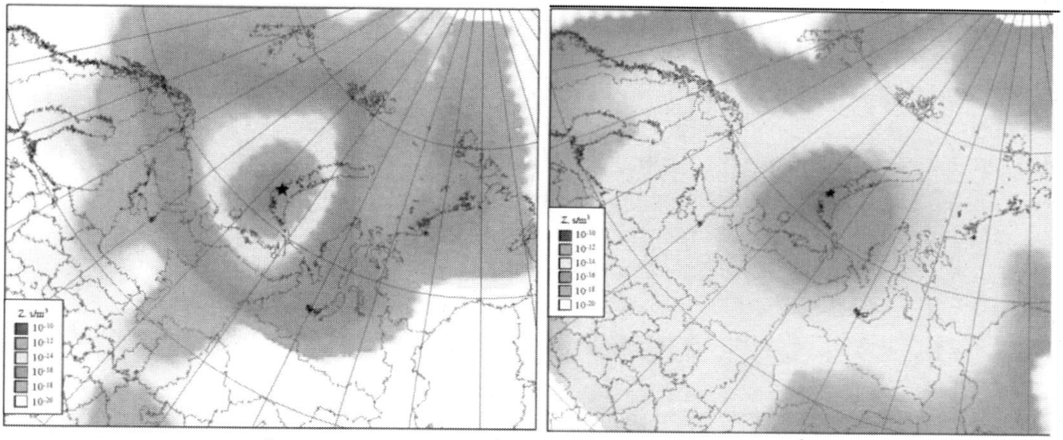

a b

Figure 5. Spatial distribution of the function Z for the territory of the Novaya Zemlya archipelago a) in July, b) in January.

As can be seen from the spatial distribution of the values of the function Z in July (Figure 5 a), the area of effective transfer to the Novaya Zemlya region does not significantly capture the continental territories where the main sources of pollution are located. At the same time, some influence of the sources of the Kola Peninsula, the coast of the Barents and Kara Seas on the territory of Novaya Zemlya, is nevertheless traced. During the winter period, the aero genic transfer from the continental territories increases (Figure 5 b), the area of the effective transfer of pollutants increases, which is associated with an increase in the sedimentation rate of impurities on the surface during the summer period.

Conclusion

It was found that on the coast of Moller Bay, in the automorphic positions of the relief, the soils of the primary trunk of the soil formation of the department of underdeveloped soils (petrozems) and the post lithogenic trunk of the lithozems (peaty lithozem, dry-peaty lithozem, gray-humus lithozema, and hard-humus lithozema), are common in trans-elluvial accumulative and trans-superaqual relief forms, the development of the soil of the post-lithogenic trunk of the gley soils department (peaty-gley) was developed. The soil cover is predominantly represented by moss-shrub vegetation, moss areas and, fragmentarily, meadow vegetation.

For the first time, maps of the spatial distribution of natural and man-made radionuclides in the surface soil layer on the coast of Moller Bay were constructed.

The obtained data are comparable with the levels of specific activity of these radionuclides obtained earlier by other authors for the soils of the east coast of Novaya Zemlya (Usacheva et al., 2016).

According to the data obtained, for ^{137}Cs and ^{40}K, the level of specific activity in the soils of the coast of Moller Bay in the west of the southern island of Novaya Zemlya is less than 3 and 2 times, respectively, compared to the east coast (Usacheva et al., 2016). The level of specific activity of ^{137}Cs in the studied soils is in 2,6 times lower than in the soils of the northern part of the new island of Novaya Zemlya (Miroshnikov et al., 2017).

In a number of soil sections on the Moller coast, the presence of short-lived isotopes ^{134}Cs and ^{60}Co was established, which indicates the continued flow of man-made radionuclides into the soils of the Arctic.

Atmospheric transport of pollutants to the territory of Novaya Zemlya is of greatest importance in winter, while in summer the main transport comes from water areas.

Acknowledgment

Support from the state funding for research No AAAA-A19-119011890018-3 is acknowledged.

References

Aleksandrov Yu. A. (2007) *Fundamentals of radiation ecology*. (Osnovy radiatsionnoy ekologii). Yoshkar-Ola: Mar. state Univ. 268 p. (in Russian).

AMAP Assessment 2015: *Radioactivity in the Arctic*. Arctic Monitoring and Assessment Programme (AMAP), (2015) Oslo, Norway. 89 pp.

Arctic Pollution Issues: A State of the Arctic Environment Report. (1997) Oslo: AMAP (Arctic Monitoring and Assessment Programme), 188 p.

Aybulatov N. A. (2000) *Ecological echo of the Cold War in the seas of the Russian Arctic* [*Ekologicheskoye ekho kholodnoy voyny v moryakh Rossiyskoy Arktiki*]. M .: GEOS. 306 p. (in Russian).

Baklanov A., Bergman R., Segerstahl B. (1996) Radioactive sources in the Kola region: Actual and potential radiological consequences for man. *International Institute for Appl. Systems Analysis*, Laxenburg, Austria, 255 p.

Baklanov A. A., Mahura A. G., Morozov S. V. (1994) The Simulation of Radioactive Pollution of the Environment After an Hypothetical Accident at the Kola Nuclear Power Plant. *J. Environmental Radioactivity*, 25. 65-84.

Chernova O. V., Beketskaya O. V. (2011) *Permissible and background concentrations of pollutants in the environmental regulation* (heavy metals and other chemical elements. Soil science. [*Dopustimyye i fonovyye kontsentratsii zagryaznyayushchikh veshchestv v ekologicheskom normirovanii* (tyazhelyye metally i drugiye khimicheskiye elementy. Pochvovedeniye) 9. 1102–1113]. (in Russian).

Demeshkin A. S. (2015) *Geoecological assessment of the state of the environment in the area of the Russian coal mining mine Barentsburg on the Svalbard archipelago*: Dis... Cand. geol. sciences. [*Geoekologicheskaya otsenka sostoyaniya prirodnoy sredy v rayone raspolozheniya rossiyskogo ugledobyvayushchego rudnika Barentsburg na arkhipelage Shpitsbergen*: dis. ... kand. geol. Nauk] St. Petersburg. 181 p. (in Russian).

Draxler R. R. and Rolph G. D., (2013) *HYSPLIT* [*HYbrid Single-Particle Lagrangian Integrated Trajectory*] *Model,* NOAA Air Resources Laboratory, College Park, MD, http://www.arl.noaa.gov/HYSPLIT.php.

EMEP Centre on Emission Inventories and Projections (CEIP). (http://webdab1.umwelt bundesamt.at/scaled_country_year.html?cgiproxy_skip=1 (25.08.2017).

Evseev A. V. (1982) Arctic tundra soils on. Aion. Soil science. [Arkticheskiye tundrovyye pochvy o. Ayon. Pochvovedeniye.] 1. 31–38. (in Russian).

Gablin V. A. (2014) *Radiation assessment of litomonitoring objects in urbanized areas (theory and methods). Theses for the degree of Doctor of Geological and Mineralogical Sciences. Russian State Geological Prospecting University named after Sergo Ordzhonikidze.* [*Radiatsionnaya otsenka ob"yektov litomonitoringa na urbanizirovannykh territoriyakh (teoriya i metody). Dissertatsii na soiskaniye uchenoy stepeni doktora geologo-mineralogicheskikh nauk. Rossiyskiy gosudarstvennyy geologorazvedochnyy universitet im. Sergo Ordzhonikidze.*] Moscow, 188 p. (in Russian).

Gablin V. A., Bolshakov M. O., Govorova L. V., Shanin S. O., Loginova A. I. (2012) About errors in the monitoring of natural environments. *Ecology and geological changes in the environment of the Northern regions.* (O pogreshnostyakh pri provedenii monitoringa prirodnykh sred. Ekologiya i geologicheskiye izmeneniya v okruzhayushchey srede Severnykh regionov) Arkhangelsk. 48-51. (in Russian).

Gudkov AB, Yermolin S. P., Popova ON, Nebuchennykh A. A. (2015) Characteristics of the functional reserves of the cardiorespiratory system in servicemen in the Arctic during the contrasting seasons of the year. *Bulletin of the Northern (Arctic) Federal University. Series: Biomedical Sciences.* (Kharakteristika funktsional'nykh rezervov kardiorespiratornoy sistemy u voyennosluzhashchikh v Arktike v kontrastnyye sezony goda. Vestnik Severnogo (Arkticheskogo) federal'nogo universiteta. Seriya: Medikobiologicheskiye nauki). 4. P. 13-22. (in Russian).

Halperin M., Sofiev M., Gusev A., Afinogenova O. (1995) *Approaches to modeling transboundary air pollution of Europe with heavy metals.* [Podkhody k modelirovaniyu transgranichnogo zagryazneniya atmosfery Evropy tyazhelymi metallami]. M., EMEP/MSC-B, Report 7/95. 85 sec. (in Russian).

Harrington E. C. (1965) The desirable function. *Industrial Quality Control*, 21, 10, 124-131.

Hirdman D., Sodemann H., Eckhardt S., Burkhart J. F., Jefferson A., Mefford T., Quinn P. K., Sharma S., Strom J., Stohl A. (2010) Source identification of short-lived air pollutants in the Arctic using statistical analysis of measurement data and particle dispersion model output. *Atmos. Chem. Phys.* 10. 2. 669-693.

Ivanov A. B., Krasilov G. A., Logachev V. A., Matuschenko A. M., Safronov V. G. (1997) Northern polygon (New Earth). *Radioecological consequences of nuclear testing.* [Severnyy poligon (Novaya Zemlya). *Radioekologicheskiye posledstviya yadernykh ispytaniy*]. M .: GIPE, 85 p. (in Russian).

Kiselev G. P., Bazhenov A. V., Zykov S. B., Kriauciunas V. V., Kiseleva I. M., Lastovsky A. M. (2006) *About radioactivity of the environment of the Arkhangelsk industrial region. Human ecology.* [O radioaktivnosti okruzhayushchey sredy arkhangel'skogo promyshlennogo rayona. Ekologiya cheloveka]. 2. 3-6. (in Russian).

Kostenkov N. M., Klyshevskaya S. V. (2014) *Impact of the pulsation process on the salt content in the soils of coastal sea landscapes.* Herald KrasGAU. (Vliyaniye protsessa impul'verizatsii na soderzhaniye soley v pochvakh pribrezhnykh morskikh landshaftov. Vestnik KrasGAU). 10. 81-84. (in Russian).

Kreshtapova V. N. (1974) *Guidelines for the evaluation of the content of trace elements in the peat deposits of the European part of the RSFSR.* (Metodicheskiye rekomendatsii po otsenke soderzhaniya mikroelementov v torfyanykh metorozhdeniyakh Yevropeyskoy chasti RSFSR). M.: Mingeo USSR. 200 p. (in Russian).

Kriauciunas V. V., Iglovsky S. A. and Kuznetsova I. A. (2018 b). *Trace Metals in Soil Catenas of the Arctic Islands* [The Svalbard and Novaya Zemlya Archipelagos]. *Biogeochemistry of trace elements. Editors Oleg S. Pokrovsky and Jerome Viers. Nova Science Publishers, Inc.* 309 - 359.

Kriauciunas V. V., Iglovsky S. A., Kuznetsova I. A., Shakhova Ev., Bazhenov A. V., Mironenko K. A. (2018 a). Spatial distribution of natural and anthropogenic radionuclides in the soils of Naryan-Mar. *Arctic Environmental Research*. 18. 3. 82-89.

Krupskaya V. V., Miroshnikov A. Yu., Dorzhieva O. V., Zakusin S. V., Semenkov I. N., Usacheva A. A. (2017) The mineral composition of the soil and bottom sediments of the bays of the Novaya Zemlya archipelago. *Oceanology*. [Mineral'nyy sostav pochv i donnykh osadkov zalivov arkhipelaga Novaya Zemlya. Okeanologiya]. 57. 238–245. (in Russian).

Kriauciunas V. V., Iglovsky S. A., Shakhova E. V., Malkov A. V. (2014) Heavy metals in the arctic soils of the west coast of the Svalbard archipelago. *Human ecology*. [Tyazhelyye metally v arkticheskikh pochvakh zapadnogo poberezh'ya arkhipelaga Shpitsbergen. Ekologiya cheloveka]. 9. 8-13. (in Russian).

Kuznetsov V. M. (2002) *The main problems and the current state of the nuclear fuel cycle enterprises of the Russian Federation*. [Osnovnyye problemy i sovremennoye sostoyaniye predpriyatiy yadernogo toplivnogo tsikla Rossiyskoy Federatsii]. M., 259 p. (in Russian).

Kuznetsov V. M., Nazarov A. G. (2006) *Cold War Radiation Legacy. The experience of historical research*. [Radiatsionnoye naslediye kholodnoy voyny. Opyt istoriko-nauchnogo issledovaniya]. M.: Klyuch-S, 720 p. (in Russian).

Laverov N. P., Velichkin V. I., Miroshnikov A. Yu., Krupskaya V. V., Asadulin E. E., Semenkov I. N., Usacheva A. A., Zakusin S., Terskaya E. A. (2016) *Geochemical structure and radiation state of coastal landscapes of the Kara Sea Bays of the New Earth. Reports of RAS*. [*Geokhimicheskaya struktura i radiatsionnoye sostoyaniye pribrezhnykh landshaftov zalivov Karskogo morya Novoy Zemli*. Doklady RAN]. 467. 3. 342–346. (in Russian).

Matishov D. G. (2001) *Anthropogenic radionuclides in marine ecosystems/Abstract. diss... Dr. Geol. sciences*. [*Antropogennyye radionuklidy v morskikh ekosistemakh/Avtoref. diss. ... d-ra geol. Nauk*]. St. Petersburg, 50 p. (in Russian).

Matishov G. G., Matishov D. G., Namyatov A. A., Zuev A. N., Kirillov E. E. (1997) *Radionuclides and oceanographic conditions of their accumulation in the Kola and Matovosky bays (Barents Sea)*. [*Radionuklidy i okeanograficheskiye usloviya ikh nakopleniya v Kol'skom i Matovskom zalivakh (Barentsevo more)*]. Murmansk: MMBI RAS, 32 p. (in Russian).

Melnik N. A., Velichko O. S., Krasheninnikov O. N., Belogurova T. P. (2005) *Radio-geoecological assessment of shale of the Peninsula Middle and Rybachy. Mineralogy in the whole space of this word*. [*Radiogeoekologicheskaya otsenka slantsev poluostrovov Sredniy i Rybachiy. Mineralogiya vo vsem prostranstve sego slova*]. Apatity: K & M, 126-128. (in Russian).

Miroshnikov A. Y., Laverov N. P., Chernov R. A., Kudikov A. V., Ysacheva A. A., Semenkov I. N., Aliev R. A., Asadulin E. E., Gavrilo M. V. (2017) Radioecological investigations on the Northern Novaya Zemlya Archipelago. *Oceanology*. 57, 1. 204-214.

Nikitin A. I., Shershakov V. M., Tsaturov Yu. S. (2011) A cooperative Russian-Norwegian study of radioactive contamination of the western Arctic seas in areas prone to exposure to local sources. [Sovmestnyye rossiysko-norvezhskiye issledovaniya radioaktivnogo zagryazneniya zapadnykh arkticheskikh morey v rayonakh, podverzhennykh vozdeystviyu lokal'nykh istochnikov. Arktika: Ekologiya i ekonomika]. *Arctic: Ecology and Economics*. 2. 26-35. (in Russian).

Novaya Zemlya: Nature. Story. Archeology. Culture. Book 1: Nature. (1998) Ed. P. V. Boyarsky: Tr. MAKE. [Novaya Zemlya: Priroda. Istoriya. *Arkheologiya. Kul'tura*. Kn. 1: *Priroda. Pod red*. P. V. Boyarskogo: Tr. MAKE]. M., 280 p. (in Russian).

Pechurov L. V. (1983) Svalbard (Shpitsbergen). M: Thought, 124 p. (in Russian).

Petriga A. A., Goryunova N. V. (2016) The composition of snow in the impacted regions of the Arctic on the example of the settlements of the Is-fjord Bay of the Svalbard archipelago. *Bulletin of science and education*. [Sostav snega v impaktnykh rayonakh Arktiki na primere poselkov zaliva Is-f'ord arkhipelaga Shpitsbergen. *Vestnik nauki i obrazovaniya*]. 2 (14). 106–110. (in Russian).

Radiation situation on the territory of Russia and neighboring countries in 2015. (2016) *Yearbook. Comp*. M. N. Katkov. Obninsk: FGBU "VNIIGMI-WDC." [Radiatsionnaya obstanovka na territorii Rossii i sopredel'nykh gosudarstv v 2015 godu. Yezhegodnik. Sost. M. N. Katkova. Obninsk: FGBU «VNIIGMI-MTSD»]. 348 p. (in Russian).

Rovinsky F. Ya., Gromov S. A., Burtseva L. V., Paramonov S. G. (1994) Heavy metals: long-range transport in the atmosphere and precipitation. *Meteorology and Hydrology*. [Tyazhelyye metally: dal'niy perenos v atmosfere i vypadeniye s osadkami. Meteorologiya i gidrologiya] 10. 5-14. (in Russian).

Salvador P., Art´ınano B., Pio C., Afonso J., Legrand M., Puxbaum H., Hammer S. (2010) Evaluation of aerosol sources at European high altitude background sites with trajectory statistical methods. *Atmosph. Environ*. 44. 2316–2329.

Shvartsman Yu. G., Bolotov I. N., Iglovsky S. A. (2008) Climate change and its impact on the environment of the European North of Russia. Environmental and climate change. Natural and related technological disasters. T. VI. *Climate change: the impact of extraterrestrial and terrestrial factors*. M.: IFA RAS. [Izmeneniya klimata i ikh vliyaniye na okruzhayushchuyu prirodnuyu sredu Yevropeyskogo Severa Rossii. Izmeneniye okruzhayushchey sredy i klimata. Prirodnyye i svyazannyye s nimi tekhnogennyye katastrofy. T. VI. *Izmeneniya klimata: vliyaniye vnezemnykh i zemnykh faktorov*. M.: IFA RAN]. 80-98. (in Russian).

Song L., Jian J., Tan D. J., Xie H. B., Luo Z. F., Gao B. (2015) Estimate of heavy metals in soil and streams using combined geochemistry and field spectroscopy in Wan-sheng mining area, Chongqing, *China International Journal of Applied Earth Observation and Geoinformation*. 34. P. 1–9.

The Arctic is on the threshold of the third millennium (resource potential and environmental problems). (*Arktika na poroge tret'ego tysyacheletiya* (resursnyy potentsial i problemy ekologii).) Ed. I. S. Gramberg, N. P. Laverov. SPb.: *Science*, 2000. 247 p. (in Russian).

The state and environmental protection of the Arkhangelsk region in 2015. (2016) Ed. S. V. Maslova: report. [*Sostoyaniye i okhrana okruzhayushchey sredy Arkhangel'skoy oblasti za 2015 god*. 2016./Pod red. S. V. Maslova: doklad]. Arkhangelsk, 432 p. (in Russian).

Usacheva A. A., Semenkov I. N., Miroshnikov A. Yu., Krupskaya V. V., Zakusin S. V. (2016) Geochemical features of the Arctic tundra landscapes of the east coast of the Novaya Zemlya. *Bulletin of Moscow University. Series 5: Geography*. 6. [Geokhimicheskiye osobennosti arktotundrovykh landshaftov vostochnogo poberezh'ya Novoy zemli. Vestnik Moskovskogo Universiteta. Seriya 5: Geografiya. 6]. 87-95. (in Russian).

Vinogradova A.A. (2000) Anthropogenic pollutants in the Russian Arctic atmosphere: sources and sinks in spring and summer. *Atm. Environ.* 34. 29-30. 5151-5160.

Vinogradova A. A. (2014) *Remote assessment of the impact of air pollution on remote areas. Geophysical processes and biosphere*. [*Distantsionnaya otsenka vliyaniya zagryazneniya atmosfery na udalennyye territoriy. Geofizicheskiye protsessy i biosfera*]. 13. 4. 5-20. (in Russian).

Vinogradova, A. A., Ponomareva, T. Ya. (2012) *Atmospheric transfer of anthropogenic impurities to the arctic regions of Russia* (1986-2010). *Optics of the atmosphere and the ocean.* [Atmosfernyy perenos antropogennykh primesey v arkticheskiye rayony Rossii (1986-2010 gg.). Optika atmosfery i okeana] 25. 6. 475-483. (in Russian).

Zharnikov V. B., Nikolaeva ON, Safonov V. V. (2013) Technogenic transformation of land and its indicators in the monitoring system. *Bulletin of SUGUGT*. [*Tekhnogennaya transformatsiya zemel' i yeye pokazateli v sisteme monitoringa.* Vestnik SGUGiT]. 2 (22). 36-43. (in Russian).

In: The Arctic: Current Issues and Challenges
Editors: Oleg S. Pokrovsky, Sergey N. Kirpotin et al.
ISBN: 978-1-53617-306-2

Chapter 11

RADIONUCLIDES IN SEA ECOSYSTEMS OF THE WESTERN SECTOR OF THE RUSSIAN ARCTIC

Sergey B. Zykov, Sergey V. Druzhinin, Evgeny Yu. Yakovlev* and Elena N. Zykova
FCIARctic - Federal Center for Integrated Arctic Research
Russian Academy of Sciences, Institute of Geodynamics and Geology,
Arkhangelsk, Russia

ABSTRACT

The relevance of assessing the contamination of the Arctic ecosystems with radioactive elements is associated with the presence of land and sea testing sites for modern weapons, radioactive waste disposal sites, the operation of nuclear power and nuclear fleet (Matishov et al., 1997; Strand et al., 1997; Kuznetsov et al., 1994). Along with natural and anthropogenic factors, transboundary transfers of radioactive components by air and water masses are of great importance for the state of the environment in the Western Sector of the Russian Arctic (Kiselev et al., 2012).

The chapter presents the results of studies the radionuclides in the seas, coastal and island territories of the Western Arctic of Russia during sea expedition on research vessel "Professor Molchanov" in 2012. In the process of research, sampling of bottom sediments, sea waters, aerosols, island soils within the Barents and White seas was performed. The measurements were carried out by alpha and gamma spectrometry using radiochemical analysis. The content of technogenic (^{134}Cs, ^{137}Cs, ^{60}Co) and natural (^{234}U, ^{238}U, ^{40}K, ^{226}Ra, ^{232}Th, ^{7}Be) radionuclides was determined. In general, it has been shown that ^{137}Cs activity in the bottom sediments of the Western Arctic seas of Russia has recently decreased, but in many samples a rapidly decaying ^{134}Cs radionuclide has been detected, which may indicate a continued flow of technogenic radioactivity into the Arctic seas. The activities of natural radionuclides in marine bottom sediments are within the clarke content and depends on the composition of sediments. In the Barents Sea, lenses of waters with an excess of ^{234}U/^{238}U isotopic ratios of more than 1.3 were

* Corresponding Author's Email: evgeny.yakovlev@fciarctic.ru.

detected. These lenses with an excess of ^{234}U isotope are areas of continental flow. A decrease in the ^{7}Be activity in air aerosols at high latitudes has been established.

The results obtained on the activity of technogenic radionuclides showed the following pattern of their distribution and accumulation: in seawater there is a significant decrease in the activity of radionuclides compared with the beginning of the 80s, but at the same time insignificant fallout continues; the deposition of radionuclides on the seabed area is uneven, the greatest accumulation of radionuclides is observed in the fine fraction of the shelf zones; ^{137}Cs radionuclide accumulation in soils is typical for the upper genetic horizons, the distribution of radionuclides over the territory depends on the conditions of regional and global fallout.

Current research results show that levels of radiation pollution are low and tend to decrease, but the picture may change due to increasing anthropogenic activity on the continental shelf and adjacent territories. Also, the active use of the Northern Sea Route with the arrangement of related infrastructure and the implementation of dredging in the mouths of large northern rivers could be cause the redistribution of natural and man-made radionuclides.

Keywords: Arctic sea, technogenic and natural radionuclides, bottom sediments, soils, Russian Arctic.

INTRODUCTION

From June 1 to July 10, 2012, the sea research expedition “Floating University” was conducted in the White and Barents Seas on the research vessel “Professor Molchanov.” The main goal of expedition was assess the current radiological situation of the Western sector of the Russian Arctic on example the sea ecosystems of the Barents and White Seas. The mission of the expedition was to conduct radiological studies in the bottom sediments, sea water and the atmosphere of the White and Barents seas. For the rapid assessment of the radioactivity of samples, a Progress-2000 gamma-spectrometer and a filtering device for the selection of aerosols on Petryanov-Sokolov filters were installed onboard the research vessel “Professor Molchanov.” In addition, radiochemical preparation of water samples for studies of uranium isotopes was organized on board. As part of the overall research, employees of the Laboratory of Environmental Radiology took samples of bottom sediments, sea water, and seaweed. Filters with aerosols were measured immediately after their removal from filtering device, which was required according to the method of work with the fast-decaying ^{7}Be cosmogenic radionuclide (El-Hussein et al., 2001).

The relevance of assessing the contamination of the Arctic ecosystems with radioactive elements is associated with the presence of land and sea testing sites for modern weapons, radioactive waste disposal sites, the operation of nuclear energy facilities and the presence of a nuclear fleet (Kjellstrom, 1986). Along with natural and anthropogenic factors, transboundary transfers of radioactive components by air and water masses are of great importance for the state of the environment in the Western sector of the Russian Arctic (Vernadsky, 1997).

The initial relevance of radioactivity research in the Russian Arctic was due to the creation in 1954 of a nuclear test site on Novaya Zemlya archipelago (Logachev, 1998). At this time, the United States of America actively conducted nuclear explosions in the atmosphere and under water in different parts of the world’s ocean, which prompted the test of nuclear weapons in the USSR (Ivanov et al., 1993). Taking into account the experience of the Semipalatinsk test site, as well as the experience of controlling radioactive contamination during nuclear explosions in other countries, a network of observation points for the spread of radioactivity was created around the test site at Novaya Zemlya archipelago based on meteorological stations of the USSR Hydromet (Katrich, 1993). The network involved the capture of radioactive aerosols from several tens to thousands of kilometers from the epicenters of future explosions (Kershaw et al., 1993). This network allowed assess the radioactive effect of test site to the territory of Russia. The first nuclear explosion was produced under water in Chernaya Bay on Novaya Zemlya archipelago in 1955. A group of radiologists from the Institute of Occupational Hygiene and Occupational Diseases of the Academy of Medical Sciences of the USSR worked to control the radioactive contamination of sea water and biota before, after and during the explosion in the Chernaya Bay. This indicates that from the very beginning of nuclear weapon tests at Novaya Zemlya archipelago radioactive research was conducted, but it should be noted that their results were not publicized. As a result of the first radiological studies, it was found that severe contamination of territories and waters in the epicenters of nuclear explosions occurs during underwater and surface tests. On Novaya Zemlya archipelago, one ground-based nuclear explosion was carried out, which left the contaminated area at the epicenter with an activity of more than 1 mR/h (Haartsen, 1995). To date, very high contamination persists in the bottom sediments of the Chernaya Bay. Atmospheric explosions ejected their decay products into the stratosphere and created conditions for global fallout throughout the Earth (Aarkrog, 1994). Part of the radioactive products spreaded out in the surface atmosphere. Winds of the eastern and southern directions caused the movement of radioactivity clouds which brought pollution to the territory of Siberia and the European North of Russia. One of the indicators of radioactive contamination from nuclear tests is the long-lived radioactive isotope ^{137}Cs. According to the model of the formation of radioactive contamination during atmospheric tests of nuclear weapons, confirmed experimentally, ^{137}Cs were released into the high atmospheric layers, where it was accumulated and distributed throughout the atmospheric envelope, followed by subsidence on the ground in the form of atmospheric fallout (Strand et al., 1993).

The second phase of active radioecological research began in the 70s of the last century (it continues at the present time) and was due to an assessment of the global environmental impact of testing nuclear weapons and the nuclear industry. A number of European countries, the USA, and the USSR have joined this program. Global studies of the atmosphere, oceans, vegetation and soil began. At this stage, it was investigated not only the state of radioactivity of natural environments, but also the sources of radionuclide release into the environment, as well as the real harm for world's population (Ivanov et al., 1995; Demin and Demina, 1999).

Particular attention was paid to research in the Arctic region in connection with the extraction of seafood and minerals. The greatest attention was paid the Barents Sea, due to the location of the nuclear testing site on Novaya Zemlya archipelago (Kershaw, 1993).

Huge maritime spaces demanded the efforts and financing of many countries, such as Norway, Finland, Germany, Poland, Sweden, Russia, England, Canada, Denmark, Netherlands, USA. The joint researchs were carried out during the Arctic Monitoring and Assessment Programme program (Strand et al., 1997). These works allowed to determine the state of radioactive contamination of sea water and bottom sediments of the northern and Arctic seas, the continental Arctic and adjacent territories (Arctic Polar Environment, 1997). The main potentially dangerous radioactive sources were identified, as well as the radioactivity of sea water, seafood, bottom sediments, soils, vegetation of the European North of Russia. In the early 90s of the last century, the tense situation of radioactive danger was created in the European North of Russia due to the massive utilization of nuclear submarines and the liquidation of naval military bases in the absence of funding and experience of such works.

The huge potential of radioactive contamination of the territory is concentrated in the European North and in the Arctic. First of all, this is the Novaya Zemlya test site itself, where 130 nuclear tests were conducted. This is followed by areas of flooding and disposal of radioactive waste in the Barents and Kara Seas (Aibulatov et al., 1994). In addition, these are the bases of the navy and nuclear fleet of Russia, underground nuclear explosions for non-military purposes, the dumping areas of nuclear waste on the continent.

In the work (Vakulovsky et al., 1988) there is evidence of cesium radioactive isotopes entering the White Sea from radioactive waste dumps of the English Sellalfield plant. The accumulated experimental data allow us to determine the possibility of such a transfer. Between 1981 and 1982, anomalous concentrations of ^{137}Cs (up to 200 Bq/m^3) were concentrated in the Northern and Irish seas. They spread out along the two branches of the Nordkapp Current along the coasts of Norway and the Kola Peninsula and reached the White and Kara Seas and left in the direction of Svalbard. In the direction of currents, ^{137}Cs activity decreases with distance from the source - the Northern and Irish Seas. In the western and southern parts of the Barents Sea, its activity became 20-40 Bq/m^3. During this period, nuclear weapons were tested at Novaya Zemlya, but there was no increase in ^{137}Cs concentration above the level of its trend in the sea water of the Barents, White and Kara Seas (Vakulovsky et al., 1988).

In the period 1992-1995, ^{137}Cs activity in Northern waters decreased by an order of magnitude. In the Northern and Irish Seas, it reached 16–20 Bq/m^3 and a plume of radioactive contamination from this area is observed along the Norwegian coast and towards Spitsbergen. The distribution character of ^{137}Cs in seawater of this period is fully consistent with its distribution in the early 80s (Yudahin et al., 1998).

The problem of transport the radionuclides in the sea attracts the attention of many researchers (Agamirov, 1963; Baturin et al., 1966, 1969). The main patterns of the accumulation the radionuclides are as follows. Along the western coast of the Novaya Zemlya archipelago, no significant sedimentation of anthropogenic radionuclides are observed in marine sediments. This is explained by the fact that along the western border of Novaya Zemlya there are no conditions for rapid sedimentation of the finely dispersed fraction of abiogenic and biogenic material that promotes the transfer of radioactive elements, and strong currents intensively dilute surface runoff from Novaya Zemlya and transport the trapped products to the World Ocean (Nilsen and Bömer, 1994). Similar dilution of surface runoff is also possible along the southern border (Kara Gate), where there is also no intensive deposition of ^{137}Cs in bottom sediments. High concentrations of ^{137}Cs and other radioactive elements along the archipelago occur in Black Bay (Matishov et al., 1994). Here is where the largest underwater nuclear explosions were made. Strong currents along Scandinavia and the Kola Peninsula also do not allow the accumulation of radioactive elements in the bottom sediments, except for the "traps" forming in the bays, as shown in the work (Matishov et al., 1997). Radioactive products captured by sea waters are carried by currents hundreds, thousands of kilometers and deposited in areas where currents slow down. This is the shelf zone of the Kara Sea, the mouths of large rivers and estuaries, that is, areas with maximum concentrations of biogenic and abiogenic impurities in sea water and bottom sediments. The areas of deposition of radioactive cesium will be susceptible to use in economic activities (drilling of wells, pipelines, geological exploration and dredging, bottom trawling of fish, and mining), which in turn will lead to the secondary emission of radioactive elements into seawater. In avoidance of re-contamination of sea water by radioactive elements during the period of economic activity requires detailed studies of concentrations of antropogenic radioactive elements in bottom sediments and localization of areas with their high concentrations (Matishov et al., 1995).

The most common natural radioactive element in continental and marine waters is uranium. The concentration of uranium in seawater was first determined in 1933 by Gerner and Dwarf. The average concentrations of uranium in the Pacific and Atlantic oceans from the surface to a depth of 3000 m range from 2.2 to 2.8 10^{-6} g/l. In seawater, the uranium content varies in the range from 0.6 10^{-6} to 3.2 10^{-6} g/l and depends on the amount of uranium introduced by rivers from the continent. Similar data are given in the works of V. P. Shvedov and S. A. Patin and later confirmed by Chen in 1986 (Shvedov et al., 1968; Chen et al., 1986). A representative indicator of river runoff from the continent is the dependence of the uranium concentration of water on its salinity for the Baltic Sea. With sea water salinity up to 10 ppm, uranium concentration can vary from $0.6 \cdot 10^{-6}$ to $5.8\text{-}8 \cdot 10^{-6}$ g/l. High and low values are due to river runoff and abrasion of coastline by sea waves, followed by dissolution of the rocks. With an increase in the salinity of the Baltic Sea from 10 to 35 ppm, which is observed when moving from bays to the open sea, the concentration of uranium increases from $0.6 \cdot 10^{-6}$ to 1.8 10^{-6} g/l.

The common data on uranium in the World ocean shows that it concentration ranges from 0.7 10^{-6} to 5.8 10^{-6} g/l. The uranium isotope ratio $^{234}U/^{238}U$ for oceans was originally studied by E. Golberg and M. Koide (Golberg 1965; Koide 1965). They proved that due to the high solubility of uranium and its long presence in the sea water of the open seas and the World ocean, it does not change and is 1.14 ± 0.014. At the same time Miyke in 1966 showed that the uranium isotope ratio $^{234}U/^{238}U$ for the northwestern Pacific Ocean equal to 1.09 ± 0.05, that is, close to the values given above (Miyke et al., 1966).

V. Cherdyntsev showed similar data on uranium isotopes in the southeastern part of the Pacific Ocean (1.167 ± 0.007) (Chalov, 1975). The values indicate a constant excess of ^{234}U in the oceans from 10 to 17%. This can be explained by the residence time of uranium in the ocean at 500,000 years – about two half-lives of ^{234}U (245,000 years) (Edvards et al., 1987). The introduction of uranium into the ocean from the continent by surface runoff increases the average excess of ^{234}U to 50% (Kiselev and Zykov, 2004).

Figure 1 shows a map of the total distribution of uranium concentration in the surface waters of the White Sea catchment area according to R. Salminen (2002). The highest concentrations of uranium are shown in the surface waters of the Belomorsko-Kuloi Plateau (Winter Coast of the White Sea).

River runoff from this area should significantly increase the uranium concentration in the coastal waters of the White Sea. However, according to the data we obtained for the waters of this coastal zone, such an increase is not observed (Table 1).

Measurements of uranium concentration in the coastal zones of Dvinsk, Onega and Kandalaksha Bays and coast of the Kola Peninsula, also did not show an increase in uranium concentrations in seawater as compared with its level in the World Ocean (Table 2).

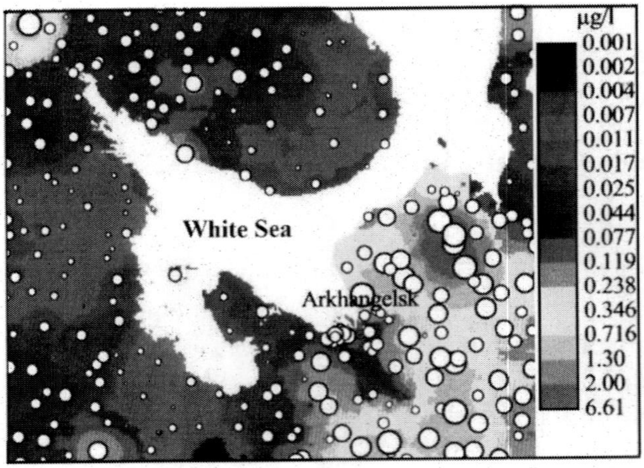

Figure 1. The distribution of uranium in the surface waters of the White Sea coast (Salminen, 2002).

Table 1. The uranium isotopic composition of the White Sea (Kiselev and Zykov, 2004)

Sample number	**234U/238U, γ**	**Concentration 238U, g/l**
Sea coastal waters (Winter Coast)		
C-18	1.16	$1 \cdot 10^{-6}$
C-22	1.25	$1 \cdot 10^{-6}$
C-25	1.27	$8 \cdot 10^{-7}$
Estuaries of the rivers(Winter Coast)		
C-30 (Trozhmariver)	1.38	$4 \cdot 10^{-7}$
C-32 (Kuyariver)	1.62	$4 \cdot 10^{-7}$
Groundwater of the White Sea basin		
PV-1	5.91	$2 \cdot 10^{-5}$
PV-2	7.20	$1 \cdot 10^{-5}$
PV-6	2.02	$5 \cdot 10^{-6}$
PV-7	3.33	$3 \cdot 10^{-7}$

It has been established that in coastal sea waters, in areas of large rivers and lakes, there are high excesses of ^{234}U (Kuptsov, 1979; Kiselev and Zykov, 2003). A particularly high uranium isotope ratio of $^{234}U/^{238}U$ was found in water of estuary the Pyalitsariver. At the same time, a decrease in the concentration of uranium in seawater is observed in the estuary areas of rivers, which is associated with its low concentrations in the surface runoff. The mechanism of the formation of an excess of ^{234}U in exogenous zones has been considered in a number of works (Chalov, 1975; Kiselev et al., 2000; Kiselev and Zykov 2003).

Studies of the uranium isotope composition in the White Sea were started in 1996 (Kiselev et al., 1996; Kiseleva, 2000). There were studied 130 water samples from different parts of the White Sea and at different depths. Also there were studied over 40 samples in coastal zones and river estuaries. The uranium isotope ratio $^{234}U/^{238}U$ in the central part of the White Sea varied for deep waters from 1.08±0.01 to 1.25±0.01 and for surface waters from 1.09±0.01 to 1.33±0.01 (Kiselev, 1996, 2005). Significant excess of ^{234}U, probably caused by lenses of continental waters penetrating into the sea in two ways. The first way is a river runoff, as evidenced by studies of Siberian rivers (Ostlund, 1993). The second way is the penetration of groundwater into the sea.

Table 2. The uranium isotopic composition in the coastal zones of the White Sea

Sample number	Brief description of the sample	Coordinates		U, g/l	$^{234}U/^{238}U$, Bq/Bq
		N	E		
KOL-1-08	sea water (estuary of Varzuga river)	$66^0$15.877'	36^0 53.604'	$4.33 \cdot 10^{-7}$	1.21
KOL-4-08	sea water	66^0 08.103'	37^0 22.062'	$1.66 \cdot 10^{-7}$	1.96
KOL-6-08	sea water (estuary of Chopoma river)	66^0 05.537'	38^0 54.936'	$17.2 \cdot 10^{-7}$	1.27
KOL-8-08	Piyalitsa river (estuary)	66^0 11.065'	39^0 31.576'	$2.27 \cdot 10^{-7}$	4.95
KOL-10-08	Sea (estuary of Pulonga river)	66^0 15.591'	39^0 59.112'	$18.1 \cdot 10^{-7}$	1.19
KOL-11-08	Lihodeyka river	66^0 19.810'	40^0 10.293'	$0.89 \cdot 10^{-7}$	1.81
KOL-13-08	Sea water (Sosnovka)	66^0 29.819'	40^0 35.357'	$15.7 \cdot 10^{-7}$	1.32
KOL-14-08	Sea water	66^0 04.844'	40^0 04.988'	$6.17 \cdot 10^{-7}$	1.30
BM09-1в	Dvina Bay	64^0 49.650'	40^0 18.570'	$4.60 \cdot 10^{-7}$	1.78
BM09-18в	Sea water (Karelian coast)	64^0 20.780'	35^0 27.080'	$13.1 \cdot 10^{-7}$	1.16
BM09-19в	Sea water (Karelian coast)	64^0 27.117'	35^0 19.166'	$19.5 \cdot 10^{-7}$	1.06
BM09-20в	Sea water (Belomorsk)	64^0 30.394'	35^0 11.112'	$10.7 \cdot 10^{-7}$	1.27

In Northeastern border of the White Sea, waters have an excess of ^{234}U, characteristic of the world ocean and in particular the Barents Sea. In coastal waters, an excess of ^{234}U increases. In some coastal areas, anomalous excesses of it were found, in particular in the Kandalaksha Bay, Dvina Bay, in the estuaries of the Ponoy and Kem rivers. The ratio $^{234}U/^{238}U$ reaches value of 2.2. Compared to the waters of the White Sea (about 1.2-1.3), this ^{234}U excess is anomalous. In the course of studies conducted in 2010, even higher values of $^{234}U/^{238}U$ were obtained (Table 2). Their nature is due to the groundwater, runoff from uranium, apatite, iron, iodine and diamond deposits into the White Sea. Regions with a large excess of ^{234}U in seawater are of interest from the standpoint of pollution of sea by groundwater containing microcomponents of ore deposits.

METHODS

Measurements of gamma active isotopes in samples of bottom water sediments were carried out using gamma spectrometer "Progress-gamma." After warming up, calibrating and background measuring of the spectrometer, a weighted 1 liter water sample was placed in a Marinelli vessel (if samples are insufficiently active, they were physically concentrated). The Marinelli vessel with the sample was inserted into the chamber of gamma spectrometer. The counting samples was measured within 1-2 hours. The results of measurements (radiation spectra) were processed using the "Progress-2000" software. Samples of soil and bottom sediments were dried in a drying cabinet at 40°C, quartered, weighed and transferred to the Marinelli vessel and then measured.

Samples of aerosols deposited on the Petryanov-Sokolov filter using filtering device, were dried in a drying cabinet and burned. The residue was weighed, transferred to a Petri dish and measured on the gamma spectrometer in Petri geometry. The obtained measurement results were recorded in the form of a protocol with an indication of all sample parameters of measured values and measurement errors.

Radiochemical extraction of uranium isotopes and measurements were carried out according to the following procedure. A sample of natural water with a volume of at least 20 liters was acidified with hydrochloric acid to pH 1-2, the necessary amount of ^{232}U isotope label was introduced so that its activity was commensurate with the activity of the natural isotopes of uranium. If the water was colored, it is necessary to add 2-3 ml of hydrogen peroxide with concentration of 28-32%. After 4 hours, using a saturated solution of hexamethylenetetramine (hexamine), the pH was adjusted to 4.5-5.5 by controlling it with universal indicator paper. After approximately 1-2 hours, a sorbent in the form of specially prepared activated charcoal BAU-A grade, was added to the sample. Sorbent preparation was as follows. Charcoal BAU-A was ground in a mill to 200 mesh particles. Then 1 kg of charcoal was poured 10 liters of a 0.5% solution of hydrochloric acid, stirred and incubated for 24 hours. Then the water was decanted using a hose, and the charcoal was filtered on a Buchner funnel and washed with distilled water, controlling the absence of chlorine ions in the wash water adding a solution of silver nitrate. Next, the coal was dried in a drying cabinet at 105°C and packaged in paper bags.

The sorbent prepared in this way (15 g) was added to a 20 liter container with water and mixed thoroughly with a stirrer for 10 minutes, after which it is allowed to settle for 24 hours.

The water was decanted with a hose, and the sorbent was filtered and dried in a drying cabinet. In the expeditionary conditions the sorbent after decanting with the remaining water was poured into plastic containers of 1-2 liters.

Filtered and dried sorbent was transferred to a porcelain crucible and burned in a muffle furnace at 500°C. The resulting residue was transferred to a one-liter flask, dissolved in 10 ml of hydrochloric acid, and brought to a liter with distilled water. After dissolution the residue was boiled under a watch glass for 20 minutes to remove the carbonates. After cooling down, 1 ml of a 25% ferric chloride solution was added to the solution, stirred, and iron hydroxides precipitated with an aqueous solution of ammonia. After the yellow-brown flakes fall out, the solution was filtered on a Buchner funnel and dissolved in 70 ml of a hot solution of 7M HNO_3. The resulting nitric acid solution free from ^{210}Po and ^{226}Ra isotopes was transferred to a separatory funnel and 15 ml of a 30% tributyl phosphate solution in toluene was added and shaken for 5 minutes. At this stage was the separation of uranium from thorium isotopes. The bottom layer was discarded, and the top layer of tributyl phosphate was washed twice for 1 minute with 15 ml of 7M HNO_3. One more time it was washed for 1 minute with 15 ml of 0.25 M HNO_3 solution in 0.04 M HF. The wash water was discarded. Next, re-extraction of uranium was carried out, for which the organic phase is washed 3 times for 1 minute with distilled water in 15 ml portions. The combined aqueous re-extract was transferred to a 100 ml quartz crucible and evaporated to dryness, 5 ml of concentrated nitric acid are added to remove traces of organic matter and evaporated again to dryness.

Electrolytic deposition of uranium isotopes was carried out as follows. The dry residue containing uranium isotopes was dissolved in 10 ml of a 2% soda solution when heated, filtered through a “blue ribbon” filter and transferred to a Teflon electrolytic cell with stainless steel disk with a diameter of 39 mm. Pre the disc were cleaned with emery paper and rubbed with acetone. Platinum was used as a positive electrode in the form of a cone with holes or a spiral-twisted wire. The crucible and filter are additionally washed with 10 ml of soda and added to the solution in the cell. Electrolytic deposition of uranium isotopes was carried out at a current of 2A for 30 minutes. The disc was removed, washed with distilled water and dried.

After warming up the semiconductor alpha spectrometer "Progress-alpha," its calibration for 40 minutes and background measurement, a thin-layer sample with uranium isotopes was introduced into the spectrometer chamber. The measurement was carried out in a vacuum for several hours. The results were processed using the “Progress-alpha” software. The measurement error was to 10-15%.

RADIOACTIVITY OF WATERS OF THE BARENTS AND WHITE SEAS

Figure 2 shows the route of the expedition “Floating University” at the research vessel “Professor Molchanov” in June – July 2012 in the Barents Sea. Large points show the stations at which sea water samples were taken. For these points, the Figure 2 shows two parameters – the concentration of uranium in water and the isotopic ratio of $^{234}U/^{238}U$. This relationship is most important for determining the genesis of seawater. Table 3 shows the detailed data for each of the studied samples – the name of the sample, the coordinates of the sampling, the activity of ^{238}U, ^{234}U, ^{235}U isotopes, the errors of their determination and the ratios $^{234}U/^{238}U$

and $^{235}U/^{238}U$. From the data obtained it follows that the concentration of uranium in sea water corresponds to the level of the world ocean and varies from $1.78 \cdot 10^{-6}$ g/l to $3.56 \cdot 10^{-6}$ g/l (Kiselev, 2010). At the same time, with respect to $^{234}U/^{238}U$, the surface waters of the Barents Sea are not uniform. This ratio $^{234}U/^{238}U$ varies from 1.18 to 1.91 (for the World Ocean this ratio is 1.14). The Northern Dvina River carries uranium with an isotopic ratio of 1.8 - 2. We think that water lenses diluted with continental waters existed in the surface waters of the Barents Sea for the period of June 2012 at latitude 71°.

Table 3. The gamma-active radionuclides in water of the Barents Sea

No	Sample name	Coordinates		Radioactivity, Bq/l					
		N	E	^{134}Cs	^{137}Cs	^{40}K	^{226}Ra	^{232}Th	^{60}C
1	B-bt-ZFI-12	80 20403	52 46 326	-	< 0.10	< 6.15	< 0.59	< 7.19	-
2	PU-st-37	75 08 980	55 23 957	-	< 0.08	8.93	< 0.10	< 0.41	-
3	B-st-21-12	77 55 014	56 52 239	-	-	2.10	< 0.01	< 0.02	-
4	B- Bar m St-14	74 30 050	33 27 490	-	< 0.06	9.38	< 0.03	< 0.81	-
5	B-Bar m St-17	75 59 653	33 28 033	-	< 0.30	< 8.3	< 0.12	< 0.81	-

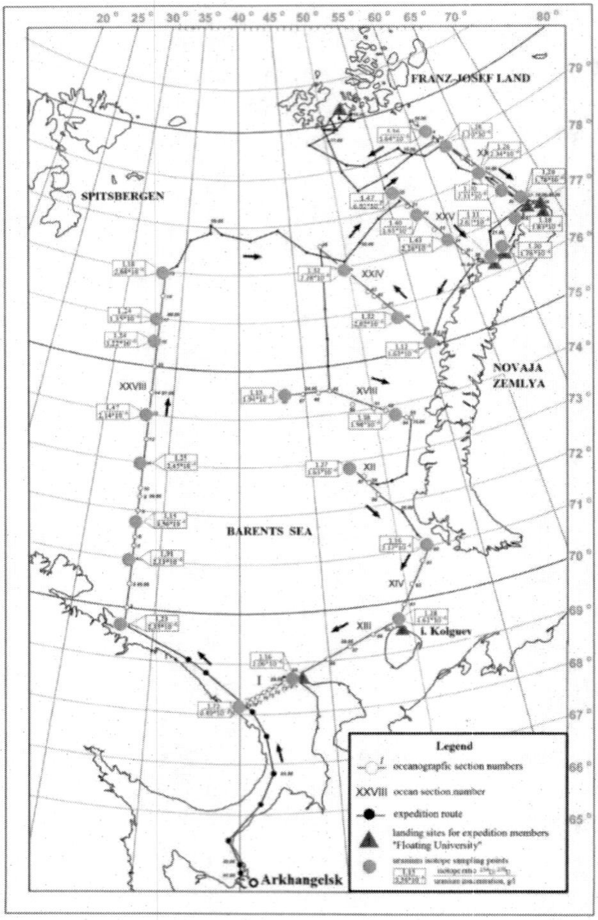

Figure 2. Map of water sampling and uranium isotope composition of water in the Barents Sea, 2012.

A runoff from the Scandinavian Peninsula could be the cause of lenses appearance. But it should be noted that such studies were conducted for the first time in the Barents Sea, for this reason it is not possible to indicate the defined source of the appearance of waters with a high excess of ^{234}U. It is necessary to continue the study of this isotopic parameter in the Barents Sea with greater detail, both along the littoral and at depth.

Figure 3 shows a map of route of the research vessel "Professor Molchanov" in the White Sea. The map shows the samples location (large circles), near which the concentration of uranium in water and the ratio of $^{234}U/^{238}U$ are shown. As well as for the Barents Sea, the concentration of uranium in the water of the White Sea corresponds to the concentration uranium in the World Ocean. It should also be noted that due to the strong influence of the Northern Dvina River, on Dvina Bay of the White Sea (the river carries water with a concentration of uranium $n \cdot 10^{-7}$ g/l), the concentration of uranium in the surface waters of the Dvina Bay during the testing period was $4.8 \cdot 10^{-7}$ g/l to $6.8 \cdot 10^{-7}$ g/l (Malov and Kiselev, 2008).

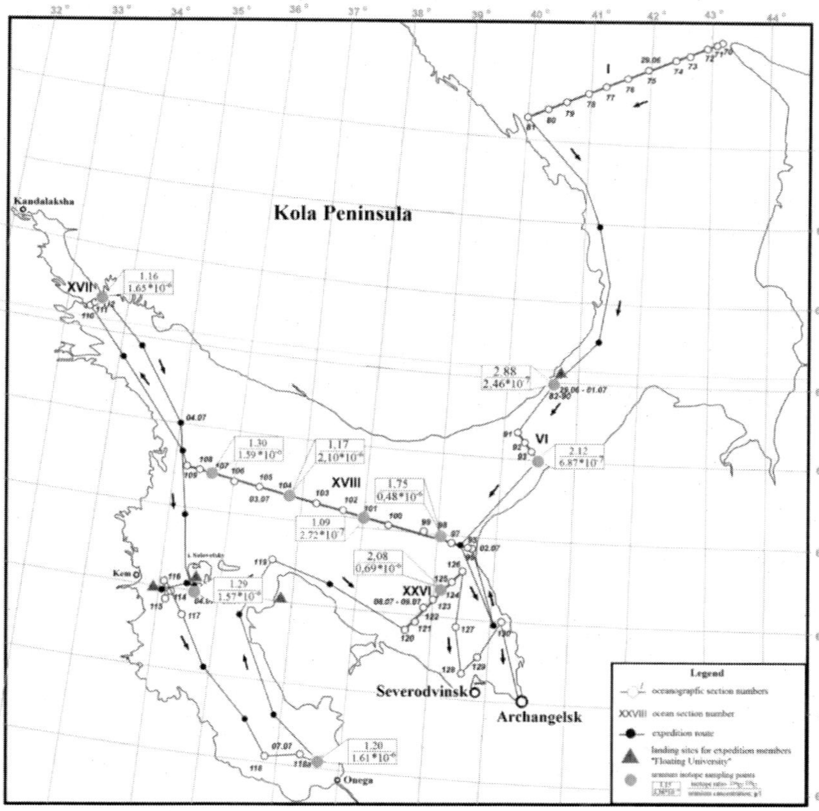

Figure 3. Map of water sampling and uranium isotope composition of water in the White Sea, 2012.

In spring of 2012, during the Northern Dvina flood, the uranium-isotope ratio $^{234}U/^{238}U$ was registered up to 3.32, with a concentration of uranium of $3.29 \cdot 10^{-7}$ g/l (Zykova and Zykov, 2017). This explains the increase $^{234}U/^{238}U$ ratio and the decrease in the concentration of uranium in coastal waters of the Dvina Bay. With distance from the influence of the Northern Dvina River on Dvina Bay, the concentration of uranium in seawater increases and the uranium isotope ratio decreases. In the center of the White Sea, these water parameters correspond to previously determined values (Malov and Kiselev, 2008).

Measurement of gamma-activity radionuclides in seawater is hampered by their low concentrations and requires high precision laboratory research. During the expedition, it was possible to enrich some seawater samples by evaporation 20 liters to 1 liter. Measurement of seawater was carried directly on board the research vessel "Professor Molchanov" using gamma-spectrometer "Progress 2000." The results of some measurements are presented in Table 4.

Table 4. Gamma-active radionuclides in the bottom sediments of the Barents and White Seas at the sea expedition on the research vessel "Professor Molchanov" in 2012

Sample No, horizon	Coordinates		Weight, g	Radionuclide activity, Bq/kg					
	N	E		^{7}Be	^{134}Cs	^{137}Cs	^{226}Ra	^{232}Th	^{40}K
DO-1	78° 06.56'	37° 17.96'	953	-	1.4	1.2	4.4	10.8	327.2
DO-2-1 0-2	76° 14.15'	62° 40.85'	320	-	-	-	6.9	17.3	508.4
DO-2-2 2-5			1056	-	-	10.6	9.4	24.2	534.3
DO-2-3 5-10			1140	-	1.3	9.6	3.9	38	407.6
DO-3	76° 37.84'	65° 42.11'	950	17.3	0.7	1.3	4.6	12.7	427.5
DO-4-1	76° 53.09'	66° 47.35'	495	-	2.2	27.6	8.6	33	675
DO-4-2			750	13.6	1.4	3.3	2.7	17.7	477
DO-4-3			-	-	1.08	1.4	9.6	31.9	627
DO-5-1	77° 03.44'	67° 42.56'	861.7	-	-	3.8	20.9	5.1	536
DO-5-2			1235	-	-	3.8	5.4	36.3	682
DO-C-65	69° 37.25'	49° 06.07'	828.1	-	1	-	6.9	13	429.2
DO-C-66-1	69° 23.70'	47° 33.08'	383.2	-	1.4	1.7	4.8	12	393.3
DO-C-66-2			1202.2	-	8.3	27.5	3.4	40.4	643
DO-C-67	69° 11.67'	46° 15.73'	-	8.2	-	4.8	7.4	14.8	505.7
DO-C-67-1 0-2			410.3	2.9	1.04	2.8	1	17.2	394
DO-C-68-1 0-1	69° 00.02'	44° 59.98'	317.3	-	-	2.3	6.7	14.4	409.4
DO-C-68-2 1-10			1289.3	-	-	-	3.2		149.2
DO-C-68-total			455.8	25.8	3.45	1.2	6.7	12.6	505.1
DO-C-69-1	68° 45.66'	43° 14.35'	79.4	-	-	0.6	4.1	9.2	372.1
DO-C-69-2			382.6	-	0.7	1	5.8	11.1	336.3
DO-C-69- total			880.8	6.1	-	2.6	7.4	13.6	354.6
DO-C-70-1 0-3	68° 38.71'	43° 10.71'	987.5	-	1.4	4	2.8	24.3	442.5
DO-C-70-2 3-7			1237.4	-	-	6.1	3.4	13.6	522
DO-C-78	68° 19.5'	40° 54.6'	335	-	2.1	7.5	6.1	32.2	622
DO-C-82	66° 28.88'	40° 39.83'	2002.8	3.3	2.5	2.8	19.1	33.1	583.3
DO-C-82			-	4.5	1.2	0.8	28.1	37.2	575.7
DO-C-83	66° 29.48'	40° 40.73'	687.7	15.8	-	3.6	9.1	16.5	375.5
DO-C-94	66° 02.33'	40° 41.86'	429.8	-	-	2.6	20.2	36.2	650
DO-C-95	65° 27.89'	39° 37.01'	1959.5	-	-	-	14	20.3	458
DO-C-97-1 0-1	65° 28.92'	39° 17.85'	743.1	-	1.2	4.1	19.6	27.6	530
DO-C-97-2 1-10			1064.7	-	-	1.4	16	22.4	460.7
DO-C-97-total			746.1	-	-	3.2	5.2	12.7	329.2

Sample No, horizon	**Coordinates**		**Weight, g**	**Radionuclide activity, Bq/kg**					
	N	**E**		**^{7}Be**	**^{134}Cs**	**^{137}Cs**	**^{226}Ra**	**^{232}Th**	**^{40}K**
DO-C-99-1 0-2	65° 30.97'	38° 49.02'	359.9	-	0.7	1.8	13.9	24.2	496.4
DO-C-99-2 2-20			664.5	-	-	-	18.2	20.5	494.6
DO-C-99-3 20-25			606.5	28.5	1.3	4.8	21.9	26.4	536.3
DO-C-99-total			792.5	-	-	4.3	9	38.4	561
DO-C-101-1 0-1	65° 33.13'	38° 01.25'	135.3	-	1.3	-	11.7	25.1	487.1
DO-C-101-2 1-10			321	-	-	-	8.6	10.4	527
DO-C-101-total			548.5	-	-	2	15.6	21.1	295.8
DO-C-107-1 0-1	65° 42.02'	35° 42.12'	503.3	-	-	-	6.1	16.9	580
DO-C-107-2 1-5			543.8	-	0.9	-	6.5	7.1	335.6
DO-C-107-3 5-15			609.5	13.8	-	4.9	4.9	23.4	434.5
DO-C-107-total			505.6	-	-	-	6	21.6	615
DO-C-108-1 0-1	65° 43.15'	35° 25.19'	268.6	-	0.6	-	7.3	18.2	536.4
DO-C-108-2 1-10			763	-	2.9	9.4	15.3	54.7	697
DO-C-108-total			1077.6	-	1.6	10.3	8	18.9	523.7
DO-C-110	66° 39.05'	33° 19.81'	-	5	2.3	3.3	3.6	23.4	468.8
DO-C-111	66° 40.07'	33° 22.11'	1202.4	13.4	2.5	2.3	9	16	443.1
DO-C-112-1 0-1	66° 41.42'	33° 25.47'	564.4	-	-	3.3	6.3	14.2	411
DO-C-112-2 1-5			1175	23.8	1.7	5.2	8.7	24.3	465
DO-C-112			621.8	-	2	4.5	7.6	32.7	535.2
DO-C-112-total			1106.7	-	-	1.1	6.7	12.3	355.9
DO-C-115-1 0-1	64° 50.55'	35° 06.58'	337	-	1.2	1.6	6.4	16.1	487.7
DO-C-115-2 1-15			1420.6	-	-	2.9	6.7	11.6	303.3
DO-C-115-total			1330.5	-	0.5	3.1	7.7	15.7	359.2
DO-C-116	64° 57.18'	35° 06.14'	1722.2	26.9	1.3	16.7	6	11.3	522

The measurement results showed that in seawater there are no abnormal or excessive concentrations of radioactive isotopes. The technogenic isotope ^{134}Cs was not detected in sea water, the presence of which would indicate the release into the environment of large quantities of modern emissions from anthropogenic sources. ^{137}Cs activity is below the determination level, which indicates its almost complete absence in these samples of sea water. The technogenic isotope ^{60}Co is also completely absent in these samples. The activity of such natural isotopes as ^{226}Ra and ^{232}Th is below the instrument measurement limits, and the gamma-active isotope ^{40}K is recorded in some samples, due to its abundance and high activity in the environment, since it is part of salts dissolved in sea water. Overall, modern radioactive contamination of the waters of the Barents Sea was below detection limits of our methods.

RADIOACTIVITY OF BOTTOM SEDIMENTS OF THE BARENTS AND WHITE SEAS

The bottom sediments during the expedition at the research vessel "Professor Molchanov" were selected with a special on-board sampler on the winch. At the same time, it was possible to sample the bottom sediments only from shallow depths and in the presence of sandy and clayey deposits. The sampling points and the results of gamma spectrometry analysis are shown in Figure 4 for the Barents Sea, and in Figure 5 for the White Sea.

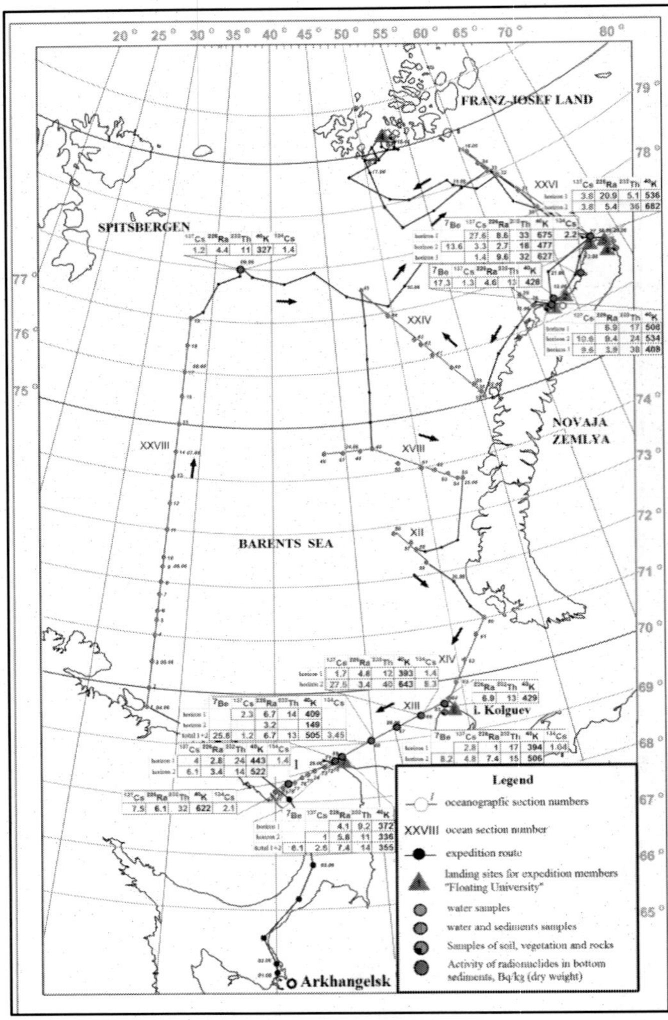

Figure 4. Map of the bottom sediments sampling in the Barents Sea during at the sea expedition on the research vessel "Professor Molchanov" in 2012.

The distribution of ^{226}Ra in the White Sea corresponds to the previously obtained data (Kiselev, 2010). In the Barents Sea, its concentration in bottom sediments does not exceed the clarke values. The activity of ^{232}Th in different horizons of the bottom sediments of these seas varies from 3 to 54 Bq/kg. The maximum value of ^{232}Th (54 Bq/kg) was established in the layer from 1 to 10 cm of sample at station No 108 in the White Sea near the Karelian coast, where its elevated concentrations were previously noted in the upper two centimeter layer (Kiselev et al., 1999). The emission of ^{232}Th to the bottom sediments of this region can be carried out by exogenous processes destroying the acidic granitoids that make up numerous islands and the Karelian coast. Increased concentrations of ^{232}Th (36.3 Bq/kg) were established near the Oranskie Islands, station No 27 at the northern end of the Novaya Zemlya archipelago, and at station No 66 near Kolgueva Island (40 Bq/kg). Obviously, these are natural concentrations and are not related to man-made pollution. In the remaining territories, the activity of ^{232}Th does not exceed the clarke values, and for the White Sea, the obtained data are consistent with previous studies (Bashenov et al., 1999).

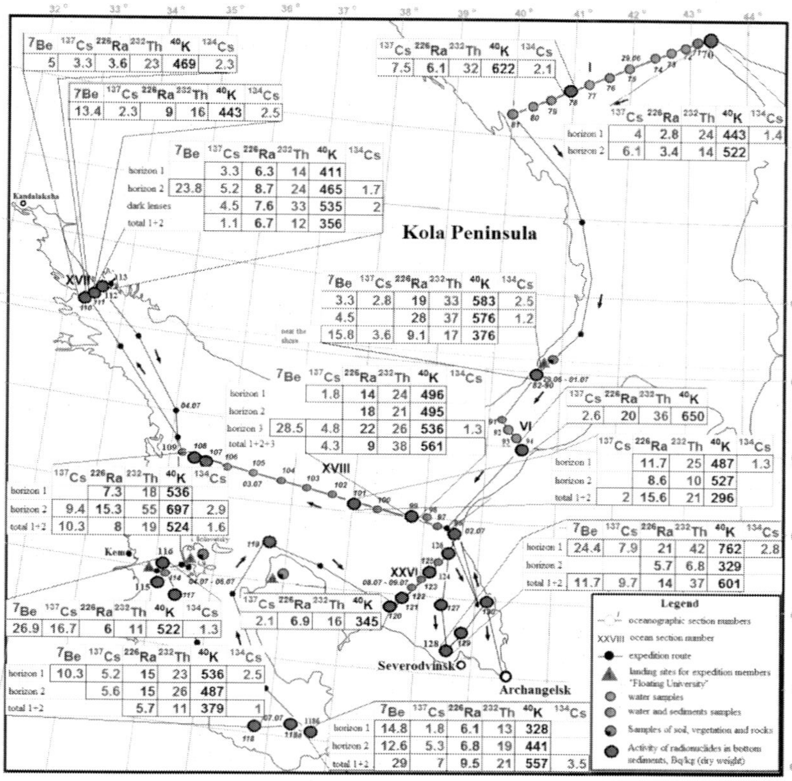

Figure 5. Map of the bottom sediments sampling in the White Sea during at the sea expedition on the research vessel "Professor Molchanov" in 2012.

In some samples, a natural radioactive isotope ^{7}Be of cosmogenic origin is observed. Its maximum recorded activity is 28.5 Bq/kg.

Technogenic radioactive isotopes ^{134}Cs, ^{137}Cs, ^{60}Co are the main indicators of the contamination by anthropogenic radioactivity of bottom sediments of territories and water areas.

The isotope ^{60}Co was not found in the bottom sediments. ^{137}Cs, having a half-life of about 30 years, was found at almost every station in a given horizon. The maximum values of its activity were 27.6 Bq/kg in one sample on the northern edge of the Novaya Zemlya archipelago and in the area of Kolguev Island (station 66). For the rest of the water area, the concentration of this isotope is low and corresponds to the previously obtained results (Kiselev, 1999, 2001).

The general patterns of distribution of radioactive isotopes are as follows. The highest activity values belong to ^{40}K, as noted earlier. Its activity depends on the composition of sediments and ranges from 147 Bq/kg to 680 Bq/kg for selected samples, which corresponds to a normal distribution of ^{40}K in marine sediments of the White Sea. The activity of ^{226}Ra in the bottom sediments of the Barents and White Seas varies from 1 to 28 Bq/kg. The highest values of 20.9 Bq/kg in the Barents Sea were found in the OranskieIslands and 28.1 Bq/kg in the Sosnowiec Island in the White Sea. These islands are composed of acidic igneous rocks which contain high concentrations of ^{226}Ra (Nikolaev et al., 1979).

^{134}Cs has a half-life of about 2.5 years. Previously, this isotope was not recorded by us in bottom sediments under the same methodological and instrumental conditions (Kiselev et al., 1999). In the samples of bottom sediments in the Barents and White Seas, it is found in quantities from 0.7 to 8.3 Bq/kg. This indicates that technogenic radioactivity is present in the Barents and the White Sea, which should be alarming.

Coastline

Studies of vegetation and algae sampling points in coastal zone of White Sea show that mosses growing on the Kuzova Archipelago have an increased gamma activity of such isotopes as ^{134}Cs and ^{137}Cs. For example, the lichen Cladonia has ^{137}Cs radioactivity of 350 Bq/kg, the activity of ^{134}Cs in the same sample reaches 19.63 Bq/kg, which indicates the entrance of modern technogenic radioactivity to this area, which lichens absorb and accumulate. The Moss (*Helodium blandowii*) is similar to the lichen Cladonia sample, but has slightly lower values of the activities of anthropogenic isotopes ^{134}Cs and ^{137}Cs (Table 5).

Table 5. The activity of gamma-active isotopes in vegetation

N	Samplename, coordinates	Radioactivity, Bq/kg					
		^{134}Cs	^{137}Cs	^{40}K	^{226}Ra	^{232}Th	^{60}Co
1	Brown algae (*Laminaria*), Barents Sea,N 76^0 14.147; E 62^0 40.852	2.6	-	1412.0	< 1.0	-	-
2	Dwarf willow (*Salix herbacea*), Barents Sea, Kolguev Island, N 69^0 37 138; E 49^0 05.697	-	-	514.0	-	20.2	-
3	Moss (*Helodiumblandowii*), White sea, Kuzova Archipelago, N 64^0 56.700; E 35^0 06.958	10.6	104.6	157.3	< 8.8	< 7.3	< 4.1
4	Lichen (*Cladonia*), White sea, Kuzova Archipelago, N 64^0 56.700; E 35^0 06.958	19.6	350.0	171.0	18.1	< 20	< 3.0

In the sample of brown algae (*Laminaria*) there is a slight presence of ^{134}Cs, which is consistent with the previously obtained data (Makhonko, 1998). There is also a high activity of ^{40}K, although its increased content is typical for algae as a whole and is associated with the physiological features of this plant. Leaves of dwarf willow (*Salix herbacea*), selected on Kolguev Island have typical values for this kind of plant and do not have elevated values of radioactivity.

^{7}Be in Atmospheric Air Above the Barents and White Seas

In the process of analyzing the data on the distribution of ^7Be in the atmosphere of the western part of the Arctic, it was established that the atmospheric pressure in the northern regions of Russia plays the main role in the flow of the cosmogenic radioactive isotope into the surface layer of atmosphere.

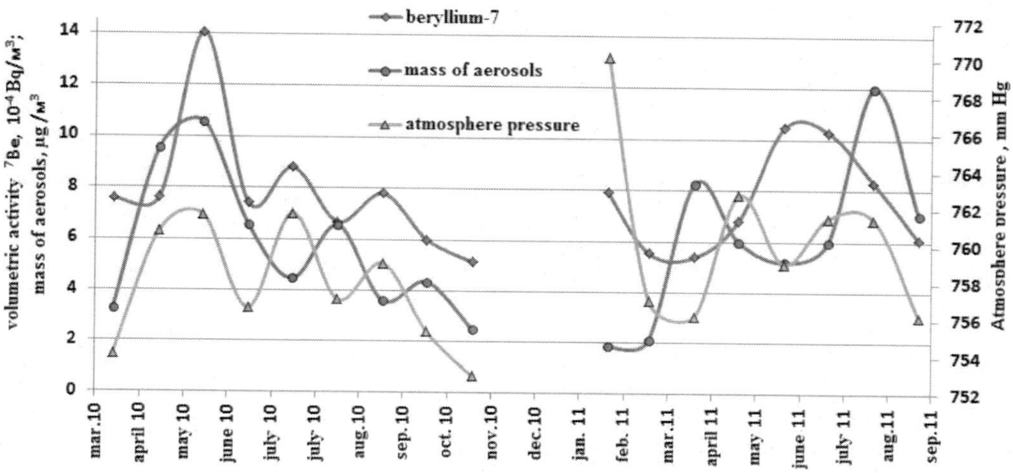

Figure 6. The average monthly distribution of the volume activity of ^{7}Be (Bq/m^3), atmospheric pressure (mm Hg) and the mass of aerosols (µg/m^3) for the Arkhangelsk region from March 2010 to September 2011.

Figure 6 presents data on average monthly atmospheric pressure in comparison with the concentration of ^{7}Be and mass of aerosols for surface layer of atmosphere for the period from March 2010 to September 2011.

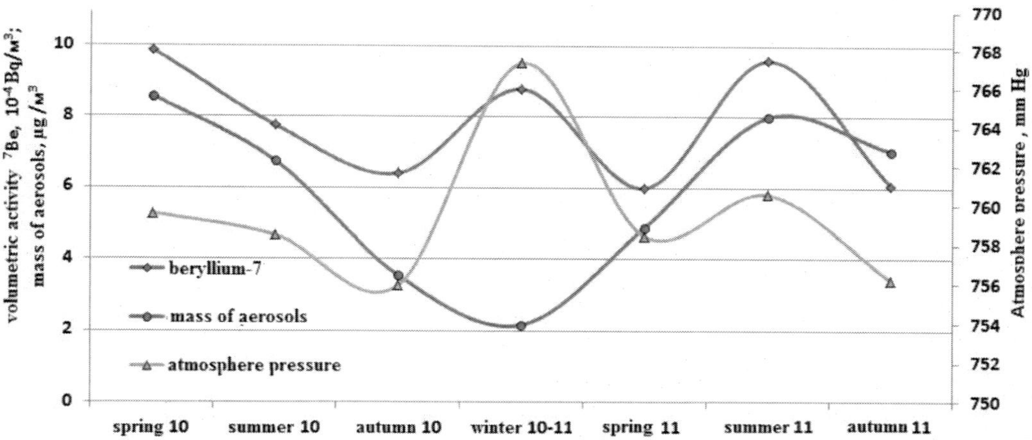

Figure 7. The average seasonal distribution of concentration of ^{7}Be (Bq/m^3), aerosol masses (µg/m^3) and atmospheric pressure (mm Hg) for the period spring 2010 - autumn 2011 for the Arkhangelsk region.

An increase in atmospheric pressure leads to an increase in the concentration of ^{7}Be in the surface atmosphere. A particularly strict correlation of these parameters was observed in 2010. In 2011, the increasing ^{7}Be concentration with increasing atmospheric pressure also takes place. High pressure in the surface atmosphere is uniquely associated with anticyclones. In the centers of anticyclones, vertical air flows arise from top to bottom, which leads to the transfer of ^{7}Be from high atmospheric horizons to the surface atmosphere. Figure 7 shows the average seasonal parameters of ^{7}Be, atmospheric pressure and the mass of aerosols for the period spring 2010. - Autumn 2011 from which it follows that in the winter time also increase

^{7}Be concentration in the surface atmosphere with increasing atmospheric pressure. The mass of aerosols in the air is significantly reduced. In the total flux of ^{7}Be to the surface atmosphere, atmospheric pressure plays a major role.

It was established experimentally that the concentration of ^{7}Be in air-dry atmospheric aerosols decreases with increasing latitude of the terrain. This phenomenon has not previously been noted. Perhaps this is due to the active leaching of ^{7}Be precipitation - snow and rain in the Arctic region. Table 6 shows the data of various authors on the content of ^{7}Be in air aerosols.

Table 6. Distribution of ^{7}Be in aerosols depending on the latitude of the area

Region	**Coordinates Northlatitude**	**Average activity of ^{7}Be, mBq /m^3**	**Source**
Barents Sea	72°	0.37	(Druzhinin and Kiselev, 2014)
Kola Peninsula, Apatity	68°	0.75	(Melnik, 2004)
Arkhangelsk	64°	0.8	(Kiselev and Druzhinin, 2013)
Moscow	55°	2.7	(Petrova et al., 2009))
Sevastopol	44°	2.8	(Arbuzov et al., 2010)
Egypt	30°	30.0	(EL-Husseinet al.,2001)

During the expedition on the research vessel "Professor Molchanov," ^{7}Be activity was meseared in aerosols of air that were taken along the entire route of the vessel. The data are presented in Figure 8.

These data show the dependence of the decrease in ^{7}Be activity in aerosols of the surface air layer from the south to the north.

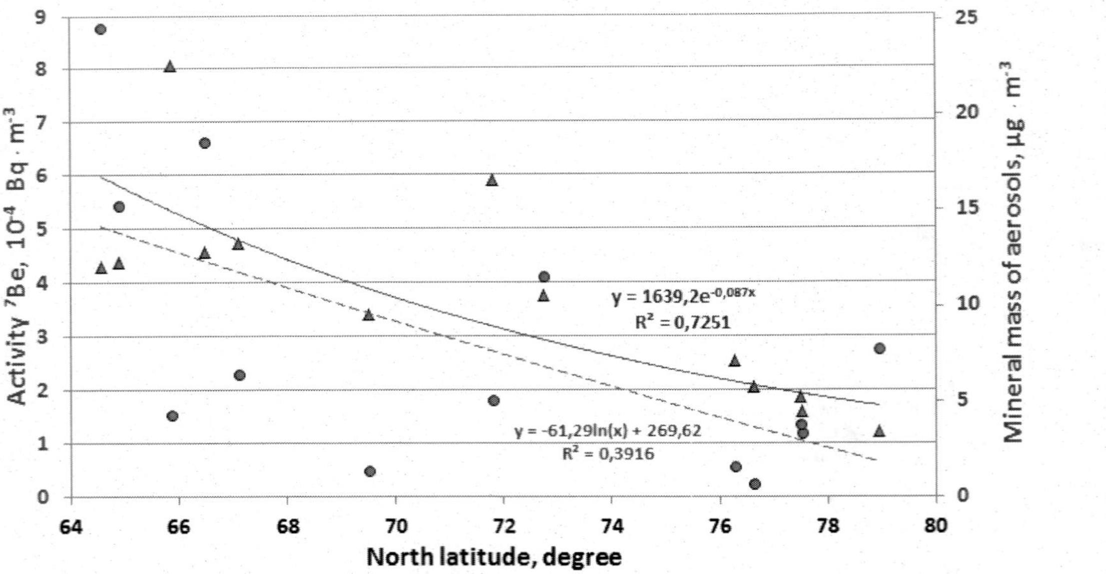

Figure 8. Distribution of ^{7}Be in aerosols at high latitudes according to the results of the expedition "Professor Molchanov."

CONCLUSION

During the expedition on the research vessel "Professor Molchanov," a radiological group of FCIARctic carried out the radiological studies in the White and Barents seas of the Western Arctic of Russia. In the process of research, sampling of bottom sediments, sea waters, aerosols, island soils within the Barents and White seas was performed. The measurements were carried out by alpha and gamma spectrometry using radiochemical analysis. The content of technogenic (^{134}Cs, ^{137}Cs, ^{60}Co) and natural (^{234}U, ^{238}U, ^{40}K, ^{226}Ra, ^{232}Th, ^{7}Be) radionuclides was determined. In general, it has been shown that ^{137}Cs activity in the bottom sediments of the Western Arctic seas of Russia has recently decreased, but in many samples a rapidly decaying ^{134}Cs radionuclide has been detected, which may indicate a continued flow of technogenic radioactivity into the Arctic seas. The activities of natural radionuclides in marine bottom sediments are within the clarke content and depends on the composition of sediments. In the Barents Sea, lenses of waters with an excess of $^{234}U/^{238}U$ isotopic ratios of more than 1.3 were detected. These lenses with an excess of ^{234}U isotope are areas of continental flow. A decrease in the ^{7}Be activity in air aerosols at high latitudes has been established.

The results obtained on the activity of technogenic radionuclides showed the following pattern of their distribution and accumulation: in seawater there is a significant decrease in the activity of radionuclides compared with the beginning of the 80s, but at the same time insignificant fallout continues; the deposition of radionuclides on the seabed area is uneven, the greatest accumulation of radionuclides is observed in the fine fraction of the shelf zones; ^{137}Cs radionuclide accumulation in soils is typical for the upper genetic horizons, the distribution of radionuclides over the territory depends on the conditions of regional and global fallout.

Current research results show that levels of radiation pollution are low and tend to decrease, but the picture may change due to increasing anthropogenic activity on the continental shelf and adjacent territories. Also, the active use of the Northern Sea Route with the arrangement of related infrastructure and the implementation of dredging in the mouths of large northern rivers could be the cause the redistribution of natural and man-made radionuclides.

REFERENCES

Aarkrog A. Radioactivity in Polar Regions - main sources. *Environmental Radioactivity in the Arctic and Antarctic*. Osteras. 1994. 388 p.

Agamirov S. Sh. Geokhimicheskiy balans radioaktivnykh elementov v basseyne Chernogo morya [Geochemical balance of radioactive elements in the Black Sea basin]. *Geochemistry*. 1963. № 6. P. 612-618.

Arctic Polar Enviroment. Pol. AMAP. Oslo. 1997. 796 p.

Aybulatov N. A., Sapozhnikov Yu. A., Plishkin A. N., Politova N. V., Sapozhnikova L. D. Cs v donnykh otlozheniyakh zapadnoy chasti Karskogo morya [Cs in the bottom sediments of the western part of the Kara Sea]. *DokladyAkademiiNauk USSR*. 1994. № 1. P. 98-100.

Bashenov A. V., Yudakhin F. N., Kiselev G. P. Raspredeleniye ^{137}Cs v pochvakh severnoy, sredney i sredney yuzhnoy taygi Arkhangel'skoy oblasti (Distribution of ^{137}Cs in the soils of the northern, middle and middle southern taiga of the Arkhangelsk region). *Geodynamics and Geoecology*. Arkhangelsk. 1999. P. 89-99.

Baturin G. N., Kochenov A. V. Migratsiya urana v rekakh i vremya yego prebyvaniya v vodakh okeanov, morey i ozer [*Migration of uranium in rivers and the time of its stay in the waters of the oceans, seas and lakes*]. *Geochemistry*. 1969. №6. P. 510-517.

Baturin G. N., Kochenov A. V., Kovaleva S. A. Nekotoryye osobennosti rasprostraneniya urana v vodakh Chernogo moray [Some features of the distribution of uranium in the waters of the Black Sea]. *DokladyAkademiiNauk*. 1966. №2. P. 89-106.

Chalov P. I. *Izotopnoye fraktsionirovaniye prirodnogo urana* [*Isotopic fractionation of natural uranium*]. Frunze: ILIM. 1975. 196 p.

Chen J. H., Edvards R. I. and Wasserburg G. J. 1986. ^{238}U-^{234}U and ^{232}Th in seawater. *Earth Planet. Sci. Lett.* Vol. 80. P. 241-251.

Demin Yu. I., Demina L. I. *Toriy v gornykh porodakh i rudakh yavlyayetsya istochnikom radiatsionnoy opasnosti v gornykh rayonakh [sootnosheniye yestestvennoy i tekhnogennoy radiatsii) (Thorium in rocks and ores is a source of radiation hazard in mining areas (the ratio of natural and man-made radiation*)]. Moscow: GEOS. 1999. 429 p.

Edvards R. L., Chen J. H., Wasserburg G. J.· ^{238}U-^{230}Th-^{232}Th over the past 500,000 years. *Earth Planet. Sci. Lett.* 1987. Vol. 81. P. 465-475.

EL-Hussein A., Mohamemed A., Abd. EL-Haby M., Ahmed A. A. DiBrasen et al., ^{7}Be and ^{210}Pb in Egipt. *Atmos. Environ. Radioactivity*. 2001. Vol. 35. P. 43-45.

Golberg, E. D., Koide M. *Questions of ocean geochemistry and geochronology*. Ed. V. M. Vdovenko. Moscow: MIR. 1965. 585 p.

Haartsen Tialda. *The North Sea Ecosystem Assessment*. Arctic Center, University of Groningen. Norway. 1995. 342 p.

Ivanov G. I., Bordukov Yu. K., Ivanov V. I., Polyak L. V., Sadikov M. A. *Radioecology of the Banks and the Sea.* Woords Hole OI. USA. 1993. P. 267-278.

Ivanov G. I., Bordukov Yu. K., Polyak L. V., Krinitski P., Nescheretov A. V., Krasnyuk A. D., Ellis K. Assessment of Radiological Contamination from the Dumpsites in the Kara Sea. *Environmental Radioactivity in the Arctic.* Osteras. 1995. 349 p.

Katrich Yu. In *the Arctic and the Arctic Ocean*. Norway. 1993. 258 p.

Kershaw P. J. Baxter A. J. Environmental Radioactivity in the Arctic and Antarctic. *Osteras.* 1993. 428 p.

Kiselev G. P., Bashenov A. V., Malov A. I., Zykov S. B. Kharakteristika radioaktivnosti pochv i donnykh otlozheniy r. S. Dvina. Ekologiya Severnoy Dviny [*Characterization of radioactivity of soils and bottom sediments of the r. S. Dvina*]. *Ecology of the Northern Dvina*. Arhangelsk: ELPA. 1999. P. 185-191.

Kiselev G. P., Ponomarev A. V., Kisileva I. M. Osobennosti rasprostraneniya radionuklidov v pochvakh Arkhangel'skoy oblasti. Pochvovedeniye na Yevropeyskom Severe Rossii [Features of the distribution of radionuclides in the soils of the Arkhangelsk region. Soil studies in the European North of Russia]. *Coll. articles and bibliographer.* pointer lit. Arkhangelsk. 1996. P. 138-146.

Kiselev G. P. Obzor radioaktivnykh issledovaniy na Yevropeyskom Severe Rossii [Overview of radioactive research in the European North of Russia]. *Ecology of the Arctic and*

Subarctic Territories: materials of the international symposium. Institute of Ecological Problems of the North UB RAS. Arkhangelsk. 2010. P. 324-325.

Kiselev G. P. *Prognoz mestorozhdeniy poleznykh iskopayemykh i zagryazneniya geologicheskoy sredy uran-izotopnymi metodami.* [*Forecast of mineral deposits and pollution of the geological environment by uranium-isotope methods*]. Arkhangelsk. 2005. 201 p.

Kiselev G. P. *Izotopy urana v geosphere* [Isotopes of uranium in the geosphere]. Ekaterinburg: UB RAS. 1999. 219 p.

Kiselev G. P., Zykov, S. B. *Izotopnyy sostav urana v Belom more.* [*The isotope composition of uranium in the White Sea*]. Geology of the seas and oceans. Abstracts of the XV International School of Marine Geology. Vol. 2. Moscow: GEOS. 2003. P. 129-156.

Kiselev G. P., Kiseleva I. M., Zykov S. B., Bashenov A. V., Malov A. I. *Radioaktivnyye izotopy v donnykh otlozheniyakh Belogo morya* [*Radioactive isotopes in bottom sediments of the White Sea*]. *North: Ecology: Sat. scientific* tr. Ekaterinburg. 2000. P. 18-30.

Kiselev G. P., Kisileva I. M., Krauchunas B. B., Bashenov A. V. Sostoyaniye radioaktivnykh issledovaniy na Yevropeyskom Severe [*The state of radioactive research in the European North*]. *Problems of biogeochemistry and geochemical ecology*. 2012. №3 (20). P. 131-140.

Kiselev G. P., Zykov S. B. *Izotopnoye stroyeniye Severa Rossii* [*Isotope structure of the North of Russia*]. Moscow. LOIRA-VII. 2004. P. 183-192.

Kjellstrom B. Design and operational features of Swedish boling water reactors. *International nuclear reactor reactor study*. Report prepared for GREENPEACE. Exergetics AB. Trosa. Sweden. September. 1986. 581 p.

Koide M., Golberg E. $^{234}U/^{238}U$ ratios in sea water. *Progress in Oceanography*. 1965. №3. P. 173-192.

Kuznetsov Y. V., Revenko Yu. A., Legin V. K., Rakov N. A., Zhidkov V. V., Egorov Yu. M. On assessment of the contribution of the Yenisei River to the total radioactive contamination of the Kara Sea. *Radiochemistry*. 1994. №6. P. 546-559.

Kuptsov V. M. Primeneniye metoda neravnovesnogo urana dlya opredeleniya doli poverkhnostnogo stoka v pritokakh shakhtnykh vod (Application of the method of non-equilibrium uranium to determine the share of surface runoff in the mine water inflows). *Isotope studies of natural waters*. 1979. №2. P. 210-213.

Logachev V. A. Vozmozhnyye dozy oblucheniya naseleniya Rossiyskoy Federatsii v rezul'tate yadernykh ispytaniy na Severnom poligone [*Possible radiation doses to the population in the Russian Federation as a result of nuclear tests at the Northern Test Site*]. *Marine Arctic Complex Expedition*. 1998. №2 (1). P. 240-258.

Makhonko K. P. Severnyy poligon: kontrol' radioaktivnogo zagryazneniya ot yadernykh ispytaniy v strane [*Northern range: control of radioactive contamination from nuclear tests in the country*]. *Marine Arctic Complex Expedition*. 1998. №2 (1). P. 205-209.

Malov A. I. Kiselev G. P. *Uran v podzemnykh vodakh Mezenskoi syneklizy. (Uranium in the underground waters of the Mezensyneclise).* Publishing House of Russian Academy of Sciences. Ekaterinburg. 2008. 238 p.

Matishov G. G. Matishov D. G., Shchipa E., Rissanen H. *Radionuklidy v ekosisteme Barentseva i Karskogo morey [Radionuclides in the ecosystem of the Barents and Kara seas]*. KSC RAS. Apatity. 1994. 367 p.

Matishov G. G., Matishov D. G., Namyatov A. A. Urovni i usloviya nakopleniya antropogennykh radionuklidov v Kol'skom i Matovskom zalivakh [*Levels and conditions for the accumulation of anthropogenic radionuclides in the Kola and Matovsky bays*]. *Doklady Akademii Nauk*. 1997. №6. P. 812-814.

Matishov GG, Matishov D. G., Namyatov A. A., Zuev A. N., Kirillov E. E. *Radionuklidy i okeanograficheskiye usloviya ikh nakopleniya v Kol'skom i Matovskom zalivakh (Barentsevo more)* [*Radionuclides and oceanographic conditions of their accumulation in the Kola and Matovsky Bays (Barents Sea)*]. Murmansk Marine Biological Institute. Murmansk. 1997. P. 196-203.

Matishov G. G., Matishov D. G., Rissanen X. Zagryazneniye donnykh otlozheniy Belogo morya iskusstvennymi radionuklidami [*Pollution of bottom sediments of the White Sea by artificial radionuclides*]. *DokladyAkademiiNauk*. 1995. №2. P. 256-258.

Miyke Y., Sugimura J, Uchida T. Ratio $^{234}U/^{238}U$ and the concentration of the northwest Pacific. *Journal of Geophysics*. 1966. Vol. 71. P. 30-83.

Nikolaev D. S., Efimova E. I., Drozhzhin V. M. Sredniye dannyye o raspredelenii urana, ioniya radiya i toriya v vodakh Tikhogo, Atlanticheskogo okeanov i ikh morey [*Average data on the distribution of uranium, ionium of radium and thorium in the waters of the Pacific, Atlantic oceans and their seas*]. Moscow: Nauka. 1979. 189 p.

Nilsen T., Bömer N. Sources to radioactive contamination in Murmansk and Arkhangelsk counties. *Bellona Report*. 1994. №1. P. 140-146.

Ostlund Gote. *The transport pattern of the Arctic and the Arctic. Antarctic*. Ed. P. Strand, E. Holm, Osteras. 1993. 233 p.

Salminen R, Glavatskikh S. P., Tomilina O., Bogatyrev I., Gregorauskiene V. BW ecogeochemistry. Ecology of Northern Territories of Russia. Problems, forecast of the situation, ways of development, solutions. Arkhangelsk. 2002. P.16-19.

Shvedov V. P., Patin S. A. Radioaktivnost' okeanov i morey [*Radioactivity of the oceans and seas*]. Moscow: Atomizdat. 1968.

Strand P. et al., *Survey of artificial radionuclides in the Kara sea.* Scientific Committee on the Arctic and Antarctic. Norway. 1993. 528 p.

Strand P., Balonov M., Aarkrog A., Bewers M., Howard B., Salio A., Tsaturov Y. Radioactive contamination in the Arctic. *AMAP International Symposium on Environmental Radioactivity in the Arctic.* Tromso. Norway. 1997. P. 434-442.

Vakulovsy S. M., Nikitin A. I., Chumichev V. B. Zagryazneniye Belogo morya radioaktivnymi otkhodami stran Zapadnoy Yevropy [*Pollution of the White Sea with radioactive wastes of Western European countries*]. *Atomic Energy.* 1988. Vol. 65. №1. P. 66-67.

Vernadsky A. I. Raboty po radiogeologii [*Works on radiogeology*]. Moscow: *Science*. 1997. 123 p.

Yudahin F. N., Kiselev G. P., Kiseleva I. M. Tekhnogennyye radioaktivnyye izotopy v moryakh Zapadnoy Arktiki v period 1972-1996 gg [*Technogenic radioactive isotopes in the seas of the Western Arctic in the period 1972-1996*]. RGO. 1998. №5. P. 13-21.

Zykova E. N., Zykov S. B. 2017. Izmeneniya v izotopnom sostave poverkhnostnykh vod r. Severnoy Dviny [*Changes in the isotopic composition of the surface waters of the Northern Dvina River*]. *Volga Scientific Herald.* №5 (69). P. 25-29.

In: The Arctic: Current Issues and Challenges
Editors: Oleg S. Pokrovsky, Sergey N. Kirpotin et al.
ISBN: 978-1-53617-306-2

Chapter 12

SEISMIC MONITORING OF THE EUROPEAN ARCTIC AND ADJOINING REGIONS

Galina Antonovskaya*, Alexey Morozov, Natalia Vaganova and Yana Konechnaya
N. Laverov Federal Center for Integrated Arctic Research, Arkhangelsk, Russia

ABSTRACT

This chapter elaborates on the results and analysis of the seismic monitoring of the European Arctic and adjoining regions obtained from the early 20th century and modern period from 2011 to 2018. We have relocated earthquakes recorded within the Arctic region beginning from the early 20th century with a view to creating a relocated Seismic catalogue. For the relocation we collected all available seismic bulletins from the global network using data acquired for the ISC-GEM project (International Seismological Centre–Global Earthquake Model), the EuroSeismos project, the Geophysical Survey of the Russian Academy of Sciences, and the Russian State Library. The relocation was performed using a modified method of generalized beamforming and the AKA135 travel time model. As a result, the new coordinates of some earthquakes turned out to be significantly different from those that were determined previously. Most of the relocated earthquake epicenters are confined to major seismic zones in the Arctic, namely, mid-oceanic ridges, the Svalbard archipelago, and the Laptev Sea shelf. With opening some new seismic stations in Russian sector of the European Arctic, we can now register small-magnitude earthquakes within the Gakkel Ridge area and Barents-Kara Sea region including the “continent-ocean” transition zone. In order to improve the accuracy of the earthquakes epicenter locations, velocity models and regional seismic phase travel-times for spreading ridges in areas within the European Arctic have been calculated. The Gakkel Ridge is seismically active, regardless of having the lowest spreading velocity among global mid-ocean ridges. Quiet periods alternate with periods of higher seismic activity. Earthquakes epicenters are unevenly spread across the area. Most of the epicenters are assigned to the Sparsely Magmatic Zone (SMZ), more specifically, to the area between 1.5°E and 19.0°E. We hypothesize that assignment of most earthquakes to the SMZ segment can be explained by the amagmatic character of the spreading of this

* Corresponding Author’s Email: essm.ras@gmail.com.

segment. The structuring of this part of the ridge is characterized by the prevalence of tectonic processes, not magmatic or metamorphic ones. The “continent-ocean” transition zone of the Barents-Kara Sea region is seismically active and we hypothesize that the prevailing geodynamic factor responsible for the occurrence of small-magnitude earthquakes is isostatic compensation of avalanche sedimentation in the “continent-ocean” transition zone. Based on the generalization and analysis of previously available and received new information of Arctic earthquakes distribution, the conclusion that the modern Arctic geodynamics is determined by tectonic processes taking place at the divergent boundary of the North American and Eurasian lithospheric plates is confirmed. The modern seismicity is indicators of geodynamic processes. In general case, the increased seismic activity of separate sites of the European Arctic and adjoining regions is defined by total influence of three main factors. First, the partial discharge of stresses generated in the axial zones of the Arctic basins, and secondly, the lithosphere reaction to the pressure of thick sediments and, thirdly, the uploading from the ancient glaciation.

Keywords: seismic monitoring, European Arctic, Gakkel Ridge, Barents and Kara Sea region, crustal velocity structure, relocation procedures, seismic catalogue

GENERAL IDEA OF THE EUROPEAN ARCTIC SEISMICITY ACCORDING TO THE ARKHANGELSK SEISMIC NETWORK

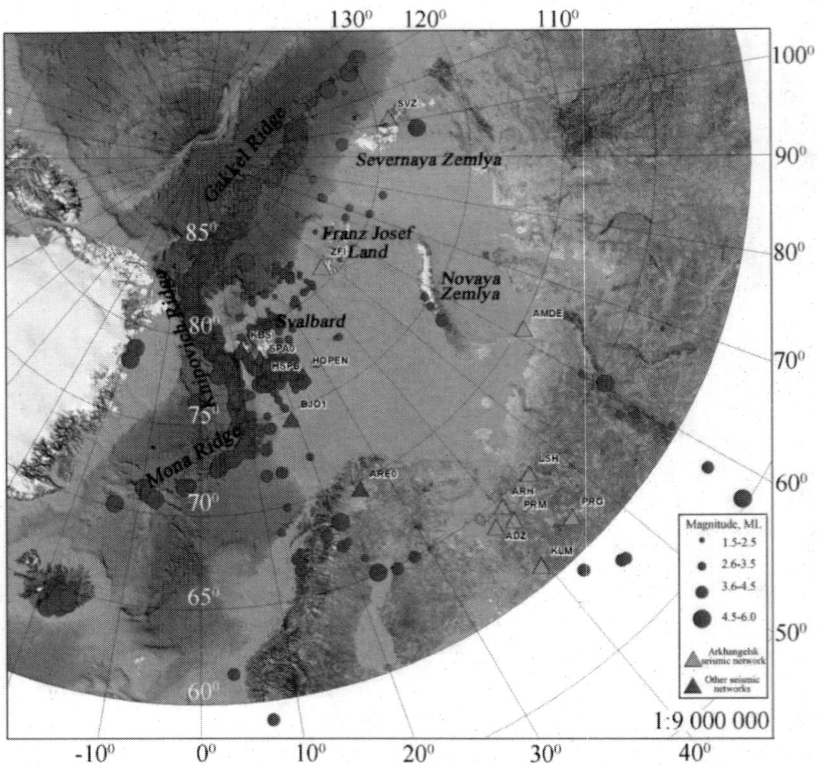

Figure 1. Seismic map according to the Arkhangelsk seismic network for 2014-2018. Green triangles are stations of the Arkhangelsk seismic network, the blue triangles are stations of other seismic networks.

The distribution of earthquakes registered by stations of the Arkhangelsk seismic network (ASN) for 2014-2018 are shown in Figure 1. Over the past five years, we have recorded about 6,000 seismic events of different nature (earthquakes and open blasts at mining enterprises). There is significant earthquake activity along the Mid-Atlantic Ridge (Knipovich Ridge, Gakkel Ridge, Mohns Ridge). The Western Barents Sea exhibits frequent earthquake activity (Figure 1). It must be emphasized that the earthquakes in the Arctic region east of 30°E are new data obtained after installation of the ASN stations. In the zone of Franz Victoria and Orel troughs, the earthquake epicenters coincide with the morphostructures of the continental slope (grabens) and the distribution of earthquake is nonuniform, with larger clusters at graben mouths. The eastern part of the Barents-Kara region is somewhat less exposed to earthquakes activity, although we recorded recent events occurred on Novaya Zemlya.

Next, we consider seismicity situation of the Arctic in the Early Twentieth Century, and more details will focus on contemporary seismicity of the Barents-Kara Sea region, and the "continent-ocean" transition zone, and Novaya Zemlya archipelago, and Gakkel Ridge.

Seismicity of the Arctic in the Early Twentieth Century: Relocation of the 1904-1920 Earthquakes

Research in seismic hazard for any area often based on an earthquake catalogue from which seismicity parameters can be calculated. The most significant events are the largest earthquakes; it is these which define the seismic potential of a seismic source zone. Large earthquakes have long recurrence times; hence the estimation of seismic hazard for an area is based on a composite earthquake catalogue that contains both historical and instrumental earthquakes, as well as including data on palaeo-earthquakes. The information on historical earthquakes for the Arctic region is extremely scanty, while the data on palaeo-earthquakes are practically inaccessible. It follows that the earthquakes that have been recorded during the period of instrumental observation are the most significant.

The development of instrumental observation in Europe in the late 19th to the early 20th century, especially since the installation of the Bergen (1904), Abisco (1906), Pulkovo (1906), and Reykjavik (1909) seismic stations, enabled the recording of seismic motion from individual larger earthquakes occurring in the Arctic. The early history of seismometry is described in detail by (Lee et al., 1988; Ferrari, 1990; Avetisov, 1996; Schweitzer, 2007; Storchak et al., 2015). The lowest magnitude of complete reporting during the earlier stage of instrumental observations for the Arctic, until 1957, was 5.5 – 6.0 (Avetisov, 1996). First reviews of Arctic seismicity can be found in (Tams, 1922; Hodgson, 1930; Gutenberg and Richter, 1941; Emery, 1949; Linden, 1959). However, the hypocenter parameters as reported by earthquake catalogues published in the earlier half of the 20th century, and which are still used today, are based on data from bulletins of some of the seismic stations operated at the time and on approximate notions of patterns that govern the propagation of seismic waves in the Earth's interior.

At present, the modern projects like the ISC-GEM (International Seismological Centre–Global Earthquake Model) catalogue and the EuroSeismos (ES) project enabled researchers to study bulletins and seismograms from stations that have been operated since the beginning of the instrumental period. This makes it possible to estimate earthquake hypocenter parameters from instrumental data, while felt data only had been available until recently for such estimation. Examples are (Bungum et al., 2009) who refined the hypocenter parameters for the Fennoscandian earthquake of October 23, 1904 (MS = 5.4) (the Oslofjord Earthquake), (Nikonov & Chepkunas, 2009) who sought to more accurately determine the parameters of the Sysola earthquake of January 13, 1939 in the northern European part of Russia, and Niemz, P. & Amorèse, D. who did a similar study for the November 10, 1935 earthquake near Montserrat (Niemz & Amorèse, 2016).

A large amount of work was done by collaborators at the International Seismological Centre (Storchak et al., 2013) to develop the standardized ISC-GEM Global Instrumental Earthquake Catalogue (1904-2014). The ISC-GEM catalogue is the result of a special effort to adapt and substantially extend and improve currently existing bulletin data for large earthquakes (magnitude 5.5 and above) to serve the requirements of the specific user group who assess and model seismic hazard and risk (Storchak et al. 2013). For earthquakes recorded during 1904-1919, they digitized body wave arrival times, as well as amplitudes and periods for various phases using a multitude of sources with a view to relocation and magnitude revision using the approaches described in (Di Giacomo et al., 2015a).

However, not all earthquakes that have occurred in the Arctic during the early 20th century are found in the ISC-GEM catalogue, with some being even absent from the Bulletin of the ISC (International Seismological Centre, 2015) and the ISS (International Seismological Summary) Bulletin. This study is a review of all data on the earthquakes that have been recorded in the Arctic for the period between 1904 and 1920 using a number of sources. We have restricted our study to the earthquakes that occurred before 1920 since errors in determining the hypocenter parameters are more likely for them due to a small number of actual seismic stations and absence of accurate seismic velocity models at that time. We refined their hypocenter parameters based on the same velocity model, the same location algorithm, and all available station bulletins.

Description of Dataset and Methods

We used data from (Tams, 1922; Gutenberg and Richter, 1941; Linden, 1959), the ISS Bulletin between 1917 and 1920, and the ISC-GEM catalogue, to make a preliminary earthquake catalogue for the period between 1904 and 1920 for events whose epicenters lie above 70° N (Table 1). The preliminary catalogue contains a total of 25 earthquakes, with most of these being borrowed from different sources and having different hypocenter parameters. Eight earthquakes are present in the ISC-GEM catalogue. Nearly all epicenters are in the Eurasian Arctic region and most of them lie at mid-oceanic ridges.

For each earthquake we looked for the times of arrival in the bulletins of the seismic stations that were operated during that time period. Seismic station bulletins were collated using data from the ISC-GEM catalogue, the EuroSeismos project, and the Geophysical Survey of the Russian Academy of Sciences, and the Russian State Library.

Table 1. Seismic catalogue of earthquakes in an area of the Arctic region from early 20th century to 1920

No	Date dd.mm.yyyy	Origin time hh:mm:ss.0	Hypocenter			Magnitude	Source of the hypocenter
			φ,°	λ,°	h, km		
1	2	3	4	5	6	7	8
1	09.10.1904	13:52:00.0	73.5	-5.6			Tams (1922)
2	19.03.1906	07:57:00.0	73.8	9.1			Tams (1922)
		07:56:59.9	71.71	-6.14	15	Mw(ISC-GEM)=6.26	ISC-GEM Catalogue
3	08.07.1908	12:50:00.0	82.9	-5.4			Tams (1922)
4	14.10.1908	14:56:00.0	81.5	28.7		MS(PAS)=6.6 M=6.25 Mw(ISC-GEM)=6.61	Tams (1922)
		14:56:18.0	82.0	30.0	35		Gutenberg and Richter (1941)
		14:56:22.0	81.5	16.0			Linden (1959)
		14:56:14.4	82.64	23.62	10		ISC-GEM Catalogue
5	10.04.1909	18:46:54.0	77.5	128.0	35	M(GUTE)=6.6 MS(PAS)=6.6 M=6.5 Mw(ISC-GEM)=6.71	Gutenberg and Richter (1941)
		18:46:58.0	78.0	128.0			Linden (1959)
		18:46:54.3	78.54	129.16	10		ISC-GEM Catalogue
6	04.12.1911	14:39:00.0	79.0	26.2			Tams (1922)
7	25.01.1912	01:37:00.0	79.9	2.6			Tams (1922)
8	19.02.1912	10:32:56.0	71.0	-158.6		M=5.0	Linden (1959)
9	13.04.1912	02:40:00.0	86.4	94.6		MS(PAS)=5.6 M=5.0	Tams (1922)
		02:39:42.0	80.0	100.0	35		Gutenberg and Richter (1941)
		02:39:36.0	78.9	107.9			Linden (1959)
10	07.06.1914	16:24:00.0	73.0	119.0		M=5.25	Linden (1959)
11	04.11.1914	12:54:00.0	73.5	-3.0		M=5.5	Tams (1922)
		12:52:55.0	74.0	-2.0			Linden (1959)
12	05.11.1914	08:00:40.0	75.5	5.0		M=5.5	Linden (1959)
13	01.06.1915	14:43:54.0	78.5	8.0	35	M(GUTE)=6.6 MS(PAS)=6.8 M=5.75 Mw(ISC-GEM)=6.54	Gutenberg and Richter (1941)
		14:43:45.0	77.0	7.0			International Seismological Summary (ISS)
		14:43:00.0	82.0	8.0			Tams (1922)
		14:43:57.0	78.5	10.0			Linden (1959)
		14:44:03.3	77.30	9.09	10		ISC-GEM Catalogue
14	02.06.1915	23:24:04.0	77.5	2.0		M=4.5	Linden (1959)
15	16.09.1915	10:21:44.0	80.0	-8.0		M=4.5	Linden (1959)
16	30.09.1915	14:31:20.0	77.0	12.0		M=4.5	Linden (1959)
17	11.05.1916	03:05:00.0	79.0	-2.0		M=4.25	Bulletin of the seismic station "Pulkovo" (PUL)
		03:05:52.0	79.4	-1.0			Linden (1959)
18	06.12.1916	22:17:00.0	81.0	61.4		MS(PAS)=5.8 M=5.25 Mw(ISC-GEM)=5.75	Tams (1922)
		22:17:12.0	87.0	48.0	35		Gutenberg and Richter (1941)
		22:17:05.0	88.0	40.0			Linden (1959)
		22:17:14.0	87.20	44.86	10		ISC-GEM Catalogue
19	14.05.1917	06:57:00.0	72.0	-2.8			International Seismological Summary (ISS)
		06:57:00.0	74.8	-6.7			Tams (1922)
20	21.08.1917	10:44:10.0	72.0	-2.8		M=5.0 Mw(ISC-GEM)=5.67	ISS
		10:43:00.0	76.1	-7.8			Tams (1922)
		10:44:13.0	71.4	-7.8			Linden (1959)
		10:44:21.4	71.43	-3.50	10		ISC-GEM Catalogue

Table 1. (Continued)

No	Date dd.mm.yyyy	Origin time hh:mm:ss.0	Hypocenter			Magnitude	Source of the hypocenter
			φ,°	λ,°	h, km		
1	2	3	4	5	6	7	8
21	27.01.1918	02:51:00.0	64.8	35.3		M=4.25	ISS
		02:51:00.0	73.2	12.2			Tams (1922)
		02:51:07.0	72.7	7.8			Linden (1959)
22	20.10.1918	05:44:55.0	72.0	-2.8			ISS
23	30.11.1918	06:48:40.0	71.0	132.0		MS(PAS)=6.2 M=6.0	Gutenberg and Richter (1941)
		06:48:38.0	71.2	134.0		MS(ISC)=6.4	Linden (1959)
		06:48:47.0	70.56	130.44	15	Mw(ISC-GEM)=6.52	ISC-GEM Catalogue
24	02.02.1919	20:02:50.0	72.0	-2.8		M=5.5 Mw(ISC-GEM)=6.07	International Seismological Summary (ISS)
		20:02:00.0	72.0	-18.5			Tams (1922)
		20:02:57.0	72.0	-8.0			Linden (1959)
		20:03:05.2	71.58	-5.05	10		ISC-GEM Catalogue
25	12.09.1919	14:26:37.0	72.0	-2.8			ISS

All detected instrumental onset times were compared with the ak135 velocity model (Kennet et al., 1995) in order to remove those onset times that were inappropriate for P and S and to identify the phase that was defined as the Rayleigh wave maximum (LRM). This had to be done because of the following three reasons. Firstly, some bulletins report only the arrival times of unidentified seismic phases and the times of the onset of "Maximum Wave." Secondly, as already mentioned in the work (Abe, 1988; Bungum et al., 2009): "Because of the low sensitivity and limited resolution of the seismographs at these early years of instrumental earthquake observations, many of the reported onsets are related to surface-wave observations and cannot be used for an instrumental location."

Thirdly, the timekeeping accuracy was low at that time. In the absence of a centralized worldwide timekeeping service at the beginning of the 1900s, early seismometers were often housed at or in the vicinity of astronomical observatories where the time was regularly determined using meridian transit observations (Storchak et al., 2015).

The amplitudes and respective periods of Rayleigh waves for some earthquakes were used to find MS magnitudes. The Rayleigh wave periods were between 10 and 60 s and the epicentral distances were 20°–160°. We used the Prague formula (Vanek et al., 1962):

$$M_S = log(\frac{A}{T})_{max} + 1.66log(\Delta) + 3.3 \quad (1)$$

Event Location Algorithm

The NAS program (Asming & Prokudina, 2016; Asming et al., 2016) performs phase association using a new grid search method based on the idea of generalized beamforming (Kvaerna & Ringdal, 1996) in combination with the minimization of origin time estimation residuals. NAS makes a grid search in a limited region of time and space around a point of a preliminary event location (prototype event). It takes a circle of a relatively large radius (value of 250 km was used in this study) around the preliminary location of the event's epicenter. This circle is considered as a search area for a more precise location. It is covered by a set of overlapping circles of smaller radii, thus forming a grid of round cells. A rating function R(c,t) is computed for each cell c of the grid, with the hypothesis that the event has occurred in this particular cell at time t. To describe this function in more detail, let a phase (P or S) arrived at the i-th seismic station at time t_i. Let $r_{i0}(c)$ and $r_{i1}(c)$ be the minimal and maximal distances from the i-th station to the cell c. If the event actually occurred in this cell, its origin time has to be within the limits $[t_i - r_{i1}(c)/v, t_i - r_{i0}(c)/v]$, where v is the wave's apparent propagation velocity. Thus, following Kvaerna and Ringdal (1996), we can define the total rating function as:

$$R(c,t) = \sum_i S(t, t_i - r_{i1}(c)/v, t_i - r_{i0}(c)/v),$$

where $S(t,t_a,t_b)$ is the step function:

$$S(t,t_a,t_b) = \begin{cases} 1, t \in [t_a, t_b] \\ 0, otherwise \end{cases}$$

It is assumed that each phase that can correspond to an event that occurred in a given cell at a given time makes unit contribution to the rating function for this cell. This would work well if all measurements $\{t_i\}$ were accurate and the knowledge of v was exact. In the present case we measure onsets with some uncertainty Δt_{onset}, and we know the velocity with uncertainty Δv. Therefore, the interval for the origin time must be expanded as:

$$[t_i - r_{i1}(c)/v - \Delta t, t_i - r_{i0}(c)/v + \Delta t],$$

where $\Delta t = \Delta t_{onset} + r \cdot \Delta v / v^2$.

Instead of $S(t, t_a, t_b)$ we introduce a trapezoidal-shaped function T:

$$T(t,t_a,t_b,\Delta t) = \begin{cases} 1-(t_a - t)/\Delta t, t \in [t_a - \Delta t, t_a] \\ 1, t \in [t_a, t_b] \\ 1-(t - t_b)/\Delta t, t \in [t_b, t_b + \Delta t] \\ 0, otherwise \end{cases}$$

And finally we define the rating function as:

$$R(c,t) = \sum_i T(t, t_i - r_{i1}(c)/v, t_i - r_{i0}(c)/v, \Delta t)$$

This is a sum of contributions of all phases that can correspond to an event occurred in a given cell at a given time. Those phases that belong to the exact time interval $\left[t_i - r_{i1}(c)/v, t_i - r_{i0}(c)/v\right]$ add 1 to the sum, whereas the ones that are outside the main interval but may correspond to the event add smaller values.

The rating functions are computed for an interval of possible origin times [t_0-ΔT, t_0+ΔT], where t_0 is a preliminary estimation of origin time of the prototype event, ΔT is some large time reserve (30 sec value was taken for the events considered in this study). Maximal values of the rating function among the times within the interval are rating estimations for the grid cells. The grid is diminished several times. Each time 75% of the cells with the smallest ratings are excluded and each remaining cell is divided into four smaller ones, to keep the same total cell number. Following, ratings are recalculated.

The grid search is performed several times with different fixed depths (from 0 to 100 km with a step 5 km for the present study). Finally, the cell with the maximal rating is selected. The time t_0 for which the rating function has reached its maximum is considered to be the event's origin time.

A weight is attributed to each phase, which is taken equal to the contribution of the phase to the rating function for an estimated origin time t_0:

$$w_i = T(t_0, t_i - r_{i1}(c)/v, t_i - r_{i0}(c)/v, \Delta t).$$

The phases with non-zero weights are taken into account for the next stage of processing. This approach enables us to screen out phases with non-realistic times. This procedure may be proven useful for processing onsets obtained from old analog seismograms.

In the next step the phases and the weights are used to improve the location by searching the point where the residual of origin time estimation computed for this point is minimal.

We briefly remind the method. Again, let t_i be an arrival time of a P or S wave to the i-th station with coordinates (φ_i, λ_i). Suppose that the event coordinates are (φ, λ, h), where φ is latitude, λ is longitude and h is depth. Origin time estimations are:

$$t_{0i}(\varphi, \lambda, h) = t_i - TT(\varphi_i, \lambda_i, 0, \varphi, \lambda, h),$$

where TT is the travel time between two points for this wave type. If (φ, λ, h) are true coordinates of the event, onsets are measured exactly and TT is known, then all t_{0i} must be the same and equal to the true origin time.

The average origin time estimation is:

$$\bar{t}_0(\varphi, \lambda, h) = \sum_i w_i \cdot t_{0i} \Big/ \sum_i w_i,$$

where w_i are weights of the phases and its standard deviation is:

$$\sigma(\varphi,\lambda,h) = \sqrt{\sum_i w_i \cdot (\bar{t}_0 - t_{0i})^2 \Big/ \sum_i w_i}$$

To locate the event we find the point (φ_{event}, λ_{event}, h_{event}) where σ is minimal:

$$\sigma_{\min} = \min_{(\varphi,\lambda,h)} \sigma(\varphi,\lambda,h) \text{ , } (\varphi_{event}, \lambda_{event}, h_{event}) = Arg \min_{(\varphi,\lambda,h)} \sigma(\varphi,\lambda,h)$$

Usually for bulletin processing the coordinates obtained by the minimization differ very slightly from those obtained by the grid search. As a rule, the final point lies within the same cell of the grid where the maximum of the rating function was reached. Thus, it would seem that this step of processing could be omitted. Nevertheless, the function $\sigma(\varphi,\lambda,h)$ appeared to be useful for the estimation of the confidence region.

We estimate the confidence region by solving the inequality (with fixed depth h_{event}):

$$\sigma(\varphi,\lambda,h_{event}) \le \sigma_0 \text{ ,}$$

where σ_0 is a border value of the standard deviation, depending on the expected errors (uncertainties) of the onset time measurements Δt_{onset} and the uncertainties of travel velocities Δv.

It is estimated for a given set of arrivals $\{t_i\}$ as: (r_i is distance from the i-th station to the event):

$$\Delta t_i = \sqrt{\Delta t_{onset}^{\ 2} + \left(\frac{r_i \cdot \Delta v}{v^2}\right)^2} \text{ (summary onset time uncertainty)}$$

$$\sigma_0 = \sqrt{\sum_i (w_i \cdot \Delta t_i)^2 \Big/ \sum_i w_i} \text{ .}$$

The region is estimated numerically and is approximated by an ellipse.

To determine the depth range a similar inequality is solved for h:

$$\min \sigma(\varphi,\lambda,h) < \sigma_0$$

The onset picking errors (Δtonset) were taken Δtp = Δts = 3 s. That is reasonable for old seismograms. The velocity uncertainties were taken Δvp = Δvs = 0.15 km/s. For the relocation was used ak135 velocity model (Kennet et al., 1995). Verification of the modified method was performed using data on four nuclear explosions that occurred in the area of the Novaya Zemlya archipelago and in the north of the European part of Russia (Morozov et al., 2018).

It is impossible to compute the focal depths reliably, because of the lack of P and S measurements at very close stations as well as the absence of teleseismic phases. For this reason the epicenters were located using a fixed depth for each earthquake. Since the magnitudes are high, the respective rupture zones cannot have been shallower than some definite depths (Hmin). We found the lowest possible depth of focus for an earthquake from its magnitude using the relation from (Riznichenko, 1979):

$$M \leq 3.3 \log h + 3.1` \tag{2}$$

All the depths are depths beneath the seafloor. Next, consider the results of our events relocation. We will focus in detail on the results for only some earthquakes.

The Earthquake of October 9, 1904

The earthquake of October 9, 1904 was the first earthquake to have been recorded by seismic stations. It is only mentioned by (Tams, 1922) with the hypocenter parameters indicated in Table 1. We collected bulletins from five seismic stations whose observed seismic onsets we could associate with the earthquake of October 9, 1904. The distribution of the stations is shown on the Figure 2.

The relocation procedure used the following data: Nstations = 8; Nphases = 13; Gap = 239°. The epicentral distances varied between 1743 and 5011 km. No onsets of the LRM phase with relevant amplitude and period measurements have been found in the bulletins, hence it has not been feasible to calculate magnitudes and the minimum depth of focus. We used a fixed depth of 10 km. The small number of stations and their great distances from the events, as well as the narrow azimuthal coverage, were unfavorable for determining the epicenter parameters accurately. According to the new location of the 1904 earthquake, the epicenter is shifted 60 km to the south-east with respect to the initial coordinates (see Figure 2 and Figure 3).

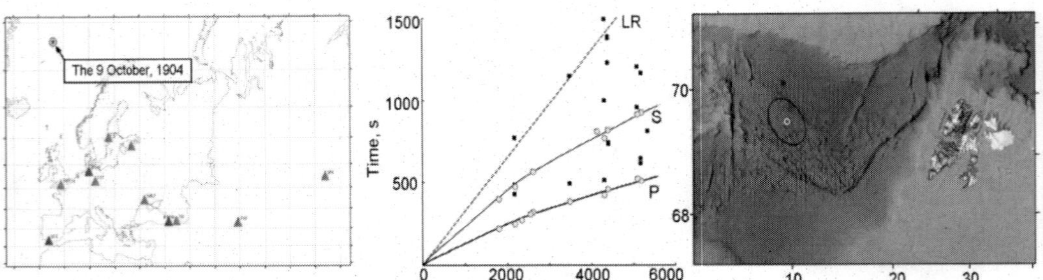

Figure 2. The map (left) of the seismic stations that recorded the October 9, 1904 earthquake. The circled dot marks the relocated epicenter; green triangles show the stations used for the location; red triangles mark the stations, which onsets were rejected due to possible large timing errors or misinterpretation. The plot in the middle shows all available onset times for the earthquake. Theoretical (ak135 velocity model) P wave travel time is plotted with a thick line, S is thin line, and surface wave is dashed line. The onsets used for location are shown as hollow circles and rejected onsets are marked as black squares. The map on the right shows the relocated epicenter of the earthquake. The solid circle indicates the position of the epicenter, calculated by (Tams, 1922); the circled dot marks the new position. The error ellipse is plotted with a black line.

The error ellipse has an area of 24 768 km^2. Since the error ellipse partly covers the area of the Mona Mid-Oceanic Ridge we can assume that the event has occurred in the area (Table 1). Tams used data of the POT station (Potsdam) to find maximum surface-wave amplitudes on the north–south component (860 μm) and on the east–west component (560 μm) at periods of T = 11–14 s. Because the arrival times of the maximum phases for which the measurements have been made were not available, we could not verify whether these were really the LRM phases. We calculated the range of possible magnitude values MS=7.3–7.6 from (1), as well as using the Rayleigh wave amplitude ratio for different components $\frac{A_Z}{A_{N,E}} = \sqrt{2}$. However, the resulting values cannot be trusted owing to the factors described above.

The Earthquake of April 10, 1909

The earthquake is mentioned by (Gutenberg & Richter, 1941; Linden, 1959) with those hypocenter parameters contained in Table 1. The epicenter parameters were revised in the ISC-GEM catalogue and the magnitude was determined as Mw(ISC-GEM)=6.71. This magnitude, Mw = 6.71, is one of the highest for this region during the entire observation period.

We collected bulletins from twelve seismic stations whose observed seismic onsets we could associate with the earthquake of April 10, 1909. The ak135 velocity model was used to select 21 P and S phases at the twelve seismic stations for the relocation procedure. Thus, the relocation procedure used the following data: Nstations = 11; Nphases = 18; Gap = 344°. The epicentral distances varied between 3600 and 5396 km. We used a fixed depth of 13 km as the minimum possible depth of focus for an earthquake of magnitude Mw(ISC-GEM) = 6.71 in accordance with (2).

According to the new location of the 1909 earthquake, the epicenter was located near epicenters from (Gutenberg & Richter, 1941; Linden, 1959), and the ISC-GEM catalogue. The initial epicenter from (Gutenberg & Richter, 1941; Linden, 1959), and the ISC-GEM catalogue falls in the error ellipse of the new epicenter. The epicenter is located in the eastern Gakkel Ridge at the boundary between the oceanic part of the ridge and the shelf (see Figure 3). The error ellipse has an area of 223 774 km^2. Since the error ellipse partly covers the area of the Gakkel Ridge we can assume that the event has occurred in the area. The error ellipse does not include other significant seismic zones (Table 2).

The Earthquake of February 2, 1919

This earthquake is mentioned by (Tams, 1922; Linden, 1959), and in the International Seismological Summary with the hypocenter parameters of Table 1. The epicenter parameters for this earthquake have been revised for the ISC-GEM catalogue and magnitude Mw(ISC-GEM)=6.07 has been determined.

We collected bulletins from 34 seismic stations whose observed seismic onsets could be associated with the earthquake of February 2, 1919. The ak135 velocity model was used to select 39 P and S onset times as recorded at 26 seismic stations for the relocation procedure. The relocation procedure thus used the following data: Nstations = 23; Nphases = 36; Gap = 166°. The epicentral distances varied between 1666 and 5214 km. We used a fixed depth of 8 km as the minimum possible depth of focus for an earthquake of magnitude Mw(ISC-GEM) = 6.07 in accordance with (2). According to the new location of the 1919 earthquake, the epicenter was located at the junction of the Jan Mayen Fracture Zone and the Mona Mid-

Oceanic Ridge (see Figure 3) at different distances from earthquakes due to other source zones. The epicenter from the ISC-GEM catalogue falls in the error ellipse of the new epicenter. The error ellipse has an area of 5 872 km^2 (Table 1).

RESULTS

The instrumental period for the Arctic region started with the recording of the October 9, 1904 earthquake. Many large earthquakes were recorded during the first two decades of the 20th century (Figure 3, Table 2). These are important both for understanding the general patterns of seismicity in the Arctic and for the assessment of earthquake hazard.

The seismic stations of the worldwide network recorded earthquakes of magnitudes equal to or larger than 6.0 in the Arctic region during the period between 1904 and 1911. Since 1912, it became possible to record 5.0+ earthquakes. Our relocation resulted in large error ellipses for most earthquakes owing to the narrow azimuthal coverage and the great epicentral distances for the seismic stations, which is characteristic for the Arctic region during that time period (Figure 3, Table 2).

However, this does not prevent us from ascribing the epicenters to definite seismic zones. Most of the earthquakes are confined to major Arctic seismic zones, which include mid-oceanic ridges, the Svalbard archipelago, and the Laptev Sea shelf. Note, there is a single "unusual" earthquake occurred on October 14, 1908, whose magnitude was Mw(ISC-GEM)= 6.6 (Figure 3). The earthquake occurred in the shelf of the Barents Sea in the continent-ocean transition zone. Since the error ellipse partly covers the area of the Franz Victoria Graben we can assume that the event has occurred in the area, because, first, the ellipse does not include other significant seismic zones in the Barents-Kara area, i.e., the Gakkel and Knipovich Mid-Oceanic Ridges and the Svalbard archipelago. Secondly, large earthquakes have been recorded during the instrumental period in the Franz Victoria Graben region (Table 3).

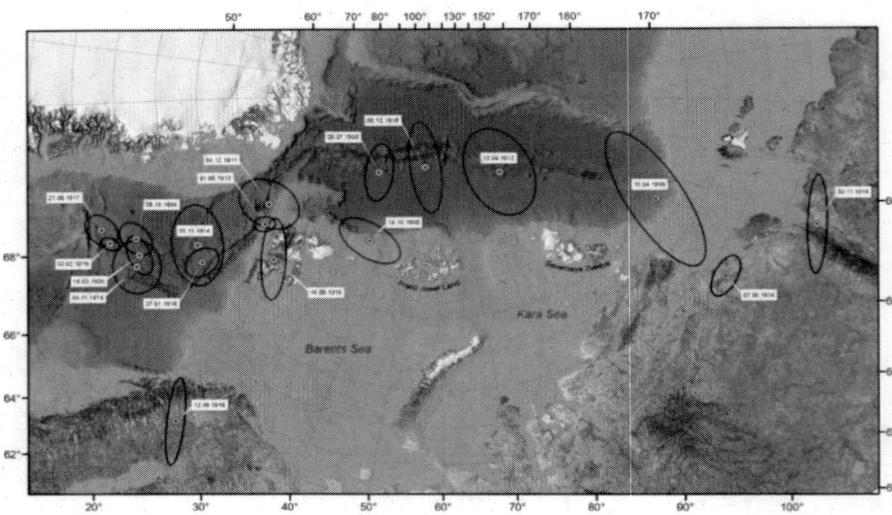

Figure 3. Map showing the relocated epicenters of earthquakes of the Arctic in the Early Twentieth Century. Symbols: circle with a dot, the new coordinates and error ellipse.

Table 2. Catalogue of relocated earthquakes in an area of the Arctic region

N	Date dd.mm.yyyy	Time hh:mm:ss.0	Hypocenter			N_{st}/N_{def}	Gap, °	Distances*, km	Error ellipse			Magnitude	Note
			φ,°	λ,°	**h, km**				**Az_{Major}**	**R_{minor}**	**R_{major}**		
1	09.10.1904	13.52:00.5	73.12	-4.24	10f	8/13	239	1743 - 5011	80	73.3	107.9	MS=7.3–7.6	Approximate determination of the MS
2	19.03.1906	07.56:47.5	73.85	-1.33	9f	14/23	244	1982 - 4928	100	59.2	120.5	Mw(ISC)=6.26	
3	08.07.1908	12.50:02.8	84.87	22.61	10f	12/18	244	2950 - 5231	150	75.9	169.9	MS=6.3	
4	14.10.1908	14.56:24.1	81.44	34.93	12f	17/26	329	2419 - 4271	100	100.7	193.0	Mw(ISC)=6.61	
5	10.04.1909	18.47:02.2	78.12	123.01	13f	11/18	343	3600 - 5396	30	156.0	456.6	Mw(ISC)=6.71	
6	04.12.1911	14.38:47.9	79.70	3.48	8f	8/15	325	2336 - 3592	70	133.1	177.8	MS=5.8 – 6.0	Approximate determination of the MS
7	25.01.1912	-	-	-	-	2/3	-	-	-	-	-	-	The relocation procedure was not carried out, not enough data
8	19.02.1912	-	-	-	-	3/6	-	-	-	-	-	-	-«-«-«-
9	13.04.1912	02.39:45.9	85.05	94.93	5f	6/8	325	3127 - 4460	10	181.0	270.3	MS=5.1	
10	07.06.1914	16.24:02.5	73.03	116.29	10f	8/13	253	2359 - 5155	80	63.4	129.7	MS=6.3	
11	04.11.1914	12.54:14.7	72.42	0.22	5f	7/11	307	2269 - 4063	120	130.9	157.6	MS=5.0	
12	05.11.1914	08.00:38.3	75.54	2.63	5f	4/5	233	2453 - 4831	120	141.2	241.3	M=5.25	
13	01.06.1915	14.43:51.2	78.88	7.45	12f	16/28	155	2228 - 5026	30	44.3	59.3	Mw(ISC)=6.54	
14	02.06.1915	-	-	-	-	2/4	-	-	-	-	-	-	-«-«-«-
15	16.09.1915	10.22:07.0	77.91	15.78	4f	3/4	238	2453 - 4831	140	67.1	244.8	M=4.5	Poor determination
16	30.09.1915	-	-	-	-	1/1	-	-	-	-	-	-	-«-«-«-
17	11.05.1916	-	-	-	-	1/2	-	-	-	-	-	-	-«-«-«-
18	06.12.1916	22.17:17.2	86.09	47.69	7f	6/8	229	2787 - 6062	160	83.5	273.7	Mw(ISC)=5.75	
19	14.05.1917	-	-	-	-	1/2	-	-	-	-	-	-	-«-«-«-
20	21.08.1917	10.44:09.1	71.83	-8.07	7f	8/11	271	1722 - 3400	70	75.2	102.4	Mw(ISC)=5.67	
21	27.01.1918	02.50:40.4	75.18	6.10	6f	8/10	302	1777 - 3283	0	74.5	102.1	MS=5.4	
22	20.10.1918	-	-	-	-	3/4	-	-	-	-	-	-	-«-«-«-
23	30.11.1918	06.48:44.5	70.38	130.80	11f	9/12	197	3492 - 6309	70	52.7	288.7	Mw(ISC)=6.52	
24	02.02.1919	20.03:00.4	71.88	-5.42	8f	23/36	166	1666 - 5214	50	35.7	52.3	Mw(ISC)=6.07	
25	12.09.1919	14.26:32.9	68.02	22.06	5f	4/5	242	1865 - 7532	150	47.8	263.7	-	Not relevant to the study area

* - the range of epicentral distances (from … to …); -«-«-«- – The relocation procedure was not carried out, not enough data.

Table 3. Seismic catalogue of earthquakes in an area of the Franz Victoria Graben

No	Date dd.mm.yyyy	Origin time hh:mm:ss.0	Hypocenter			Magnitude	Source of the hypocenter (International Seismological Centre, 2015)
			φ,°	λ,°	h, km		
1	18.02.1948	20:29:47.0	82.50	41.50	-	M=6.5 by Linden (1959)	ISC Bulletin
2	26.09.1948	05:51:12.0	82.50	41.50	-	-	ISC Bulletin
3	22.11.1948	23:32:48.0	82.50	41.50	-	-	ISC Bulletin
4	13.03.1967	21:44:07.7	82.23	39.60	33.0f	m_b(ISC)=4.3	ISC-EHB Bulletin
5	14.03.1967	07:50:14.9	82.38	39.10	13.0f	m_b(ISC)=4.7	ISC-EHB Bulletin
6	25.06.1975	10:14:57.6	82.35	38.72	33.0f	m_b(ISC)=4.6	Reviewed ISC Bulletin (Prime hypocentres)

Modern studies of the low magnitude seismicity in the “continent-ocean” transition zone corroborate seismic activity in the Franz Victoria Graben (Morozov et al., 2015; Antonovskaya et al., 2018). The prevailing geodynamic factor responsible for the occurrence of weak earthquakes is likely the isostatic compensation of avalanche sedimentation in the continent-ocean transition zone (Zayonchek et al., 2010).

A reverse situation concerns the earthquake of April 13, 1912, with magnitude MS=5.1. According to (Tams, 1922; Gutenberg & Richter, 1941; Linden, 1959) the epicenter of the earthquake was in the Severnaya Zemlya Archipelago. However, as a result of the relocation, it was obtained that the epicenter was to the north, in the area of the Gakkel Ridge. It is possible that the earthquake took place exactly in the area of the Gakkel Ridge because during the entire instrumental period strong earthquakes from the region of the Severnaya Zemlya archipelago were not recorded (Morozov et al., 2018; Antonovskaya et al., 2018).

We have not succeeded in finding onsets in available bulletins of seismic stations for seven earthquakes of twenty three. It therefore remains an unresolved issue whether these earthquakes actually occurred in the Arctic region. The earthquakes call for more study involving an analysis of actual records rather than bulletins. We thus hope that the present results improve our knowledge of seismic activity in the Arctic. Registration of each earthquake from the Arctic at the beginning of the twentieth century was a unique event. Because the information on historical earthquakes for the Arctic region is extremely scanty, while the data on paleo-earthquakes are practically inaccessible. It follows that the earthquakes that have been recorded during the period of instrumental observation are the most significant.

SEISMICITY OF THE BARENTS-KARA REGION: RELOCATED EVENT CATALOGUE FROM EARLY 20TH CENTURY TO 1989

The Barents-Kara region, fringing the northern rim of the Eurasian continent, cover one of the widest continental shelves in the world (Figure 4). The shelf of the Barents and Kara Seas is the most studied in the Arctic region. However, the shelf has not been studied enough to elucidate the crustal dynamics and evolution in continent-ocean transition zones, requiring further deep investigations in this region. Previous geological and geophysical research in the Barents and Kara sea region indicated its oil and gas potential created opportunity to the discovery of several large fields (Shtokmanovskoye, Ledovoye, Prirazlomnoye and others on the Barents Shelf; Rusanovskoye, Leningradskoye on the Kara Shelf) (Ivanova et al., 2006).

The shelf of the Barents and Kara Seas is characterized by low seismicity. Tectonic earthquake's activity is observed mostly at its margins, in particular in the Svalbard archipelago, and along the "continent-ocean" transition zone. There are significant difficulties in the seismic monitoring of the Barents-Kara region. Widely spaced seismic stations cannot register small-magnitude earthquakes, and poor station coverage makes it difficult to determine earthquake hypocenters and focal mechanisms accurately. According to the studies (Assinovskaya, 1994; Avetisov, 1996), the completeness threshold of earthquakes (M) for the Barents-Kara region in the period before 1950 was M = 5.5-6.0, in the period 1951-1990 M varied from 5.0 to 3.9. The value M even decreased to 2.7 due to development of seismic networks NORSAR (NOA network), Institute of Seismology (University of Helsinki) (ISUH, HEL network), Sodankylä Geophysical Observatory (University of Oulu) (FN network), Kola Branch of GS RAS (KOLA network), Federal Center for Integrated Arctic Research (AH network doi.org/10.7914/SN/AH) and other in last decades (Figure 4) (Morozov et al., 2018).

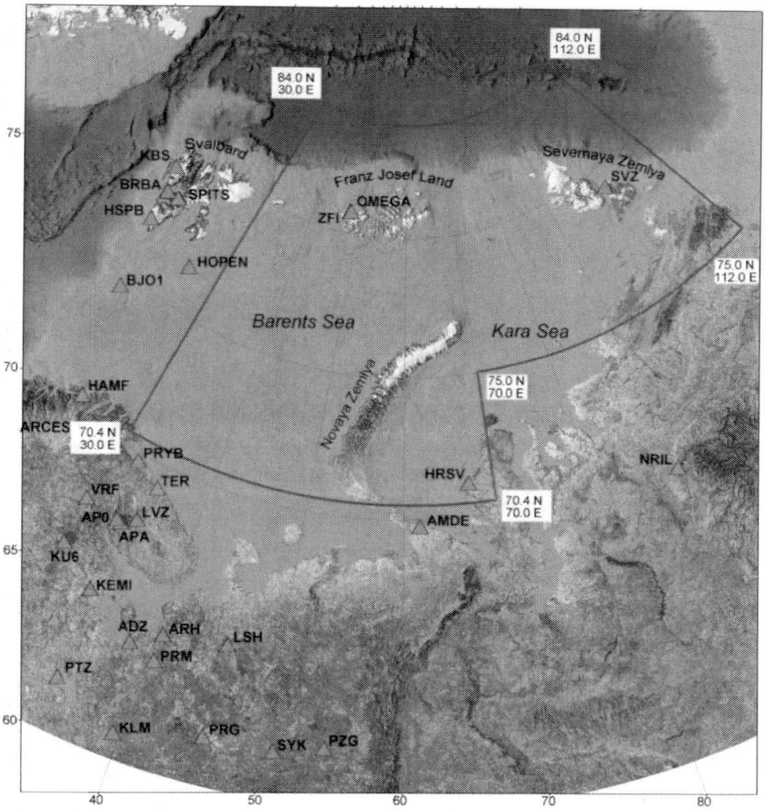

Figure 4. Map showing the location of the Barents-Kara region, combined with a bathymetric data (www.ngdc.noaa.gov). The red line indicates the area of our study. Location of seismic stations NORSAR (NOA network), Institute of Seismology (University of Helsinki) (ISUH, HEL network), Sodankylä Geophysical Observatory (University of Oulu) (FN network), Kola Branch of GS RAS (KOLA network), Federal Center for Integrated Arctic Research (AH network) and other.

A common practice in seismology is a posterior reprocessing of the events when data of all seismic stations are collected, velocity models are updated and location techniques are worked out for the purpose of creating a relocated catalogue. We collected all available seismic bulletins from the global network including Soviet seismic stations. The location was performed by applying a modified method of generalized beamforming (Kvaerna & Ringdal, 1996). We have considered several travel time models and selected one with the best location accuracy for ground truth events. Verification of the modified method and selection of the travel time model were performed using data on four nuclear explosions that occurred in the area of the Novaya Zemlya archipelago and in the north of the European part of Russia. Similar studies were carried out earlier for the Novaya Zemlya archipelago (Morozov et al., 2017). Here we will continue our research for the entire Barents-Kara region.

From a practical point of view, our research studies are focused on ensuring the environmental safety in the Arctic zone of Russia and clarifying the seismic hazard map of the Barents and Kara sea region. The existing world experience shows that insufficient consideration of geodynamic factors within the process of the development and field exploitation leads to sufficiently large economic losses.

Description of Dataset and Methods

The area of our study is shown in Figure 4. The area includes the central and eastern part of the Barents Sea, the Kara Sea and their continental shelf, the archipelagos Novaya Zemlya, Severnaya Zemlya and Franz Josef Land. We did not specifically include seismically active areas, such as the Svalbard, Knipovich and Gakkel Ridges because we believe that they are well covered by seismic observation and have been well studied. The region considered in this work, in opposite, always had a weak instrumental representativity compared to western part of the Barents Sea region. A catalogue of seismic events was created for the Barents and Kara sea region using the data of International Seismological Centre (ISC; International Seismological Centre, 2015) from early 20th century to 1989 (Table 4).

A Soviet nuclear test site was operated on the Novaya Zemlya archipelago until 1990. At Novaya Zemlya, a total of 130 tests were carried out high in the atmosphere, at low levels above water, at the water/air interface, below water, and underground (Khalturin et al., 2005). Nuclear explosions were excluded from the catalogue using the data of Databases of Nuclear Tests (Yang et al., 2003), as well as seismic events that occurred in the area of the nuclear test site (Table 4).

Seismic station bulletins from the global network were collated using data from the ISC Bulletin (International Seismological Centre, 2015), the ISC-GEM project (http://storing.ingv.it/bulletins/ISC-GEM/) and the EuroSeismos (ES) project (http://storing.ingv.it/es_web/).

The ISC-GEM Global Instrumental Earthquake Catalogue (1904-2014) is the result of a special effort to adapt and substantially extend and improve currently existing bulletin data for large earthquakes (magnitude 5.5 and above) to serve the requirements of the specific user group who assess and model seismic hazard and risk (Storchak et al., 2013). The ES project is oriented to research, recovery, restoration, reproduction and scientific valorization of the historical assets of the instrumental observation of earthquakes in the whole Euro-Mediterranean area.

Table 4. Seismic catalogue of events in an area of the Barents and Kara sea region from early 20th century to 1989

No	Date dd.mm.yyyy	Origin time hh:mm:ss.0	Hypocenter			Magnitude	Source of the hypocenter
			φ,°	λ,°	h, km		
1	14.10.1908	14:56:18.0	82.00	30.00	35.0	M_S(PAS)=6.6	ISC Bulletin
2	13.04.1912	02:39:42.0	80.00	100.0	35.0	M_S(PAS)=5.6	ISC Bulletin
3	24.09.1922	12:26:00.0	75.00	100.0	-		ISC Bulletin
4	11.11.1923	14:00:24.0	84.00	100.0	-		ISC Bulletin
5	19.10.1924	15:34:45.0	79.00	110.0	-		ISC Bulletin
6	27.03.1928	17:46:20.0	84.00	100.0	-		ISC Bulletin
7	02.06.1928	20:12:54.0	83.00	70.00	-		ISC Bulletin
8	25.08.1935	04:00:00.0	74.00	36.00	-		ISC Bulletin
9	18.02.1948	20:29:47.0	82.50	41.50	-		ISC Bulletin
10	26.09.1948	05:51:12.0	82.50	41.50	-		ISC Bulletin
11	22.11.1948	23:32:48.0	82.50	41.50	-		ISC Bulletin
12	04.03.1956	03:18:10.0	83.5	112.0	-		Reviewed ISC Bulletin (Prime hypocentres)
13	11.01.1962	12:32:36.0	78.00	30.00	-		ISC Bulletin
14	22.06.1962	07:47:11.0	78.00	48.00	-		ISC Bulletin
15	04.08.1963	18:43:00.0	75.00	38.00	-		ISC Bulletin
16	08.10.1963	10:52:07.0	75.60	87.80	-	M(CGS)=4.7	ISC Bulletin
17	13.03.1967	21:44:07.7	82.23	39.60	33.0f	m_b(ISC)=4.3	ISC-EHB Bulletin
18	14.03.1967	07:50:14.9	82.38	39.10	13.0f	m_b(ISC)=4.7 M(MOS)=5.5	ISC-EHB Bulletin
19	19.08.1967	16:27:58.0	77.00	32.00	0.0		ISC Bulletin
20	20.08.1967	11:31:44.0	84.00	97.00	33.0	M(LAO)=3.7	ISC Bulletin
21	31.03.1968	03:03:59.0	78.90	64.30	33.0	M(LAO)=3.9	ISC Bulletin
22	06.05.1970	07:09:50.0	78.00	31.00	0.0f		Reviewed ISC Bulletin (Prime hypocentres)
23	16.05.1970	13:19:25.0	71.00	40.00	0.0		ISC Bulletin
24	12.12.1974	21:19:46.4	71.96	49.53	0f	m_b(ISC)=4.4	Reviewed ISC Bulletin (Prime hypocentres)
25	03.03.1975	05:44:39.7	81.40	111.1	0.0f	m_b(ISC)=4.1 M(LAO)=4.0	Reviewed ISC Bulletin (Prime hypocentres)
26	25.06.1975	10:14:57.6	82.35	38.72	33.0f	m_b(ISC)=4.6 m_b(NEIS)=4.6 M(LAO)=4.8	Reviewed ISC Bulletin (Prime hypocentres)
27	11.12.1975	05:32:15.0	72.00	45.00	-		ISC Bulletin
28	24.01.1976	19:09:08.6	80.61	88.11	33.0f	m_b(ISC)=4.0	ISC Bulletin
29	27.05.1977	17:50:37.0	77.00	30.00	-		ISC Bulletin
30	07.08.1978	12:48:23.1	79.60	98.85	33.0f	m_b(ISC)=4.5 M_S(ISC)=3.5 m_b(NEIS)=4.5	ISC-EHB Bulletin
31	07.08.1978	13:35:19.7	80.12	97.28	33.0f	m_b(ISC)=4.3 m_b(NAO)=4.4	Reviewed ISC Bulletin (Prime hypocentres)
32	15.11.1978	08:30:04.9	72.57	52.84	0f	m_b(NAO)=4.3	Reviewed ISC Bulletin (Prime hypocentres)
33	19.05.1986	13:52:30.8	75.57	111.4	10.0f	m_b(ISC)=4.6 M_S(ISC)=3.7	ISC-EHB Bulletin
34	01.08.1986	13:56:37.0	72.93	56.06	0f	m_b(ISC)=4.8 M_S(ISC)=3.7 m_b(NAO)=4.5	ISC-EHB Bulletin
35	29.10.1988	02:12:23.0	76.80	31.60	-		ISC Bulletin
36	09.12.1989	12:55:28.2	70.77	33.87	12.2	m_d(BER)=3.9	ISC Bulletin

The ES project coordinated by Graziano Ferrari (SGA – Storia Geofisica Ambiente) and Nicola Alessandro Pino (Istituto Nazionale di Geofisica e Vulcanologia, INGV). For each seismic event placed in Table 4, the bulletins of the Soviet seismic stations (Table 5) were viewed and, in case of presence such seismic event, were involved in processing. Bulletins were obtained from the Geophysical Survey of the Russian Academy of Sciences (GS RAS).

Seismic event location and depth estimation were carried out using the NAS (New Association System) program, which is a part of the NSDL (New System for Detection and Location) system developed in the Kola Branch of GS RAS. The software was used to carry out automated seismicity monitoring of any region using an arbitrary network (Asming & Prokudina, 2016; Asming et al., 2016). We have relocated only seismic events with at least 5 phases from 3 stations. Calculation of magnitudes and earthquake-explosion discrimination of events from Table 4 was not carried out.

Table 5. Parameters of Soviet seismic stations

Station			Lat.	Lon.	elevation/	Start	Instrument	Status
Code	**Network**	**Name, Country**	(°)	(°)	**depth, m**			
APA	KOGSR	Apatity, Russia	67.569	33.405	182	1956 2010	- Guralp-3ESP	open
ARU	II	Arti, Russia	56.429	58.562	260	1976 1999	- Streckeisen STS-1/BB Seismometer	open
KHE		Kheis, Russia	80.617	58.050	22	1957	-	Close (1992)
OBN	II	Obninsk, Russia	55.114	36.569	160	1964	Guralp CMG-3T Seismometer	open
PUL	GE	Pulkovo, Russia	59.773	30.322	89	1906	STS-2	open
SVE	OBGSR	Sverdlovsk, Russia	56.827	60.637	275	1906	-	open

(-) – parameters of stations are not known.

Verification of the Algorithm, Selection of Travel Time Model, Location Error Estimation

The NAS program (Asming & Prokudina, 2016, Asming et al., 2016) performs phase association using a new grid search method based on the idea of generalized beamforming (Kvaerna & Ringdal, 1996) in combination with the minimization of origin time estimation residuals. The algorithm is described in more detail in previous chapter.

Two Novaya Zemlya nuclear explosions on November 2, 1974 and October 24, 1990 and two peaceful nuclear explosions in the north of the European part of Russia on July 18, 1985 and September 6, 1988 were chosen for verification of the algorithm, selection of the travel time model and location error estimation. The 1974 explosion is denoted as event GT5. The 1985, 1988 and 1990 explosions are denoted as event GT1 (Yang et al., 2000). The parameters of the hypocenters were obtained from (Yang et al., 2003).

Several 1D velocity models were tested: BARENTS (Kremenetskaya et al., 2001), BAREY, and BAREZ from (Schweitzer & Kennett, 2007) and NORP (Morozov & Vaganova, 2011). The BARENTS model has been developed for Fennoscandia, the Baltic shield and adjacent areas. In its upper layers, the model is a simplified average of various models developed for parts of the region. Below the Moho the model uses the IASPEI91 model layers (Kennett & Engdahl, 1991).

The BAREY and BAREZ models were adapted from a model developed by (Kremenetskaya et al., 2001) by making minor P-velocity adjustments and varying the VP/VS ratio in the upper mantle. The BAREZ model uses a VP/VS ratio of 1.73 in the crust and a ratio of 1.72 below the Moho. The BAREY model uses a VP/VS ratio of 1.77 in the uppermost mantle (Schweitzer & Kennett, 2007). The NORP model has been developed based on calculation of the crustal velocity structure of two areas of the Russian Plate using inversion of receiver functions (Morozov & Vaganova, 2011).

Seismic station bulletins for the nuclear explosions were collated using data from the ISC and GS RAS bulletins. All available onsets were collected, and several datasets from stations located at certain distance intervals from the event's location were prepared: 0–10°, 0–15°, 0–20°, 0–30°, and 0–90°. The results are summarized in Figure 5. The BARENTS and BAREY models showed sufficient location accuracy for all the datasets. The focal depth was estimated to be equal to zero for all of these variants. We chose the BARENTS model for the relocation because the distance between each obtained location and the true coordinates for this model were slightly smaller than for the other models (Table 6).

Thus, the application of the algorithm implemented in the NAS program together with the BARENTS model gives the most accurate parameters of hypocenters. This allows us to apply the algorithm and model to relocate seismic events registered in the area of the Barents-Kara region.

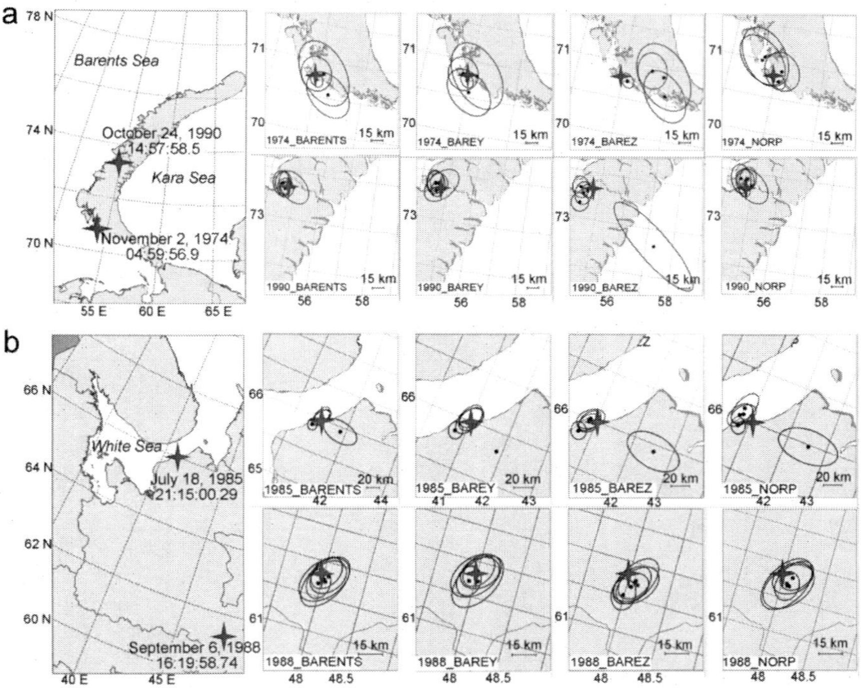

Figure 5. Checking the efficiency of the modified generalized beamforming and the velocity models using the underground nuclear explosions. Maps showing the computed (circles) and true (asterisks) epicenters of the underground nuclear explosions.

Table 6. Checking the efficiency of the modified generalized beamforming and the BARENTS velocity models using the underground nuclear explosions (Yang et al., 2003)

Ground truth* event: underground nuclear explosions					Result of the relocated procedure										
					R**,°	Nst	Nphases	Time hh:mm:ss.0	Hypocenter			Error ellipse			D***, km
Date dd.mm.yyyy	Time hh:mm:ss.0	Hypocenter													
		φ,°	λ,°	h, km					φ,°	λ,°	h, km	Az_{Major},°	R_{minor}, km	R_{major}, km	
02.11.1974	04:59:56.9	70.81	53.91	0	0-90	275	305	5:00:01.3	70.73	53.99	(0) 0-7	140	8.6	12.5	9.4
					0-30	65	86	5:00:00.5	70.81	53.93	(0) 0-11	150	23.3	27.3	0.7
					0-20	12	19	4:59:55.5	70.55	54.58	(0) 0-75	130	26.7	40.9	38.0
					0-15	8	14	4:59:56.8	70.83	54.24	(0) 0-99	140	33.0	54.4	12.3
18.07.1985	21:15:00.29	65.99	41.04	0.8	0-45	125	135	21:15:00.4	66.04	40.96	(0) 0-13	160	8.1641	9.6875	6.6
					0-30	121	131	21:14:59.9	66.08	41.04	(0) 0-15	30	9.3359	15.117	10.0
					0-20	23	33	21:15:00.2	66.04	41.01	(0) 0-11	30	9.6484	14.805	5.7
					0-15	16	24	21:15:00.7	66.06	40.99	(0) 0-16	20	10.27	18.32	8.1
					0-10	9	13	21:14:55.8	65.93	41.80	(0) 0-99	100	16.60	30.70	35.1
06.09.1988	16:19:58.74	61.33	47.98	0	0-90	173	187	16:20:01.4	61.27	47.969	(0) 0-9	10	7.2266	9.84	6.7
					0-30	127	141	16:20:00.9	61.33	48.03	(0) 0-10	30	7.656	11.719	2.7
					0-20	38	49	16:20:00.4	61.32	48.06	(0) 0-11	40	8.71	12.62	4.4
					0-15	16	26	16:20:00.9	61.31	48.02	(0) 0-19	40	10.82	17.77	3.1
					0-10	8	14	16:20:01.5	61.29	48.04	(0) 0-25	30	10.39	19.18	5.5
24.10.1990	14:57:58.5	73.33	54.76	0	0-90	518	534	14:58:03.0	73.30	54.63	(0) 0-8	150	6.3	7.9	5.3

Ground truth* event: underground nuclear explosions					**Result of the relocated procedure**											
					R,°**	**Nst**	**Nphases**	**Time hh:mm:ss.0**	**Hypocenter**			**Error ellipse**			**D***, km**
Date dd.mm.yyyy	Time hh:mm:ss.0	**Hypocenter**													
		φ,°	**λ,°**	**h, km**					**φ,°**	**λ,°**	**h, km**	**Az_{Major},°**	**R_{minor}, km**	**R_{major}, km**	
					0-30	74	85	14:58:02.2	73.36	54.53	(0) 0-7	170	9.7	13.4	8.1
					0-20	24	31	14:58:01.1	73.31	54.64	(0) 0-8	160	9.7	11.1	4.4
					0-15	10	14	14:58:01.5	73.35	54.48	(0) 0-16	160	10.4	14.9	9.2
					0-10	4	7	14:58:01.2	73.33	54.95	(0) 0-100	130	14.2	22.2	6.1

* - GT categories. GTX refers to events with location accuracy better than X km (Yang et al., 2000); ** - epicentral distances; *** - the distance between the calculated and truth coordinates.

DISCUSSIONS

From 36 seismic event placed in Table 4, were compiled bulletins and conducted relocation procedures for 31 events, including the events on October 14, 1908 and April 13, 1912 (Table 7). The event on November 11, 1923 did not meet our requirements (at least 5 phases by 3 stations). The phases were found in the bulletins of five stations, but only 3 phases by 2 stations participated in the calculation. The remaining phases were removed by the NAS program because of a large deviation from the theoretical entry times of the BARENTS travel time model due to either erroneous identification of the wave type or due to inappropriate to considered event. No entry in the bulletins of seismic stations for the event on 25.08.1935 was found. This fact allows us to exclude this event from the Barents-Kara region catalogue.

For seismic events on December 11, 1975, May 27, 1977 and October 29, 1988, the bulletins of seismic stations could not be detected. The seismic events on December 12, 1974, on November 15, 1978 and on August 1, 1986 occurred in the Novaya Zemlya archipelago were considered in details earlier in the work (Morozov et al., 2017). The results of relocation procedures are presented in Table 7 and on Figure 6. It is impossible to compute the focal depths reliably for most of the seismic events due to the small number of available stations and phases. The 0-km maximum depth of the rated function is inaccurate and must not be taken into account. Other maximums of the rating function are obtained at depths within the Earth's crust.

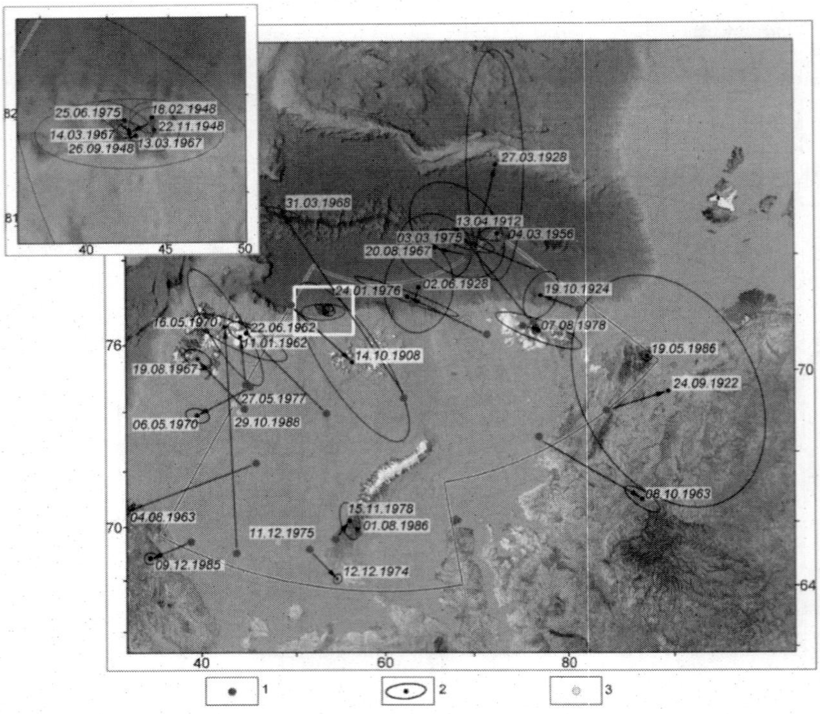

Figure 6. Maps showing the relocated epicenters of seismic events in the area of the Barents and Kara sea region: 1 – initial coordinates; 2 – new coordinates and error ellipses; 3 – the epicenters were not relocated due to the lack of bulletins.

Within analyzing the seismic events of the early 20th century, it become clear that most of the seismic phase entries have a large deviation from the theoretical entry times of the BARENTS travel time model due to either erroneous identification of the wave type or due to inappropriate to considered event. Events for which were conducted relocation procedures have large ellipses of errors due to the wide azimuthal gap and the narrow range of epicentral distances. It is impossible to determinate the coordinates more accurately. However, it was important for us to ensure that considering events are relevant to our research area.

After the refinement procedure, the thesis on the aseismicity of the Barents and Kara seas shelf was further confirmed. The seismic events, which previously belonged to the shelf, after the relocated procedure are referred to the margins of the shelf, such as the Svalbard archipelago and along the "continent-ocean" transition zone, or to the Gakkel Ridge. Exception event on May 6, 1970, the episenter is shifted to the south-west to Bjornoya Basin of the Barents Sea shelf (Figure 7).

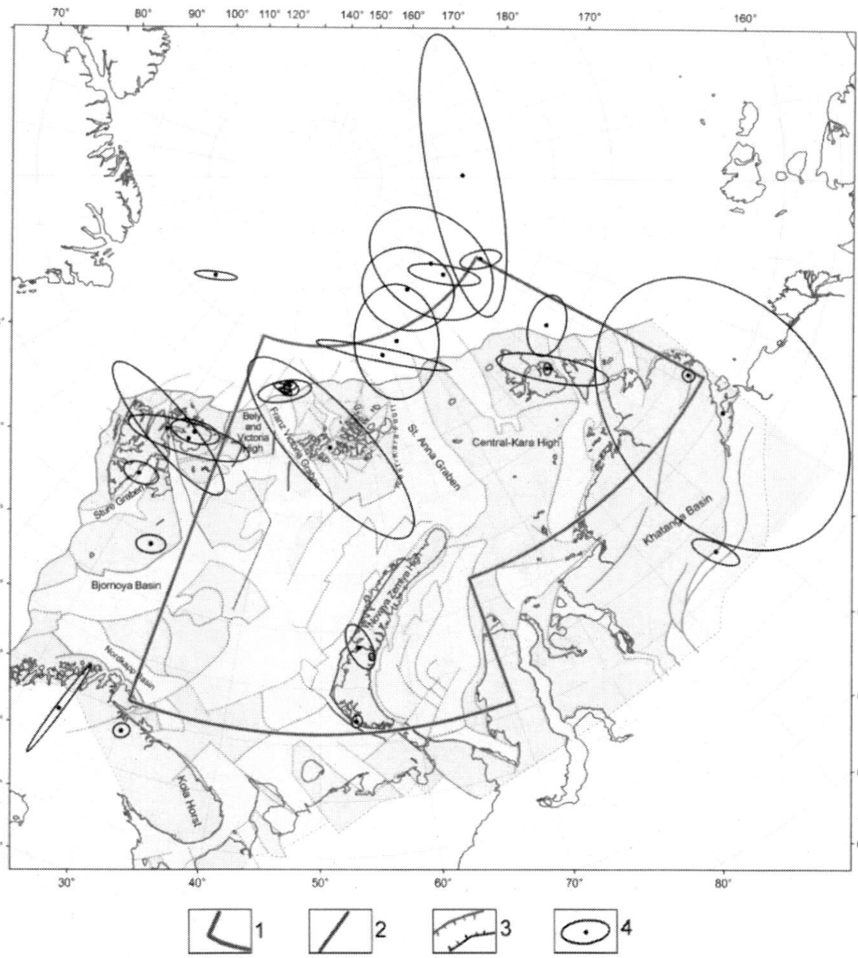

Figure 7. Maps showing the relocated epicenters of seismic events in the area of the Barents-Kara region, combined with the neotectonic structures and active faults (Atlas, 2004): 1 - boundaries of the study area; 2 - main neotectonic faults; 3 - borders of neotectonic structures and dislocation with a break of continuity; 4 - relocated epicenters.

Table 7. Catalogue of relocated seismic events in an area of the Barents-Kara region

N	Date dd.mm.yyyy	Time hh:mm:ss.0	Hypocenter			N_{st}/N_{def}	Gap, °	Distances*, km	Error ellipse			Magnitude	Note
			φ,°	λ,°	h, km				Az_{Major}	R_{minor}	R_{major}		
1	14.10.1908	14:56:17.7	80.44	51.81	(0) 0-100	6/10	335	2534-3912	140	143.7	454.5	M_S(PAS)=6.6	
2	13.04.1912	02:39:44.8	85.18	100.87	(0) 0-99	5/7	326	3195-4529	0	188.6	266.0	M_S(PAS)=5.6	
3	24.09.1922	12:25:43.6	73.80	109.25	(0) 0-100	5/5	340	4650-5578	20	414.0	597.3		
4	19.10.1924	15:34:49.5	80.64	106.13	(0) 0-100	5/7	181	3050-5794	70	73.2	122.2		
5	27.03.1928	17:46:14.1	85.18	141.00	(0) 0-100	4/4	283	3509-3874	80	134.4	565.1		
6	02.06.1928	20:13:00.7	83.65	73.81	(0) 0-100	3/5	272	2844-3620	20	163.0	226.4		
7	18.02.1948	20:29:52.8	82.53	41.42	(0) 0-16	92/142	85	1931-7951	80	18.0	24.4		
8	26.09.1948	05:51:18.4	82.30	40.22	(4) 0-99	10/11	174	2877-6663	70	40.5	104.7		
9	22.11.1948	23:32:48.0	82.42	41.90	(1) 0-65	40/48	97	2507-6730	90	25.8	38.1		
10	04.03.1956	03:18:16.4	83.83	112.14	(15) 0-99	14/15	231	2322-7222	140	34.3	81.0		
11	11.01.1962	12:32:30.5	79.64	23.40	(0) 0-100	5/8	344	1125-2118	80	61.7	242.3		
12	22.06.1962	07:47:05.5	79.94	23.98	(0) 0-99	8/14	334	1131-2146	80	44.0	96.3		
13	04.08.1963	18:42:28.2	69.18	23.97	(0) 0-100	4/4	284	197-947	10	14.5	213.4		
14	08.10.1963	10:52:37.0	70.97	96.18	(0) 0-100	11/14	252	2003-5429	160	35.4	99.1	M(CGS)=4.7	
15	13.03.1967	21:44:07.8	82.30	40.71	(0) 0-32	46/51	77	340-7059	100	12.3	21.7	m_b(ISC)=4.3	
16	14.03.1967	07:50:18.1	82.33	40.12	(4) 0-36	84/94	63	349-8054	100	10.2	20.3	m_b(ISC)=4.3 M(MOS)=5.5	

N	Date dd.mm.yyyy	Time hh:mm:ss.0	Hypocenter			N_{st}/N_{def}	Gap, °	Distances*, km	Error ellipse			Magnitude	Note
			φ,°	λ,°	h, km				Az_{Major}	R_{minor}	R_{major}		
17	19.08.1967	16:27:53.7	77.77	19.01	(0) 0-100	7/14	333	909-1551	80	40.3	70.4		
18	20.08.1967	11:31:45.5	85.07	86.66	(5) 0-100	6/6	318	2037-6398	150	160.7	180.7	M(LAO)=3.7	
19	31.03.1968	03:03:58.1	84.89	3.25	(0) 0-100	9/9	193	783-2733	50	15.6	82.8	M(LAO)=3.9	
20	06.05.1970	07:10:32.8	75.76	25.36	(0) 0-100	9/14	312	674-1438	70	34.7	56.9		
21	16.05.1970	13:20:01.2	79.60	19.02	(0) 0-100	4/5	347	1353-2602	110	82.3	327.4		
22	12.12.1974	21:19:39.7	70.83	53.76	(0) 0-40	16/23	220	790-7020	20	19.5	24.6	m_b(ISC)=4.4	probably an explosion
23	03.03.1975	05:45:01.7	84.58	100.73	(0) 0-99	9/9	220	2015-6359	150	36.3	141.3	m_b(ISC)=4.1 M(LAO)=4.0	
24	25.06.1975	10:14:57.9	82.42	39.54	(0) 0-43	38/47	107	360-8059	100	14.1	35.5	m_b(ISC)=4.6 m_b(NEIS)=4.6 M(LAO)=4.8	
25	24.01.1976	19:09:28.7	83.35	68.61	(15) 0-99	10/10	227	1763-5368	120	25.7	265.8	m_b(ISC)=4.0	
26	07.08.1978	12:48:22.7	79.68	99.14	(0) 0-26	52/56	83	771-7253	50	15.7	23.5	m_b(ISC)=4.5 M_S(ISC)=3.5 m_b(NEIS)=4.5	
27	07.08.1978	13:35:16.1	79.54	99.66	(10) 0-99	25/25	251	2183-5335	150	49.5	220.0	m_b(ISC)=4.3 m_b(NAO)=4.4	
28	15.11.1978	08:30:01.7	73.44	54.69	(0) 0-99	6/10	336	980-2170	160	43.8	93.5	m_b(NAO)=4.3	probably an explosion
29	19.05.1986	13:52:33.9	75.50	111.30	(0) 0-99	53/54	120	1280-7563	20	23.9	25.8	m_b(ISC)=4.6 M_S(ISC)=3.7	
30	01.08.1986	13:56:40.1	73.05	56.01	(0) 0-16	52/64	11	1030-8430	160	11.0	15.4	m_b(ISC)=4.8 M_S(ISC)=3.7 m_b(NAO)=4.5	
31	09.12.1989	12:55:55.3	69.32	30.30	(0) 0-100	6/11	312	283-1286	50	27.5	32.0	m_d(BER)=3.9	probably an explosion

* - the range of epicentral distances (from … to …).

The epicenters within Franz Victoria Graben and Severnaya Zemlya archipelago did not significantly change their location. Their initial epicenters are located within the error ellipses of the new epicenters. Thus, both solutions for these events are correct. Exception event on April 13, 1912, the epicenter is shifted to the Gakkel Ridge.

For the seismic events on December 12, 1974 and on November 15, 1978, the epicenters are located in the vicinity of a nuclear test site, which could imply a possible anthropogenic nature. According to the new location of the 1989 seismic event, the epicenter located in the area of operation large industrial quarry, which could imply a possible anthropogenic nature too.

We deliberately did not try to relate the recalculated epicenters to the active tectonic structures of the Barents-Kara region. Future research will aim to identify the relationship between geological and tectonic processes. The thesis on the low seismicity of the shelf of the Barents and Kara Seas has received additional confirmation after the relocation procedure. The seismic events, which previously belonged to the shelf, after the relocated procedure are referred to the margins of the shelf, such as the Svalbard archipelago and along the "continent-ocean" transition zone, or to the Gakkel Ridge. Epicenters of twelve seismic events have shifted beyond the area of the study after relocation procedure. The Franz Victoria Graben is the main seismically active zone in the Barents-Kara region within the area of the study.

Thus, the present results improve our knowledge of seismic activity in the area of the Barents and Kara Seas. The relocated event catalogue will make it possible to clarify the seismic hazard map of these.

SEISMICITY OF THE NOVAYA ZEMLYA ARCHIPELAGO: RELOCATED EVENT CATALOGUE FROM 1995 TO 2014

The Novaya Zemlya archipelago is the largest in the Russian Arctic and is located in the eastern part of the Barents Sea (Figure 8). It includes two large islands, Severniy and Yuzhniy, which are separated by the narrow Matochkin Shar Strait (2–3 km) and numerous of smaller islands. A Soviet nuclear test site was operated on the archipelago until 1990. At Novaya Zemlya, a total of 130 tests were carried out high in the atmosphere, at low levels above water, at the water/air interface, below water, and underground (Khalturin et al., 2005).

Novaya Zemlya is characterized by low seismicity, and most of the detected seismic events from the region resulted from nuclear tests. There are significant difficulties in the seismic monitoring of the Novaya Zemlya archipelago. Widely spaced seismic stations cannot register small-magnitude earthquakes, and poor station coverage makes it difficult to determine earthquake hypocenters and focal mechanisms accurately. According to the studies (Assinovskaya, 1993; Avetisov, 1996), the completeness threshold of earthquakes for the Barents Sea region and, in particular, the Novaya Zemlya archipelago in the period 1971-1980 was mb = 4.3-4.6, and in the period 1981-1990 was mb = 3.9-4.2. As a result, understanding of the archipelago's seismic activity is based on 18 seismic events that are not identified as nuclear explosions but are probably tectonic in nature (Table 8).

Two new seismic stations were installed in the region between 2010 and 2011. The AMD station (69.76°N/61.68°E) was placed in the settlement of Amderma on the coast of the Kara

Sea, and the ZFI station (80.81°N/47.66°E) was installed on Alexandra Land Island in the Franz Josef Land archipelago (Morozov & Konechnaya, 2013). The ZFI and AMD seismic stations belong to the Arkhangelsk seismic network (AH network). The threshold of earthquakes for the Barents Sea region and, in particular, the Novaya Zemlya archipelago is now 2.7 (Morozov & Konechnaya, 2013). Only two earthquakes have occurred after the installation of the stations, one on 11.10.2010 at 22:48:27.8 (mb (ISC) = 4.7) and one on 04.03.2014 at 04:42:36.0 (ML (AH) = 3.3). The uncommon occurrence of tectonic activity requires a detailed survey for each earthquake detected. A list of seismic stations is presented in Figure 9.

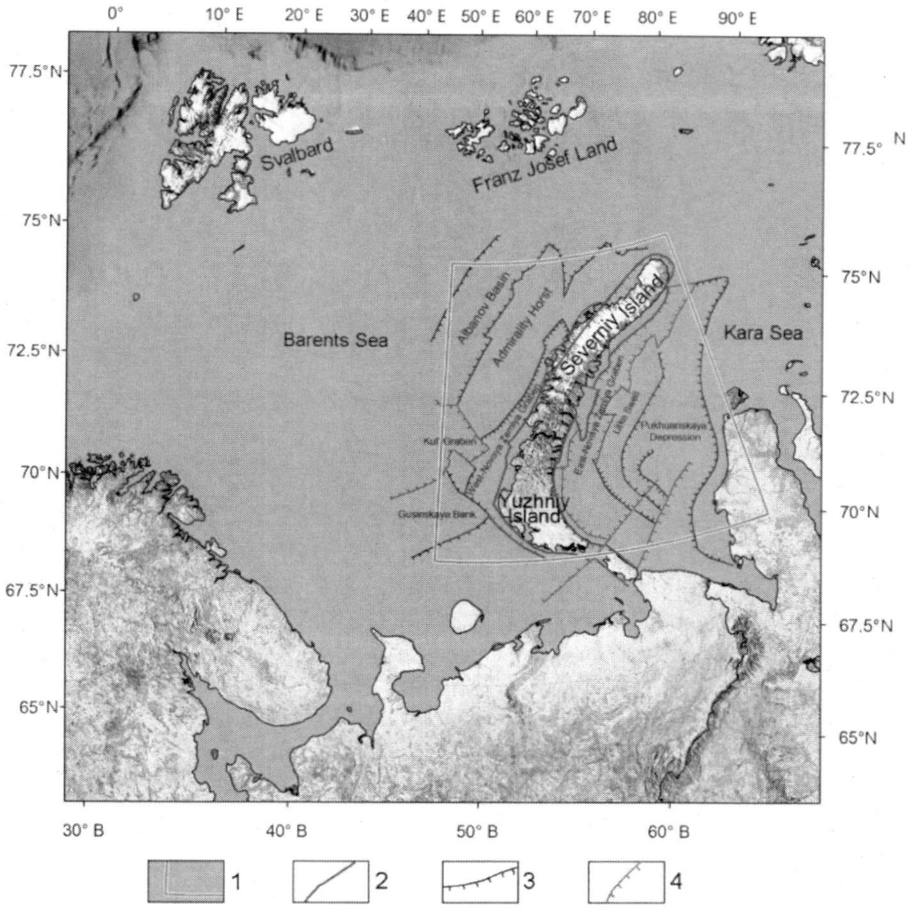

Figure 8. Map showing the location of the Novaya Zemlya Archipelago, the neotectonic structures and active faults (Atlas, 2004), combined with a bathymetric data (www.ngdc.noaa.gov): 1 – boundaries of the study area; 2 – main neotectonic faults; 3 – borders of neotectonic structures; 4 – dislocation with a break of continuity.

Table 8. Combined seismic catalogue of events in an area of the Novaya Zemlya Archipelago and adjacent territories over the instrumental observation period

No	Date	Origin time	Hypocenter			Magnitude				Source of the hypocenter
	dd.mm.yyyy	hh:mm:ss.0	φ,°	λ,°	h, km	mb(ISC)/ MS(ISC)	mb(EIDC)/ md(NEIC)	mb(NAO)/ ML(NAO)	ML(HEL)/ ML(AH)	
1	31.12.1992	09:29:24.0	73.61	55.18	0f			2.7/-		(Ringdal, 1997)
2	13.06.1995	19:22:37.9	75.26	56.88	0f			3.5/-		(Ringdal, 1997)
3	13.01.1996	17:17:23	75.13	56.73	0f			2.4/-		(Ringdal, 1997)
4	16.08.1997	02:10:59.2	72.62	56.94	10.0f		3.9/3.2	3.5/-	3.8	Reviewed ISC Bulletin (Prime hypocentres)
5	23.02.2002	01:21:12.2	74.27	55.83	10.0f	3.0/-		-/3.0	3.1/-	Reviewed ISC Bulletin (Prime hypocentres)
6	10.11.2002	11:04:41.5	70.40	48.64	10.0f			-/2.0	2.4/-	Institute of Seismology (University of Helsinki)
7	08.10.2003	23:07:19.7	75.64	62.89	-			-/2.6		NORSAR
8	05.03.2006	23:17:35.7	76.80	64.30	0f			-/2.7		NORSAR
9	14.03.2006	20:57:02.4	75.07	53.06	-			-/2.2		NORSAR
10	30.03.2006	10:46:02.4	70.89	51.01	0			-/2.3	2.6/-	Reviewed ISC Bulletin (Prime hypocentres)
11	26.06.2007	03:19:19.4	73.39	53.39	-			-/2.8		NORSAR
12	11.11.2009	04:18:20.2	71.42	47.07	10.0f				3.2/-	Institute of Seismology (University of Helsinki)
13	11.06.2010	16:46:25.6	70.81	49.42	0					Kola Regional Seismic Centre
14	11.10.2010	22:48:28.8	76.30	64.27	19.4	4.7/3.5	/4.8			Reviewed ISC Bulletin (Prime hypocentres)
15	04.03.2014	04:42:36.0	74.72	56.72	5.0f				-/3.3	Arkhangelsk seismic network
16	02.05.2017	13:35:42.0	75.29	57.71	10.0f				-/3.4	Arkhangelsk seismic network

Figure 9. Location of seismic stations.

A catalogue of seismic events was created for the Novaya Zemlya archipelago using the data of International Seismological Centre (ISC; International Seismological Centre, 2015), NORSAR (NOA network), Institute of Seismology (University of Helsinki) (ISUH, HEL network), Kola Regional Seismic Center (KOLA network), and the Arkhangelsk seismic network (AH network) (Table 8). The area of interest was limited to the following coordinate ranges: latitude from 70.4°N to 77.5°N and longitude from 47.0°E to 70.0°E.

Nuclear explosions were excluded from the catalogue using the data of Databases of Nuclear Tests (Yang et al., 2003), as well as seismic events that occurred in the area of the nuclear test site, thus considering only seismic events of tectonic origin (Table 8). Bulletins and waveform data from the Soviet and Russian seismic stations were obtained from the Geophysical Survey (OBN-network) of the Russian Academy of Sciences (RAS), the Kola Regional Seismic Center, and the Arkhangelsk seismic network. Seismic station bulletins from the global network were collated using data from the ISC, NORSAR, and the ISUH. The waveform data were obtained from the GEOPHON Seismological Archive (www.geofon.gfzpotsdam.de/geofon/).

The goal of analyzing the waveform data from seismic events was phase identification. Onset determination was performed using the WSG (Windows Seismic Grafer) program, a joint product from the Geophysical Survey of the RAS and Geotech, a limited liability company (www.ceme.gsras.ru). Seismic event location and depth estimation were carried out using the NAS (New Association System) program, which is a part of the NSDL (New System for Detection and Location) system developed in the Kola branch of the Geophysical Survey of RAS (Asming & Prokudina, 2016; Asming et al., 2016). The algorithm is described in more detail in the paragraph "*Event location algorithm.*"

We have not found additional seismic phases in the bulletins and waveform data for the 31.12.1992, 13.01.1996, 08.10.2003, 05.03.2006, 14.03.2006, 26.06.2007, and 11.06.2010 seismic events shown in Table 8. These events also did not meet our requirements (at least 8 phases by 4 stations), so a relocation procedure was not performed. We also did not find additional seismic phases in the bulletins and waveform data for the 16.08.1997 and 11.10.2010 seismic events in Table 8. (Schweitzer & Kennett, 2007; Gibbons et al., 2016) have obtained satisfactory results based on all available bulletins and waveform data, so a relocation procedure was not performed in this case as well.

Discussions

Thus, the relocation procedure was applied for 9 seismic events registered in the area of the Novaya Zemlya archipelago (Figure 10, Table 9). For the 13.06.1995, 23.02.2002 and 11.11.2009 events, the initial epicenters are located within the error ellipses of the new epicenters, so, both solutions for these events are correct. It is impossible to compute the focal depths reliably for most of the seismic events due to the small number of available stations and phases. The 0-km maximum depth of the rated function is inaccurate and must not be taken into account. Other maximums of the rating function are obtained at depths within the Earth's crust.

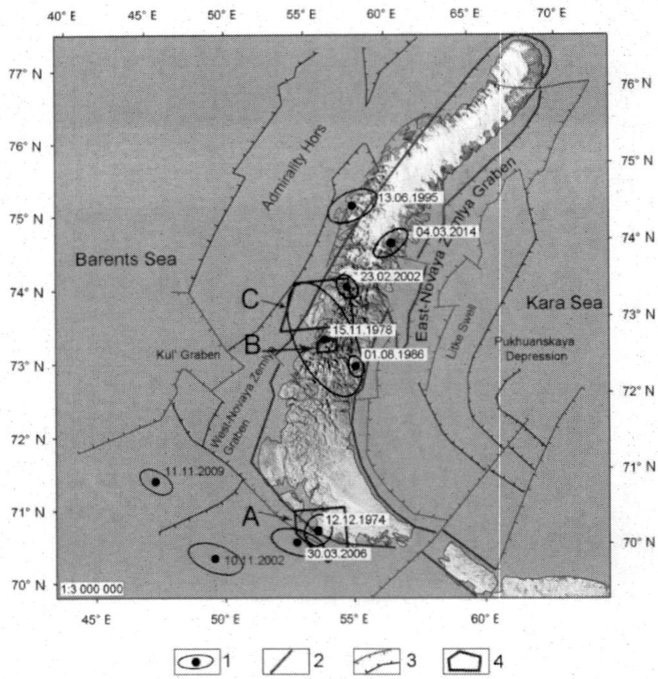

Figure 10. Map showing the relocated epicenters of seismic events in the area of the Novaya Zemlya archipelago: 1 – new coordinates; 2 – main neotectonic faults; 3 – borders of neotectonic structures and dislocation with a break of continuity; 4 – A, B, and C denote three main areas (zones) of military activity (Khalturin et al., 2005): A = Guba (Bay) Chernaya. B = Guba Mityushikha, south bank of Matochkin Shar. C = Sukhoy Nos Cape and its vicinity.

Table 9. Catalogue of relocated seismic events in an area of the Novaya Zemlya archipelago and adjacent territories over the instrumental observation period

N	Date	Time	Hypocenter			Error ellipse			Magnitude			
	dd.mm.yyyy	hh:mm:ss.0	φ,°	λ,°	h, km	AzMajor	Rminor	Rmajor	mb(ISC)/ MS(ISC)	mb(EIDC)/ md(NEIC)	mb(NAO)/ ML(NAO)	ML(HEL)/ ML(AH)
1	13.06.1995	19:22:37.9	75.21	56.70	(0) 0-99	60	17.1	33.3			3.5/-	
2	16.08.1997	02.11:00.2	72.47	58.02	(0) 0-25	40	16.4	11		3.9/3.2	3.5/-	3.8
3	23.02.2002	01:21:17.6	74.13	56.01	(0) 0-39	150	14.4	20.0	3.0/-		-/3.0	3.1/-
4	10.11.2002	11:04:41.7	70.47	49.62	(99) 2-99	0	12.4	20.2			-/2.0	2.4/-
5	08.10.2003	23:07:14.9	75.77	65.36	(5) 0-100	70	19.8	51.5			-/2.6	
6	30.03.2006	10:46:02.1	70.68	52.88	(35) 0-99	110	19.0	40.0			-/2.3	2.6/-
7	11.11.2009	04:18:20.2	71.52	47.06	(0) 0-99	120	15.6	26.6				3.2/-
8	04.03.2014	04:42:34.4	74.58	58.32	(40) 0-99	70	13.1	22.3				-/3.3
9	02.05.2017	13:35:42.0	75.31	58.05	(25) 0-99	70	40.9	11.3				-/3.4

We deliberately did not try to relate the recalculated epicenters to the active tectonic structures of the Novaya Zemlya archipelago. Future research will aim to identify the relationship between geological and tectonic processes. The regional seismicity will be based on the overall set of earthquakes that occurred during historic and instrumental periods, as well as paleoearthquake data. The present results improve our knowledge of seismic activity in the area of the Novaya Zemlya archipelago. Using for all relocated events the same velocity model, the same methodological approach, and all the currently available waveform data and bulletins allowed us to locate the earthquakes more realistically.

New Data About Small Magnitude Earthquakes of the Ultraslow-Spreading Gakkel Ridge

Before the opening new island stations in the Barents-Kara region, information on the seismic activity of the Gakkel Ridge was supplemented by data from relatively remote (Δ>700km) polar seismic stations and during Arctic expeditions (Schlindwein et al., 2007).

The Gakkel Ridge forms the North America-Eurasia plate boundary in the Arctic (Figure 11). The ridge strikes approximately for 1800 km from 7° W near the intersection of the Gakkel Ridge and the Lena Trough to 125° E near the Siberian continental margin. The Gakkel Ridge is the slowest spreading section of the global mid-ocean ridge system (Karasik, 1968; Vogt et al., 1979; DeMets et al., 1994). It was revealed that the rate of spreading of the ridge decreases from 12.8 mm/year to 6.5 mm/year from West to East.

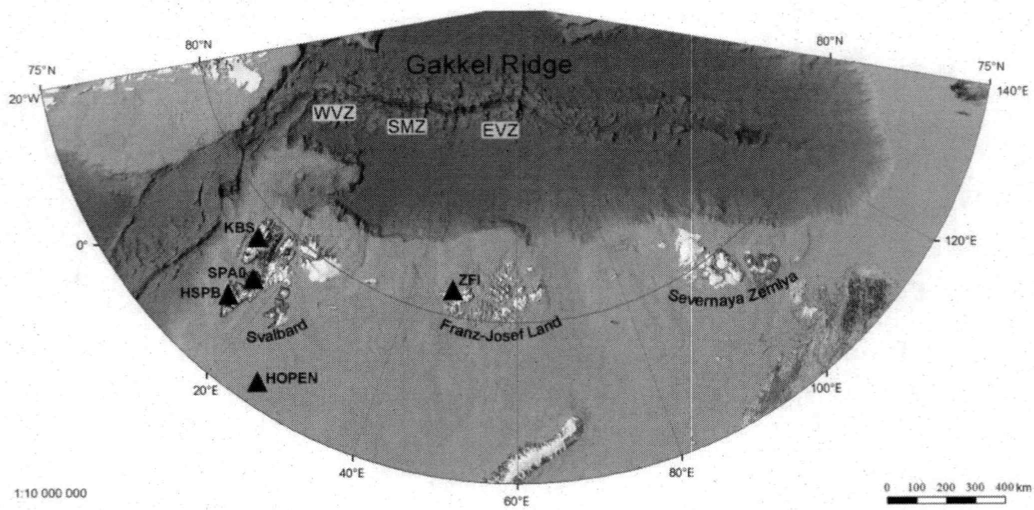

Figure 11. The study area is the Gakkel Ridge. Dotted line divides the Gakkel Ridge into segments: Western Volcanic Zone (WVZ), the Sparsely Magmatic Zone (SMZ) and the Eastern Volcanic Zone (EVZ). Black triangles mark the position of the station ZFI and the stations operating in the Svalbard archipelago. Bathymetry is from the National Geophysical Data Centre (NOAA - www.ngdc.noaa.gov).

The extreme environmental and geographical conditions impose limitations on scientific research conducted at the Gakkel Ridge. More recently, however, new geophysical data sets have become available. The SCence ICe EXercises (SCICEX) program utilized U.S. Navy nuclear submarines for a series of unclassified scientific research cruises to the Arctic Ocean between 1993 and 1999. In particular, USS (United States ship) Hawkbill carried out a systematic survey of the Gakkel Ridge axis utilizing the SCAMP (Seabed Characterization and Mapping Pods) instrument package during the 1998 and 1999 cruises. These cruises obtained nearly complete bathymetry, sidescan and gravity coverage to 50 km from the ridge axis, for 600 km along the Gakkel Ridge, from 8° E to 73° E, with coverage of the axis to 96° E (Edwards et al., 2001; Cochran et al., 2003; Cochran, 2008).

In 2001, the international Arctic Mid-Ocean Ridge Expedition (AMORE2001) onboard the icebreakers RV Polarstern and USCGS (US Coast Guard Ship) Healy studied the bathymetry, petrological and geophysical characteristics of the crust and the hydrothermal activity of Gakkel Ridge (Thiede et al., 2002). The obtained geophysical data allowed for the studying of the structure and features of the crustal accretion within the Gakkel Ridge area.

Geological Setting

The Western Volcanic Zone (WVZ), the Sparsely Magmatic Zone (SMZ) and the Eastern Volcanic Zone (EVZ) are distinguished in the Gakkel Ridge on the basis of morphological, geological, and geophysical features (Dick et al., 2003; Cochran et al., 2003) (Fig 11). The WVZ extends for 220 km from 7° W to 3° E. The spreading rate varies here from 15 to 13.5 mm/yr. The bottom of the rift valley is located at a depth of 3.8–4.0 km. The crustal thickness is 2.5 - 4.9 km. The rift valley hosts elongated volcanic ridges, the seafloor is covered with basalts and several hydrothermal plumes were discovered (Michael et al., 2003; Jokat & Schmidt-Aursch, 2007; Dubinin et al., 2013).

The SMZ from 3° to 30° E extends at about 300 km. The spreading rate varies from 13.5 to 12.7 mm/yr. The bottom of the rift valley is located at a depth of 5.0–5.5 km and consists of a series of elongated basins. The crustal thickness in the central amagmatic segment of the Gakkel Ridge is only 1.3–2.5 km. Only one large volcanic center at 19° E has been noted over the entire extent of this segment. Gabbro and peridotite dominate in the rift valley (Cochran et al., 2003; Michael et al., 2003; Jokat & Schmidt-Aursch, 2007; Dubinin et al., 2013).

The EVZ extends for 600 km from 30° to 94° E. The spreading rate varies here from 12.7 to 6 mm/yr. The bottom of the valley is located at a depth of 4.5-5.0 km and its width is 12-15 km. The thickness of the crust in the adjacent basins is limited to 2.5-3.3 km (Michael et al., 2003; Dubinin et al., 2013). The largest volcanic centre is situated near 85°E in the eastern part of Gakkel Ridge, it showed massive volcanic activity in 1999 with a submarine eruption and a swarm of teleseismically recorded earthquakes (Edwards et al., 2001; Tolstoy et al., 2001; Riedel & Schlindwein, 2010).

Based on the analyses of bathymetry and geophysical features, the researchers determine additional segments in the Gakkel Ridge. The paper of (Gurevich et al., 2004) defines 7 segments. At the same time, based on seismic data for the ridge, the paper of (Engen et al., 2003) determined there are 8 fracture zones, which divide the ridge into small segments.

The morphological and geological-geophysical characteristics of the considered ultraslow spreading ridge and the simulation results allowed (Dubinin et al., 2013) to establish the main structure forming factors that act in rift zones. The Gakkel Ridge is distinguished by the lowest spreading rate and a significant thickness of the lithosphere. The degree of mantle heating and geometry of spreading vary along the strike of the ridge. The relationships between endogenic processes also change along the strike. A prevalent tectonic factor is combined with locally developed magmatic and metamorphic processes.

Seismicity of the Gakkel Ridge

Thanks to the geophysical expeditions of 1999-2001, nowadays there are the detailed bathymetry, gravimetric, magnetometer, petrological, and seismic (mb>4) data for the Gakkel Ridge. However, so far we only have a vague idea on the character of small magnitude earthquakes (or microearthquakes) distribution within the Gakkel Ridge area. Most of the seismicity of the Gakkel Ridge occurs at magnitudes well below the completeness and detection thresholds of the Global Seismological Network. As a result, our knowledge of geodynamic processes in the Gakkel Ridge is incomplete and shows real situation only unilaterally. Because data on small magnitude earthquakes are often absent, studying them can provide much information on active processes at the rift axis on the style of magmatism and tectonic faulting and on the mechanism of hydrothermal circulation (Schlindwein et al., 2005, 2007).

There are some objective difficulties in the seismic monitoring of Arctic areas. First, widely spaced seismic stations cannot register small magnitude earthquakes. Second, poor station coverage makes it difficult to determine earthquake hypocenters and focal mechanisms. As a result, the understanding of the Gakkel Ridge seismic activity is based on the large magnitude earthquakes recording performed by a large number of seismic stations. Works by (Fujita et al., 1990; Avetisov, 1996; Engdahl et al., 1998; Engen et al., 2003; Schlindwein, 2012) are dedicated to the generalization of information on earthquakes for the entire period of instrumental seismological observations within the Arctic Region. The detection threshold of earthquakes is ranged from 3 to 5 at different parts of Gakkel Ridge. In particular, (Avetisov, 1996) noted, the seismic belt tracing the Gakkel Ridge has a nearly persistent strike and an average width not exceeding 20-30 km. The most remarkable offset of 100-120 km of epicenters is located at the site between 40 and 80°E where nearly linear fragment some 300 km long is shifted to the north in its western part, and farther east it fairly smoothly approaches the general axial line.

At different times expeditions for small magnitude earthquake recording were carried out in the Arctic Region, including the Gakkel Ridge. From 1968 to 1988 researchers for the VNIIOkeangeologia (Research Institute for Marine Geology and Mineral Resources of the World Ocean) conducted the field works on installation of temporal seismic stations across the Arctic territories and the Frantz Josef Land archipelago, Novaya Zemlya, and the New Siberian Islands (Avetisov, 1993, 1996). The duration of measurements varied from a month to two years. Based on the acquired information some new assumptions have been made regarding geodynamics of this region, which to a certain extent have changed the previous views.

In 1979 during the drift of the FRAM I ice station Kristoffersenet al. (1982) recordings of the microseismicity of the Arctic Mid-Ocean Ridge were made. They recorded and located about 21 regional earthquakes at a distance of 15-100 km and magnitudes in the ranging between mb 1 and 4. (Sohn & Hildebrand, 2001) used U.S. Navy hydrophones (Spinnaker Array) detect earthquakes of magnitude mb=3-4 at Gakkel Ridge at a distance of 700-1000 km by means of their hydroacoustic T-phase.

During AMORE2001 attempts by (Schlindwein et al., 2007) were made at recording the microearthquake activity of the Arctic Mid-Ocean Ridge. The arrays consisted of four three-component seismometers installed on ice floes. Three such arrays were deployed in different rift segments and recorded the seismic activity continuously for 5-11 days. All of the arrays recorded numerous microearthquakes in the central rift valley and on its flanks, which (Schlindwein et al., 2007) interprets as tectonic earthquakes. A swarm of microearthquakes was localized with high accuracy underneath the crest of a volcanic ridge in the rift valley and is proposed to have magmatic origin.

The paper of (Schlindwein et al., 2015) analyzed all of the information about small and large magnitude earthquakes in context with the spreading processes of the Gakkel Ridge and conclude the following, "To further advance our understanding of these processes, it is necessary to gain datasets that close the gap in magnitude coverage between the very local and short seismicity studies on ice floes and the teleseismic record of the Arctic ridge system. Long-term OBS (Ocean Bottom Seismometers) studies are needed that explore the evident differences in seismicity between the WVZ and the SMZ covering also events with magnitude 2-4." The renewal of instrumental seismological observations in 2011 (station ZFI) on Alexandra Land Island in the Franz Josef Land archipelago combined with other modern regional stations has created favorable conditions for the seismic monitoring of small magnitude earthquakes the Gakkel Ridge, and studying of the differences in seismicity between the WVZ and the SMZ.

According to (Schlindwein et al., 2007; Schlindwein, 2012) seismic activity on the Gakkel Ridge reflects itself in the form of individual earthquakes and swarm sequences. Swarms of earthquakes have been recorded near volcanic centers and are possibly related to the rise of magma. However, the previously analyzed data practically did not contain weak seismicity (ML<3), which is necessary both for the identification of the Omori Law sequential shake sequences and for swarms. With the addition of seismic information from the Arctic stations "Franz Josef Land," "Omega" and "Severnaya Zemlya" to the on-going Gakkel Ridge regional monitoring, it became possible to include weak seismicity in the analysis (Rogozhin et al., 2016). Figure 12 shows the graphs of the representative magnitude of the Gakkel Ridge according to the ISC (International Seismological Center) before the opening of the island stations of the Arkhangelsk network (Figure 12 a) and taking into account their contribution (Figure 12 b). We see that the new stations allowed to reduce the level of representative magnitude to 2.9.

The Figure 13 shows the results of registration of earthquakes within the Gakkel Ridge for 2013 recorded by station ZFI of the Arkhangelsk seismic network. In order to improve the accuracy of the small magnitude earthquakes epicenters locations, the travel-time for regional phases within the ridge was determined. To minimize errors in locating earthquakes, data from boadband stations operating in the Svalbard archipelago were added, including KBS (GE network, GEOFON- Global Seismic Network), HSPB, (PL network, Polish Seismic Network), HOPEN (NS network, Norwegian National Seismic Network, University of

Bergen, Norway), station SPA0, member of the seismic array SPITS (NORSAR, Norway) (Figure 13). Access to the seismic waveforms was carried out using the GEOFON program (www.geofon.gfz-potsdam.de/geofon/). The analysis of seismic data to earthquake hypocenters was completed with the help of the WSG program, a joint product of the Geophysical Survey of the Russian Academy of Science and "RPE Geotech" (www.ceme.gsras.ru). To determine local magnitude ML the WSG calculation method was used, based on an averaged northern Eurasia calibration function (Gabsatarova, 2006).

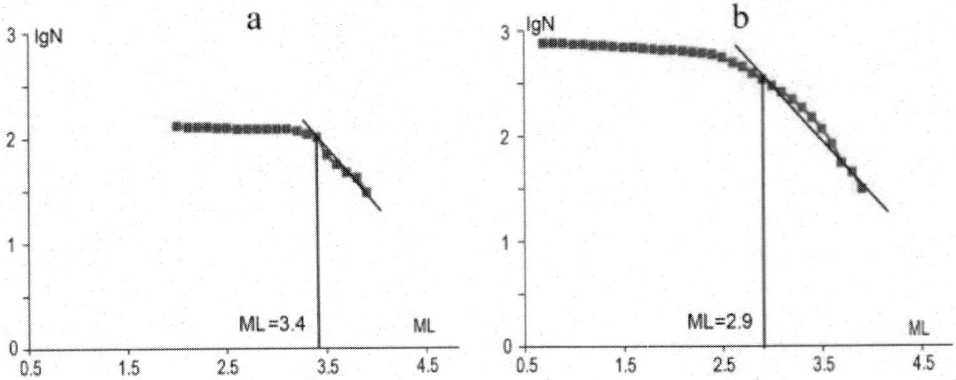

Figure 12. Graphs of the representative magnitude of the Gakkel Ridge according to the ISC data: before opening stations in the Russian sector of the Arctic from 2007 to 2011 (a) and taking them into account from 2012 to 2016 (b).

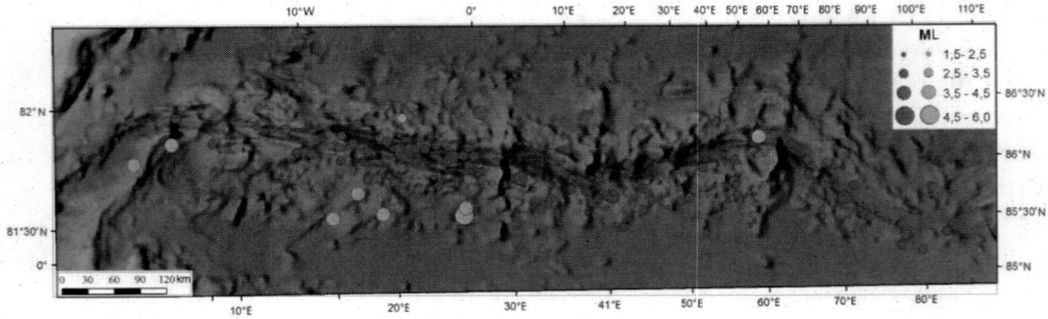

Figure 13. Comparison of the results of registration of earthquakes within the Gakkel Ridge for 2013, according to the seismic station ZFI (red circles) of the Arkhangelsk seismic network and stations by seismic agency NORSAR (green circles). The epicenters of earthquakes map the combined bathymetric data (www.ngdc.noaa.gov).

Velocity Model for Location Earthquakes on the Gakkel Ridge

In this paragraph we consider the topical issue of creation of velocity model and regional seismic phase travel-times for spreading ridges for areas within the Euro-Arctic Region. This is especially important, and necessary, for locating epicenters of small magnitude earthquakes from the Gakkel and Knipovich Ridges, which are registered by a small number of seismic stations operating in the Svalbard archipelago and Franz Josef Land. Figure 14 is given as an example to illustrate the problem described above.

The Figure 14 shows epicenters of earthquakes that occurred on the Gakkel Ridge during the period of 2011 to 2014. These epicenters were registered by Arkhangelsk seismic network stations and the stations operating in the Svalbard archipelago (SPA0, KBS, HOPEN and HSPB). Each epicenter was located with the help of joint processing of records of at least three stations by using the BARENTS velocity model (Kremenetskaya et al., 2001; Hicks et al., 2004). The figure clearly shows that the epicenters are located along the Gakkel Ridge, though slightly shifted to the south with respect to its axial part. The chain of epicenters repeats all morphological bends of the Gakkel Ridge, but there is an "undershoot" of epicenters on latitude. This may be due to a discrepancy between model seismic wave velocities in this region and real-time velocity. The next paragraph is devoted to the calculation of travel-times for regional phases of the Euro-Arctic region and verification of their effectiveness.

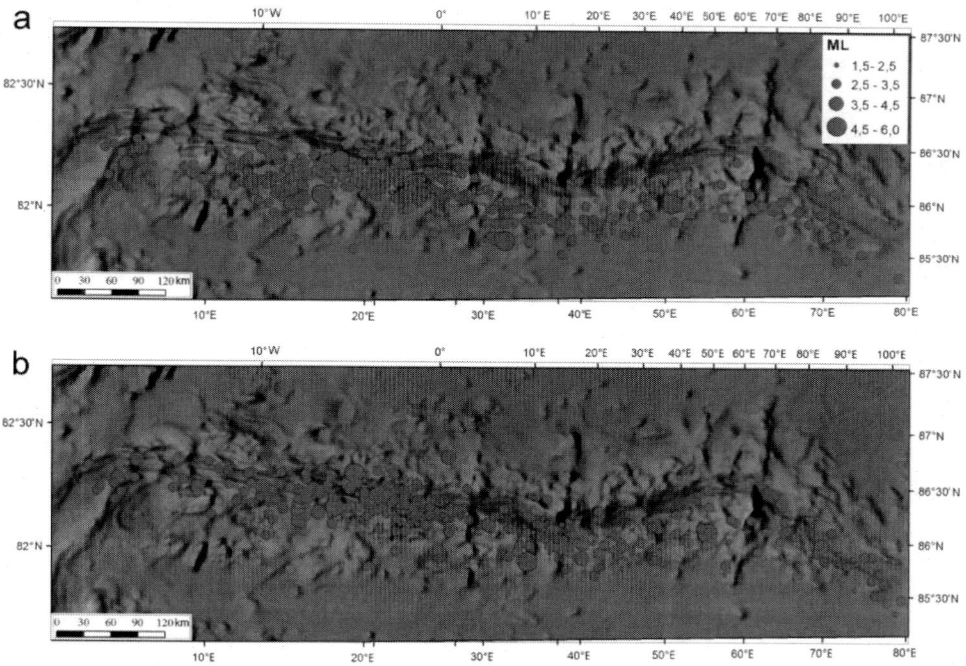

Figure 14. The distribution of earthquakes which occurred on the Gakkel Ridge during the period from 2011 to 2014 and registered by the seismic station ZFI and the stations operating in the Svalbard archipelago. (a) Each epicenter was located with the help of joint processing of records of at least three stations by using the BARENTS velocity model; (b) Each epicenter was located by using the NOES velocity model. The epicenters of earthquakes map the combined with a bathymetric data (www.ngdc.noaa.gov).

Calculation of Travel-Time for Regional Phases

Calculation of P and S regional seismic phase travel-time was carried out as follows. For the determination of crustal velocity structure for two areas of the Euro-Arctic region - the Svalbard archipelago and Franz Josef Land, the PRF receiver functions technique was employed (Vinnik, 1977). The PRF technique is aimed at the registration, selection and interpretation of converted Ps waves, which spread from the source area as "primary" waves (P waves) before being converted into "secondary" waves (S waves) at the edges of the seismic station area. The crustal velocity structure was determined beneath seismic stations SPA0 (SPITS array, NORSAR) and ZFI (Arkhangelsk seismic network), in the substation areas. Previously, determination of crust and upper mantle structure in the Svalbard archipelago was carried out in the works of (Ottemöller & Midzi, 2003, Wilde-Piorko et al., 2009), but based on the data from other stations. The crust and upper mantle structure in the Franz Josef Land archipelago was obtained earlier by the authors in the (Morozov et al., 2015).

Broadband three-component records of teleseismic earthquakes from 2011 and 2013 were used as source data for the receiver function study. A total of 120 earthquake data records were selected in all azimuthal directions and epicentral distances from 35° to 90°. Earthquake magnitudes ranged from 5.8 to 8.3, with focus depths extending down to 300 km. The records have clear P-wave arrivals, with the first arrival amplitude exceeding that of noise by more than 3 times.

Processing of the selected earthquake waveforms was carried out by the Seismic Handler software package (www.seismic-handler.org) and included the following procedures: filtering with band pass 0.03 – 1 Hz; rotation of the primary axis components Z, N, E to the coordinate system L,Q,T, where the L axis corresponds to the principal direction of the P wave particle motion in the wave-propagation plane, the Q axis is perpendicular to L in the same plane, and the converted Ps wave is included in the form of T – the transverse component. Further standardization of the components was also performed, as well as the acquisition of the individual receiver functions for each earthquake. The Q-component describes the displacement in the converted Ps phase and is also known as a receiver function. This function is free from source influence and characterizes the structure within the limits of the lithosphere area located under the SPA0 and ZFI stations. Furthermore, we have summarized the individual PRF from several sources (Figure 15).

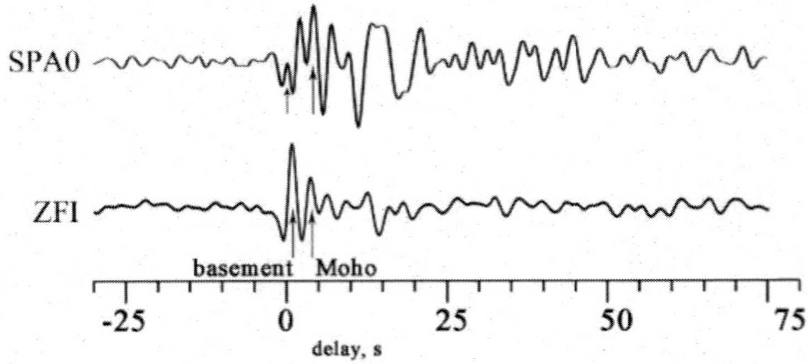

Figure 15. The summary P-receiver function of ZFI and SPA0 stations. Time zero refers to the direct P phase. The Ps conversion phases from the basement and Moho discontinuities are marked by arrows.

In order to carry out inverse modeling and determination of the crustal velocity structure beneath the SPA0 and ZFI stations the regularization method (Tikhonov & Glasko, 1965) was selected (Figure 16). In this method the problem is reduced to solving a system of linear equations, containing parameters for the velocity of P and S waves (Vp, Vs), as well as layer densities (ρ) and layer thickness. For the inversion a horizontally-layered model was employed, lying on a half-space, with constant velocities inside the 13 layers, gradually increasing with depth from 0 to 52 km. We used a fixed velocity ratio Vp/Vs=1.732 (in the crustal) and Vp/Vs=1.8 (in the mantle).

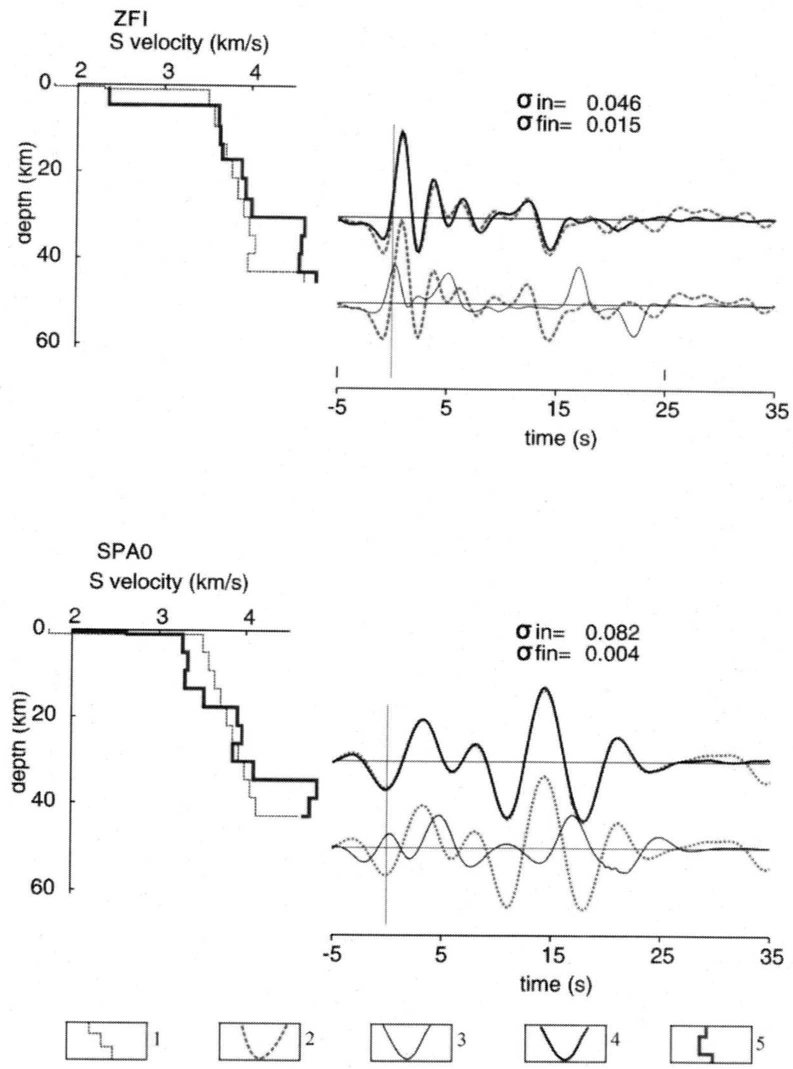

Figure 16. 1-D velocity models of the crust and upper mantle beneath ZFI and SPA0 stations: 1 - initial velocity model; 2 – experimental P-receiver function; 3 – synthetic P-receiver function calculated using the initial model; 4 – synthetic P-receiver function calculated using the final model; 5 – final model.

The density was as specified by (Birch, 1961). The search for the optimal parameters of the medium in the velocity model is based on the general method employed to solve poorly-posed inverse problems, and is determined by the selection and iterative minimization of discrepancies between the synthetic receiver functions and experimental functions. Synthetic receiver function is calculated using the (Haskell 1962) method. Calculations were performed using the programs developed by the Institute of Physics of the Earth of the Russian Academy of Sciences (Kosarev et al., 1987).

The optimal velocity model parameters are determined by the iterative minimization smoothing functional F (X):

$$F(X) = \frac{1}{T}\int_0^T \left(Q(t) - Q_{syn}(t, X, c)\right)^2 dt + \alpha \sum q_i \left(x_i - x_i^0\right)^2$$

where $X=\{x_i\}$ is x_i model parameter vector; $Q(t)$ is experimental P-receiver function, Q-component; Q_{syn} is synthetic P-receiver function calculated using the Haskell method (Haskell, 1962); x_i^0 is the parameter value in the initial model; q_i is weight coefficient; α is the regularization parameter, $\alpha=\alpha_n \cdot \Delta\alpha$, where α_n is the regularization parameter value in the preceding step to minimize. $\Delta\alpha<1$.

We select the time interval from –5 to 25 s for receiver function calculation to include the major converted phases. The iterations number determined by the condition that the calculated deviation synthetic Q-component from the experimental approximate noise level or next iterations do not reduce the deviation. Obtained velocity models were called ‘NOES‘ (North of Eurasian shelf) and ‘Sval‘ (Svalbard) for the sake of convenience (Table 10). A common feature of cuts received is evidence of several borderlines. The following elements are most vividly separated in the earth's crust (from top to bottom): top of crystalline basement, boundary within the crust at the depth of ~ 17 km, and Moho discontinuity. Sval velocity model is consistent with the model obtained earlier in the work of (Ottemöller & Midzi, 2003).

Figure 17 presents a comparison of various Vp and Vs velocity models - NOES, Sval, BARENTS (Kremenetskaya et al, 2001; Hicks et al, 2004), IASPEI91 and BARENTS3D (http://www.norsar.no/seismology/barents3d/). The BARENTS3D model for the seismic velocity and density structure of the crust and upper mantle in the greater Barents Sea Region has been developed by a Norwegian-US team as a composite of two separate models, one for the crust BARENTS50 (Ritzmann et al., 2006) and one for the upper mantle BARMOD (Levshin et al., 2007). We only used the model of the crust and upper mantle at the point of seismic stations ZFI location. Also, there is a model EUNAseis (Artemieva & Thybo, 2013) for the Euro-Arctic region, but it we did not use in the calculations.

Table 10. The NOES and Sval crustal and upper mantle models

NOES			**Sval**		
Depth, km	**Vp, km/s**	**Vs, km/s**	**Depth, km**	**Vp, km/s**	**Vs, km/s**
0	4.3	2.36	0	5.77	3.33
4	6.1	3.6	17	6.75	3.9
17	6.8	3.94	35	8.4	4.6
30	8.15	4.52	71	8.48	4.6
43	8.25	4.75	271	8.523	4.628
71	8.35	4.81			
210	8.373	4.556			
271	8.523	4.628			

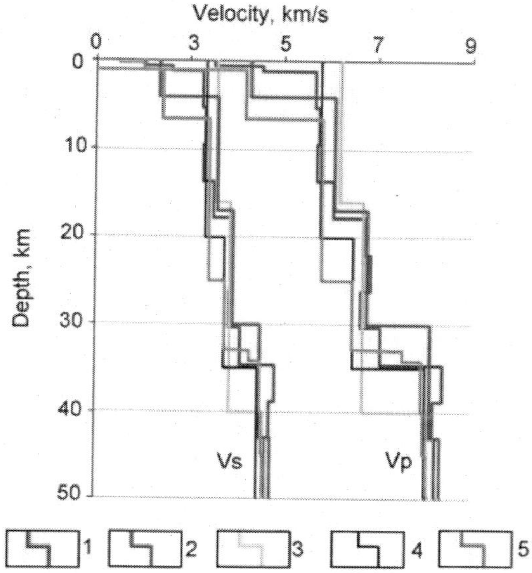

Figure 17. The comparison of various Vp and Vs velocity models: 1 – NOES; 2 – Sval; 3 – BARENTS; 4 - IASPEI91; 5 – BARENTS3D. Note: we only used the BARENTS3D model of the crust and upper mantle at the point of seismic stations ZFI location.

When comparing models, the distribution of velocities is observed similarly for all models with the difference not exceeding 0.5 km/s. What makes it distinctive is a different depth of crystalline basement and Moho discontinuity, due to regional particularities and varied thicknesses of sedimentary cover and crust under the Svalbard archipelago, Franz Josef Land and averaged model for the whole Earth.

Validation of NOES and Sval Travel-Times

Based on received NOES and Sval velocity models, regional travel-time curves were calculated for P and S seismic phases, which were given the same name. Calculation of travel-time was carried out in "Layers_Travel_Time" program, developed by V.E. Asming from the Kola Branch of the Geophysical Service RAS (Kremenetskaya et al., 2004). Travel-times are calculated for epicentral distance values ranging from 0 to 30° with steps of 0.1° and for the following focal depths values: 0, 5, 10, 20 and 30 km.

To check the effectiveness of received NOES and Sval travel-times, the following work was done. On the basis of the International Seismological Center's catalogue (http://www.isc.ac.uk), selection of earthquakes was made with mb ≥4.0 which occurred in the Arctic region during the period from 2002 to 2012. According to the ISC catalogue, epicenter parameters of each selected earthquake were calculated on the basis of data from at least 100 seismic stations at various epicentral distances, and parameters of particular earthquakes were received on the basis of data received from 2500 stations. This being the case, it is assumed that location of epicenters was determined with sufficient accuracy and can be considered as "true." Table 11 shows selection of earthquakes from ISC catalogue, intended for checking the NOES and Sval travel-times. Figure 18 shows spatial distribution of earthquake epicenters from Table 11.

According to the original data of seismic stations SPA0, KBS, HSPB, ZFI, BJO1 and HOPEN, the epicentral distances for each of the earthquakes were calculated on the basis of difference in times of P and S seismic phases (Figure 18).

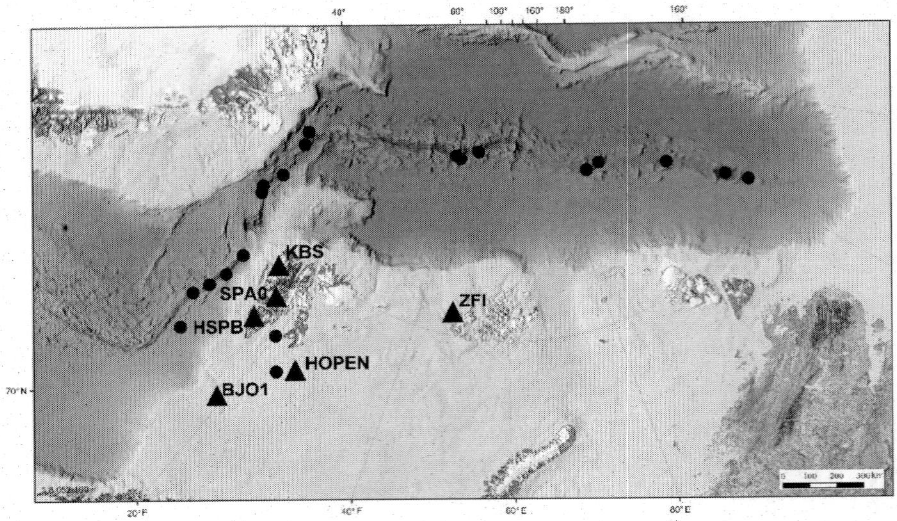

Figure 18. The distribution of earthquake epicenters from Table 11 and seismic stations, the epicentral distances for each of the earthquake were calculated on the basis of difference in times of P and S seismic phases.

Table 11. The seismic earthquake catalogue for validation of NOES and Sval travel-times

Date yyyy-mm-dd	Time hh:mm:ss.00	Hypocenter			Magnitude, mb
		φ,°	λ,°	h, km	
2002-05-03	11:20:51.75	86.0010	31.5120	10.0f	5.2
2003-07-04	07:16:43.45	76.3603	23.3677	0.4	5.7
2005-03-18	04:24:49.73	85.2492	91.7329	10.0f	5.2
2005-08-06	04:02:30.42	85.2061	97.5848	7.1	5.3
2008-08-13	18:30:57.57	83.5319	114.5420	10.0f	5.2
2009-03-05	19:41:41.29	80.2067	-1.4032	14.4	5.4
2009-03-06	10:50:30.16	80.3143	-1.9637	14.2	6.4
2009-05-17	19:24:21.40	82.5186	-8.7914	16.0	5.4
2009-06-22	19:05:17.49	76.3002	6.8595	15.2	5.1
2009-06-22	18:15:40.09	76.3311	6.8091	13.1	5.3
2009-10-07	20:52:13.82	81.5856	120.0240	15.0	5.6
2011-10-02	20:54:42.10	77.3035	20.0068	2.7	4.0
2011-10-19	22:35:09.06	75.2577	9.4035	10.0f	4.1
2011-11-19	21:04:21.76	81.1045	-2.2170	10.0f	4.5
2011-11-26	23:59:52.97	82.2287	-6.2379	12.3	5.0
2011-12-24	22:39:06.85	77.5941	7.5859	10.0f	4.5
2012-02-07	23:37:05.23	76.9170	7.2679	12.2	4.8
2012-05-26	03:27:01.40	85.9810	32.1040	9.0	5.1
2012-06-18	20:14:53.50	86.4070	41.5480	10.0	5.1
2012-09-02	12:44:03.46	78.4300	6.7640	10.0f	5.4
2012-12-13	23:03:38.53	80.8040	121.3830	10.0f	5.7

The epicentral distances were calculated by using different NOES, Sval, BARENTS, BARENTS3D and IASPEI91 travel-times. Calculated values of epicentral distances were compared with "true" distances between stations and epicenters from Table 11. Access to the data from SPA0, KBS, HSPB, BJO1 и HOPEN stations was carried out using the GEOFON program (www.geofon.gfz-potsdam.de/geofon/).

Figure 19 shows the results of comparison of travel-times, where the x-axis represents epicentral distance between the seismic station and the epicenter in degrees and the vertical axis shows δΔ° value, defined as the difference between calculated and "true" epicentral distances:

$$\delta\Delta^{\circ} = \Delta^{\circ}_{calculate\_dist} - \Delta^{\circ}_{true\_dist}$$

Thus the value of δΔ° allows us to determine distribution of points in the graph with such terms as "undershoot" and "overshoot", i.e., when calculated epicentral distance is less or more than the "true" distance from the station to the epicenter, according to the data from Table 11.

The results of comparing NOES, Sval, BARENTS, BARENTS3D and IASPEI91 travel-times are, in our opinion, quite interesting (Figure 19). Thus, within epicentral distances up to 2.2° travel-times demonstrate almost the same inaccuracy, which is mainly "undershoot". In this case, BARENTS and BARENTS3D travel-times show that seismic waves at these distances have much higher velocity compared to other travel times.

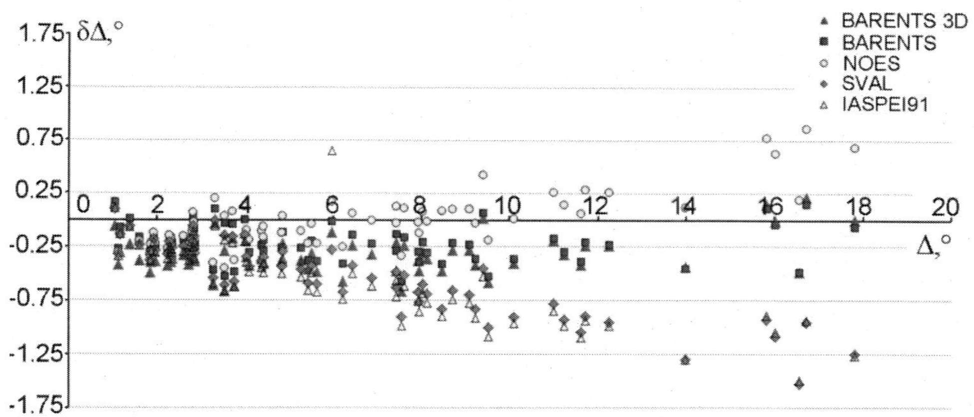

Figure 19. The results of comparison of travel-times NOES, Sval, BARENTS, BARENTS3D and IASPEI91.

Starting from the distance of 2.2°, travel-times behave differently. A big "undershoot" is typical for IASPEI91 and Sval travel-times. With increasing epicentral distance, variation increases and may reach 1.5°, i.e., almost 170 km. BARENTS and BARENTS3D travel-times, within the whole considered range of distances (from 0° to 18°), is characterized by a small "undershoot", but values of the "undershoot" do not exceed 0.5°. NOES travel-time is characterized by slight variation at zero point up to epicentral distance of 10°, then the "overshoot" starts. In this case, variation values do not exceed 0.25° up to 14° of epicentral distance. After the distance of 14°, a big "overshoot" is registered. Thus, it has been determined empirically that the use of NOES travel-time in calculating parameters of arctic earthquakes gives smaller errors in location of epicenters than when using other travel-times.

Efficiency of computed NOES travel-time can be illustrated by Figure 14b, which shows epicenters of earthquakes for the period from 2011 to 2014. Earthquake source parameters were recalculated using received NOES travel-time. It is evident that epicenters of up to 30° are clearly distributed along the centerline of the Gakkel Ridge. Since 30° the epicenters are slightly shifted to the south with respect to its axial part. The reason for this is likely in the absence of seismic stations in the east. However, this allows us to analyze seismic activity along the Gakkel Ridge.

Location of Low-Magnitude Earthquakes

We have processed 417 earthquakes of magnitude from 1.5ML to 5.7ML from the Gakkel Ridge (from 7° W to 80° E) for the period from December 2011 to January 2015 (Figure 20). In this, about 70% of earthquakes are of a magnitude, which does not exceed 3.0.

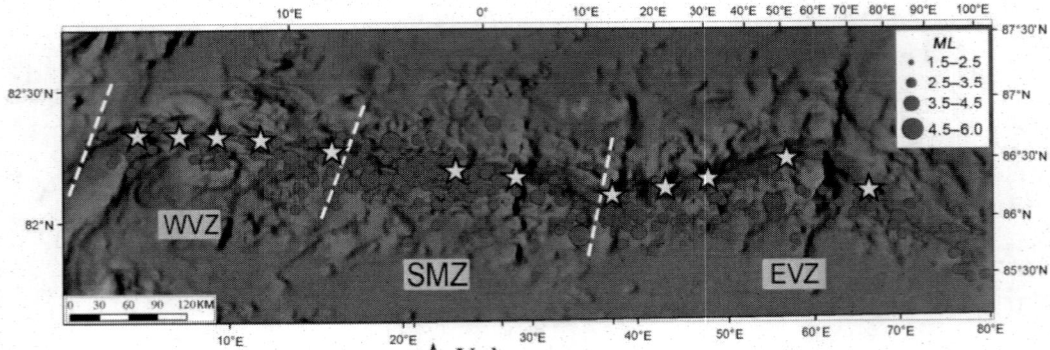

Figure 20. The location of the epicenters of earthquakes of the Gakkel Ridge for the period 01.12.2011–31.01.2015.

The accuracy of the location of the epicentre of the Gakkel Ridge earthquake was made by calculating error ellipses. This method allows determining the direction with the highest and lowest measurement error. Calculations were performed in the NAS program (Asming & Prokudina, 2016), the following errors were made for the calculations: phase input time was 0.3s, velocity model 0.15 km/s) within period 01.01.2004–01.01.2011 and for regional monitoring (according to NOES hodograph) for the period 01.12.2011–31.01.2015. The results are shown in Table 12 and Figure 21.

The values of error ellipses according to regional monitoring data are quite large due to the imperfection of the configuration of recording stations. After the opening of the SVZ station, the average error in the location of EVZ earthquakes with ML>3 decreased and became: Rminor=15.8 km, Rmajor=40.1 km. As shown in Table 12, it almost corresponds to the error according to the ISC data. Thus, for events with a magnitude ML>3 recorded by a large number of seismic stations, ISC processing remains more accurate. For low-magnitude earthquakes, the use of the NOES hodograph remains the most effective for location.

Table 12. The average error of the epicenters location of the earthquakes of the Gakkel Ridge according to regional data (01.12.2011–31.01.2015) and teleseismic monitoring (01.01.2004–01.01.2011)

Catalogue	R_{minor}, км			R_{major}, км		
	WVZ	SMZ	EVZ	WVZ	SMZ	EVZ
	ML>3					
FCIARctic RAS	26.5	26.8	36.9	51.4	48.3	50.5
ISC	6.2	12.9	14.9	10.2	33.6	36.9
	ML<3					
FCIARctic RAS	25.2	26.4	41.2	53.3	55.0	63.4

The distribution of earthquakes by magnitude in each zone from 01.12.2011 – 01.10.2017 is shown in Figure 22. The value of ML for WVZ is 3.1, for SMZ is 2.5, for EVZ is 3.3. It can be seen that the western and most central zone is least seismically active, which is most likely due to differences in earthquake generation mechanisms. There are practically no active volcanic centers in the central zone, but earthquakes with ML≥4 are relatively small. The obtained result shows that the process of spreading Gakkel Ridge weakens from west to east not evenly, but segments in which different seismic mechanisms work. This is also indicated by differences in the spectral composition of the earthquake that we have previously identified (Rogozhin et al., 2016).

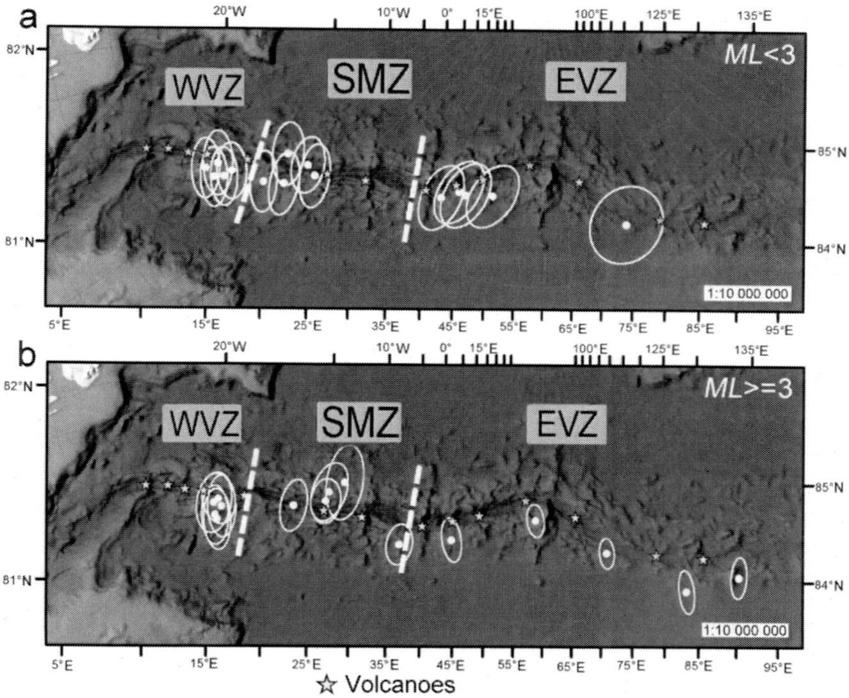

Figure 21. Error ellipses of earthquakes of the Gakkel Ridge according to the regional monitoring based on the NOES hodograph: a – *ML*<3, b – *ML*>3.

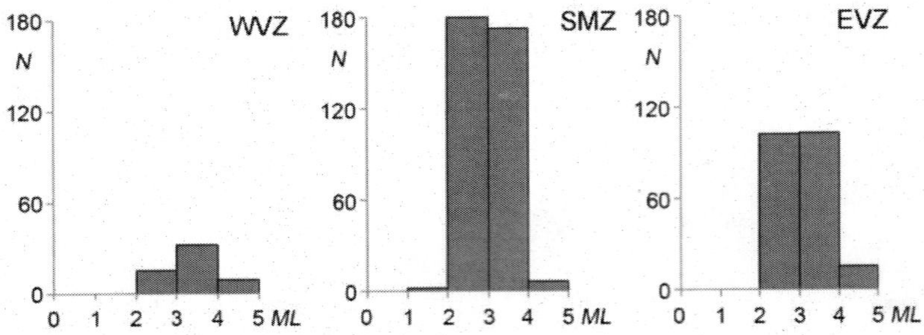

Figure 22. Histograms of magnitudes of earthquakes of the Gakkel Ridge for period 01.12.2011–01.10.2017 according to regional monitoring.

The spatial-temporal distribution of earthquakes is an indicator for the geodynamics of the region. The earthquakes epicenters are distributed along the ridge unevenly. The large number of epicenters are assigned to the Sparsely Magmatic Zone, particularly to the area from 1.5°E to 19.0°E (Figure 23). This area accounts for up to 50% of all recorded earthquakes. The left border of this area merges with the break of the WVZ и SMZ. The right border coincides with the location of the only one volcanic center in the SMZ. Further along the ridge (up to 26°E) earthquakes occur rarely. Only from 26°E, the number of epicenters of the recorded earthquakes starts to increase again. We can distinguish the area from 26°E to 42°E within which there is also a significant number of earthquakes.

Within the WVZ and EVZ the number of recorded earthquakes is fewer, as compared with the SMZ, and the epicenters are distributed unevenly. Within the WVZ we can define the area that lies between 6° W and 3° W, which most epicenters are assigned to (Figure 23). Within the EVZ small number makes it impossible to identify certain regularities in the distribution of epicenters.

Earthquake swarms are characteristic of volcanic regions, shocks for destruction of tectonic nature (Radziminovich & Ochkovskaya, 2013). During the entire operation of the seismic observation points "Franz Josef Land" and "Severnaya Zemlya" in the Gakkel Ridge area, 4 (four) swarms of earthquakes were detected. Table 13 shows their parameters. All are located near volcanic centers, but location errors are quite large.

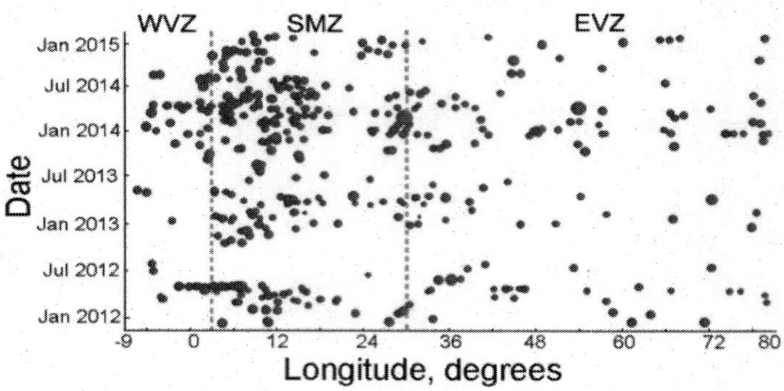

Figure 23. Temporal earthquake distribution along the Western Volcanic Zone and Sparsely Magmatic Zone of the Gakkel Ridge during the period from December 2011 to August 2014 and registered by the seismic station ZFI and the stations operating in the Svalbard archipelago. Dotted line divides the Gakkel Ridge into segments: Western Volcanic Zone (WVZ), Eastern Volcanic Zone (EVZ) and the Sparsely Magmatic Zone (SMZ).

Table 13. Parameters of earthquakes swarms of the Gakkel Ridge according to regional monitoring data

Period	District	*N*	*f*, Hz	ML_{min}	ML_{max}	Average value		
						R_{minor}, км	R_{major}, км	Az_{major}, °
29.04.2012–30.04.2012	1°E	9	1.5–10	2.8	3.3	25.8	58.0	120
20.04.2012–21.04.2012	40°E	6	1.5–12	2.0	2.9	37.0	64.0	14
31.12.2013	75°E	6	1.5–9	2.0	3.1	61.9	74.0	60
28.10.2017–29.10.2017	55°E	19	1.5–10	3.6	4.8	16.8	25.7	137

Assignment of most earthquakes to the SMZ segment can be explained by the amagmatic character of this segment's spreading. Within the segment the crustal thickness is minimal, and the mantle's temperature is lower. The structuring of this part of the ridge is characterized with the prevalence of tectonic processes, not magmatic or metamorphic ones (Goldstein, 2008; Dubinin, 2013). However, for large magnitude earthquakes (above mb>4.4) the assignment of epicenters to the area from 1.5°E to 19.0°E has not been observed for the total observation period (Engen et al., 2003).

A feature of the frequency content in waveforms of the ZFI station was also identified for a part of earthquakes within the SMZ segment (Figure 24). These earthquakes are characterized with both low and high frequencies in spectral-time diagrams. The rest of the recorded earthquakes are specified only with low frequencies in spectral-time diagrams. We have not determined the correlation between the character of frequency content of such earthquakes and their distances and magnitude. We suppose that the reason lies in processes within the very focal points of such earthquakes.

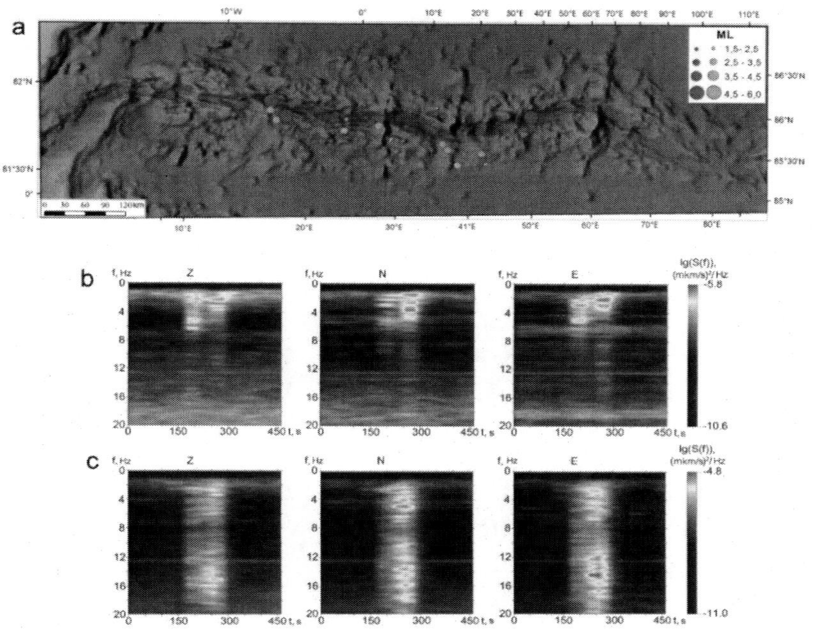

Figure 24. The distribution of earthquakes of the Gakkel Ridge (a): red circles – these earthquakes are characterized only with low frequencies in spectral-time diagrams (b); green circles – these earthquakes are characterized with both low and high frequencies in spectral-time diagrams (c). The epicenters of earthquakes map the combined with a bathymetric data (www.ngdc.noaa.gov).

The number of recorded earthquakes within the Gakkel Ridge has varied during the year (Figure 24). During summer months (from July to September) the reduction of the recorded earthquakes is connected to the seasonal variations of the ZFI seismic station's response. But this cannot be the case for the total lack of earthquakes from May to October 2012 within the entire length of the WVZ and SMZ segments. Perhaps, here we had a seismic quiescence, which occurred after a large series of earthquakes in April 2012. That month 39 earthquakes were recorded. And in the area between 1.5°W and 7.0°E, 13 earthquakes were recorded within two days. Moreover, for the WVZ segment the quiescence period lasted the next year until November 2013.

The part of epicenters can be combined into the groups due to their close proximity to each other (Figure 24). They either precede an earthquake of larger magnitude (foreshocks), or have substantially similar magnitudes. They seem to reflect some active process, proceeding at the given time and place. As mentioned above, basing on the bathymetric and geophysical features analysis (Engen et al., 2003; Gurevich et al., 2004) there are additional segments and fracture zones (FZ) in the Gakkel Ridge. The positional relation of earthquake epicenters (Figure 23) allows for determining, where we can define the border between segments (or FZ) clearly, and where it is impossible (Table 14). FZs with coordinates of 3-4°W, 3°E, 23-25° and 63°E are clearly defined (comparatively clear) with the positioning pattern of earthquakes epicenters, and FZs with coordinates of 7°E, 14°E, 34°E and 39°E cannot be defined at all.

Table 14. List of identified or not identified boundaries defined by the geophysical data on the distribution of earthquakes in Figure 23

Name of the boundaries (or FZ) by (Gurevich et al., 2004) (*) and (Engen et al., 2003) (**)	Latitude,°	Description
I-II*	~3°-4° W	Identified
II-III*, FZ5**	~3° E	Identified
III-IV*	~7° E	Not identified
FZ6**	~14° E	Not identified
IV-V*, FZ7**	~23°-25°E	Identified
V-VI*, FZ8**	~34° E	Not identified
VI-VII*, FZ9**	~39° E	Not identified
FZ10**	~63° E	Identified

Discussions

The results of the seismic monitoring at the Gakkel Ridge from December 2011 to January 2015 allow for the determination of the character of small magnitude earthquakes distribution. The article presents the following results:

i. The crustal velocity structures of two areas of Euro-Arctic Region Svalbard and Franz Josef Land were determined using receiver functions. This allowed us to calculate two travel-times for regional phases there are NOES and Sval. Validation of NOES and Sval, as well as IASPEI91, BARENTS and BARENTS3D travel-times showed that NOES travel-time was the best for calculating the location of earthquakes in the area of spreading Arctic ridges.

ii. The Gakkel Ridge is seismically active, despite its lowest spreading velocity among global Mid-Ocean Ridges. The epicenters of recorded earthquakes within the ridge area are distributed unevenly. The most number of epicenters are assigned to the Sparsely Magmatic Zone (SMZ), and particularly to the area within the coordinates between 1.5°E and 19.0°E. The assignment of the most earthquakes to the SMZ segment can be explained by the fact, that spreading of this segment is amagmatic. The structuring of this part of the ridge is characterized with the prevalence of tectonic processes, not magmatic or metamorphic ones. The only exclusion is the single volcanic center area (19°E) in the SMZ, where we can register practically no seismic activity. For a number of earthquakes within the SMZ both low and high frequencies are observed in the spectral-time logs. The rest of the earthquakes are characterized only with low frequencies in the spectral-time logs.

iii. The number of recorded earthquakes within the Gakkel Ridge has varied during the year. There were periods of quiescence and seismically active periods. Some of the ridge's sections are characterized with earthquake swarms, which seem to reflect some active process, proceeding at the given time at the place.

iv. The positional relation of earthquake epicenters allowed us to determine, where we can define the border between segments (or FZ) clearly, and where it is impossible. FZs with coordinates of 3-4°W, 3°E, 23-25° and 63°E are clearly defined (comparatively clear) with the positioning pattern of earthquakes epicenters, and FZs with coordinates of 7°E, 14°E, 34°E and 39°E cannot be defined at all.

NEW DATA ABOUT SEISMICITY AND CRUSTAL VELOCITY STRUCTURE OF THE "CONTINENT-OCEAN" TRANSITION ZONE OF THE BARENTS-KARA REGION

A great deal of emphasis is currently being placed on conducting basic and applied scientific research examining the shelf and continental slope of the Arctic region. This is the result of several factors, including the discovery of large oil and gas deposits in the region and the necessity of their commercial development, as well as a general lack of knowledge regarding the tectonic and geodynamic issues surrounding the interaction of continental and oceanic crust. Most important in this context is information about current seismicity of the studied areas, which forms the basis of seismic hazard studies and data on modern geodynamic processes. This paragraph presents the results of a study of seismic of the "continent-ocean" transition zone of the Barents-Kara region and determination the crustal structure in the area of the ZFI seismic station.

Tectonics and Geomorphological Description of the Study Area

Fundamental data concerning the tectonics of the Arctic basin have previously been set out in (Khain, 2001). Special aspects of the geological and tectonic structure of the Arctic Ocean "continent-ocean" zone are discussed in (Pogrebitskiy, 1984), with the results of recent research aimed at clarifying the structure of individual sections of this zone presented in

(Zayonchek et al., 2010; Moroz et al., 2010). The evolution of the Barents-Kara region is examined in (Baturin, 1988; Verba, 2007), with geomorphic data associated with the largest morphostructures of the Barents Sea shelf available in (Dibner, 1978; Milanovskiy, 1996). The Barents-Kara region is characterized by the activated and recently uplifted (up to + 500-1500 m) blocks of the continental marginal plate.

The continental slope of the Arctic Ocean's Eurasian sub-basin can be classified as Atlantic type based on morphological indices. According to (Pogrebitskiy, 1984), the tectonic structure of the "continent-ocean" transition zone under consideration is represented by a wedge-shaped crustal block around 150 km in width, with the thickness of 28-30 km at the shelf edge and around 10 km at the coupling with the oceanic plate, constrained by the bottom surfaces and Moho with generalized slopes of 1.5° and 6°, respectively. According to regional gravimetrical surveys, the wedge-shaped block is isostatically balanced. The continental slope includes the edge of the shelf, the slope itself and the continental rise.

The unique aspect of the marginal areas of the Barents and Kara Seas is the existence of large negative morphostructures, which are referred to as trough valleys and/or trenches. In the north these trenches include the Orle, Franz Victoria, St. Anna, Voronin, Shokalski, Vilitski and Komsomolskaya Pravda. Some trenches fall deep into the shelf and diverge into branches; further investigation into their extensions has revealed more shallow grabens, resembling river valleys in size. Tectonically they are associated with grabens formed as a result of multiple tectonic and geomorphic processes. Verba (Verba, 2007) describes in detail the morphology, genesis and age of these shelf edge grabens, showing their rhegmatogenous nature and association with rifting processes taking place during the "break-up" stage of Eurasian sub-basin generation.

Theories surrounding the seismicity of the "continent-ocean" transition zone of the Barents-Kara region are rather schematic, which reflects the remoteness of the recording seismic stations. Based on the analysis of all available data concerning the Arctic earthquakes, (Avetisov, 1996) found that the continental slope is marked by rare unevenly-distributed epicentres; these form individual groups, gravitating towards the marginal-shelf grabens and faults.

Analysis of the "Continent-Ocean" Transition Zone of the Barents-Kara Region

Study area (latitude from 80.5° to 84.5°, longitude from 5.0°E to 115.0°E) is shown on the Figure 25. We considered period from October 2011 to January 2014. The map of the seismic stations witch we used for monitoring is shown in Figure 20. Access to the data was carried out using the GEOFON program (www.geofon.gfz-potsdam.de/geofon/).

The analysis of seismic data to earthquake hypocenters was completed with the help of the WSG program (www.ceme.gsras.ru). Earthquakes were located using the body wave model BARENTS developed by the Kola Branch, GS RAS (Kremenetskaya et al., 2004). To determine local magnitude ML the WSG calculation method was used, based on an averaged northern Eurasia calibration function (Gabsatarova, 2006).

Table 15. The seismic catalogue of recorded earthquakes and the calculated parameters of earthquakes for period 2011-2014

No	Date dd.mm.yyyy	Origin time hh:mm:ss.s	Lat., φ,°	Lon. λ,°	Magnitude ML	Stations
1	09.12.2011	04:56:29.1	81.23	35.92	2.1	ZFI, SPA0
2	10.12.2011	11:04:16.0	80.96	29.36	1.4	ZFI, SPA0
3	21.12.2011	23:53:33.6	82.66	33.65	1.4	ZFI, SPA0
4	27.01.2012	09:05:46.0	81.66	35.77	2.2	ZFI, SPA0, KBS
5	04.03.2012	21:53:57.4	81.33	31.60	1.3	ZFI, SPA0
6	06.03.2012	18:13:11.6	82.55	34.22	1.3	ZFI, SPA0
7	08.03.2012	00:46:56.0	83.57	23.83	2.1	ZFI, SPA0
8	13.03.2012	04:48:43.0	83.99	42.50	1.8	ZFI
9	17.03.2012	15:17:37.0	84.05	49.75	2.1	ZFI
10	08.04.2012	21:51:09.3	81.69	70.47	2.1	ZFI, SPA0, KBS
11	29.04.2012	10:07:34.0	82.16	40.33	2.2	ZFI, SPA0, ARE0, SFJD
12	03.05.2012	08:51:10.8	82.97	64.53	2.2	ZFI
13	07.05.2012	10:52:41.0	80.50	20.16	2.2	ZFI, SPA0, KBS
14	07.06.2012	00:55:06.0	82.38	68.75	2.1	ZFI
15	08.06.2012	00:55:40.0	82.08	38.41	1.4	ZFI
16	21.06.2012	14:10:55.4	84.06	87.85	2.5	ZFI
17	30.06.2012	18:55:12.4	84.37	55.30	2.4	ZFI
18	11.10.2012	09:28:51.2	83.43	71.14	-	ZFI
19	27.10.2012	04:48:55.3	80.80	108.48	-	ZFI
20	07.12.2012	02:47:01.4	82.93	50.21	2.9	ZFI, SPA0, KBS, KEV, HSPB, HOPEN, ARA0
21	28.01.2013	13:13:28.0	81.93	38.91	-	ZFI
22	30.01.2013	10:58:53.0	80.59	33.65	1.4	ZFI, SPA0
23	30.01.2013	12:07:30.5	81.75	39.46	0.9	ZFI
24	31.01.2013	23:37:23.0	81.90	12.79	2.2	ZFI, SPA0
25	03.02.2013	20:55:38.0	81.29	36.12	1.9	ZFI, SPA0, KBS
26	09.02.2013	07:28:19.0	81.52	09.21	2.6	ZFI, SPA0
27	20.02.2013	07:10:52.0	82.34	38.89	1.6	ZFI
28	20.02.2013	15:14:56.7	83.40	63.69	2.0	ZFI
29	09.03.2013	12:35:47.6	79.33	114.12	-	ZFI
30	10.03.2013	21:15:21.4	82.40	93.25	2.3	ZFI
31	11.03.2013	05:38:22.7	81.92	37.29	1.3	ZFI, KBS
32	12.03.2013	01:05:24.6	83.02	64.35	1.8	ZFI
33	12.03.2013	14:26:09.8	82.85	68.96	1.7	ZFI
34	16.03.2013	20:07:50.4	81.58	32.13	1.3	ZFI, SPA0
35	17.03.2013	02:30:16.0	80.77	38.50	0.8	ZFI, SPA0
36	19.03.2013	00:20:11.0	82.24	39.62	1.2	ZFI
37	19.03.2013	00:21:20.0	82.51	46.54	1.2	ZFI
38	21.03.2013	01:53:53.0	82.04	38.84	1.4	ZFI, SPA0, KBS
39	26.03.2013	14:42:18.0	80.58	13.10	2.1	ZFI, SPA0, KBS
40	01.04.2013	16:11:24.6	82.70	29.52	2.9	ZFI, SPA0
41	02.04.2013	03:40:42.9	83.03	40.89	1.7	ZFI, SPA0
42	03.04.2013	12:40:38.1	83.08	69.33	2.0	ZFI, SPA0
43	04.04.2013	23:13:59.7	80.59	41.88	1.7	ZFI, SPA0
44	07.04.2013	10:09:27.9	82.09	50.56	0.7	ZFI
45	07.04.2013	11:09:48.2	81.82	42.03	0.7	ZFI
46	07.04.2013	15:58:28.0	83.50	66.01	3.3	ZFI, KBS
47	08.04.2013	10:40:20.3	81.39	36.23	0.9	ZFI
48	25.05.2013	17:45:36.0	80.54	39.20	2.0	ZFI, SPA0, KBS
49	18.06.2013	04:09:26.5	84.13	86.93	2.2	ZFI
50	27.06.2013	01:33:10.0	82.18	39.50	0.9	ZFI
51	12.07.2013	01:31:36.8	83.64	65.44	2.1	ZFI, SPA0
52	25.10.2013	01:25:56.0	80.33	40.06	1.9	ZFI, SPA0, KBS

As the main station working in considered region, we chose the seismic station ZFI. However, not all earthquakes in the continental slope region registered by ZFI were detected by other stations, even when using filtering procedures. In this case, single station locations were determined using backazimuths and epicentral distances (Kremenetskaya & Asming, 2002). The backazimuths are obtained by P wave polarization analysis and the distances are computed from S-P time differences. In this case depth is impossible to estimate so it is assumed to be 0. Such estimation is rough: errors of backazimuth can amount to 10 degrees. To minimize erroneous locations, only earthquakes having P and S phases with clear onsets and high signal to noise ratios of were analyzed. The final seismic catalogue is given in Table 15.

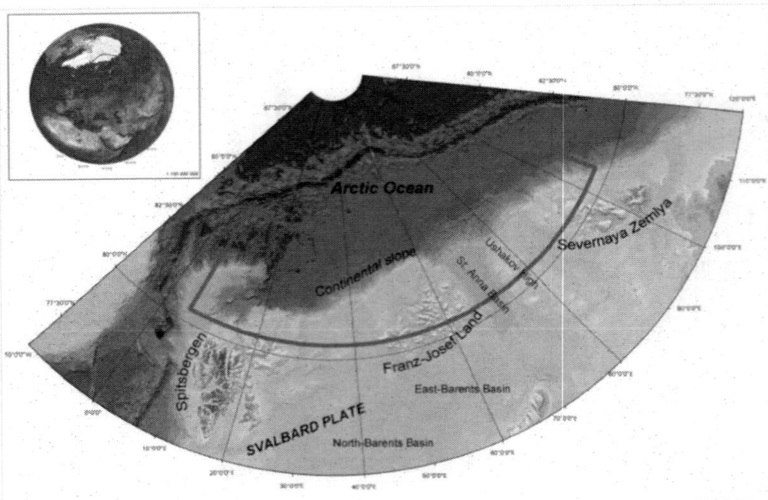

Figure 25. Map showing the boundaries of the study area.

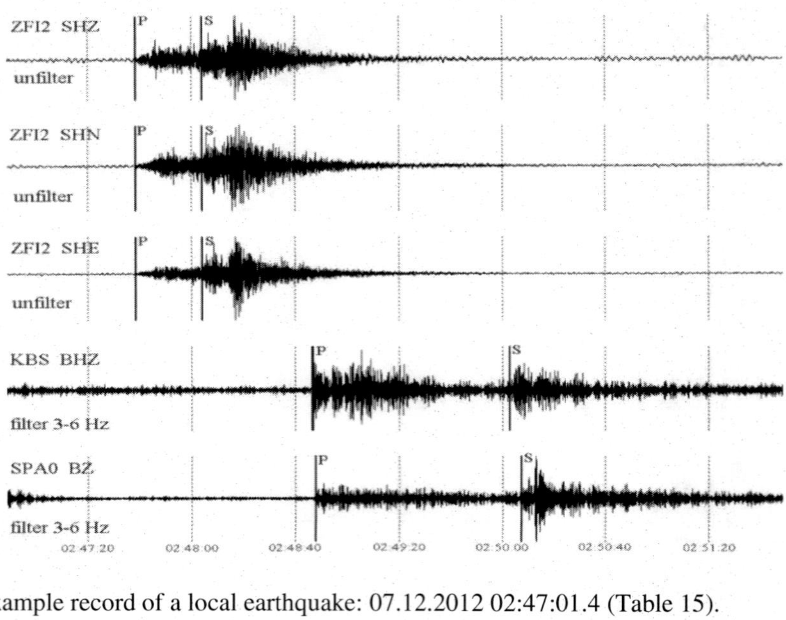

Figure 26. Example record of a local earthquake: 07.12.2012 02:47:01.4 (Table 15).

We registered 52 earthquakes within the "continent-ocean" transition zone of the Barents-Kara region during the period from October 2011 to January 2014. Magnitudes ranged from 0.7 ML to 3.3 ML. An example wave forms of local earthquake recorded by some stations is shows in Figure 26.

The spatial distribution of the epicenters exhibits a range of specific features (Figure 27), including their clear confinement to the negative morphostructures of the continental slope (grabens). Most of the located earthquakes are confined to the Franz Victoria and St. Anna Grabens. The earthquake epicenters are also located unevenly in the area of the grabens themselves.

In the area of the Franz Victoria Graben, the earthquake epicenters gravitate towards several different sections, with most located in the estuarial part of the graben, right on the continental slope. Small groups of epicenters lie on the border between the Bely and Victoria uplifts in the northern and central parts of the graben, with several epicenters of weak earthquakes also confined to the continental rise in front of the graben estuary.

The earthquake epicenters registered in and around the St. Anna Graben gravitate towards the latter's wide estuarial part in the continental rise area. In this part of the graben, an earthquake with a magnitude of 3.3 was recorded on April 07, 2013, the strongest of all the considered events. Two additional earthquakes took place right in the graben estuary, closer to the shelf edge. To the north of the Franz Josef Land archipelago, in the region of the Britannia Graben, two weak earthquakes were registered with ML<2.0, as well as one earthquake with ML=2.9 on the continental slope.

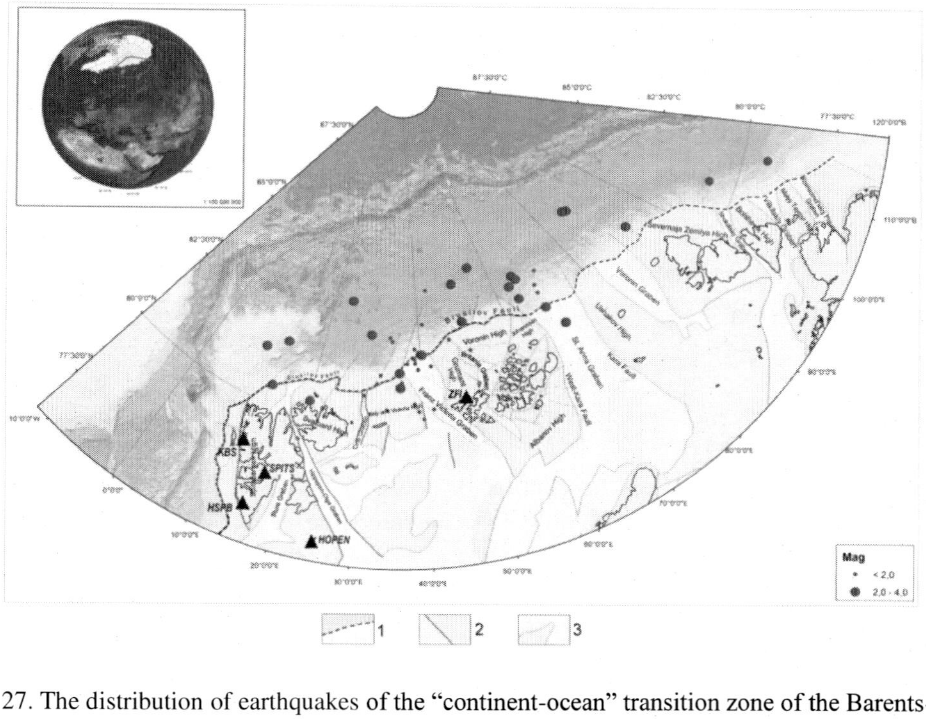

Figure 27. The distribution of earthquakes of the "continent-ocean" transition zone of the Barents-Kara region during the period from October 2011 to January 2014. The epicenters of earthquakes map the neotectonic structures and active faults (Atlas, 2004), combined with a bathymetric data (www.ngdc.noaa.gov): 1 – shelf edge, steep of flexure-fault zone; 2 – main neotectonic faults; 3 - dislocation with a break of continuity.

Almost all earthquakes recorded within the regions of the Franz Victoria, Britannia and St. Anna Grabens were characterized by local magnitudes of up to 2.0, unlike in those grabens located to the east. This pattern can be explained by the remoteness of the ZFI seismic station. However, the absence of minor seismicity within the Orle graben region, in which only one weak earthquake (in its estuarial part) was observed during the whole investigation period, cannot be explained by its distance from the station.

Three earthquakes were registered in the vicinity of the Voronin graben: one on the slope and two on the continental rise in front of the graben estuary. Two weak earthquakes were recorded on the continental rise in front of the estuarial parts of the Shokalski and Komsomolskaya Pravda grabens. As already stated however, their remoteness from the station did not enable comprehensive information regarding the seismic events occurring in these regions to be acquired.

Several earthquakes were also registered in the region of the continental rise to the north of the Franz Josef Land archipelago, as well as in the Yermak plateau area. These events are not directly connected to the large morphostructures of the continental slope. The spatial distribution of earthquakes in the study region is of specific scientific interest. Firstly, due to its geographical and climatic conditions, the considered region poorly studied in terms of seismicity. Secondly, the obtained data, together with the available geophysical information enable a fresh look to be taken at the geodynamics of the region. An analysis of the registered earthquakes can help to identify the geodynamic processes responsible.

The formation of rift structures in the Barents-Kara region is a consequence of crustal extension processes that took place during the formation of the Eurasian sub-basin, with the development of the grabens occurring during the basin's pre-oceanic development or "break-up" stage. According to the data presented in (Verba, 2007), this "break-up" stage lasted for around 100 million years, from the second half of the Jurassic to the end of the Cretaceous. Rift formation then went through a complex phase with the greater involvement of fluid- rather than rupture-driven failure. As a result, the period of active rifting has now finished and thus should not make a substantial contribution to the seismicity of the region.

This conclusion has been confirmed by research examining heat flow in the Franz Victoria graben, which is generally characterized by typical Barents Sea background values (Khutorskoy et al., 2009). Geo-acoustic profiling data have also been obtained (Verba, 2007) along the Franz Victoria and St. Anna grabens, with no evidence observed of active graben development. However, heat flow research performed in the vicinity of the Orle graben has revealed anomalously high values ranging from 300 to 520 mW/m^2, almost 10 times higher than background levels (Khutorskoy et al., 2009). Such high flow is typical for both the entire Orle graben and its extension on the continental slope, right up to the 1200 m isobar. The authors of this study assumed within the graben area the presence of crust destruction throughout its entire thickness, with the injection of hot mantle material inside the foundation already providing evidence of the graben active development phase. The almost complete absence of registered earthquakes from the Orle graben area can be explained by the high temperatures in the lithosphere.

From 2006 to 2009, the staff of the Geological Institute of the Russian Academy of Sciences, in collaboration with the Norwegian Petroleum Directorate, undertook three expeditions aimed at clarifying the structure of individual sections of the "continent-ocean" transition zone of the Barents Sea's northwestern margins (Zayonchek et al., 2010). These expeditions revealed a system of large landside on the continental slope of the Arctic Ocean.

Landslide had previously been found on the Norwegian continental slope, in the Litke graben and at the extension of the Hinlopen Strait (Hjelstuen et al., 2007; Vanneste et al., 2006; Winkelmann & Stein, 2007).

During one of the expeditions, signs of the intensive offset of large rock fragments of turbidite fracture and alluvial cone formation were detected inside the Orle graben formation. Similar phenomena were also observed on the western side, mainly occurring due to the isostatic compensation of avalanche sedimentation. It is therefore possible to assume that the eastern grabens, starting from the Franz Victoria, also act as channels for sedimentary material transport.

Staff at the Kola Branch, GS RAS (Vinogradov & Baranov, 2013) studied the nature of seismic events in the western part of the Barents Sea shelf, within the continental slope area. Regarding minor seismicity ($ML \leq 2.2$), seasonality was observed within the earthquakes registered in the area from the Strait of Storfjorden (Svalbard archipelago) to Bjornoya Island. In accordance with the suggested hypothesis, most of the weak earthquakes are the result of the landslide processes taking place in steep shelf areas, with their seasonality reflecting rapid changes in alluvial material mass flow during the warm season. In particular, by analyzing the records of the SPITS seismic group, the authors were able to identify waveforms made by landslide events in the north of Spitsbergen.

Therefore, based on an overall analysis of the available geophysical, geotectonic and new seismic data, it can be assumed that the prevailing geodynamic factor responsible for the occurrence of weak earthquakes is the isostatic compensation of avalanche sedimentation in the "continent-ocean" transition zone.

Determination of the Crustal Velocity Structure

For the determination of crustal velocity structure in the substation area, the PRF receiver functions technique was employed (Vinnik, 1977). Broadband three-component records of teleseismic earthquakesfrom 2012 and 2013were used as source data for the receiver function study. A total of 41 earthquake data records were selected, in all azimuthal directions and epicentral distances from 35o to 90°. Earthquake magnitudes ranged from 5.8 to 8.3, with focus depths extending down to 300 km. The records have clear P-wave arrivals, with the first arrival amplitude exceeding that of noise by more than 3 times. Processing of the selected earthquake waveforms was carried out by the Seismic Handler software package (www.seismic-handler.org) and included and included a similar procedure described in paragraph "*Calculation of travel-time for regional phases*".

The receiver function maxima, clearly defined in the resulting Q-component, can be explained by converted Ps phase formation on the sedimentary sheath base, with a delay time of tps=0.8 s and a Moho discontinuity of tps=3.8 s (Figure 28). Large amplitudes between 10 and 15 seconds are conditioned by the presence of multiples Ppps and Ppss phases, formed by the conversions-reflections of the free surface and at the Moho discontinuity.

In order to carry out inverse modelling and determination of the crustal velocity structure beneath the ZFI station, the regularization method (Tikhonov & Glasko, 1965) was selected. In this method the problem is reduced to solving a system of linear equations, containing parameters for the velocity of P and S waves (Vp, Vs), as well as layer densities (ρ) and layer thickness. For the inversion a horizontally-layered model was employed, lying on a half-space, with constant velocities inside the layers, gradually increasing with depth. The density was as specified by (Birch, 1961). The search for the optimal parameters of the medium in the velocity model is based on the general method employed to solve poorly-posed inverse problems, and is determined by the selection and iterative minimization of discrepancies between the synthetic receiver functions and experimental functions. Synthetic receiver function is calculated using the (Haskell, 1962) method. Calculations were performed using the programs developed by the Institute of Physics of the Earth of the Russian Academy of Sciences (Kosarev, 1987).

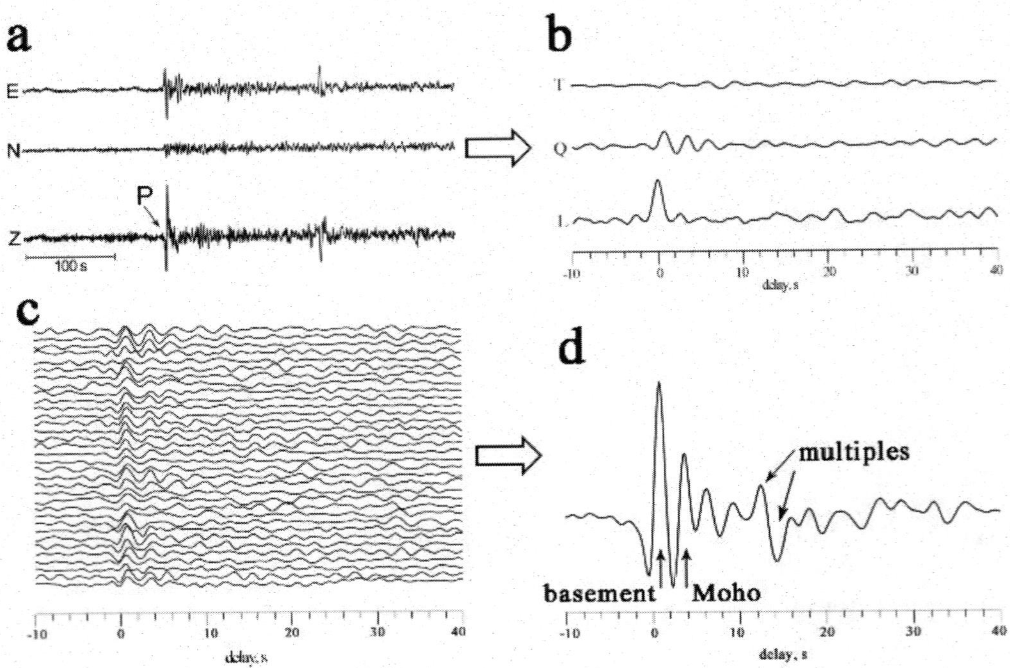

Figure 28. The processing steps by receiver functions for ZFI station: the initial three-component recording (a), P-receiver functions of one earthquake (L, Q, T components) (b), P-receiver functions (Q components) several earthquakes (c), summary P-receiver function of 41 earthquakes (Q component) (d). Time zero refers to the direct P phase. The Ps conversion phases from the basement, Moho and associated multiples from the Moho and reflection at the free surface (Moho multiples) are marked by arrows.

The results of mathematical modeling, synthetic receiver function and its comparison with experimental (right), and the final one-dimensional velocity model Vs (left) are shown in Figure 29. Several interfaces can be observed in depth section, distinguished by their varying degrees of converted phases traceability. The following features are most obviously identifiable (bottom - up): the Moho discontinuity at a depth of 30 km, with the velocities in the underlying mantle being Vp=8.15 km/s and Vs=4.5 km/s, and the discontinuity of a sedimentary layer having a total thickness around 4 km with Vp=4.3 km/s and Vs=2.36 km/s. The resulting model also indicates a mid-crustal boundary at a depth of about 17 km with a velocity contrast separating the two layers: the upper layer of crust has Vp=6.1 and Vs = 3.6 km/s; lower crustal layer has Vp=6.8 and Vs = 3.9 km/s. Velocity Vp was recalculated to Vp based on the ratio Vp/Vs between 1.7 and 1.82.

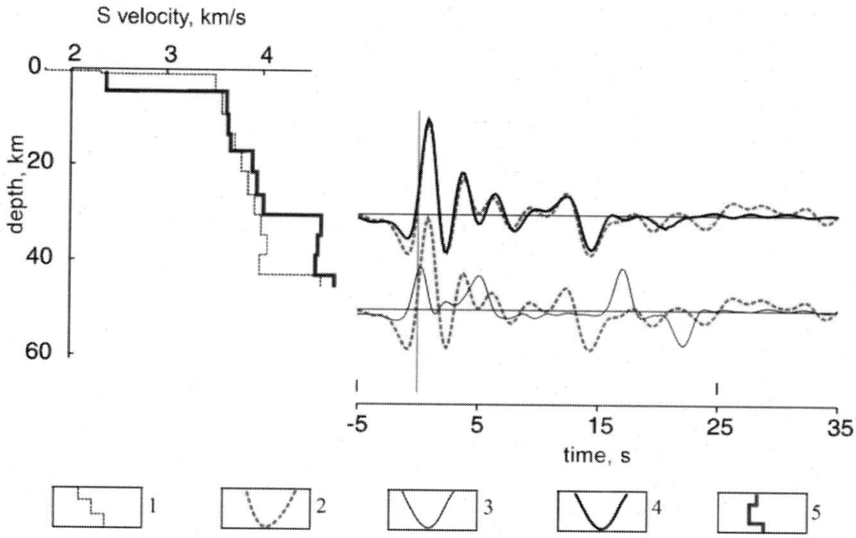

Figure 29. 1-D velocity models of the crust and upper mantle beneath ZFI station: 1 - initial velocity model; 2 – experimental P-receiver function; 3 – synthetic P-receiver function calculated using the initial model; 4 – synthetic P-receiver function calculated using the final model; 5 - final model.

One unique feature visible in the obtained depth section is a small depth of "crust-mantle" discontinuity (H=30 km). The observed thinning of the consolidated crust to 26 km (excluding sedimentary cover) is the result of rift structure formation related to crustal extension under conditions of continental rifting, as well as the transition from continental to oceanic crust. The obtained velocities of P and S waves inside the crustal velocity structure and upper mantle are here not considered abnormal since they possess values typical of continental crust.

The recorded boundary and velocity data match the results of previous research examining the deep structure of Franz Josef Land, published in (Avetisov & Bulin, 1974; Sakoulina et al., 2000). Detailed crust dissections performed at the Franz Josef Land archipelago (crustal thickness 23-25 km) detected the Conrad discontinuity at a depth of 15-17 km, with a Mesozoic rock thickness of 1.5-4 km.

General Discussions

Let's summarize the research. The opening some island stations in the Russian sector of the Arctic has created opportunities for ongoing monitoring the Barents-Kara region and the Gakkel Ridge. Obviously, the opening of new stations in areas with a low density of seismic study improves the earthquakes location accuracy. We made a theoretical calculation of the minimum magnitudes (Mmin) for Barents-Kara region and Gakkel Ridge area by using V.Y.Burmin's program (Burmin, 2012) for two options. First option is the calculation for existing seismic stations, the second one is with opening of new ones, for example, on the Ushakova island, Severnaya Zemlya and Novosibirsk Islands (Figure 30). The theoretical distribution of Mmin makes assess not only the possibilities of individual stations but also to understand the geography of seismic monitoring. According to the maps (Figure 30) the opening of new stations does not contribute a significant change in the Gakkel Ridge monitoring but allows us to improve the quality of the Barents-Kara region seismic monitoring. At present this is especially important since there are widely known cases of induced seismicity in the oil and gas fields (Melnikov & Kalashnik, 2011; Adushkin & Turuntaev, 2015 and others).

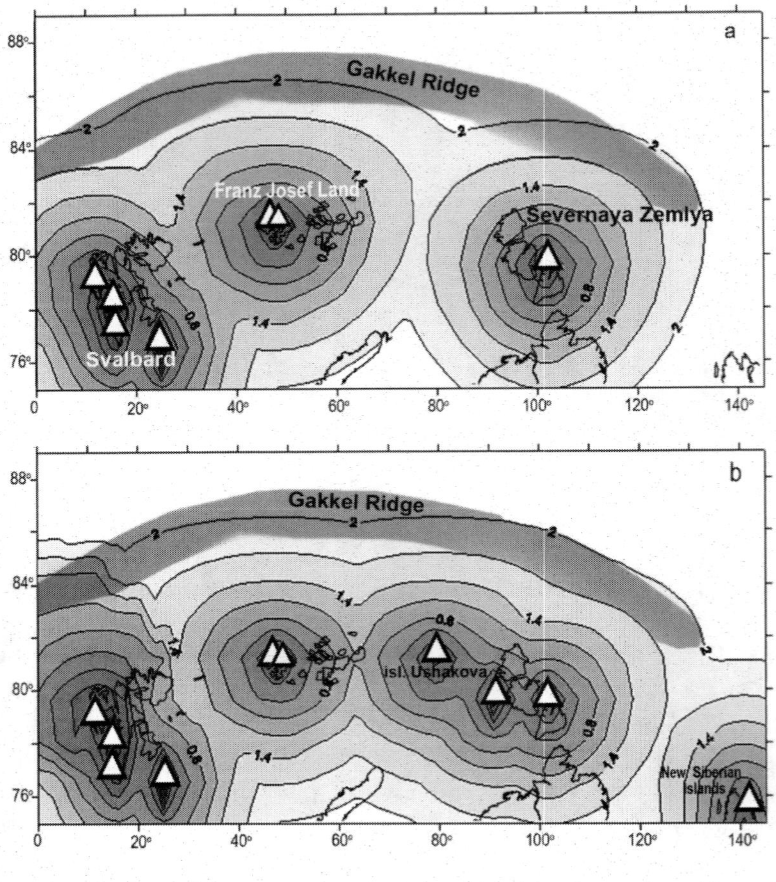

Figure 30. The theoretical Mmin values maps for Barents-Kara region and Gakkel ridge area by using program (Burmin, 2012) for: a - the current location of stations above 76 ° N, b - taking into account the opening of new stations, triangles - seismic stations.

The opening of some seismic stations in the Russian Arctic allow us to expand the geography of seismic monitoring to 100°E with an increase in magnitude sensitivity not lower than 3.5 for Barents-Kara region. Despite the progress, further improvement the seismic network is necessary.

CONCLUSION

We have relocated seismic events registered within the Barents-Kara region and in the area of the Novaya Zemlya archipelago with a view to creating a relocated catalogue. The location was performed by applying a modified method of generalized beamforming. We have considered several travel time models and selected one with the best location accuracy for ground truth events. Verification of the modified method and selection of the travel time model were performed using data on four nuclear explosions that occurred in the area of the Novaya Zemlya archipelago and in the north of the European part of Russia. The modified method and the Barents travel time model provide sufficient accuracy for event location in the region.

Increase of seismic monitoring sensitivity of Arctic shelf is extremely important as for ensuring safety in the development of offshore fields as for improving fundamental knowledge of the geodynamics of the area. In particular, the results of the seismic monitoring at the Gakkel Ridge from December 2011 to January 2015 allow for the determination of the character of small magnitude earthquakes distribution.

The chapter presents the following results and conclusions.

The crustal velocity structures of two areas of European Arctic Svalbard and Franz Josef Land were determined using receiver functions. This allowed us to calculate two travel-times for regional phases there are NOES and Sval. Validation of NOES, Sval, as well as IASPEI91, BARENTS and BARENTS3D travel-times showed that NOES travel-time was the best for calculating the location of earthquakes in the area of spreading Arctic ridges.

The Gakkel Ridge is seismically active, despite its lowest spreading velocity among global mid-ocean ridges. The epicenters of recorded earthquakes within the ridge area are distributed unevenly. The most number of epicenters are assigned to the Sparsely Magmatic Zone (SMZ), and particularly to the area within the coordinates between 1.5°E and 19.0°E. The assignment of the most earthquakes to the SMZ segment can be explained by the fact, that spreading of this segment is amagmatic. The structuring of this part of the ridge is characterized with the prevalence of tectonic processes, not magmatic or metamorphic ones. The only exclusion is the single volcanic center area (19°E) in the SMZ, where we can register practically no seismic activity. For a number of earthquakes within the SMZ both low and high frequencies are observed in the spectral-time logs. The rest of the earthquakes are characterized only with low frequencies in the spectral-time logs.

The number of recorded earthquakes within the Gakkel Ridge has varied during the year. There were periods of quiescence and seismically active periods. Some of the ridge's sections are characterized with earthquake swarms, which seem to reflect some active process, proceeding at the given time at the place.

The positional relation of earthquake epicenters allowed us to determine, where we can define the border between segments (or FZ) clearly, and where it is impossible. FZs with coordinates of 3-4°W, 3°E, 23-25° and 63°E are clearly defined (comparatively clear) with the positioning pattern of earthquakes epicenters, and FZs with coordinates of 7°E, 14°E, 34°E and 39°E cannot be defined at all.

On the basis of the performed seismic monitoring of the "continent-ocean" transition zone of the Barents-Kara region for period 2011-2014, the following can be deduced:

i. The obtained seismic data have enabled the nature of minor seismic events taking place within the "continent-ocean" transition zone of the Barents-Kara region to be explored in detail for the very first time. The region can be characterized as seismically active, with a total of 52 earthquakes of ranging from ML 0.7 to 3.3 registered during the observation period.

ii. The earthquake epicentres are clearly confined to the negative morphostructures of the continental slope (grabens). These epicentres are also distributed unevenly within the graben zone and gravitate towards the latter's estuarial parts.

iii. The prevailing geodynamic factor responsible for the occurrence of weak earthquakes is likely the isostatic compensation of avalanche sedimentation in the "continent-ocean" transition zone.

iv. The use of the receiver functions technique enabled the crustal structure to be determined in the area of the ZFI seismic station on Franz Josef Land archipelago. The Moho discontinuity is located at a depth of 30 km, with the velocities in the underlying mantle equal to Vp=8.15 km/s and Vs=4.5 km/s; a discontinuity inside the consolidated crust is located at a depth of 17 km. Crustal velocities are Vp=6.8 km/s and Vs=3.9 km/s below and Vp=6.1 and Vs = 3.6 km/s above the discontinuity; the sedimentary layer has a total thickness of around 4 km, Vp=4.3 km/s and Vs=2.36 km/s.

Funding

This research was partially funded by the project no. AAAA-A18-118012490072-7 and by the Russian Foundation for Basic Research the project no 18-05-70018.

Data and Resources

Seismic station bulletins of the ISC are available from http://www.isc.ac.uk (last accessed May 2019); of the ISC-GEM project are available from http://storing.ingv.it/bulletins/ISC-GEM (last accessed May 2019); of the EuroSeismos (ES) project are available from http://storing.ingv.it/es_web/(last accessed May 2019). Bulletins of the Soviet seismic stations were collected using data from the Geophysical Survey of RAS (http://www.ceme.gsras.ru/new/eng/ssd_news.htm; last accessed March 2019) and from the Russian State Library (https://www.rsl.ru/en; last accessed May 2019).

References

Abe K. Magnitudes and origin times from Milne seismograph data: earthquakes in China and California, 1898-1912. *Historical seismograms and earthquakes of the world.* 1988. P. 37–50.

Adushkin V.V. and Turuntaev S.B. *Technogenic processes in Earth crust (hazards and catastrophes).* Moscow. 2005. 252 p.

Antonovskaya G.N., Basakina I.M. and Konechnaya Y.V. Distribution of Seismicity and Heat Flow Anomalies in the Barents Sea Region. *Geotectonics*. 2018. Vol. 52(1). P. 45–55.

Antonovskaya G.N., Kovalev S.M., Konechnaya Y.V., Smirnov V.N., Danilov A.V. New seismic information of the Russian Arctic according to the seismic stations "Severnaya Zemlya". *Problems of the Arctic and Antarctic*. 2018. T. 64. Vol. 2 (116). P. 170–181.

Artemieva I. and Thybo H. EUNAseis: A seismic model for Moho and crustal structure in Europe, Greenland, and the North Atlantic region. *Tectonophysics*. 2013. Vol. 609. P. 97–153.

Asming V. and Prokudina A. System for automatic detection and location of seismic events for arbitrary seismic station configuration NSDL. ESC 2016-373, *35th General Assembly of the European Seismological Commission*. 2016.

Asming V.E., Fedorov A.V., Vinogradov A.N., Vinogradov Yu.A., Goryunov V.A., Evtyugina Z.A. Automatic system for seismic monitoring of Northwest Russia and the western sector of the Arctic. *Proceedings of the XI International Seismological Workshop*. Kyrgyzstan. 2016. P. 34–37 [in Russian].

Atlas: *Geology and mineral resources of the shelf of Russia*. 2004. Moscow: Scientific world. 108 p.

Avetisov G. *Seismically active zones of the Arctic*. St. Petersburg: VNIIOkeangeologia. 1996. 186 p. http://www.ngdc.noaa.gov/mgg/avetisov/avetisov.htm.

Avetisov G. and Bulin N. Deep structure of the Franz Josef Land on the seismological data. *Geophysical methods of exploration in the Arctic*. 1974. Vol. 9. P. 26–32.

BARENTS3D velocity model. http://www.norsar.no/seismology/barents3d/.

Baturin D. Structure and evolution of the continental margin of the Eurasian basin between Svalbard and Franz Josef Land archipelagos. *Reports of Academy of Sciences of the USSR*. 1988. Vol. 299. P. 412–423.

Birch F. The velocity of compressional waves in rocks in 10 kilobars, part 2. *Journal of Geophysical Research*. 1961. Vol. 66. P. 2199–2224.

Bungum H., Pettenati F., Schweitzer J., Sirovich L. and Faleide J.I. The 23 October 1904 MS 5.4 Oslofjord earthquake: Reanalysis based on macroseismic and instrumental data. *Bulletin of the Seismological Society of America*. 2009. Vol. 99(5). P. 2836–2854.

Burmin V. Inverse kinematic problems of seismology: New approaches and results. 2013. LAP LAMBERT Academic Publishing. 140 p. ISBN-13: 978-3659344824.

Cochran J.R. Seamount volcanism along the Gakkel Ridge, Arctic Ocean. *Geophys. J. Int.* 2008. Vol. 174. P. 1153–1173.

Cochran J.R., Kurras G.J., Edwards M.H., and Coakley B. The Gakkel Ridge: Bathymetry, Gravity Anomalies and Crustal Accretion at Extremely Slow Spreading Rates. *J. Geophys. Res*. 2003. Vol. 108. P. 2116–2137.

DeMets C., Gordon R. G., Argus D. F. and Stein S. Effect of recent revisions to the geomagnetic reversal time scale on estimates of current plate motions. *Geophys. Res. Lett.* 1994. Vol. 21. P. 2191–2194.

Demuth A., Ottemöller L., Keers H. Ambient noise levels and detection threshold in Norway. *Journal of Seismology*. 2016. 20(3). P. 889–904. Doi: 10.1007/s10950-016-9566-8

Di Giacomo D., J. Harris, A. Villaseñor, D.A. Storchak, E.R. Engdahl, W.H.K. Lee and the Data Entry Team, ISC-GEM: Global Instrumental Earthquake Catalogue (1900-2009), I. Data collection from early instrumental seismological bulletins. *Phys. Earth Planet. Int.* 2015. Vol. 239. P. 14–24.

Dibner V. *The morphological structure of the Barents Sea shelf.* 1978. 211 p.

Dick H.J.B., Lin J. and Schouten H. An ultraslow-spreading class of ocean ridge. *Nature.* 2003. Vol. 426. P. 405–411.

Dubinin E.P., Kokhan A.V. and Sushchevskaya N.M. Tectonics and magmatism of ultraslow spreading ridges. *Geotectonics.* 2013. Vol. 47 (3). P. 131–155.

Edwards M. H., Kurras G. J., Tolstoy M., Bohnenstiehl D. R., Coackley B. J. and Cochran J. R. Evidence of recent volcanic activity on the ultraslow-spreading Gakkel Ridge. *Nature.* 2001. Vol. 409. P. 808–812.

Emery K.O. Topography and Sediments of the Arctic Basin. *The Journal of Geology.* 1949. Vol. 57(5). P. 512–521.

Engdahl E.R., R. van der Hilst, and Buland R. Global teleseismic earthquake relocation with improved travel times and procedures for depth determination. *Bull. Seismol. Soc. Am.*, 1998. Vol. 88, P. 722–743.

Engen O., Eldholm O., Bungum H. The Arctic plate boundary. *Journal of Geoph. Research.* 2003. Vol. 108, B2. P. 5.1–5.17.

Ferrari G. (Ed.) Gli strumenti sismici storici Italia e contesto europeo – historical seismic instruments Italy and the European framework. Istituto Nazionale di Geofisica, Bologna. 1990.

Fujita K., Cook D.B., Hasegawa H., Forsyth D., Wetmiller R. Seismicity and focal mechanisms of the Arctic region and the North American plate boundary in Asia. *The Arctic Ocean Region: The Geology of North America.* 1990. Vol. 50. P. 79–100.

Gabsatarova I. Introduction into routine practice divisions of the Geophysical Survey of RAS calculation procedure of the local magnitude. *Modern methods of processing and interpretation of seismic data.* 2006. Geophysical Survey. Saint-Peterburg. 2–6 October. P. 49–53.

GEOFON. GEOFON program. http://geofon.gfz-potsdam.de/geofon.

Geophysical Survey of Russian Academy of Science. http://www.ceme.gsras.ru.

Gibbons S. J., Antonovskaya G., Asming V., Konechnaya Y.V., Kremenetskaya E., Kværna T., Schweitzer J. and Vaganova N. The 11 October 2010 Novaya Zemlya Earthquake: Implications for Velocity Models and Regional Event Location. *Bulletin of the Seismological Society of America.* 2016. Vol. 106. No 4. P. 1470–1481.

Goldstein S.L., Soffer G., Langmuir C.H. et al. Origin of a “Southern Hemisphere” geochemical signal in the Arctic upper mantle. *Nature.* 2008. Vol. 453. P. 89–94.

Gurevich N.I., Astafurova E.G., Glebovskiy V.U. and Abel’skaya A.A. Some features of crustal accretion of the western part of the ultraslow-spreading Gakkel Ridge, Arctic ocean. *Geological-geophysical characteristics of the lithosphere of the Arctic Region.* 2004. St. Peterburg, VNIIOkeangeologia. Vol 5. P. 87–97.

Gutenberg B. and Richter C.F. *Seismicity of the Earth.* 1941. 126 p.

Gutenberg B. and Richter C. Magnitude and energy of earthquakes. *Ann. Di Geofisica.* 1956. Vol. 1, No 9. P. 1–15.

Haskell N. A. Crustal reflection of plane P and SV waves. *Jornal of geophysical research.* 1962. Vol. 67. P. 4751–4767.

Hicks E.C., Kværna T., Mykkeltveit S., Schweitzer J. and Ringdal F. Travel times and attenuation relations for regional phases in the Barents Sea region. *Pure and applied geophysics.* 2004. Vol. 161. P. 1–19.

Hjelstuen B., Eldholm O. and Faleide J. Recurrent Pleistocene mega-failures on the SW Barents Sea margin. *Earth and Planetary Science Letters*. 2007. Vol. 258. Issues 4-3. P. 605–618.

Hodgson E. A. The Seismicity of the Arctic. *Journal of the Royal Astronomical Society of Canada*. 1930. Vol. 24. P. 201–210.

International Seismological Centre, On-line Bulletin, http://www.isc.ac.uk, Internatl. Seismol. Cent., Thatcham. United Kingdom. 2015.

Ivanova N.M., Sakoulina T.S. and Roslov Y.V. Deep seismic investigation across the Barents–Kara region and Novozemelskiy Fold Belt (Arctic Shelf). *Tectonophysics*. 2006. Vol. 420(1). P. 123–140.

Jokat W. and Schmidt-Aursch M. Geophysical Characteristics of the Ultraslow Spreading Gakkel Ridge, Arctic Ocean. *Geophys. J. Int*. 2007. Vol. 168. P. 983–998.

Karasik A. M. Magnetic anomalies of the Gakkel Ridge and origin of the Eurasia Subbasin of the Arctic Ocean. *Geophys. Methods Prospect. Arctic*. 1968. Vol. 5. P. 8–19.

Kennett B. and E. R. Engdahl Traveltimes for global earthquake location and phase identification. *Geophys. J. Int*. 1991. Vol. 105(2). P. 429–465.

Kennett B. L. N., Engdahl E. R. & Buland R. Constraints on seismic velocities in the Earth from travel times. *Geophysical Journal International*. 1995. Vol. 122(1). P. 108–124.

Khain V. *Tectonics of continents and oceans*. 2001. Moscow: Scientific world. 606 p.

Khalturin V.I., Rautian T.G., Richards P.G. & Leith W.S. A review of nuclear testing by the Soviet Union at Novaya Zemlya, 1955–1990. *Science and Global Security*. 2005. Vol. 13(1-2). P. 1–42.

Khutorskoy M., Leonov Yu., Ermakov A. and Ahmedzyanov V. Anomalous heat flow and the nature of troughs in the Northern part of the Svalbard plate. *Doklady Earth Ssciences*. 2009. Vol. 424, 2. P. 227–233.

Kosarev G.L., Makeyeva L.I. and Vinnik L.P. Inversion of teleseismic P-wave particle motions for crustal structure in Fennoscandia. *Physics of the Earth and Planetary Interiors*. 1987. Vol. 47. P. 11–24.

Kremenetskaya E., Asming V., Vinogradov Yu. Seismic profiling and calibration of seismic travel-time models for Barents region. *Geophysical Research Abstracts*. 2004. Vol. 6, 00881. SRef-ID: 1607-7962/gra/EGU04-A-00881 c European Geosciences Union 2004.

Kremenetskaya E. and Asming V. Problems of regional seismic event location and depth estimation in the European Arctic. *Workshop on IMS Location Calibration and Screening* 4. 2002. Oslo. Norway.

Kremenetskaya E., V. Asming and F. Ringdal Seismic location calibration of the European Arctic. *Pure Appl. Geophys*. 2001. Vol. 158(1). P. 117–128.

Kristoffersen Y., Husebye E. S., Bungum H. and Gregersen S. Seismic investigations of the Nansen ridge during the FRAM I experiment. *Tectonophysics*. 1982. Vol. 82. P. 57–68.

Kvaerna T. and Ringdal F. Generalized beamforming, phase association and threshold monitoring using a global seismic network. *Monitoring a Comprehensive Test Ban Treaty*. 1996. P. 447–466.

Lee W. H. K., Meyers H. and Shimazaki K. Historical seismograms and earthquakes of the World. *Academic Press*. 1988. 513 p.

Levshin A., Schweitzer J., Weidle Ch., Shapiro N. and Ritzwoller M. Surface wave tomography of the Barents Sea and surrounding regions. *Geophys. J. Int*. 2007. Vol. 170(1). P. 441–459.

Linden N.A. *Seismic chart of the Arctic*. Seismology and seismological researches during the IGY. 1959. No. 2. Academy of Science. Moscow. P. 7–17.

Melnikov N.N. and Kalashnik A.I. Shelf oil and gas workings out of the western sector of the Russian Arctic: geodynamic risks and safety. *The gas industry*. 2011. P. 46–53.

Michael P.J., Langmuir C.H., Dick H.J.B., Snow J.E., Goldstein S.L., Graham D.W., Lehnert K., Kurras G., Jokat W., Mühe R. and Edmonds H.N. Magmatic and amagmatic seafloor generation at the ultraslow-spreading Gakkel Ridge, Arctic Ocean. *Nature*. 2003. Vol. 423. P. 956–961.

Milanovskii E. Geology of Russia and CIS countries (North Eurasia). 1996. Moscow: Moscow State University. 448 p.

Moroz E., Mazarovich A., Abramova A., Efimov V., Zaraiskaya Yu. and Sokolov S. Neotectonics of the North-West of the Barents sea. *Geology and Geoecology of the continental margins of Eurasia*. 2010. Vol. 2. P. 161–173.

Morozov A. and Konechnaya Ya. Monitoring of the Arctic region: contribution of the Arkhangelsk seismic network. *Journal of Seismology*. 2013. Vol. 17. P. 819–827.

Morozov A.N. and Vaganova N.V. The Travel Times of Regional P and S for Spreading Ridges in the European Arctic. *Journal of Volcanology and Seismology*. 2017. Vol. 11, No. 2. P. 156–163.

Morozov A.N., Asming V.E., Vaganova N.V., Konechnaya Y.V., Mikhaylova Y.A. and Evtyugina Z.A. Seismicityof the Novaya Zemlya archipelago: relocated event catalogue from 1974 to 2014. *Journal of Seismology*. 2017. Vol. 21(6). P. 1439–1466.

Morozov A.N., Vaganova N.V., Konechnaya Y.V. and Asming V.E. New data about seismicity and crustal velocity structure of the "continent-ocean" transition zone of the Barents-Kara region in the Arctic, *Journal of Seismology*. 2015. Vol. 19(1). P. 219–230.

Morozov A.N. and Vaganova N.V. Godografy sejsmicheskih voln dlya Severa Russkoj plity po dannym Arhangel'skoj sejsmicheskoj seti. (*The travel-time of seismic waves for the north of the Russian Plate according to the data from the Arkhangelsk seismic network). Prospect and protection of mineral resources*. 2011. No 12. P. 48–51 [in Russian].

Morozov A.N., Vaganova N.V., Asming V.E., Konechnaya Y.V. and Evtyugina Z.A. The instrumental seismicity of the Barents and Kara sea region: relocated event catalogue from early twentieth century to 1989. *Journal of Seismology*. 2018. Vol. 22(5). P. 1171–1209.

Morozov A.N., Vaganova N.V., Konechnaya Y.V. and Asming V.E. New data about seismicity and crustal velocity structure of the "continent-ocean" transition zone of the Barents-Kara region in the Arctic. *Journal of Seismology*. 2015. Vol. 19(1). P. 219–230.

National Geophysical Data Centre. National Oceanic and Atmospheric Administration (NOAA). http://www.ngdc.noaa.gov/.

Niemz P. and Amorèse D. Relocalizing a historical earthquake using recent methods: The 10 November 1935 Earthquake near Montserrat, Lesser Antilles. *Journal of South American Earth Sciences*. 2016. Vol. 66. P. 166–179.

Nikonov A.A. and Chepkunas L.S. The Sysol earthquake of January 13, 1939 on Russian plate- new approach. *Seismic Instruments*. 2009. Vol. 36(4). P. 25–41.

Ottemöller L. and Midzi V. The crustal structure of Norway from inversion of teleseismic receiver functions. *Journal of Seismology*. 2003. Vol. 7. P. 35–48.

Radziminovich N.A., Ochkovskaya M.G. Vydeleniye aftershokovykh i royevykh posledovatel'nostey zemletryaseniy Baykal'skoy riftovoy zony. *Geodynamics and Tectonophysics*. 2013. No 4(2). P. 169–186.

Riedel C. and Schlindwein V. Did the 1999 earthquake swarm on Gakkel Ridge open a volcanic conduit? A detailed teleseismic data analysis. *J. Seismol*. 2010. Vol. 14. P. 505–522.

Ringdal F. Study of Low-Magnitude Seismic Events near the Novaya Zemlya Nuclear Test Site. *Bulletin of the Seismological Society of America*. 1997. Vol. 87(6). P. 1563–1575.

Ritzmann O., Maercklin N., Faleide J.I., Bungum H., Mooney W.D. and Detweiler S.T. A 3D geophysical model for the crust in the greater Barents Sea region: Database compilation, model construction and basement characterization. *Geophys. J. Int*. 2006. Vol. 170. P. 417–435.

Riznichenko Yu.V. Seismic Shaking of the USSR Territory. *Nauka*. 1979. 192 p. [in Russian].

Rogozhin E.A., Antonovskaya G.N., Kapustyan N.K. and Fedorenko I.V. Features of seismicity of the Euro-Arctic region. *Doklady Earth Ssciences*. 2016. 467(2). P. 389–392.

Sakoulina T., Telegin A. and Tichonova I., et al. The results of deep seismic investigations on geotraverse in the Barents Sea from Kola Peninsula to Franz Joseph Land. *Tectonophysics*. 2000. Vol. 329. Issues 1-4. P. 319–331.

Schlindwein V. Teleseismic earthquake swarms at ultraslow spreading ridges: indicator for dike intrusions. *Geophys. J. Int*. 2012. Vol. 190(1). P. 442–456. doi:10.1111/j.1365-246X.2012.05502.x.

Schlindwein V., Demuth A., Korger E., Läderach C. and Schmid F. Seismicity of the Arctic mid-ocean ridge system. *Polar Science*. 2015. Vol. 9(1). P. 146–157, doi:10.1016/j.polar.2014.10.001.

Schlindwein V., Muller C. and Jokat W. Seismoacoustic evidence for volcanic activity on the ultraslowspreading Gakkel Ridge, Arctic Ocean. *Geophysical Research Letters*. 2005. Vol. 32. L18306, http://dx.doi.org/18310.11029/12005GL023767

Schlindwein V., Muller C. and Jokat W. Microseismicity of the ultraslow-spreading Gakkel ridge, Arctic Ocean: a pilot study. *Geophys. J. Int*. 2007. Vol. 169. P 100–112.

Schweitzer J. The birth of modern seismology in the nineteenth and twentieth centuries. *Earth Sci*. Hist. 2007. Vol. 26 (2). P. 263–280.

Schweitzer J. and Kennett B.L. Comparison of location procedures: The Kara Sea event of 16 August 1997. *Bulletin of the Seismological Society of America*. 2007. Vol. 97(2). P. 389–400.

Seismic Handler - seismic waveform analysis tool. http://www.seismic-handler.org.

Sohn R.A. and Hildebrand J.A. Hydroacoustic earthquake detection in the Arctic Basin with the Spinnaker Array. *Bull. seism. Soc. Am*. 2001. Vol. 91(3). P. 572–579.

Storchak D.A., Di Giacomo D., Engdahl E.R., Harris J., Bondár I., Lee W. .K., Bormann P. and Villaseñor A. The ISC-GEM Global Instrumental Earthquake Catalogue (1900-2009): Introduction. *Phys. Earth Planet. Int*. 2015. Vol. 239. P. 48–63, doi: 10.1016/j.pepi.2014.06.009.

Storchak D.A., Di Giacomo D., Bondár I., Engdahl E.R., Harris J., Lee W.H.K., Villaseñor A. and Bormann P. Public Release of the ISC-GEM Global Instrumental Earthquake Catalogue (1900-2009). *Seism. Res. Lett*. 2013. Vol. 84, 5. P. 810–815.

Tams E. Die seismischen Verhältnisse des europäischen Nordmeer. zentralblatt f. Mineralogie. *Geologie und Palaentologie*. 1922. No 13. P. 385–397.

Thiede J. and the Shipboard Scientific Party Polarstern Arktis XVII/2 Cruise Report: AMORE2001 (Arctic Mid-Ocean Ridge Expedition). *Rep. Polar Mar. Res*. 2002. Vol. 421. 397 p.

Tikhonov A. and Glasko V. Use of the regularization method in non-linear problems. *USSR Computational Mathematics and Mathematical Physics*. 1965. Vol. 5. P. 93-107.

Tolstoy M., Bohnenstiehl D.R., Edwards M.H. and Kurras G.J. Seismic character of volcanic activity at the ultraslow-spreading Gakkel Ridge. *Geology*. 2001. Vol. 29. P. 1139-1142.

Vanek J., Zatopek A., Karnik V., Kondorskaya N.V., Riznichenko Y.V., Savarensky E.F., Soloviev S.L. and Shebalin N.V. Standardization of magnitude scales. *Bull. Acad. Sci. USSR Geophys. Ser*. 1962. P. 108–111.

Vanneste M., Mienert J. and Bunz S. The Hinlopen Slide: Agiant, submarine slope failure on the northern Svalbardmargin, Arctic Ocean. *Reports of Earth and Planetary Science Letters*. 2006. V. 245, 1-2. P. 373–388.

Verba M. Modern bilateral tension of the earth's crust in the Barents-Kara region and its role in the evaluation of the hydrocarbon potential. *Oil and gas Geology. Theory and practice*. 2007. Vol. 2. http://www.ngtp.ru/rub/4/026.pdf.

Vinnik L. Detection of waves converted from P to S in the mantle. *Physics of the Earth and Planetary Interiors*. 1977. Vol. 15. P. 39–45.

Vinogradov A. and Baranov S. The possible impact of landslide processes on the seismicity of the North-Western part of the Barents Sea. In: *Modern methods of processing and interpretation of seismic data, Geophysical Survey*. 2013. Gelendzghik. P. 99–103.

Vogt P.R., Taylor P.T., Kovacs L.C. and Johnson G.L. Detailed aeromagnetic investigations of the Arctic Basin. *J. geophys. Res*. 1979. Vol. 84(B3). P. 1071–1089.

Wilde-Piorko M., Grad M., Wiejacz P. and Schweitzer J. HSPB seismic broadband station in Southern Spitsbergen: First results on crustal and mantle structure from receiver functions and SKS splitting. *Polish Polar Research*. 2009. Vol. 30. P. 301–316.

Winkelmann D. and Stein R. Triggering of the Hinlopen/Yermak Megaslide in relation to paleoceanography and climate history of the continental margin north of Spitsbergen. *Reports of Geochem.Geophys. Geosyst.* (G3). 2007. Vol. 8, 6. P. 1–15. (*An electronic journal of the earth sciences* doi:10.1029/2006GC001485).

Yang X., Bondár I., and Romney C. PIDC Ground truth event (GT) database (Revision 1). *Center for Monitoring Research*. 2000.

Yang X., North R., Romney C. and Richards P.G. Worldwide nuclear explosions. *International Geophysics*. 2003. Vol. 81. P. 1595–1599.

Zayonchek A.V., Brekke H., Sokolov S.Yu., Mazarovich A.O., Dobrolyubova K.O., Efimov V.N., Abramova A.S., Zaraiskaya Yu.A., Kokhan A.V., Moroz E.A., Peive A.A., Chamov N.P. and Yampol'sky K.P. Structure of continent–ocean transition zone in the northwestern framework of the Barents Sea from the data of the 24th, 25th, and 26th cruises of R/V Akademik Nikolaj Strakhov, 2006–2009. In *Structure and Evolutionary History of the Lithosphere: Contribution of Russia to the International Polar Year* (Paulsen, Moscow). 2010. Vol. 4. P. 111–157 [in Russian].

In: The Arctic: Current Issues and Challenges
Editors: Oleg S. Pokrovsky, Sergey N. Kirpotin et al.
ISBN: 978-1-53617-306-2

Chapter 13

DIAMOND DEPOSITS OF THE EUROPEAN ARCTIC OF RUSSIA: PROBLEMS OF THE KIMBERLITES EXPLORATION, LIMITS OF CONVENTIONAL METHODS AND NEW APPROACHES

Konstantin B. Danilov**, PhD, Evgeniy Yu. Yakovlev, PhD and Nikita Yu. Afonin

N. Laverov Federal Centre for Integrated Arctic Research of Russian Academy of Sciences, Arkhangelsk, Russia

ABSTRACT

The Arctic is a unique region of the planet, where huge amount of mineral deposits are located. The extraction of them is the main direction of economic development of the Arctic territories. Along with hydrocarbon and metal deposits, diamonds are one of the most important minerals (Bogatikov et al., 1999). Currently, on the European Arctic territory, only two developed industrial diamond deposits are exist – named after Lomonosov and named after Grib. These deposits located within the Arkhangelsk region (Russia) and include more than 25% of the total diamond reserves of Russia (Rosnedra, 2016). These deposits make a great contribution to the economic development of the country and the most important link in the development of the Arctic. However, according to the Russian state balance, nowadays there is a tendency for a general decline in diamond reserves in primary deposits. A gradual reduction of diamond reserves without its proper reproduction leads to a deterioration in the economic indicators of the mining industry (VSEGEI, 2017). The explored diamond reserves of primary deposits can be fully extracted in a little over 20 years. Exploration work in the Arkhangelsk region over the past 15 years has not led to the discovery of new deposits, and the fund of high-amplitude local magnetic anomalies has exhausted itself by the beginning of this century. A fundamental breakthrough in the identification of low contrast anomalies, associated with kimberlite pipes by geophysical methods, has not yet been achieved. The possibilities of searching for kimberlites by heavy mineral sampling method in this area,

* Corresponding Author's Email: danilov_kostea@mail.ru.

are also very limited. However, the opportunity for discovery the new kimberlite pipes in the Arkhangelsk region is very high (VSEGEI, 2017). One of the solutions for this problem is developing new innovative technologies for exploration of low-contrast diamond kimberlite bodies, hidden in the depth and to strengthen the forecasting-prospecting works. The main importance is identifying diagnostic signs and criteria of diamond deposits associated with alkaline-ultrabasic magmatism in closed areas. In the chapter considers the structural features of kimberlites pipes and the enclosing rocks considered as diagnostic signs. The characteristic criterion of kimberlite pipes is sub-vertical structure of an inverted cone shape, confined to the intersection of faults and fractured zones in near-pipe space (Kutinov, Chistova, 2004; Ignatov et al., 2015). Some radioisotopic and passive seismic methods, that have not been previously used for these purposes, can be applied to identification kimberlite pipes by this criteria. As a passive seismic method, it is efficient to joint using of the microseismic sounding and passive seismic interferometry methods. Both radioisotopic and passive seismic methods have been separately and joint tested on several kimberlite pipes and other geological structural elements on the Arctic territory. The joint using of these methods allowed significantly increase the measured data informativity and the quality of the obtained results. The main advantage of the microseismic sounding method, is possibility for localization sub-vertical objects (pipes and faults). The passive seismic interferometry allows to localize the main sub-horizontal boundaries. Radioisotopic methods are use highly sensitive detectors, which allow identifying features of the distribution the radioactive isotopes. The isotopes, together with rare earth elements, are the subtlest indicators of petrogenetic and geochemical processes. That is why, they can provide important information about the radio geochemical specialization of rocks, associated with kimberlite magmatism. The using of the emanometric method in the radioisotope complex allows us to judge the permeability of the medium for the flow of gases and the presence of structures that control the location of the kimberlite pipes.

Keywords: kimberlites, faults, near-pipe space, exploration methods, structural features, radioactive isotopes, emanometry, ambient noise, microseismic sounding, passive seismic interferometry

INTRODUCTION

Since the discovery of the first kimberlite pipe in the Arkhangelsk diamondiferous province (ADP) (North of the East European Craton, Russia) many traditional geophysical methods have been tested with the purpose to search new kimberlite bodies (Bezborodov et al., 2003; Kontarovich & Babayants 2011). Aeromagnetic methods have shown the most effective (Mwenifumbo & Kjarsgaard 1999; Reed & Witherly 2007). However, at the present time crisis of the efficiency aeromagnetic survey to searching kimberlite bodies influences on the searches results increasingly. Thus, in the early eighties of the last century the efficiency verification magnetic anomalies was about 20%; at the end of the eighties, 12.5%; in the nineties it was already below 1% (Stognij & Korotkov 2010). Today, this figure is less than 0.5% for the whole territory of the Arkhangelsk diamondiferous province. It is connected with a decrease of contrast magnetic anomalies which are tending to geological noise.

The strong variations of physical properties of rocks, from which kimberlite pipes consist and enclosing environment make it difficult to use geophysical methods for exploration of them (Milashev 1984; Kutinov and Chistova 2004). For example, kimberlite pipe named after

M. Lomonosov, which is part of the Arkhangelsk diamondiferous province, has the variations of p-wave seismic velocities are about 30% (Verzhak et al. 1987).

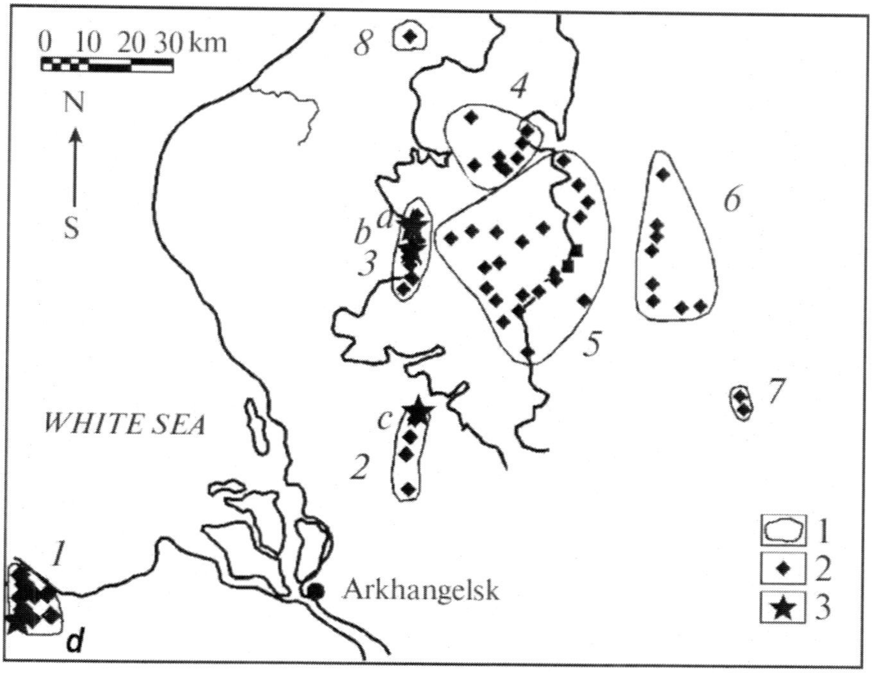

Figure 1. A sketch map showing the distribution of ADPigneous rocks (Eremenko and Nenakhov, 2002), finalized. (1) kimberlite fields ((1) Nenoksa, (2) Izhmozero, (3) Zolotitsa, (4) Verkhotin, (5) Kepino, (6) Tur'ino, (7) Oltino, (8) Mel'skoe); (2) explosion pipes; (3) explosion pipes that have been studied by microseismic sounding (a Lomonosov pipe; b Pionerskya pipe; c Chidviya pipe; d S10 pipe).

Therefore, development of new approaches for exploration and prediction of kimberlite pipes are an important task. We selected the methods for joint using because of the next features of each method. The microseismic sounding method allows identifying contrasts of seismic velocities in studied medium with relatively high resolution on horizontal.

The typical occurrence of kimberlite rocks is in the form of various pipe bodies (explosion pipes). Explosion pipes are cone-shaped bodies whose apexes look downward. An explosion pipe is flattened out with increasing depth, becoming a dike. Pipes are found in very different shapes (Milashev, 1984). As consequently the microseismic sounding method can be used for pipe's imaging.

Method H/V aimed to study horizontal boundaries inside the medium. Spectral analysis of empirical Greens functions, obtained by passive seismic interferometry, allows calculating dispersion curve of surface waves, which used for calculation of depths in the microseismic sounding method. Moreover, inversion of dispersion curve allows getting velocity model of the studied medium. The velocity model is used in the H/V method for calculation of depths to boundaries. Emanation and gamma-spectrometry mapping allow extrapolating locations of inhomogeneities, obtained by the microseismic sounding method. Therefore, joint using of these methods make it possible to improve the quality of obtained information about the structure and physical properties of the studied medium.

Furthermore, as an example at Arkhangelskaya kimberlite pipe it was found that secular radioactive equilibrium of the even uranium isotopes is disturbed in enclosing rocks of the near-contact zone. Similarly, the radioactive elements concentrate inside that zone (Yakovlev et al., 2016).

METHODS

Radon Method and Gamma-Spectrometry

Radon is a decay product in the uranium-238 chain. It is the best indicator for various geological studies, including the search for ore deposits (as emanation survey of soil air) because of its physical and chemical properties. Radon concentration in soil sediments can has significant variations over a short distance, which may indicate the presence of faults or tectonic zones (Neznal et al. 1991).

According to the data summarized in the work (Miklyaev 2015) for the Russian territory, the background values of the volumetric activity of radon (VAR) in the near-surface air sandy-clay sediments vary from 5 to 100 kBq·m-3 at a depth of 0.5-1.0 m. Higher values are rare. For areas with uranium deposits, VAR can reach values of 30,000 kBq·m-3 or more. The minimum values of radon activity (the first units of kBq·m-3) are characteristic of soils with the lowest radium concentration and rate permeability. With an increase in these parameters the radon activity in the soil air increases (Novikov 1989; Gulabiants and Zabolotsky 2004).

There are few published data about the radon content in the soil air of the East European Platform in Russia. According to published data, we can conclude that the most characteristic for the East European platform range of concentrations of radon in the near-soil air is from 5 to 50 kBq·m-3 (Makarov et al. 2003; Berdnikov and Gorky 2007). Higher values, up to 100 kBq·m-3, and even up to 250 kBq·m-3, should be considered anomalous, associated either with increasing concentration of radium in the soil or with anomalous conditions of radon transfer in the rock massif.

The mapping of the territory of Estonia by soil radon activity showed that almost a third of the continental part of the country has a high (> 50 kBq·m-3) or very high (> 250 kBq·m-3) level of radon risk. In these areas, there is a relatively high concentration of uranium (> 3.5 ppm in soils. Risk zones are classified according to the radon content in the soil air as low (<10 kBq·m-3), normal (10-50 kBq·m-3), high (50-250 kBq·m-3) and very high (> 250 kBq·m-3) (Petersell et al. 2016).

According to nearly 1,700 measurements of radon concentration in the soil horizons in Prague (Czech Republic), it was found that the radon activity values, in general, correspond to the geological situation of the region and on average fluctuate between 20-70 kBq·m-3 (Neznal et al. 1991; Neznal et al. 1996). Higher values (> 100 kBq·m-3) of soil-gas radon concentration were observed only within a wide tectonic zone with transverse faults in the contact of heterochronous rocks. According to work (Huxtable et al. 2016), the concentration of 222Rn in soil air usually ranges from 2 to 50 kBq·m-3. The average concentration of radon in the soil air of some Canadian cities varies from 11–36 kBq·m-3 (Chen and Ford 2016).

Emanation survey is traditionally considered a method that allows identifying faults hidden under deposits, but currently permeable for underground gases (Seminsky et al. 2009; Seminsky et al. 2014). In view of the sufficiently long decay time of radon, the emanation method is most effective in mapping the internal structure of faults, determining the width of fracture zones, and evaluating the degree of tectonic activity of faults (Bobrov 2009).

The emanation survey is potentially useful to identify kimberlite bodies because of two factors. Firstly, kimberlite bodies are controlled by faults (Milashev 1979; Kutinov and Chistova, 2004). Secondly, fractured zones are usually located in the near-pipe space. These fracturings are associated with the process of diatreme formation as a result of the significant mechanical effect of incipient gases and melts on the host rocks (Nikitin 1980; Milashev 1984; Khazanovich-Wulf 2007). The result of this is a system of radial and concentric fractures in the near-pipe space, with crushing and vertical displacement of the blocks of surrounding rocks. The largest area of tectonic disturbances has noneroded or slightly eroded pipes, and as the depth of the erosional truncation increases, the fracturing around them decreases (Khazanovich-Wulf 2007). S.I. Kostrovitsky (1976) studied the fracturing around the small pipe "Sputnik", located on the side of the quarry of the Mir kimberlite pipe. Concentric fracturing is developed most intensively there: up to 100 fractures per 1 m. The width of the fracturing zone around the "Sputnik" pipe is about 5-10 m. The morphometric studies of M.A. Milashev (1979) for some regions of Siberia showed that, in fact, diatreme fields typically feature increased dense tectonic fracturing. These anomalous tectonic zones were formed simultaneously with the formation of diatremes mainly in the upper part of the rocks surrounding them. The structure-forming role of the processes associated with the formation of diatremes is an obvious fact. The conclusion that follows from this regularity is that diatreme magmatism creates areas of tectonic disturbances, increased fracturing and permeability within the fields of its development (Khazanovich-Wulf 2007).

Studies of the radon activity in soil sediments were carried out on kimberlite pipes of various diamondiferous provinces. Similar emanation studies were performed on kimberlite pipes in South and West Botswana, where the prospects of the emanation method in the prediction of kimberlite bodies were demonstrated (McDowall and Koketso 1991). As a result of field radon exploration conducted on the territory of the Chetlassky and Volsko-Vymsky uplifts of Middle Timan (Magomedova et al. 2015), it was found that faults are characterized by increased values of the volume activity of radon. The increased activity of radon is also characteristic of the investigated kimberlite pipes: Umbinskaya, Vodorazdelnaya, Sredninskaya (Magomedova and Udoratin 2016).

In our own previous studies, it was found that in soils over the kimberlite pipes of the Arkhangelsk diamondiferous province, local anomalous of radon activity is also observed (Kiselev et al. 2016). Studies conducted on the Chidvinskaya pipe, which has the smallest thickness of overlying sediments among the Arkhangelsk pipes (~ 1.5 m), showed that an abnormal increase in radon concentration is observed at the boundaries of the pipe with enclosing rocks. In the course of microseismic studies on the Chidvinskaya pipe, we showed that the anomalous radon concentrations observed at the boundaries of the Chidvinskaya pipe correspond to the contrast linear low-speed zones according to the microseismic data, which are the zones of increased fracturing in the endo- and exocontacts of the pipe (Kiselev et al. 2017).

Airborne gamma spectrometry in the Yakutia has shown that many kimberlite pipes stand out against the background of host rocks by increased radioactivity, represented by secondary radiogeochemical halos (Babayants et al. 2006). The radiometric and gamma spectrometric surveys have acquitted themselves as efficient methods for prospecting for diamond placers in the Timan–Ural province, which are distinguished by elevated radioactivity (Rybalchenko et al. 2011). The promise of gamma spectrometry for detecting kimberlite bodies was demonstrated in diamond districts in Canada and India (Mwenifumbo and Kjarsgaard 1999; Ramadass et al. 2015).

There is not still clear relationship has been detected of the haloes of radioactive elements in the overlying sediments and the kimberlite pipes in the Arkhangelsk diamond province. The main reasons are impeding the use of gamma spectrometry in this area are complex landscape-geological conditions, the development of Quaternary fluvioglacial sediments and a humid climate with excessive moisture, which complicates the accumulation of radioactive elements in the upper horizons. The actual data on the content of natural radioactive elements in the upper horizons of the overlying sediments of the Arkhangelsk diamond province were obtained by predecessors in 80s mainly by the method of airborne gamma-ray survey (Babayants et al. 2006). Due to the low concentrations of radioactive elements in the overlying rocks and the low sensitivity of the equipment of those years, the weak contrast anomalies associated with kimberlites could not be identified. An airborne gamma-ray survey was subsequently declared ineffective for searching the pipes in the Arkhangelsk diamond province.

However, surface gamma-spectrometric surveys of recent years using modern high-resolution and highly sensitive detectors have shown that kimberlite pipes are distinguished by low-contrast anomalies of total radioactivity, mainly on concentrations potassium and thorium (Kiselev et al. 2016; Yakovlev, 2017). At the same time, there are many anomalies associated with glacial sediments of moraines, but, as a rule, they are distinguished by high concentrations of uranium.

Passive Seismic Methods

In our investigation, we used three passive seismic methods: microseismic sounding, H/V method and passive seismic interferometry.

The passive seismic interferometry method based on estimation of an empirical Greens function by cross-correlation or convolution of ambient noise, recorded simultaneously in different locations (Shapiro and Campilo, 2004; Wapenaar 2008). This method is widely used in surface wave tomography (Shapiro et al. 2005; Lin et al. 2007, etc.), body wave imaging (Poli et al. 2012) for study of faults structure (Afonin et al., 2017) and for study of shallow subsurface structure (Cheng et al. 2015; Le Feure 2015; Lin et al. 2013). In our study, we used improved method of passive seismic interferometry, which allow decreasing time of measurement and improving quality of obtained dispersion curve (Afonin et al. 2019). The field measurements require simultaneously records of ambient seismic noise at least in three points. It is necessary for removing effect of azimuthal distribution of ambient noise sources.

Method H/V allows estimating depth to the borders inside the medium by analysis of resonance frequencies of seismic noise. The method based on the analysis of three-component seismic noise records (Lane et al. 2008; Ibs-von Seht and Wohlenberg, 1999). In opposite to passive seismic interferometry, this method does not require time synchronization of records in different points. That it is possible to use less sensors for measurements.

In our work, for calculation of depth in microseismic sounding method, we used dispersion curve obtained by passive seismic interferometry. Moreover, we processed the data, measured along profiles, relative to one station, as one profile. This technique allows to compare results, obtained for different profiles, as one profile.

An Overview of the Microseismic Sounding Method

Microseisms are the result of interference of seismic waves of various types moving as a separate wave train of limited duration. On the one hand, regularities of moving wave trains are well studied. On the other hand, proportions of each type of waves, amplitudes, initial phases and duration of these wave trains are unknown. As a result, there are two approaches to the analysis of microseisms. The first one is based on deterministic properties of seismic waves, whereas the second one is based on statistical properties (Gorbatikov and Tsukanov 2011; Gorbatikov et al., 2013). The methods pertaining to the first group are usually aimed at experimental determination of dispersion characteristics (Asten and Henstridge 1984; Horike 1985). The drawbacks of the methods are the difficulties to identify the velocity heterogeneity without sufficient wave travel and labour intensity. The methods pertaining to the second group are based on correlative correspondence between stable parameters of microseisms and geologic features. Those transient-free parameters may be frequencies and amplitudes at spectral peaks (Kanai and Tanaka 1954; Asten 1978; Katz and Bellon 1978). It should be noted that these methods are somewhat less objective due to the need of making a number of assumptions about the properties of microseisms.

The microseismic sounding method (MSM) used in the present work belongs to the second group of methods. The method was developed by Gorbatikov et al. (2008), Gorbatikov and Tsukanov (2011) based on the results of the experiments carried out in the period between 2000 and 2007. The purpose of those experiments was to study space distribution of spectral amplitude of microseisms. As the outcome of experiments, the influence of geological heterogeneities on the intensity variation of microseisms at certain frequencies was revealed. These intensity variations were registered at distances smaller than a wavelength. The important fact is that a similar effect was observed for both the near-surface geological heterogeneity and the heterogeneity at a depth of ten kilometres.

Further, the observed regularities were studied using test geological objects and mathematical models. Surface-wave was found to be the dominant microseism mode at a certain distance from the signal source (Nogoshi and Igarashi 1971; Bath 1974; Monakhov 1977). As a consequence, microseisms were conceptualised as a Rayleigh wave field. However, the composition of the natural microseisms is a challenging issue. Vinnik (1968) found that the natural microseismic field may contain up to 60% of body waves. In an ideal case, the dominant mode must be controlled in every points of measurement. However, the body wave amplification is dependent on many factors. Consequently, the effect of the body wave on the MSM method should be studied separately. In the work, system of saline's domes, faults and gradual sinking of the foundation were investigated. According to the results of the investigations, two regularities were identified (Gorbatikov et al. 2008):

1. power spectrum amplitudes of waves of Rayleigh increases when passing through low-velocity heterogeneities and decreases when passing through high-velocity heterogeneities;
2. the greatest change in intensity is observed for wavelengths twice the size of the cover thickness of the heterogeneity. This can be explained by the fact that a shear stress zone of Rayleigh wave is located at the depth roughly equivalent to half of the length of a wave. As this takes place, the shift zone is located closer to surface.

In the microseismic sounding method it is necessary to define indicative spectrum of microseisms. To be able to apply the method correctly, the sources of microseisms and their intensity were stable during signal accumulation. In a work by Gorbatikov and Stepanova (2008), it was shown that the period of stationarity of microseisms for the frequency range of 0.12-1.1 Hz was normally equal to 1.5 hours.

More recently, a detailed mathematical modelling of Rayleigh wave propagation in a heterogeneous environment was carried out (Gorbatikov and Tsukanov 2011). Based on the results of the modelling, the following facts were identified: the rate of minimal marked heterogeneity was equal to 3-4% of the wavelength; horizontal resolution was equal to a quarter of the wavelength; vertical resolution was equal to a third of the wavelength. The study of the heterogeneity was carried out for the velocities contrast of 10%. It should be noted that the variation of velocities contrast may affect the rate of minimal marked heterogeneity. It was shown that vertical borders of the objects comparable with the length of a wave could be identified with high precision. In addition, the dependence of the coefficient of the depth adjustment on the contrast in velocity was investigated. The obtained values essentially lie in the range of 0.4 -0.7. For the zero velocity contrast, the depth adjustment coefficient for the top of the structure was equal to ca. 0.51, and for the centre of the inclusion it was equal to ca. 0.43. These values agree well with the previous experimental results (Gorbatikov et al. 2008). Thus, using average dispersive dependences for the object of interest it is fair to use a coefficient of the depth adjustment of 0.4. In the case of existence of contrast objects, it is possible to correct the coefficient. The dispersive dependences characteristics of the studied region can be obtained, for example, from the data from Deep Seismic Sounding (DSS), the method of the receiving functions for the Ps waves, cross-correlation methods, etc.

The experimental results (Gorbatikov et al. 2008; Gorbatikov et al. 2013; Danilov, 2017; Francuzova, Danilov, 2018) showed that in some cases the resolution of the method increases. According to the results obtained from the mathematical models (Gorbatikov et al. 2013), the resolution of the method increases considerably if Poisson coefficient in the inclusion and the S-wave velocity contrast are approaching zero. In turn, the proximity of Poisson coefficient to zero means that the medium in the inclusions must be fractured (Gorbatikov et al. 2013). The effect of "super resolution" made it possible to reveal two large intrusive bodies beneath the El Hierro Island (Gorbatikov et al. 2013). Similar to the results obtained for the El Hierro Island, the vertical high-intensity zone was also found in the Onega Peninsula Arkhangelsk region of Russia using MSM (Danilov, 2017). These low-velocity zones can also be ancient volcanoes.

The main advantage of the method is its high precision of determining vertical borders. The high lateral resolution is a result of the use of amplitude information. The waves of different types are developed as a result of Rayleigh wave scattering on inhomogeneity.

However, these different types of waves cannot be distinguished near the heterogeneity. In this case, the amplitude is the source of heterogeneity information available for the analysis.

It should be noted that the similar statement have been made by independent researches more recently. Taylor et al. (2009) shown that the shape and amplitude of individual station resonance peaks appeared to be correlated with local geology. The research conducted by Lin et al. (2012) and Eddy and Ekström (2014) shown that amplification of earthquake Rayleigh waves has the following features:

- amplification of waves primarily due to changes in the velocity properties of the environment;
- amplification of Rayleigh waves increases when passing through low-velocity heterogeneities and decreases when passing through high-velocity heterogeneities;
- amplification due to local geology can be used to refine current elastic models of the crust and upper mantle;
- surface-wave amplification is more sensitive than phase velocity.

The above mentioned recent findings reported by Taylor et al. (2009), Lin et al. (2012) and Eddy and Ekström (2014) support earlier studies conducted by Gorbatikov et al. (2008, 2009, 2011, 2013), Gorbatikov and Tsukanov (2011). The key advantage of research by Lin et al. (2012) and Eddy and Ekström (2014) is that the Rayleigh wave was analyzed directly. The distinctive feature of researches of the Gorbatikov et al. (2008, 2009, 2011, 2013) is that in the surveyed environment there was no velocity properties smoothing. Good agreement between the results of independent earthquakes (Lin et al. 2012; Eddy and Ekström 2014) and microseisms (Gorbatikov et al. 2008, 2011, 2013) studies confirmed their authenticity. The advantage of microseisms analysis is that Rayleigh waves amplification is observed directly above the heterogeneity (Gorbatikov and Tsukanov 2011). It is also important to note that the microseismic sounding method avoids the necessity to consider the effects of focusing/defocusing and attenuation.

Since the method is sensitive to the temporal variations of microseism, a reference station should be used to eliminate the effects of the temporal variations (Gorbatikov et al. 2008a). Additionally, the station records can be used as a "reference" for rejecting noise.

Procedure of the data processing consists of the power spectra computation for each mobile and the reference measurements points, and calculate the relative intensity (I_fi) for each point and each frequency. The relative intensity is ratio of mobile station spectral amplitude to reference station spectral amplitude expressed in decibel (Gorbatikov et al. 2008a; Gorbatikov and Tsukanov 2011; Gorbatikov et al. 2013).

The next step (Gorbatikov et al. 2013) is determination of the depth (h) to be sounded as: $H \approx KV_R(f) / f$ (here, K is a numerical factor < 1).

The result of the processing is a geophysical cross section that is a distribution of relative intensity of microseisms along the profile and at a certain depth. Zones with higher relative microseism intensity represent an area with relatively reduced velocity properties, and vice versa.

Stability Assessment of Microseism

It is necessary to correct application of the microseismic sounding method (MSM) that intensity of microseisms were stable during accumulation of a signal (Gorbatikov et al. 2008).

Knowledge of the spectral amplitude accuracy allows to reasonably use the various methods of microseisms analysis. In particular MSM accuracy depends on its parameters.

In the current work the data of three regular seismic stations of the Arkhangelsk seismic net-work were analyzed: Franz Joseph Land (ZFI), Tamitsa (TMC), Klimovskaya (KLM). As shown in Figure 1, the stations are distributed throughout Arkhangelsk region. ZFI seismic station is located on the Alexandra Land island of Franz Josef Land archipelago, and is the northernmost regular seismic station in Russia. TMC seismic station is located on the coast of the White Sea. KLM is situated in the continental part of Arkhangelsk region.

Examples of temporal variations of the spectral amplitudes of ZFI station shows that the microseismic oscillation occurs mostly in a particular band of the spectral amplitudes. The average level of spectral amplitudes is smoothly variable of 40% for 10 hours. The fact says that the variation in the microseisms intensity happens with a certain regularity. As consequence it is necessary to assess the relative error of the spectral amplitude measurement depending on microseism accumulation period.

2011-2012 recording data were analyzed. It was selected 30 intervals of records for microseisms analysis. The interval recording duration was 4 hours, which is the maximum period of stationarity (Gorbatikov and Stepanova 2008), above which determination of the spectral amplitude accuracy is meaningless. The vertical channel records were invoked in the analysis. Normal operation of the equipment and the lack of visible noise impacts on wave forms in the majority duration of a record were chosen as an interval selection criteria. In the chosen intervals fluctuations in the ranges of frequencies of 0.5-0.6, 0.6-0.7, 0.7-0.8, 0.8-0.9, 0.9-1, 1-1.2, 1.2-1.4, 1.4-1.6, 2.2-2.4, 8.4-8.6 Hz were filtered. Average values of spectral amplitudes were calculated for each 5 minutes in each frequency bandwidth. Calculations were made using the software package DAK (Popov et al. 2014). For the sake of simplicity of the material presenting we will continue to speak about the same frequency band, meaning that the similar actions are performed for all bands. From henceforth the periods of accumulation of 30, 60, 90 and 240 min. were considered. A mean error and a mean square error of the spectral amplitude were calculated for each accumulation periods. Thus, the values of the relative error of the spectral amplitudes for 30, 60, 90 and 240 minutes accumulation for each interval were obtained. Gaps in each interval were excluded from the received values. For each station, the remaining set of values were averaged for each period of accumulation. Thus, the most probable values of a relative error of the spectral amplitudes for the periods of accumulation of a signal of 30, 60, 90 and 240 min. were received. Results are presented in Figure 2.

The result of the microseismic sounding method (MSM) is a distribution diagram of relative intensity of microseisms. The relative intensity is calculated by the formula 1.

$$\Delta I = 17.37\partial A \quad (1)$$

The absolute error of the relative microseisms intensity (ΔI) is directly proportional to the relative error of the spectral amplitude (∂A). The calculated errors of the relative intensity are presented in Figure 5.

Figure 2 shows that the increase in the signal accumulation period up to 4 hours reduces the measurement error up to 3%. The largest errors occur when there is the signal accumulation for 30 minutes at frequencies higher than 2 Hz. It should be noted that microseisms at frequencies up to 1.6 Hz differ from microseisms at higher frequencies.

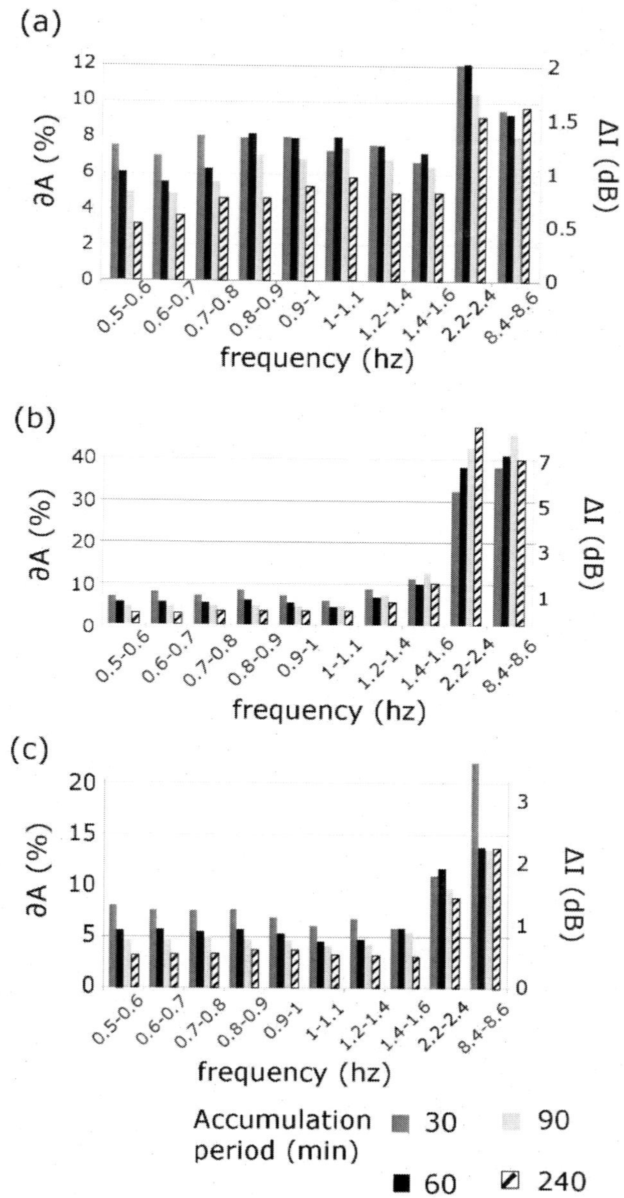

Figure 2. Diagrams of relative errors in determination by spectral amplitude of microseism (∂A) and by relative microseism intensity (ΔI, dB) in different frequencies bands with different signal accumulation in time. Diagrams constructs for permanent station of the Arkhangelsk seismic stations network: a – Klimovskaya (KLM); b – Tamica (TMC); c – Franz Josef Land (ZFI) (Danilov, 2017).

Values of a relative error at frequencies of 0.5-1.6 Hz are close for all points and are equal to 3-8%. This fact is observed regardless of the distance from the sea areas and the availability of sources of industrial noise.

Error level at frequencies over 1.6 Hz is twice as many as the one at low frequencies, thus certain features for each point are observed. Diagrams for KLM and ZFI points which were obtained for frequencies 1.4-2.4 Hz (Figures 2a and c) are very similar in nature and values. The level of errors increases with increasing accumulation period of the signal from 30 to 60 minutes. The error decreases significantly with further increasing of the accumulation period up to 4 hours. At lower frequencies similar character of charts is observed only for "continental" KLM station at frequencies over 0.8 Hz. This character of the diagrams indicates the presence in the continent sufficiently stable oscillation of microseisms intensity with a period about an hour. The error values of the ZFI point are slightly smaller than for the KLM points.

Obtained for mid frequencies in TMC point diagrams differ significantly from the other ones. This fact is caused by prevalence of technogenic sources of microseisms over natural one, even at rather uniform wave forms. In favor of the last findings speaks the following two issues. Firstly, the TMC points error value is much higher than the ones in the other observation points. Secondly, the errors increases with increasing accumulation period. This distribution can be observed only due to unpredictable changing of the set of intense sources. This fact indicates once again that microseisms at frequencies up to 1.6 Hz and above vary in their "stability" regardless of the their source. As man-made sources of microseisms in TMC point can be considered the following: industrial site with a sawmill, a federal highway, human activities in the nearest villages.

The diagrams of the ZFI and KLM points have the differences in patterns. Also there are no anthropogenic influences in the points. The ratio of the error in KLM point to the corresponding values in ZFI point was calculated to compare the results. Figure 3 shows that the value of errors in KLM point exceed the corresponding values in ZFI point. These differences are amplified by increasing the accumulation period. The biggest difference is observed at frequencies of 1-1.2 Hz. The difference at frequency of 0.5 Hz and 2.3 Hz is practically not observed. From the above it can be concluded that the frequency range of 0.8-1.6 Hz affected as "marine" and "continental" microseisms. The dominance of one or another source depends on the distance from the sea areas.

In spite of the fact that the spectral amplitudes determination accuracy can be close to 3%, in most cases the most optimum period of accumulation are 1.5 hours. So the accumulation of the signal within 1.5 hours allows almost to achieve the precision limit with substantially smaller time consuming. The errors are 5% and 9% for microseism at frequency band 0.5-1.6 Hz and 1.6- 8.4 Hz respectively.

In view of the direct relationship between each other, the accuracy of the relative intensity microseisms is characterized by the same rules that have been described for the relative error. The Figure 2 shows that the relative intensity error is close to 1 dB for the accumulation of signal 1.5 hours for the frequencies below 1.6 Hz. At the KLM and ZFI stations at frequencies above 1.6 Hz the errors are smaller than 2 dB. The increase of the accumulation period up to 4 hours will reduce the error to 0.5-0.7 dB. Most part of errors (8 dB) are observed in TMC point at frequencies higher than 2 Hz.

Description of the Field Measurements

This study used the following test objects of the ADP: the nonproductive S10 pipe in the Nenoksa field (Popov et al., 2014) and the low-diamond Chidviya pipe in the Chidviya–Izhmozero field (Kiselev et al., 2016).

Pipe C10 included in Nenokskoe field, located in north-east part of Onega ripheien aulacogen. Pipe C10 confined to the submeridional fault of the second order. The C10 depth exceed known dpeths of other pipeis of Nenoksa field. Wherein C10 depth similar as depth of pipes of Izhemsky, Suksomsky and Chidva groups (Eremenko, 2004b).

The pipe was drilled as far as 150.4 m depth (well S10/1). The area is 19.1 Ha. The overlying deposits are 47.4 m thick (Leshukov et al., 2010). The pipe is of the alkaline ultramafic type of the magnesium aluminiferous series; the composition is alkaline or ultra-alkaline, the type of alkalinity is sodium, unlike anything else worldwide. The Riphean deposits are 100 m thick, and the Vendian deposits are 300 to 500 m thick (Leshukov et al., 2010).

We measured microseismic ground motion along two lines for studying the S10 explosion pipe in the Nenoksa field. One line went from west to east (900 m, ten measurement sites) and the other from north to south (1100 m, five measurement sites) (Figure 1a). The measurement lasted 1 hour at each site.

Chidvinskaya pipe is part of the Chidvinsko- Izhmozerskoe kimberlite field, located 30 kilometers north-east from Arkhangelsk, and belonging to a membership of six pipes, constituting a chain with about 20 kilometers length to the North-North-East (Figure 1).

Chidvinskaya pipe is 1810x580 m. The pipe has low-grade ores without commercial value. However, a small amount of tests casts doubt on the accuracy of these estimates.

The microseisms at the Chidviya Pipe were measured along a line traversing the northern column of the pipe. The line included 19 measurement sites. The measurement lasted 2 hours at each site, the intersite distance varied between 50 and 150 m. The line had a total length of 1414 m.

The complex overburden rock presented quaternary loose sediments, made mainly of sands, loams, peat and sand-pebble formations. Overburden thickness change from 1.5 to 35.3 m northern and southern part of the pipe. The average thickness of quaternary sediments is about 9.4 m.

The vertical section of Chidvinskaya pipe is two-root diatreme with a common trumpet in the surface portion. The upper part of the pipe is formed by rocks of crateral facies which are divided into two bench, the top and bottom bench with total capacity of about 123.5 m.

Top bench is composed of tuffaceous sedimentary rocks represented by coarse sandstones with an admixture of magmatic material, tuff-sandstones, tuffites and tuffs. Bottom bench is distributed in the northern part of the pipe and is made by tuffites with rare interlayers of tuffs and tuff-sandstones.

Diatreme facies is composed by autolit breccias, tuff breccias that form two ore pillars. The boundary between them is held in the middle bottleneck zone. Contacts between the diatreme rocks and the surrounding sedimentary deposits are quite clear.

The results were compared with the data on the Lomonosov and Pionerskaya pipes of industrial diamond reserves in the Zolotitsa field (Frantsuzova and Danilov, 2016) and with the G23 spurious magnetic anomaly.

Lomonosovo and Pionerskaya pipes are included in diamond deposit M.V. Lomonosova. This pipe are consist of interstitial phases. Pionerskaya pipe have two-roots. The overlying deposits are 40-70 m thick. Pionerskaya pipe is 1810x580 m. Lomonosovskaya pipe is 600x300 m. The measurement lasted 1.5-2 hours at each site, the intersite distance varied between 50 and 100 m.

The G23 geophysical anomaly is understood in this study as an area identified by magnetic exploration and MMS. The G23 magnetic anomaly which is situated near to the Lomonosov explosion pipe in the ADP Zolotitsa field, will be referred to as a “spurious” anomaly, because it has a shape similar to those due to explosion pipes. This spurious magnetic anomaly was studied from data acquired by measurements on a line across it. The line includes 22 sites of microseism measurement at a step of 40 m. The line has a total length of 840 m.

We analyzed the Lomonosov explosion pipe using measurements of microseismic ground motion along a nearly line that passed nearer to the root of the pipe (Frantsuzova and Danilov, 2016). The interval between adjacent measurement sites was 50 m. The microseismic motion at a site was measured during 90 min. The line had a total length of 2700 m (Frantsuzova and Danilov, 2016). We measured microseisms during our field work using both Russian-made seismic instruments (UGRA with a 16-digit ADC and velocimetr SM3-OS) and manufactured by Guralp, UK (CMG-3ESP, CMG-40TDE and two sets of CMG-6TD broadband stations).

THE RESULT OF MICROSEISMIC SOUNDING METHOD

The C10 pipe was imaged as contras high-velocity anomaly within the following distances along the lines of measurement: 620 to 880 m for the west–east direction and 0 to 450 m for the north–south direction (Figure 3 a, b). The C10 pipe is adjacent from the east to a fault that is represented in the diagram (see Figure 3a) as a wide (up to 500 m) low-velocity band of fragmented host rocks.

The results of MSM are shown in Figure 3 were the high-velocity pipe body is allocated at observation points 6 - 15. The border dividing the western and eastern blocks of a pipe has an inclination of about 45 degrees at depths up to 400 m. Deeper than 400 m all borders remain almost vertical. Eastern block is traced up to the depths more than 1000 m despite the small sizes in deep part of a section. The pipe differ from environment in 2-7 dB by relative microseism intensity. The western and east blocks of pipe differ from each over about 2-5 dB by relative microseism intensity. According to (Gorbatikov & Tsukhanov 2011) these relative microseism intensity changes correspond to velocity properties changes in 1.5-2 times and 1.2-1.5 times.

The allocated pipe is limited by contrast linear, vertical low-velocity zones from the West and the East. These zones are likely caused by faults in endocontact rock. These faults are traced at the depths of 200-1000 m under points 5 from the West and 15-16 from the East. Also, the vertical low-velocity zones can be traced for 300-400 m off pipe to the west and east.

In near-surface area the most contrast low-velocity body is observed under points 2-5 at depths from 40 to 200 m and has a cone-shaped structure. According to (Gorbatikov & Tsukhanov 2011) this contrast approximately correspond to velocity decreasing of 2-3 times.

Also vertical low-speed zones are traced on removal from a pipe body on 300-400 m to the West and the East.

The changes in the configuration consistent with the changes in the intensity of the allocated zones allow to place the layer in the crust at about 200 m depth, and corresponds to boundary of Padun and Mezen suites of the top of Vend.

A low-velocity zone was identified in the area of the G23 magnetic anomaly based on microseismic data (see Figure 3e). The part with the highest contrast in that zone lies at depths down to 130 m within measurement sites 9 through 12.

The above comparison between the MMS results and the map of magnetic anomalies shows that the G23 anomaly occurs in a high-contrast low-velocity zone that involves some upper layers (between 50 and 200 m).

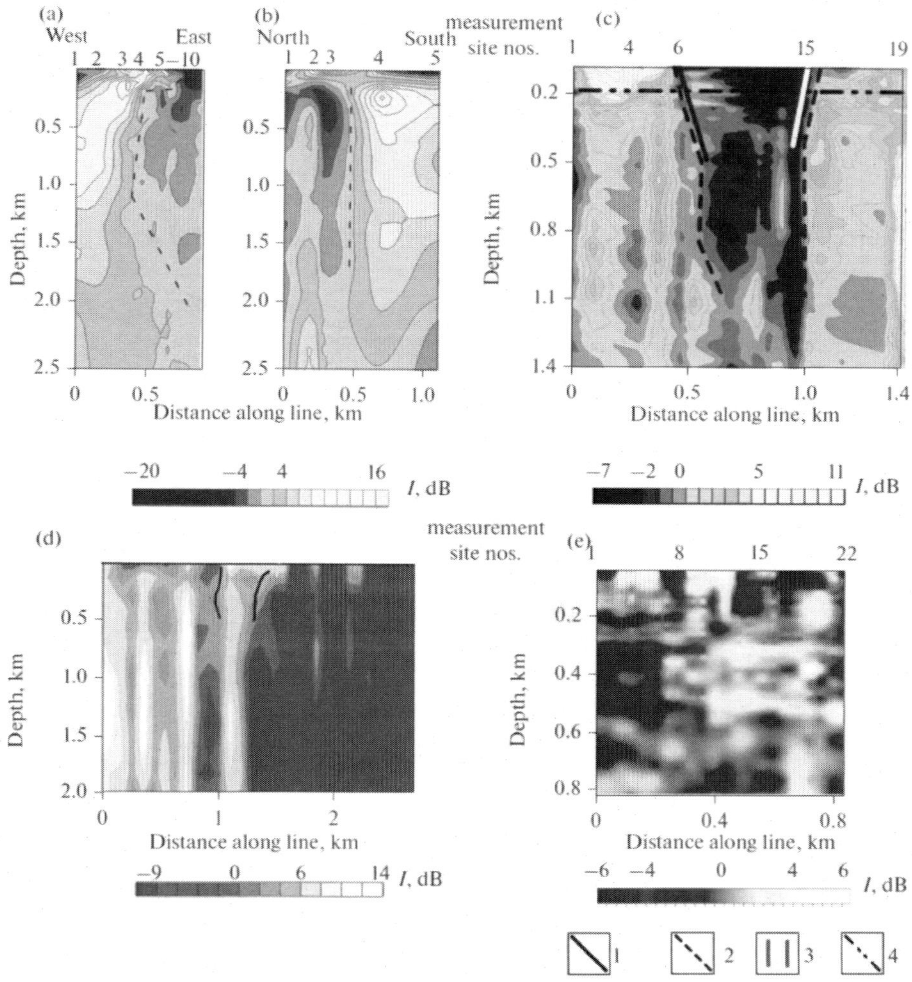

Figure 3. Vertical sections of velocity (Frantsuzova and Danilov, 2018) contrasts along lines that traverse the S10 explosion pipe (a, b), the Chidviya explosion pipe (c), Lomonosov explosion pipe (Frantsuzova and Danilov, 2016) (d), G23 spurious magnetic anomaly (e). (1) known contours of explosion pipes (Verzhak et al., 1987; Egorkin, 2004); (2) explosion pipe contours based on MMS data; (3) boundaries of magnetic anomaly (Verzhak et al., 1987); (4) boundary of Paduan and Mezenian Formations.

An analysis of the diagrams revealed the presence of four vertical low-velocity zones in the vertical cross section (see Figure 3d). The first three westernmost zones have the highest values of contrast, practically vertical boundaries, and linear structures. The fourth low-velocity zone, which is found to lie in the area that corresponds to the actual position of the pipe, has a cone-shaped structure with a downward-looking vertex, which is typical of volcanic pipes (Milashev, 1984; Verzhak et al., 1987).

Analysis of the diagrams (Figure 3) shows the presence of three high-contrast low-velocity zones westward of the pipe body; these zones can be followed downward for more than 2 km and are very likely discontinuities. This hypothesis is supported by some facts, viz., that volcanic pipes are accompanied by parallel and radial discontinuities (Kutinov and Chistova, 2004). In addition, there is certain independent geophysical evidence resulting from studies of the pipe using electri cal prospecting techniques (Stognii and Korotkov, 2010).

The leading feature that occurred in all the three pipes was the presence of a pipe body identifiable at depths greater than 1 km.

The main difference among the pipes studied here was that the S10 and Chidviya pipes are seen as highvelocity zones (see Figure 3a, 3b, 3c), in contrast to the low-velocity signature for the Lomonosov explosion pipe body (see Figure 3d). Accordingly, we tried to find an explanation of this discrepancy.

The observed differences can be related both to the material composition of pipe material and to some peculiarities of the host rocks. An analysis of the data on the composition of the typomorphic material in the pipes encountered in the Nenoksa, Izhmozero, and Zolotitsa fields of the Arkhangelsk Diamond Province suggested the following inference: the typomorphic compositions of the Nenoksa and Izhmozero pipes are nearly identical, but differ from the Zolotitsa pipes, in particular, from the Lomonosov explosion pipe (Eremenko and Nenakhov, 2002).

Joint Using of Microseismic Sounding Method and Passive Seismic Interferometry on the Example of the "Pionerskaya" Pipe

We tested the scheme of joint using microseismic sounding method (MSM), passive seismic interferometry and horizontal to vertical spectral ratio method (H/V) (Danilov et al., 2017).

The main result of our study, is number of important boundaries of the test object. So the first phase is imaged as a low-velocity zone. South root imaged at the depths from 600 m to 1200 m. The lower border of the first phase picked by microseismic sounding method and by H/V method equally.

The low-velocity zone picked by passive seismic interferometry picked at the same depths as first phase which imaged by MSM. Thus, the data of each method complement each other.

This fact proves the principle applicability of the proposed scheme of joint using of methods for study of structure and velocity characteristics of diamond pipes.

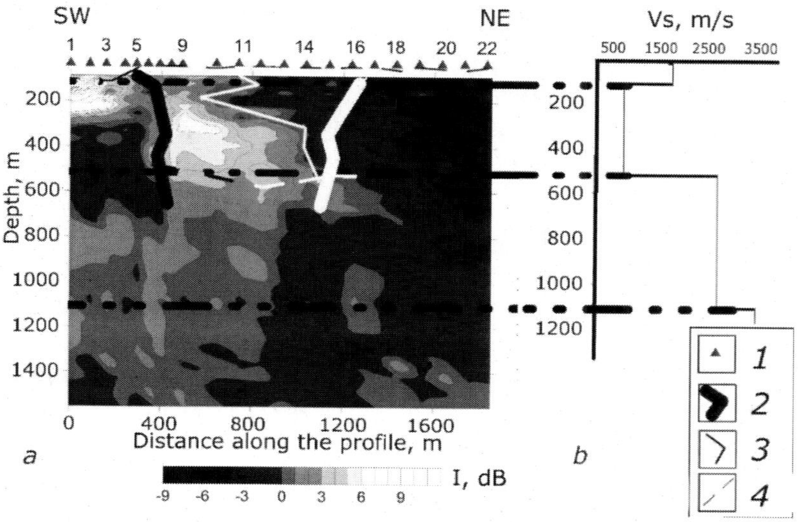

Figure 4. The Pionerskaya diamond pipe's imaging obtained by passive seismic methods (Danilov et al., 2017): (a) microseismic sounding and H/V methods (b) velocity model, calculated by inversion of dispersion curve, obtained by passive seismic interferometry (4): 1 – point of microseismic measurement, 2 – boarders of the pipe (Verzhak et al., 1987), 3 – the board of different intrusions phase (Verzhak et al., 1987).

COMPLEXING OF RADIOMETRIC AND MICROSEISMIC SOUNDING METHODS ON THE EXAMPLE OF THE CHIDVINSKAYA PIPE

Gamma-Spectrometric Studies

Figure 5 shows the map of total gamma-ray distribution of near surface layer above Chidvinskaya pipe. The total radioactivity of rocks varies widely from 172 to 1585 Bq m-3. The distribution of high gamma-field has a slightly wide configuration than the pipe contour. The most contrast and the biggest area of increased gamma-ray values is confined to the northern part of the pipe and localized within the north pole, where thickness overburden of quaternary sediments is minimal and is about 1.5 m. Value of total radioactivity reached 1500 Bq m-3.

Spatial distribution of high values of total gamma-ray radiation of near surface layer above Chidvinskaya pipe forms a linear anomaly of northeast trending which goes beyond contour of the pipe. Average values of radioactivity are about 800-900 Bq m-3 outside this area.

Concentration of thorium in Chidvinskaya pipe area varies within 0.5 - 14.0 μg/g. The maximum values of thorium are localized at small areas and are arranged substantially within the contour of the pipe. The potassium content ranges from 0.3 to less than 3.8%. The spatial distribution picture of potassium content in general is identical to the index of total gamma-ray radiation and forms an area of high values within pipe contour, especially in the northern part of the pipe where thickness overburden of quaternary sediments is minimal. Uranium concentration ranges from less than 0.4 to 4.6 μg/g. Distribution picture of uranium isn't unambiguous. Explicit spatial confinement to the contour of the pipe wasn't observed.

In general, it can be noted that Chidvinskaya pipe have increased values of gamma radiation, as well as potassium. Distribution of thorium and uranium concentration doesn't have areal disposition that apparently is a consequence of erosional feature of top bench craterial facies Chidvinskaya pipe.

Radon Emanation Results

Traverse of radon method included 16 observation points. Points of observation are shown in Figure 5. The results are shown in Figure 6.

Radon radioactivity in the soil air ranges from 114 to 1312 Bq/m3 on the profile that crosses the north pole of Chidvinskaya pipe. Variation of values radon emanation are insignificant within the contour of the pipe from 185 to 312 Bq/m3. Anomalous increase of radon concentrations was observed at the boundaries of Chidvinskaya pipe.

Maxima of radon radioactivity in soil air were obsewrved at the north-western and south-eastern borders of the profile, 1156 Bq/m3 and 1312 Bq/m3, respectively. 222Rn radioactivity falls sharply and decreases to a minimum beyond the boundaries of the pipe.

Revealed regularities of radon changes on the profile can be elucidated by the fact that the contacts of kimberlite pipe with the enclosing rocks are fractured faults zones with high permeability.

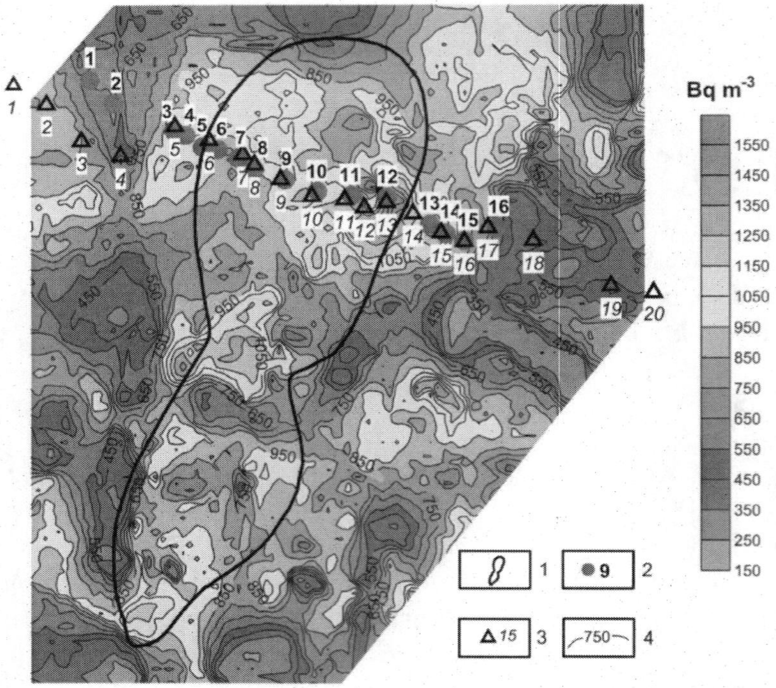

Figure 5. Distribution of total gamma-ray radiation above Chidvinskaya pipe (Kiselev et al., 2017). 1.- Chidvinskaya pipe contour; 2.- point measurement of radon emanation in the soil air; 3.– MSM measuring points; 4.- contour lines of gamma-ray radiation intensity.

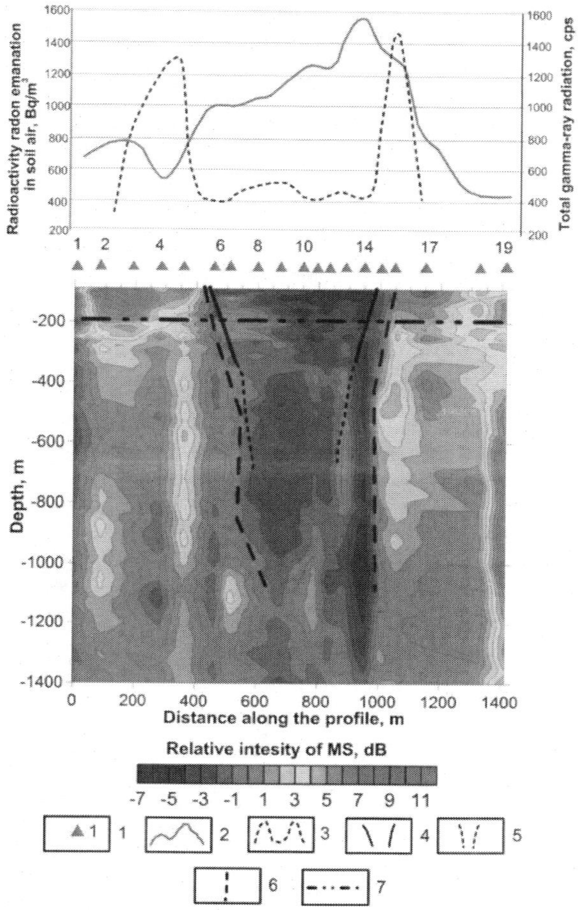

Figure 6. The depth section of relative intensity distributions of microseisms and radiometric parameters along the profile crossing Chidvinskaya pipe (Kiselev et al., 2017). 1.- point measurements of microseism; 2.- curve of change the total intensity of the gamma-ray radiation, Bq m-3; 3.- curve of change of radon emanation in soil air, Bq/m3; 4.- Chidvinskaya pipe border for geological exploration data; 5.- proposed structure of diatreme facies; 6.- side pipe body according to MSM; 7.- border Padun and Mezen suites of Vendian.

When comparing the results we observe an increase of the total gamma gamma-ray radiation consistent with MSM, that corresponds to a high-velocity anomaly. Intensity of gamma radiation on the high-speed anomaly is greater than 2-4 times for background values of enclosing rocks. At the same time, there is a decrease of total gamma radiation from western to eastern blocks of the pipe body. Such a significant change of the gamma radiation level and high-speed characteristics may indicate different material composition of the Western and Eastern pipe blocs. This difference can indicate either different implementation phases, or that data block was subjected to various (physical, metamorphic, erosive) processes after the formation of the tube.

Anomalous radon concentrations at the boundaries of Chidvinskaya pipe are appropriate contrasting linear low-velocity zones. As a dedicated fault structures, there is a possible influx of radon and other gases. High-velocity nature of the pipe body indicates a consolidated structure and a minimum influx of 222Rn. High velocity nature of the pipe body can be explained by the sharply different composition of the pipe and enclosing medium.

The proposed set of methods can be used in further studies, since the methods showed a good convergence of results.

Conclusion

The microseismic sounding method allows imaging diamond pipes of the Arkhangelsk diamond province and their rots as subvertical inhomogeneity at the depth greater than 1 km. As consequently, microseismic sounding method can be used for classify geophysical anomalies. Diamond pipes located near faults zones. The poor pipes are located insight faults zone. This criterion should be investigated further.

Our research showed, that microseism's amplitude usually stable enoughintensity. So, error of microseismic sounding method is close to 1 dB for the duration of seismic record of 1.5 hours for frequencies lower than 1.6 Hz. Errors for frequencies higher than 1.6 Hz, usually smaller than 2 dB. Increasing duration of seismic record up to 4 hours can reduce errors to 0.5-0.7 dB.

Joint using of microseism sounding, passive seismic interferometry and H/V methods allows getting more precise image of diamond pipe and host medium.

Joint using of microseism sounding and radiometric's methods allows to increase reliability of interpretation.

Presented result show, that using of passive seismic and radiometric methods can increase effectiveness of prospecting and exploration of diamond pipes of the Arkhangelsk diamond province.

Acknowledgments

The investigation by passive seismic methods was supported by the Russian Ministry of Education and Science (project 0409-2015-0135, no. AAAA-A18-118012490072-7). The investigation by radiometrics methods was supported by the Russian Ministry of Education and Science (project no. AAAA-A19-119011890018-3) and the Russian Foundation for Basic Research (projects no. 20-35-70060).

References

Afonin, Nikita., Kozlovskaya, Elena., Kukkonen, I. & DAFNE/FINLAND Working Group. (2017). Structure of the Suasselkä postglacial fault in northern Finland obtained by analysis of local events and ambient seismic noise. *Solid Earth.*, *8*, 531-544. https://doi.org/10.5194/se-8-531-2017.

Afonin, N., Kozlovskaya, E., Nevalainen, J. & Narkilahti, J. (2019). Improving quality of empirical Greens functions, obtained by cross-correlation of high-frequency ambient seismic noise. *Solid Earth*. 10, 1621–1634, https://doi.org/10.5194/se-10-1621-2019

Asten, M. W. (1978). Geological control on the three-component spectra of Rayleigh-wave microseisms, *Bulletin of the Seismological Society of America*, *68*, 1623-1636.

Asten, M. W. & Henstridge, J. D. (1984). *Array estimators and the use of microseisms for of sedimentary basins*, *Geophysics*, *49*, 1828-1837.

Babayants, P. S., Blokh, Y. I. & Zubov, E. I. (2006). Interpretation of aerogeophysical data in the search for deposits. *Prospect and protection of mineral resources.*, *5*, 18-26.

Bath, M. (1974). Spectral analysis in geophysics. Amsterdam. Elsivier., pp. 432-443.

Berdnikov, P. V. & Gorky, A. V. (2007). Experience in studying the radon danger of St. Petersburg and the Leningrad region. Abstracts nauchn.-practical. Workshop "Radon in exploration and ecology." Moscow.

Bezborodov, S. M., Verzhak, V. V. & Gerasimchuck, A. V. (2003). Diamond exploration and mining in north-west of Russia: 4th Fennoscandian exploration and mining. Final Program and Event Document. - *Rovaniemi*, 279-304.

Bobrov, A. A. (2009). Mapping of some features of fault zones of Priolkhon and the South Angara in radon emanations. *Tectonophysics and Actual Problems of Earth Sciences.*, *2*, 5-9.

Chen, J. & Ford, K. L. (2016). A study on the correlation between soil radon potential and average indoor radon potential in Canadian cities. *Journal of Environmental Radioactivity.*, *166*, 152-156.

Cheng, F., Xia, J., Xu, Y., Xu, Z. & Pan, Y. (2015). A new passive seismic method based on seismic interferometry and multichannel analysis of surface waves. *Journal of Applied Geophysics.*, *117*, 126-135.

Danilov, K. B. (2011). Application of microseismic sounding for the study of Lomonosov diatremes (Arkhangelsk diamondiferous province), Bulletin of the Kamchatka regional organizations educational and scientific center. *Series: Earth Science*, *17*, 231-237. http://www.kscnet.ru/kraesc/2011/2011_17/2011_17.html.

Danilov, K. B. (2017). The structure of the Onega downthrown block and adjacent geological objects according to the microseismic sounding method. *Pure and Applied Geophysics*, *174*(7), 2663-2676.

Eddy, C. L. & Ekström, G. (2014). Local amplification of Rayleigh waves in the continental United States observed on the US Array, *Earth and Planetary Science Letters*, *402*, 50-57.

Eremenko, A. V. & Nenakhov, V. M. (2002). *Geology and geodynamic model of the formation of the volcanic pipes of the Arkhangelsk diamond province*. Proceedings of Voronezh State University, *1*, 36-42.

Frantsuzova, V. I. & Danilov, K. B. (2016). The Structure of the Lomonosov Volcanic Pipe in the Arkhangel'sk Diamond Province from Anomalies of the Microseismic Field. *Journal of Volcanology and Seismology.*, *10*(5), 339–346.

Francuzova, V. I. & Danilov, K. B. (2018). The location of the S10 and Chidviya Kimberlite Pipes from Inhomogeneities in the Host Rock Structure of the Nenoksa and Chidviya–Izhmozero Fields in the Arkhangelsk Diamond Province as Detected by Background Microseisms. *Journal of Volcanology and Seismology.*, *12*(5), 332–340.

Gorbatikov, A. V., Larin, N. V., Moiseev, E. I. & Belyashov, A. V. (2009). The microseismic sounding method: Application for the study of the buried diatreme structure, *Doklady Earth Sciences*, *428* (1), 1222-1226. DOI: 10.1134/S1028334X0907040X.

Gorbatikov, A. V., Montesinos, F. G., Arnoso, J., Stepanova, M. Y., Benavent, M. & Tsukanov, A. A. (2013). New features in the subsurface structure model of El Hierro Island (Canaries) from low-frequency microseismic sounding: An insight into the 2011 seismo-volcanic crisis. *Surveys in Geophysics*, *34*(4), 463-489.

Gorbatikov, A. V. & Tsukanov, A. A. (2011). Simulation of the Rayleigh waves in the proximity of the scattering velocity inhomogeneities. Exploring the capabilities of the microseismic sounding method. *Izvestiya*, *Physics of the Earth*, *4*, 354-369. DOI:10.1134/S1069351311030013.

Gulabiants, L. A. & Zabolotsky, B. Yu. (2004). Seasonal variation of the radon flux from the soil and the estimation of the radon danger of the built-up area. *ANRI*, *4*, 46-50.

Horike, M. (1985). Inversion of phase velocity of long-period microtremors to the S-wave-velocity structure down to the basement in urbanized areas, *Journal of Physics of the Earth*, *33*, 59-96.

Huxtable, D., David, Read. & George, Shaw. (2016). Measuring radon-222 in soil gas with high spatial and temporal resolution. *Journal of Environmental Radioactivity*. http://dx.doi.org/10.1016/j.jenvrad.2016.11.003.

Ibs-von Seht, M. & Wohlenberg, J. (1999). Microtremor measurements used to map thickness of soft sediments. *Bulletin of the Seismological Society of America*, *89*(1), 250-259.

Kanai, K. & Tanaka, T. (1954). Measurement of the microtremor. Bulletin of the Earthquake Research Institute. *The University of Tokyo*, *32*, 199-209.

Katz, L. J. & Bellon, R. S. (1978). Microtremor site analysis study at Beatty, Nevada, *Bulletin of the Seismological Society of America*, *68*, 757-765.

Khazanovich-Wulff, K. K. (2007). *Diatreme plumes of astroblem or "bolide model" of the formation of kimberlite pipes*. GGTS GEOMASTER. Petrozavodsk.

Kiselev, G. P., Druzhinin, S. V. & Yakovlev, E. U. (2016). *Experimental-methodical and research work on the predictive-search assessment of Lomonosov deposit areas that are promising for the search for kimberlite pipes using radiometric methods*. FCIARctic. Arkhangelsk

Kiselev, G. P., Danilov, K. B., Yakovlev, E. U. & Druzhinin, S. V. (2017). Radiometric and seismic study of Chidvinskaya kimberlite pipe (Arkhangelsk diamondiferous province, North of the East European Craton, Russia). *Geofísica Internacional.*, *2*, 147-155.

Kontarovich, R. S. & Babayants, P. S. (2011). Problems of modern advanced geophysical support for prediction and exploration operations. *Prospect of mineral resources.*, *7*, 18-22.

Kostrovitsky, S. I. (1976). *Physical conditions, hydraulics and kinematics of filling kimberlite pipes*. Nauka. Novosibirsk.

Kutinov, Y. G. & Chistova, Z. B. (2004). *Hierarchical series of manifestations of alkaline-ultrabasic magmatism in the Arkhangelsk diamondiferous province*. Their reflection in geological and geophysical materials. Pravda Severa. Arkhangelsk.

Lane, J. W., White, E. A., Steele, G. V. & Cannia, J. C. (2008). January. Estimation of bedrock depth using the horizontal-to-vertical (H/V) ambient-noise seismic method. In *Symposium on the Application of Geophysics to Engineering and Environmental Problems*, 2008, (pp. 490-502). Society of Exploration Geophysicists.

Le Feuvre, M., Joubert, A., Leparoux, D. & Cote, P. (2015). Passive multi-channel analysis of surface waves with cross-correlations and beamforming. Application to a sea dike. *Journal of Applied Geophysics.*, *114*, 36-51.

Leshukov, S. I., Belov, A. V., Prusakova, N. A., et al., (2010). Poiskovye raboty na almazy na Onezhskoi ploshchadi (Arkhangel'skaya oblast') [Prospecting Surveying for Diamonds in the Onega Area, Arkhangel'sk Region], Report for Object 1-75/07. Gov. contract no. 1. License ARKH01210 KP, Arkhangel'sk: ZAO Arkhangel'skie Almazy.

Lin, F. C., Li, D., Clayton, R. W. & Hollis, D. (2013). High-resolution 3D shallow crustal structure in Long Beach, *California: Application of ambient noise tomography on a dense seismic array Geophysics.*, *78*(4), Q45-Q56.

Lin, F. C., Ritzwoller, M. H., Townend, J., Bannister, S. & Savage, M. K. (2007). Ambient noise Rayleigh wave tomography of New Zealand. *Geophysical Journal International.*, *170*(2), 649-666.

Lin, F. C., Tsai, V. C. & Ritzwoller, M. H. (2012). The local amplification of surface waves: A new observable to constrain elastic velocities, density, and anelastic attenuation, *J. Geophys. Res.*, *117*, B06302, doi:10.1029/2012JB009208.

Magomedova, A. S., Udoratin, V. V. & Ezimova, E. E. (2015). Reflection of fault zones and explosion tubes of Middle Timan in geophysical fields. *Proceedings of Institute of geology Komi UB RAS.*, *10*, 28-34.

Magomedova, A. S. & Udoratin, V. V. (2016). Radon activity of explosion pipes and magnetic anomalies of the Middle Timan. *Proceedings of the Ural Youth Scientific School on Geophysics.*, *1*, 110-112.

Makarov, V. I., Babak, V. I., Dorozhko, A. L., Bondarenko, V. M. & Demin, N. V. (2003). The influence of structural and geological features on the distribution of concentrations of the subsurface radon and radon in the basements of residential buildings (on example, the Chertanovo district of Moscow). *Geoecology.*, *2*, 139-152.

McDowall, G. & Koketso, H. (1991). Radon emanometry over some kimberlites and lamproites in Southern and Western Botswana. 53rd EAEG Meeting. Florence.

Miklyaev, P. S. (2015). Scientific basis for assessing the potential radon hazard of platform areas. PhD Thesis, Sergeev institute of environmental geoscience RAS, Moscow.

Milashev, V. A. (1979). Structure of kimberlite fields. Nedra. Leningrad.

Milashev, V. A. (1984). Pipes. Nedra, Leningrad.

Monakhov, F. I. Low-frequency seismic noise of the earth. (M.: "Nauka". 1977).

Mwenifumbo, C. J. & Kjarsgaard, B. A. (1999). Gamma-ray logging and radioelement distribution in the Fort à la Corne kimberlite pipe 169. *Exploration and Mining Geology*, *8*, 12, 137–147.

Neznal, M., Neznal, M. & Smarda, J. (1991). Radon infiltration risk from the ground in Chaby. Geological Survey of Prague. Prague.

Neznal, M., Neznal, M. & Smarda, J. (1996). Assessment of radon potential of soils e a five years experience. *Environment International.*, *22*, 819-828.

Nikitin, B. M. (1980). Deformation of host rocks during formation of kimberlite pipes. *Petrology.*, *11*, 41-49.

Nogoshi, M. & Igarashi, T. (1971). On the Amplitude Characteristics of Microtremor (Part 2) (in Japanese with English abstract). *Journal of the Seismological Society of Japan*, *24*, 26-40.

Novikov, G. F. (1989). Radiometric prospecting. L. Nedra.

Petersell, V., Taht-Kok, K., Karimov, M., Milvek, H., Nirgi, S., Raha, M. & Saarik, K. (2016). Radon in the soil air of Estonia. *Journal of Environmental Radioactivity.* http://dx.doi.org/10.1016/j.jenvrad.2016.08.004.

Popov, D. V., Danilov, K. D., Zhostkov, R. A., Dudarov, Z. I. & Ivanova, E. V. (2014). Processing the digital microseism recordings using the Data Analysis Kit (DAK) software package. *Seismic Instruments*, *50*, 75-83. DOI: 10.3103/S074792391401006X.

Poli, P., Campillo, M., Pedersen, H. & LAPNET Working Group (2012). Body-wave imaging of Earth's mantle discontinuities from ambient seismic noise. *Science.*, *338*(6110), 1063-1065.

Reed, L. E. & Witherly, K. E. (2007). 50 Years of Kimberlite Geophysics, A Review. Proceedings of Exploration 07: Fifth Decennial International Conference on Mineral Exploration" edited by B. Milkereit., 679-689.

Ramadass, G., SubhashBabu, A. & Udaya Laxmi, G. (2015). Structural analysis of airborne radiometric data for identification of kimberlites in parts of Eastern Dharwar Craton. *International Journal of Science and Research.*, *4*, 2375- 2388.

Rybalchenko, A. Ya., Rybalchenko, T. M. & Silaev, V. I. (2011). Theoretical Foundations of Forecasting and Searching for Tuffisite-Type Indigenous Diamond Fields. *Proceedings of Institute of geology Komi UB RAS.*, *1*(5), 54-66.

Seminsky, K. Z., Cheremnykh, A. V. & Bobrov, A. A. (2009). Fault zones of the Baikal region: internal structure and geophysical fields. *Tectonophysics and Actual Problems of Earth Sciences.*, *2*, 151-156.

Seminsky, K. Z., Bobrov, A. A. & Demberel, S. (2014). Variations of volume activity of radon in fault zones of the earth's crust: spatial features. *Physics of the Solid Earth.*, *6*, 80-98.

Shapiro, N. M. & Campillo, M. (2004). Emergence of broadband Rayleigh waves from correlations of the ambient seismic noise. *Geophysical Research Letters.*, *31*, L07614. https://doi.org/10.1029/2004GL019491.

Shapiro, N. M., Campillo, M., Stehly, L. & Ritzwoller, M. H. (2005). High-resolution surface-wave tomography from ambient seismic noise. *Science.*, *307*(5715), 1615-1618.

Sobisevich, A. L., Gorbatikov, A. V. & Ovsuchenko, A. N. (2008). Deep structure of the Mt. Karabetov Mud volcano, *Doklady Earth Sciences*, *422* (1), 1181-1185. DOI: 10.1134/S1028334X08070428.

Stogniy, V. V. & Korotkov, Y. V. (2010). Search of the kimberlite bodies by transients. Malotirazhnaya tipografiya 2D. Novosibirsk.

Taylor, S. R., Gerstoft, P. & Fehler, M. C. (2009). Estimating site amplification factors from ambient noise, Geophys. *Res. Lett.*, *36*, L09303, doi:10.1029/2009GL037838.

Verzhak, V. V., Medvedev, V. A. & Verichev, E. M. (1983). *Report on the results of the exploration of kimberlite pipes on the territory of Lomonosov diamond deposit in 1983–1987*. Geological Survey. Arkhangelsk.

Vinnik, L. P. (1968). *The structure microseisms and some questions of grouping methodology*. (M.: "Nedra" 1968).

Wapenaar, K., Draganov, D. & Robertsson, J. O. eds. (2008). Seismic interferometry: History and present status. Society of Exploration Geophysicists. Tulsa.

Yakovlev, E. Y., Kiselev, G. P. & Druzhinin, S. V. (2016). Non-equilibrium uranium in kimberlites and host rocks Pionerskaya pipe of the M.V. Lomonosov deposit. Vestnik of Northern (Arctic) *Federal University*, *1*, 19-28.

Yakovlev, E. Y. (2017). *The use of isotope-radiogeochemical methods for diamond deposits prospecting in Arkhangelsk diamondiferous province*. PhD Thesis. MGRI-RGGRU. Moscow.

In: The Arctic: Current Issues and Challenges
Editors: Oleg S. Pokrovsky, Sergey N. Kirpotin et al.
ISBN: 978-1-53617-306-2

Chapter 14

DISPERSED SEDIMENTARY MATERIAL IN THE SNOW AND ICE COVER OF THE CENTRAL ARCTIC AND ITS FLUXES TO THE BOTTOM

Alexander N. Novigatsky*
Shirshov Institute of Oceanology, Russian Academy of Sciences, Moscow, Russia

ABSTRACT

The monthly, seasonal and annual quantity estimates of vertical fluxes of sedimentary matter from the surface layer of the Arctic Ocean, performed out over the years by various researchers, are the basis for direct calculations of incoming chemical components, minerals, and various pollutants to the surface layer of bottom sediments. Dispersed sedimentary material in the snow and ice cover is directly determined near the North Pole. The composition of sedimentary material in the snow and ice cover was studied at the maximum distance from continents in the high latitudes of the Northern Hemisphere, and the vertical fluxes of material from the underice layer were determined. These data were obtained using sediment traps; for snow and ice, using large volumes of melt water. It was found that dispersed aeolian sedimentary material is of primary importance in snow and sea ice.

Keywords: Arctic Ocean, sedimentation, vertical sediment fluxes, sediment traps, suspended particulaite matter

INTRODUCTION

Dispersed matter is widespread in the environment. It occurs in all the Earth's geospheres: aerosol particles in the atmosphere and suspended particulate matter in the water column, ice, and snow. It is present in areas of deep-sea hydrothermal activity ("black smoker" plumes) and occurs as space dust in a small amount. An important role in the

* Corresponding Author's Email: novigatsky@ocean.ru.

formation of dispersed sedimentary matter belongs to organisms (diatoms, coccolithophores, etc.), which yield the biogenic suspension, and organisms-consumers (zooplankton and benthos), which feed on this fine suspension and small biogenic detritus (Lisitzin, 2004). The dispersed continental sedimentary matter is due to its origin from physical and chemical weathering of rocks and, to a significant degree, to the microorganisms, as well as plants (spores and pollen, plant debris).

A detailed study of dispersed forms of sedimentary material is extremely important, because these particles have a great deal of surface area, i.e., they are sorbents, and their composition and concentration reflect all facets of the sedimentation process; i.e., they are a novel and very rich source of information. The sedimentation process in the Arctic is characterized by a particular type of sedimentogenesis: sea ice. Here, terrigenous sediments predominate, which are specific in the manner of their preparation, transport, and deposition. In addition, sedimentary material is strongly dispersed in the atmosphere, snow, and sea ice (Darby et al., 1974; Stein, 2008; Lisitzin, 2010; Lisitzin, Shevchenko, 2016).

Between the appearance of sedimentary material and its accumulation in the form of bottom sediments, there is a dispersion phase up to imperceptible contents; existence in such a dispersed state during interaction with the medium and climate and biological processes; and then thickening up to seafloor sediment material, followed by diagenesis. Sedimentologists have hardly studied this stage at all.

The Arctic Ocean is a basin with a strongly expressed seasonal variation in sedimentation. Parameters such as the area of marine ice spreading, dissolved and particulate river runoff, coastal abrasion, primary production, and the temperature of surface water masses and of the near-water layer of the atmosphere demonstrate a strongly marked seasonality.

The rates of sedimentation in the bottom sediments of the World Ocean vary widely from less than 1 to 1,000 mm/year and more. The continental margins of the Arctic Ocean are characterized by values of ~10 mm/year (in deposition centers of river and oceanic water mixing zones, the Fram Strait, and some fjords). Zero and even negative values are typical of zones of bottom erosion (areas of the White, Barents, Kara, and other seas) at 0.01 mm/year – an average sedimentation rate in the pelagic zone of the Arctic Ocean (Rachold et al., 2004; Stein, 2008; Levitan, 2015).

Fundamentally new data on sedimentation were obtained for direct determination of the vertical (and inclined) fluxes of dispersed sedimentary matter by use of sediment traps, ADCP, isotopic methods, etc. (Lisitzin, 2010; Lein et al., 2012). Sediment traps are the most interesting for sedimentology (Lukashin et al., 2011). These are cones or cylinders with samplers, which capture the dispersed sedimentary material. The samplers are managed by the microprocessor, i.e., the trapping exposure may vary from 1 day up to months, seasons, or years (Honjo et al., 1988; Bauerfeind et al., 2009).

Thus, we have obtained continuous temporal samples with monthly (differential fluxes) and annual (integral fluxes) exposures in the Arctic and various oceanic zones.

The authors of this study have not investigated the ultimate product of the sedimentation process – bottom sediments – but rather the entire sedimentation process, which occurs mainly in indiscernible form as negligibly small particles dispersed in very insignificant amounts in all natural Arctic geospheres: atmosphere, cryosphere, hydrosphere, biosphere, and sedimentosphere.

METHODS

Within the Russian federal program of the International Polar Year (IPY), we have studied the snow and ice cover at the Barneo circumpolar station from 2007 through 2008. The aim of operations was to study the lithology and geochemistry of dispersed sedimentary material in snow, drift ice, underice water, and vertical underice sediment fluxes. For this, the expedition sampled the snow cover, ice cores, and underice water and placed sediment traps under the ice cover (Novigatsky, Lisitzin, 2019).

Sediment traps with 12 or 24 samplers are used to study the monthly, seasonal, or interannual variability, which is important for investigation of the cycles of the biogenic components. Traps with one sampler are used for short-period observations (up to one month) during a long period to obtain a greater total sample for the sedimentary studies. The traps can operate either at moored stations (usually with submersed buoyancy) or as drifters; i.e., they drift at a specific depth catching matter setted in the water column (Lukashin et al., 2011).

The preliminary treatment of the samples was performed in the laboratory. It included water (melted snow and ice) filtration by a vacuum through membrane and fiberglass filters. For determining the mass concentration of the particulaite matter (mg/l), water samples were filtered using the standard method under the vacuum of 400 mbar through membrane nuclear filters (pore diameter 0.45 μm, filter diameter 47 mm, production of the Joint Institute of Nuclear Research, Dubna, Russia). For determining concentrations of C_{org} the samples were filtered by a vacuum of 200 mbar through GF/F fiberglass filters produced by the Whatman Company (filter diameter 47 mm, effective pore diameters 0.7 μm) ignited at t = 450°C. Electron microscopic definitions were obtained on a TESCAN scanning electron microscope in IO RAS (Kravchishina et al., 2015).

The dispersed sedimentary matter was largely composed of biogenic and lithogenic components. The biogenic components were represented by organic matter (OM), carbonate particles, and biogenic opal. The lithogenic components included clastic and clay minerals with subordinate quantity of volcanic ash. Indicators of these components for organic matter (OM) are C_{org} ($C_{org*}2$ = OM), for siliceous organic remains $SiO_{2amorph}$, and for carbonate organic remains $CaCO_3$, while Al and Si_{total} are indicators of lithogenic particles.

The samples were analyzed to determine the composition of their major and minor elements by culonometric (C_{tot}, C_{org}) and photometric (Si, Al) methods (IO RAS), as well as C_{carb} ($C_{tot} - C_{org}$), which serves for calculating the carbonate contents. This allows the general composition of sedimentary material dispersed in water and the integrity of biogenic (triad $SiO_{2amorph}$ + OM + $CaCO_3$) as well as terrigenous (Si/Al) components to be established.

The biogenic siliceous component (diatom frustules) represents amorphous silica (Si_{amorph}) determined by the method of the terrigenous matrix using the Al module or by direct measurement of Si_{amorph} (Lukashin et al., 2003). The composition of clays and shales from the East European Platform is accepted to serve as the terrige nous matrix. The lithogenic component is calculated proceeding from the Al content measured in sediments from snow, ice and water.

RESULTS AND DISCUSSIONS

Vertical Sediment Fluxes

Based on the amounts of vertical sediment fluxes, two zones with high values are distinguished in the Arctic (>1000 mg m^{-2} day^{-1}): the marginal filters of rivers (Lisitsyn et al., 1995), which is typical of all climatic zones, and the melt zones of annual and especially perennial sea ice, which is typical only of glacial zones (Bauerfeind et al., 2005); i.e., it is a feature unique to them, just like the lowest flux values beneath the ice cover (Figure 1). These two patterns are typical of the surface layers of the Arctic water column (Lisitzin, 2002).

These values strongly contrast with the as of yet scarce data on fluxes under the perennial ice cover: all available values range from 1–150 mg m^{-2} day^{-1} (Lalande et al., 2014). This is an actual ice desert; the flux value corresponds to low, minimal for the World Ocean, SPM contents under the pack ice cover and to minimal plankton development. Such sharp reductions in fluxes have also been noted for Arctic shelf seas in the winter months (Zernova et al., 2000; Fahl, Nöthig, 2007).

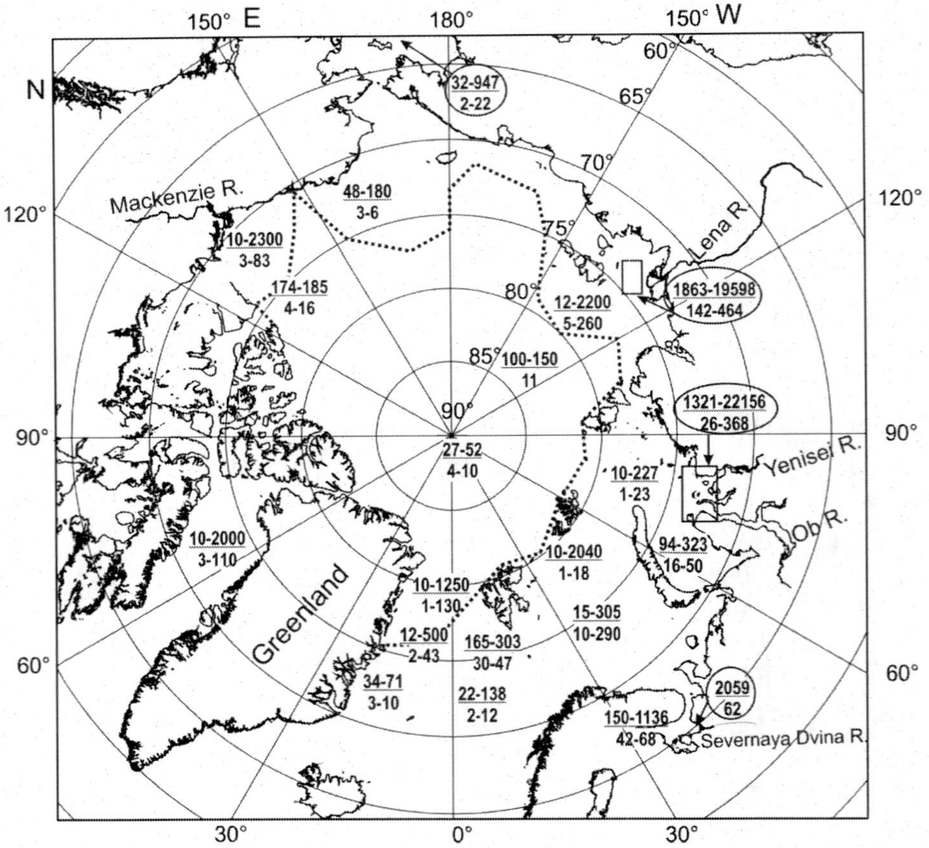

Figure 1. The vertical fluxes of sedimentary material in the Arctic and Subarctic region: the total annual flux (mg m^{-2} day^{-1}) is above the line and C_{org} flux (the biogenic component, mgC m^{-2} day^{-1}) is under the line, according to published data, for the method of sedimentation traps (Table). The dashed line signifies the mean long-term position of the ice edge in summer.

Table 1. Distribution of values of vertical fluxes of sedimentary material (total flux mg m^{-2} day^{-1} and C_{org} flux mgC m^{-2} day^{-1}) in Arctic and Subarctic according to our and literature data obtained using sediment traps

Latitude; Longitude	Date	Horizon, m	Total flux, mg m^{-2} day^{-1}	C_{org} flux, mgC m^{-2} day^{-1}	Source
Fram Strait					
N 79°00´; E 06°50´	1987-1990	150	100-1250	10-130	Hebbeln, 2000
N 72°-75°; E 04°-10°	09.1988-10.1995	500	12-500	2.1-43	Ramseier et al., 1999
N 79°01´; E 04°20´	08.2000-08.2005	260-300	10-510	1.5-52	Bauerfeind et al., 2009
Central Arctic					
N 89°37.02´; W 08°37.12´	04.2007 04.2008	20	52	10	Novigatsky, Lisitzin, 2018
		30	27	4.4	
		70	31	7.8	
N 85°17´ E 122°32´	09.2012	5	100	-	Lalande et al., 2014
		25	150	11	
Svalbard shelf					
N 81°32.0´; E 30°35.0´	07.1991	100	165	30	Andreassen et al., 1996
N 80°20.0´ E 29°09.0´	07.1991	60	303	47	
Greenland Sea					
N 73°00´; W 10°00´	1988-1991	500	34-63	2.9-10	Bodungen et al., 1995
N 80°27´; W 13° 41´	06-07. 1993	130	-	75	Pesant et al., 2002
N 75°57.7´; E 12°58.7´	04.1995	240	71	4.9	Bauerfeind et al., 2005
Norwegian Sea					
N 70°00.0´ E 00°05.0´	1986-1989	500	72-93	8.3-12	Bodungen et al., 1995
N 75°51´; E 11°28´	08.1984-07.1985	400	22-138	1.8-17	Honjo et al., 1988
N 73°42´; E 13°11´	03-08. 1993	743	59	8.5	Lukashin et al., 2000
Barents Sea					
N 73.767; E 31.876´	04.1998	200	-	30-70	Olli et al., 2002
	05.1998	200	-	500-1500	
	07.1999	150	-	150-300	
N 75°11.78′; E 12°29.21′	03-07. 1991	610	15-305	10-290	Thomsen et al., 2001
N 70°-78°; E 58°-62°	09-10. 1997	104-380	10-2040	1.1-18	Shevchenko et al., 1998; 2000
White Sea					
N 66°20.2′; E 33°39.9′	2006 -2007	30	150-1136	-	Mityaev et al., 2012
N 66°29.30'; E 34°10.43'	06.2000	55	622	42	Lukashin et al., 2003
		270	1063	68	

Table 1. (Continued)

Latitude; Longitude	Date	Horizon, m	Total flux, mg m^{-2} day^{-1}	C_{org} flux, mgC m^{-2} day^{-1}	Source
Kara Sea					
N 76°12.08´; E 75°45.30´	09.2001-08.2002	73	80-1320	4.2-77	Gaye et al., 2007
N 70°-74°; E 64°-83°	09.1993	20-116	10-1321	1.5-26	Lisitsyn et al., 1995
Laptev Sea					
N 81°04.5´; E 138°54.0´	04.1995-04.1996	150	12-196	5.1	Fahl, Nöthig et al., 2007
N 79°55.7´; E 142°21.8´	08.2005-08.2006	175	166	11	Lalande et al., 2014
Chukchi Sea					
N 72°43.3´; W 158°24.2´	05.2004	247	-	44	Lalande et al., 2007
	08.2004	247	-	129	
Canadian Arctic Archipelago					
N 71°32.23´; W 127°01.4´	2004	210	48	3.4	Forest et al., 2010
	2005		180	6.1	
	2006		66	6.4	
N 71°04.8´; W 133°37.8´	09.2005-10.2006	200	185	4.4	Lalande et al., 2009
N 76°19.6´; W 71°11.9´	08.2005-08.2006	200	174	16	
N 70°0´; W 126°0´	02-06. 2004	1-25	-	15-188	Juul-Pedersen et al, 2008
N 69°-71°; W 127°-138°	04.1988	100-500	10-2300	3.5-83	O'Brien et al., 2006
Baffin Sea					
N 76°00´; W 75°00´	08.1997-07.1999	200	10-2000	8-68	Hargrave et al., 2002
N 76°23´; W 77°24´	08.1997-07.1999	205	-	3-110	Sampei et al., 2004
Bering Sea					
N 53.5°; W 177.0°	1990-2000	600	32-947	2.4-22	Takahashi et al., 2002
N 55°0´; W 155°0´	06-09. 1988	36	-	253-654	Fukuchi et al., 1993
Subarctic					
N 50°; W 145°	1982-1993	200	45-1000	6.5-98	Wong et al., 1999

Underice fluxes of sedimentary material in the winter in Central Arctic of 2007–2008 averaged 37 mg m^{-2} day^{-1} total flux and 7.4 mgC m^{-2} day^{-1} C_{org} flux. These are quite consistent with the winter values of underice fluxes of dispersed sedimentary material measured in other regions of the Arctic Basin, which on average were 50 mg m^{-2} day^{-1} total flux and 4.5 mgC m^{-2} day^{-1} C_{org} flux (Table).

The monthly seasonal and annual fluxes of dispersed sedimentary matter can differ from each other by many times (Table), especially in spring, when the ice regime of the water area exerts a significant influence on the value of the flux: the snow–ice cover unloading of cryosol sedimentary matter, river runoff, abrasion, and phytoplankton bloom. The fluxes of sedimentary matter down to the bottom depend directly on the interaction of matter coming from the outer geospheres.

Composition of Dispersed Sedimentary Material in the Snow and Ice Cover

A low concentration of dispersed sedimentary material is characteristic of first-year ice that forms in the central parts of the Arctic Basin; perennial ice accumulates higher concentrations of insoluble particles in its sequence. The ice that forms in shelf regions, which is impacted by river runoff, coastal abrasion, and turbidity of bottom sediments during storms, picks up an even larger amount of sedimentary material (Lisitzin, 2002).

In the distribution of cryosols, two sediment accumulation zones are distinguished in the ice sequence of the central Arctic: in the roof of the ice field, where insoluble particles transported by snow accumulate, and at the foot of the ice, where newly formed ice picks up SPM from underice water, because SPM serves as crystallization nuclei for intrawater ice (Figure 2a). The upper ice horizons also contain higher values of organic carbon (C_{org}) than the lower horizons (Figure 2b), which is related to the influence of the snow cover (Aeolian material) and is characteristic of ice massifs in the central Arctic in winter (Nemirovskaya, Novigatsky, 2003).

Using Al as an indicator of lithogenic material (Shevchenko, 2003), we determined the content of terrigenous components in the snow and ice cover in the circumpolar region. For the snow cover, the lithogenic component averages 20%; for the ice cover, 25%. Thus, the biogenic component predominates both in the snow (80%) and ice cover (75%) (Figure 2c). The lithogenic component is poorly expressed in water: no more than 15%. At the foot of ice and in underice water, there is a large amount of amorphous silica, which is a structure-forming element in phytoplankton tests. These relations are confirmed by the results of microscopic studies.

Based on electron microscopy data, insoluble particles contained in the Arctic's snow and ice cover consist mainly of organic matter (phytoplankton test fragments, spores, pollen, diatomic algae, singlecelled animals, etc.) and mineral particles (mineral grains and clay aggregates) (Figure 2d).

For the snow cover, terrestrial soils are the main source of mineral particles. In winter, when the surface of the Arctic is covered in ice and snow, the main contribution is made by long-range transport, including anthropogenic particles; their aeolian transport has been noted in nearly all areas of the Arctic up to the North Pole (Shevchenko et al., 2016).

The snow cover of the Earth's high latitudes is an auspicious collector of both mineral and biogenic particles. Accumulating on the surface of the snow cover, they are preserved well and become fossilized with time. The lithogenic component in the snow cover is subordinate to the biogenic component and is represented by individual mineral grains.

In the upper part of the ice, in contrast to the rest of the sequence, a high content of mineral particles is observed (up to 30%), which is related to the influence of the snow cover and accumulation of material on the ice surface (aeolian source). The biogenic component is dominated by diatomic algae, just like in the middle of the ice, 50% of which consists of diatomic algae and silicoflagellates. The middle of the ice also hosts pollen and single-cell animals, which is also typical of the lower part of the ice (Figure 2d).

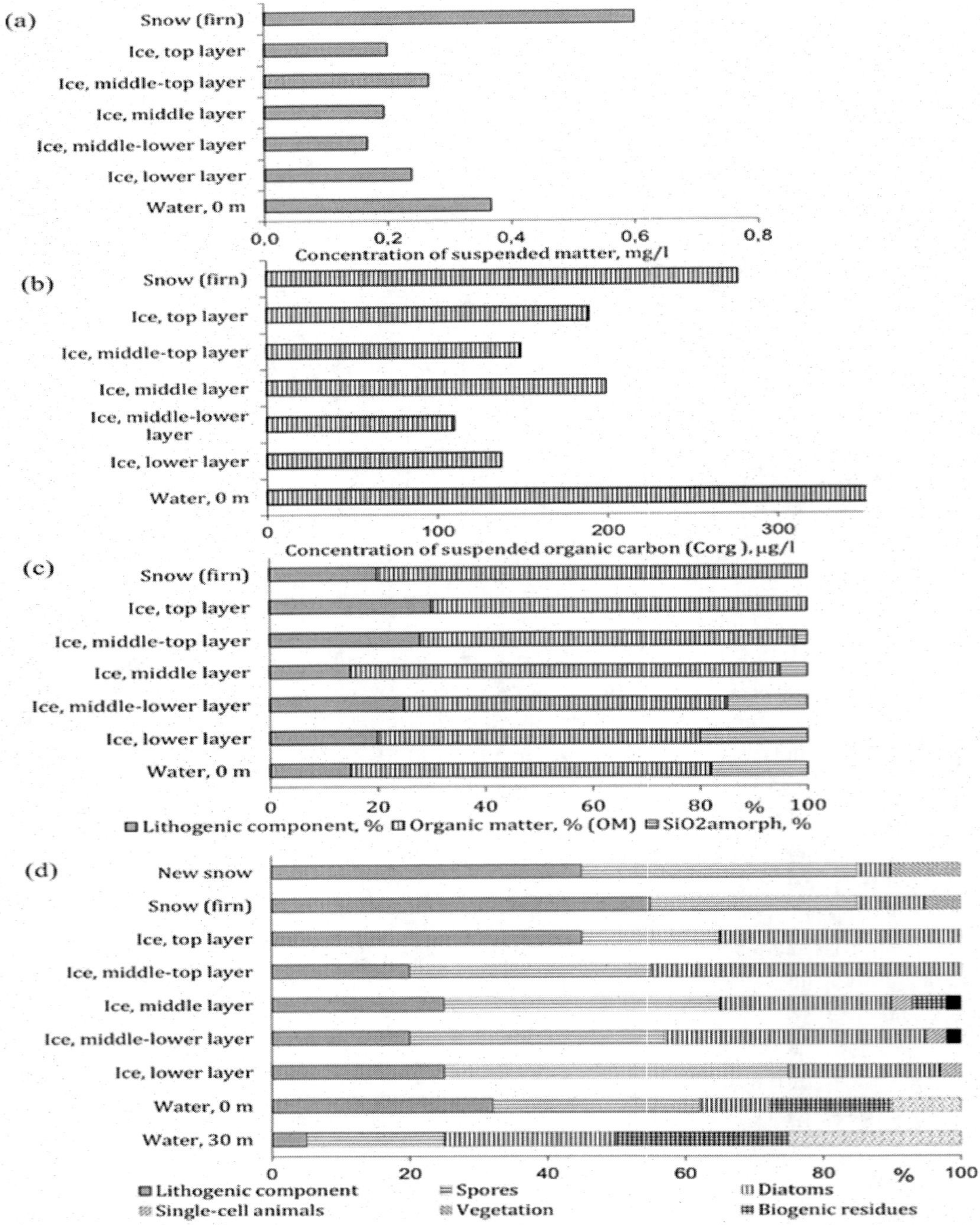

Figure 2. Distribution pattern of material on the vertical section of snow–ice–seawater: (a) concentration of suspended particles in melt water, mg/l; (b) concentration of suspended organic carbon, µg/l; (c) percentage of the lithogenic component, biogenic organic matter, and amorphous biogenic silica $SiO_{2amorph}$; (d) the quantitative and qualitative composition of sedimentation material in the snow–ice–seawater system.

In water, the lithogenic component is poorer than in the snow and ice cover; the biogenic component predominates, and a high occurrence of bacterial associations and biogenic remains is observed in the form of test and shell fragments, diatomic algae, and their spores (Figure 2d).

CONCLUSION

The maximal values of the particle fluxes (mg m^{-2} day^{-1}) are typical for the ice free spring–summer period; the minimal ones, for the winter period when the sea and the feeding catchment are covered by a snow–ice mantle and the river runoff and aeolian input are insignificant.

The sharp maxima of the fluxes are confined to the areas of estuaries and bays, where the continental runoff exerts a dominant influence on the delivery of sediment material.

Underice water is characterized by the maximum contents of dispersed sedimentary material and particulate organic carbon in the snow–ice–underice water system; the snow cover and roof of the ice cover are characterized by elevated contents of the studied components, which indicates a significant contribution from aeolian sources of sedimentary material at high latitudes. The snow–ice–underice water system is characterized by high values of the biogenic component (from 65 to 85%); the lithogenic component is characterized by lower values (from 15 to 35%) and predominates in the snow cover and the ice cover surfase, which points to an aeolian source of lithogenic material. The composition of sedimentary material is dominated by valves and spores of phytoplankton and different forms of organic matter; mineral particles are subordinate.

ACKNOWLEDGMENTS

The research was carried out under the state assignment of Minobrnauki Russia (topic nos. 0128-2019-0011). Processing of materials was supported by the Russian Foundation for Basic Research (project no. 19-05-00022).

REFERENCES

Andreassen, I., Nöthig, E. M., Wassmann, P. (1996). Vertical particle flux on the shelf off northern Spitsbergen, Norway. *Marine Ecology Progress Series*, 137: 215-228.

Bauerfeind, E., Leipe, T., Ramseier, R. O. (2005). Sedimentation at the permanently ice-covered Greenland continental shelf (74° 57.7′ N/12° 58.7′ W): significance of biogenic and lithogenic particles in particulate matter flux. *Journal of marine systems*, 56 (1-2): 151-166.

Bauerfeind, E., Nöthig, E. M., Beszczynska, A., Fahl, K., Kaleschke, L., Kreker, K., Wegner, J. (2009). Particle sedimentation patterns in the eastern Fram Strait during 2000–2005: Results from the Arctic long-term observatory HAUSGARTEN. *Deep Sea Research Part I: Oceanographic Research Papers*, 56 (9): 1471-1487.

Bodungen, B., Antia, A., Bauerfeind, E., Haupt, O., Koeve, W., Machado, E., Voss, M. (1995). Pelagic processes and vertical flux of particles: an overview of a long-term comparative study in the Norwegian Sea and Greenland Sea. *Geologische Rundschau*, 84 (1): 11-27.

Darby, D. A., Burckle, L. H., Clark, D. L. (1974). Airborne dust on the Arctic pack ice, its composition and fallout rate. *Earth and Planetary Science Letters*, 24 (2): 166-172.

Fahl, K., Nöthig, E. M. (2007). Lithogenic and biogenic particle fluxes on the Lomonosov Ridge (central Arctic Ocean) and their relevance for sediment accumulation: Vertical vs. lateral transport. *Deep Sea Research Part I: Oceanographic Research Papers*, 54 (8): 1256-1272.

Forest, A., Bélanger, S., Sampei, M., Sasaki, H., Lalande, C., Fortier, L. (2010). Three-year assessment of particulate organic carbon fluxes in Amundsen Gulf (Beaufort Sea): Satellite observations and sediment trap measurements. *Deep Sea Research Part I: Oceanographic Research Papers*, 57 (1): 125-142.

Fukuchi, M., Sasaki, H., Hattori, H., Matsuda, O., Tanimura, A., Handa, N., & McRoy, C. P. (1993). Temporal variability of particulate flux in the northern Bering Sea. *Continental Shelf Research*, 13 (5-6): 693-704.

Gaye, B., Fahl, K., Kodina, L. A., Lahajnar, N., Nagel, B., Unger, D., Gebhardt, A. C. (2007). Particulate matter fluxes in the southern and central Kara Sea compared to sediments: Bulk fluxes, amino acids, stable carbon and nitrogen isotopes, sterols and fatty acids. *Continental Shelf Research*, 27 (20): 2570-2594.

Hargrave, B. T., Walsh, I. D., Murray, D. W. (2002). Seasonal and spatial patterns in mass and organic matter sedimentation in the North Water. *Deep Sea Research Part II: Topical Studies in Oceanography*, 49 (22-23): 5227-5244.

Hebbeln, D. (2000). Flux of ice-rafted detritus from sea ice in the Fram Strait. *Deep Sea Research Part II: Topical Studies in Oceanography*, 47 (9-11): 1773-1790.

Honjo, S., Manganini, S. J., Wefer, G. (1988). Annual particle flux and a winter outburst of sedimentation in the northern Norwegian Sea. *Deep Sea Research Part A. Oceanographic Research Papers*, 35 (8): 1223-1234.

Juul-Pedersen, T., Michel, C., Gosselin, M., Seuthe, L. (2008). Seasonal changes in the sinking export of particulate material under first-year sea ice on the Mackenzie Shelf (western Canadian Arctic). *Marine Ecology Progress Series*, 353: 13-25.

Kravchishina, M. D., Lein, A. Y., Sukhanova, I. N., Artem'ev, V. A., & Novigatsky, A. N. (2015). Genesis and spatial distribution of suspended particulate matter concentrations in the Kara Sea during maximum reduction of the Arctic ice sheet. *Oceanology*, 55 (4): 623-643.

Lalande, C., Forest, A., Barber, D. G., Gratton, Y., Fortier, L. (2009). Variability in the annual cycle of vertical particulate organic carbon export on Arctic shelves: Contrasting the Laptev Sea, Northern Baffin Bay and the Beaufort Sea. *Continental Shelf Research*, 29 (17): 2157-2165.

Lalande, C., Grebmeier, J. M., Wassmann, P., Cooper, L. W., Flint, M. V., Sergeeva, V. M. (2007). Export fluxes of biogenic matter in the presence and absence of seasonal sea ice cover in the Chukchi Sea. *Continental Shelf Research*, 27 (15): 2051-2065.

Lalande, C., Nöthig, E. M., Somavilla, R., Bauerfeind, E., Shevchenko, V., Okolodkov, Y. (2014). Variability in under-ice export fluxes of biogenic matter in the Arctic Ocean. *Global Biogeochemical Cycles*, 28 (5): 571-583.

Lein, A. Y., Kravchishina, M. D., Politova, N. V., Savvichev, A. S., Veslopolova, E. F., Mitskevich, I. N., Ivanov, M. V. (2012). Transformation of particulate organic matter at the water-bottom boundary in the Russian Arctic seas: Evidence from isotope and radioisotope data. *Lithology and Mineral Resources*, 47 (2): 99-128.

Levitan, M. A. (2015). Sedimentation rates in the Arctic Ocean during the last five marine isotope stages. *Oceanology*, 55 (3): 425-433.

Lisitsyn, A. P., Shevchenko, V. P., Vinogradov, M. E., Severina, O. V., Vavilova, V. V., Mitskevich, I. N. (1995). Particle uxes in the Kara Sea and Ob and Yenisey estuaries. *Oceanology*, 34 (5): 683-693.

Lisitzin, A. P. (2002). *Sea-Ice and Iceberg Sedimentation in the Ocean Recent and Past.* Springer, 563.

Lisitzin, A. P. (2004). Sediment fluxes, natural filtering, and sedimentary systems of a "living ocean" *Russian Geology and Geophysics*, 45 (1): 15-48.

Lisitzin, A. P. (2010). Marine ice-rafting as a new type of sedimentogenesis in the Arctic and novel approaches to studying sedimentary processes. *Russian Geology and Geophysics*, 51 (1): 12-47.

Lisitzin, A. P., Shevchenko, V. P. (2016). Glacial-marine sedimentation. *Encyclopedia of Marine Geosciences*, *Springer*, 288-294.

Lukashin, V. N., Bogdanov, Y. A., Shevchenko, V. P., Rusakov, V. Y., Isaeva, A. B. (2000). A study of vertical fluxes of sedimentary material and its composition in the Norwegian Sea in the summer periods of 1991-1995. *Geochemistry International*, 38 (2): 169-183.

Lukashin, V. N., Kosobokova, K. N., Shevchenko, V. P., Shapiro, G. I., Pantyulin, A. N., Pertzova, N. M., Deev, M. G., Klyuvitkin, A. A., Novigatsky, A. N., Prego, R., Latche, L. (2003). Results of multidisciplinary oceanographic studies in the White Sea in June 2000. *Oceanology*, 43 (2): 224-239.

Lukashin, V. N., Klyuvitkin, A. A., Lisitzin, A. P., Novigatsky, A. N. (2011). The MSL-110 small sediment trap. *Oceanology*, 51 (4): 699-703.

Mityaev, M. V., Gerasimova, M. V., Druzhkova, E. I. (2012). Vertical particle fluxes in the coastal areas of the Barents and White Seas. *Oceanology*, 52 (1): 112-121.

Nemirovskaya, I. A., Novigatsky, A. N. (2003). Hydrocarbons in the snow and ice cover and waters of the Arctic Ocean. *Geochemistry International*, 41 (6): 585-594.

Novigatsky, A. N., Lisitzin, A. P. (2018). The North Pole Region: First Data on the Snow–Sea Ice–Ice Water Sedimentation System. *Doklady Earth Sciences,* 483 (2): 1534-1538.

Novigatsky, A. N., Lisitzin, A. P. (2019). Concentration, Composition, and Fluxes of Dispersed Sedimentary Material in the Snow and Ice Cover of the Polar Arctic. *Oceanology*, 59 (3): 406-410.

O'Brien, M. C., Macdonald, R. W., Melling, H., Iseki, K. (2006). Particle fluxes and geochemistry on the Canadian Beaufort Shelf: Implications for sediment transport and deposition. *Continental Shelf Research*, 26 (1): 41-81.

Olli, K., Riser, C. W., Wassmann, P., Ratkova, T., Arashkevich, E., Pasternak, A. (2002). Seasonal variation in vertical flux of biogenic matter in the marginal ice zone and the central Barents Sea. *Journal of Marine Systems*, 38 (1-2): 189-204.

Pesant, S., Legendre, L., Gosselin, M., Bauerfeind, E., Budéus, G. (2002). Wind-triggered events of phytoplankton downward flux in the Northeast Water Polynya. *Journal of Marine Systems*, 31 (4): 261-278.

Rachold, V., Eicken, H., Gordeev, V. V., Grigoriev, M. N., Hubberten, H.-W., Lisitzin, A. P., Shevchenko, V. P., Schirrmeister, L. (2004). Modern terrigenous organic carbon input to the Arctic Ocean. *In The organic carbon cycle in the Arctic Ocean,* 33-55.

Ramseier, R. O., Garrity, C., Bauerfeind, E., Peinert, R. (1999). Sea-ice impact on long-term particle flux in the Greenland Sea's Is Odden-Nordbukta region, 1985–1996. *Journal of Geophysical Research: Oceans*, 104 (C3): 5329-5343.

Sampei, M., Sasaki, H., Hattori, H., Fukuchi, M., & Hargrave, B. T. (2004). Fate of sinking particles, especially fecal pellets, within the epipelagic zone in the North Water (NOW) polynya of northern Baffin Bay. *Marine Ecology Progress Series*, 278: 17-25.

Shevchenko, V. P., Burovkin, A. A., Zernova, V. V., Shanin, S. S., Ivanov, G. I., Polyak, L. V., Dzhinoridze, R. N. (1998). Sedimentary material flows in the St. Anna Trough and eastern Barents Sea. *In Doklady earth sciences* 359 (3): 400-403.

Shevchenko, V. P., Lisitzin, A. P., Zernova, V. V., Lukashin, V. N., Politova, N. V., Rusakov, V. Y., Shanin, S. S. (2000). Vertical particle fluxes in seas of the western Russian Arctic. *Proceedings of the Humanity and the World Ocean: Interdependence at the Dawn of the New Millennium (PACON'99)*, 239-249.

Shevchenko, V. P. (2003). The influence of aerosols on the oceanic sedimentation and environmental conditions in the Arctic. *Berichte zur Polar-und Meeresforschung (Reports on Polar and Marine Research)*, 464.

Shevchenko, V. P., Vinogradova, A. A., Lisitzin, A. P., Novigatsky, A. N., Panchenko, M. V., Pol'kin, V. V. (2016). Aeolian and ice transport of matter (including pollutants) in the Arctic. *In Implications and Consequences of Anthropogenic Pollution in Polar Environments*, *Springer*, 59-73.

Stein, R. (2008). Arctic Ocean sediments: processes, proxies, and paleoenvironment. *Developments in Marine Geology*, 2: 592.

Takahashi, K., Fujitani, N., Yanada, M. (2002). Long term monitoring of particle fluxes in the Bering Sea and the central subarctic Pacific Ocean, 1990–2000. *Progress in Oceanography*, 55 (1-2): 95-112.

Thomsen, C., Blaume, F., Fohrmann, H., Peeken, I., Zeller, U. (2001). Particle transport processes at slope environments—event driven flux across the Barents Sea continental margin. *Marine Geology*, 175 (1-4): 237-250.

Wong, C. S., Whitney, F. A., Crawford, D. W., Iseki, K., Matear, R. J., Johnson, W. K., Timothy, D. (1999). Seasonal and interannual variability in particle fluxes of carbon, nitrogen and silicon from time series of sediment traps at Ocean Station P, 1982–1993: relationship to changes in subarctic primary productivity. *Deep Sea Research Part II: Topical Studies in Oceanography*, 46 (11-12): 2735-2760.

Zernova, V. V., Nöthig, E. M., Shevchenko, V. P. (2000). Vertical microalga flux in the northern Laptev Sea (from the data collected by the yearlong sediment trap). *Oceanology*, 40(6), 801-808.

EDITORS' CONTACT INFORMATION

Oleg S. Pokrovsky

CNRS, University of Toulouse, France
Email: Oleg.Pokrovski@get.omp.Eu.

Sergey N. Kirpotin

Bio-Clim-Land Center of Excellence National Research
Tomsk State University, Tomsk, Russia

Alexander I. Malov

N.P. Laverov Federal Centre for Integrated Arctic Research,
Russian Academy of Science, Arkhangelsk, Russia
Email: malovai@yandex.ru.

INDEX

#

A

B

C

D

E

F

G

H

I

K

O

P

R

S

T

U

V

W

Y